OXFORD MONOGRAPHS ON GEOLOGY AND GEOPHYSICS

OXFORD MONOGRAPHS ON GEOLOGY AND GEOPHYSICS

1. DeVerle P. Harris: *Mineral resources appraisal: mineral endowment, resources, and potential supply: concepts, methods, and cases*
2. J.J. Veevers (ed.): *Phanerozoic earth history of Australia*
3. Yang Zunyi, Cheng Yuqi, and Wang Hongzhen (ed.): *The geology of China*
4. Lin-gun Liu and William A. Bassett: *Elements, oxides, and silicates: high-pressure phases with implications for the Earth's interior*
5. Antoni Hoffman and Matthew H. Nitecki (ed.): *Problematic fossil taxa*
6. S. Mahmood Naqvi and John J.W. Rogers: *Precambrian geology of India*
7. Chih-Pei Chang and T.N. Krishnamurti (ed.): *Monsoon meteorology*
8. Zvi Ben-Avraham (ed.): *The evolution of the Pacific Ocean margins*
9. Ian McDougall and T. Mark Harrison: *Geochronology and thermochronology by the* $^{40}Ar/^{39}Ar$ *method*
10. Walter C. Sweet: *The conodonta: morphology, taxonomy, paleoecology, and evolutionary history of a long-extinct animal phylum*
11. H.J. Melosh: *Impact cratering: a geologic process*
12. J.W. Cowie and M.D. Brasier (ed.): *The Precambrian–Cambrian boundary*
13. Charles S. Hutchison: *Geological evolution of South-East Asia*
14. Anthony J. Naldrett: *Magmatic sulfide deposits*
15. D.R. Prothero and R.M. Schoch (ed.): *The evolution of perissodactyls*
16. M. Menzies (ed.): *Continental mantle*
17. R.J. Tingey (ed.): *Geology of the Antarctic*
18. Thomas J. Crowley and Gerald R. North: *Paleoclimatology*
19. Gregory J. Retallack: *Miocene paleosols and ape habitats in Pakistan and Kenya*
20. Kuo-Nan Liou: *Radiation and cloud processes in the atmosphere: theory, observation, and modeling*
21. Brian Bayly: *Chemical change in deforming materials*
22. A.K. Gibbs and C. N. Barron: *The geology of the Guiana Shield*
23. Peter J. Ortoleva: *Geochemical self-organization*
24. Robert G. Coleman: *Geologic evolution of the Red Sea*
25. Richard W. Spinrad, Kendall L. Carder, and Mary Jane Perry: *Ocean optics*
26. Clinton M. Case: *Physical principles of flow in unsaturated porous media*
27. Eric B. Kraus and Joost A. Businger: *Atmosphere–ocean interaction*, second edition
28. M. Solomon and D. I. Groves: *The geology and origins of Australia's mineral deposits*
29. R. L. Stanton: *Ore elements in arc lavas*
30. P. Wignall: *Black shales*
31. Orson L. Anderson: *Equations of state for solids in geophysics and ceramic science*
32. J. Alan Holman: *Pleistocene amphibians and reptiles in North America*
33. P. Janvier: *Early vertebrates*
34. David S. O'Hanley: *Serpentinites: recorders of tectonic and petrological history*
35. Maarten J. de Wit and Lewis D. Ashwal (ed.): *Greenstone belts*
36. Charles S. Hutchison: *South-East Asian oil, gas, coal, and mineral deposits*

South-East Asian Oil, Gas, Coal and Mineral Deposits

Charles S. Hutchison

Former Professor of Applied Geology
University of Malaya
Kuala Lumpur, Malaysia

CLARENDON PRESS · OXFORD

This book has been printed digitally and produced in a standard specification in order to ensure its continuing availability

OXFORD
UNIVERSITY PRESS

Great Clarendon Street, Oxford OX2 6DP

Oxford University Press is a department of the University of Oxford.
It furthers the University's objective of excellence in research, scholarship,
and education by publishing worldwide in

Oxford New York

Auckland Cape Town Dar es Salaam Hong Kong Karachi
Kuala Lumpur Madrid Melbourne Mexico City Nairobi
New Delhi Shanghai Taipei Toronto
With offices in
Argentina Austria Brazil Chile Czech Republic France Greece
Guatemala Hungary Italy Japan South Korea Poland Portugal
Singapore Switzerland Thailand Turkey Ukraine Vietnam

Oxford is a registered trade mark of Oxford University Press
in the UK and in certain other countries

Published in the United States
by Inc., New York

Reprinted 2004

ISBN 0-19-852649-0

Antony Rowe Ltd., Eastbourne

To Ann with thanks for always being there

Preface

This is a companion volume to my 1989 ***Geological Evolution of South-East Asia***, published as Number 13 of the same series. I was unhappy that page constraints prohibited the inclusion of a review of the economic deposits in that volume, but delighted that Oxford University Press later encouraged me to complete the coverage presented herein. It was originally to be entitled 'Economic geology of South-East Asia', but the publishers were advised that *oil* must appear in the title to encourage sales. The manuscript was accordingly modified to emphasize hydrocarbons.

In the preface to the previous volume, it was stated that my geological work in South-East Asia would have terminated by the publication date. It was, however, my great fortune to have returned to active work, firstly as Overall Co-ordinator of the SEATAR [Studies in East Asian Tectonics and Resources] transects, attached to the CCOP Secretariat in Bangkok. In May 1994 the meaning of the CCOP acronym was officially changed from 'Committee for Co-ordination of Joint Prospecting for Mineral Resources in Asian Offshore Areas' to 'Co-ordinating Committee for Coastal and Offshore Geoscience Programmes in East and South-East Asia'. But there will be no major change in either activities or direction. From August 1989, I resumed for three years my former chair of applied geology at the University of Malaya, which had remained unfilled since my departure in August 1987. From March 1993 my continuation in the region has been made possible for eighteen months as 'research professor' and 'director' of the now extinct one-man SE Asian branch of the Earth Sciences & Resources Institute [ESRI], which maintains addresses within the campuses of the universities of South Carolina (Columbia) and Utah (Salt Lake City).

My continuing involvement in the region gave me the opportunity to complete the present volume. In all probability this will be the last geology book that I will write and I would like to think that I have kept the 'best till the last'. Hopefully, however, it will prove to be no worse than my other books. Certainly writing has become progressively more difficult for me.

It is in the very nature of our human condition that the passage of time brings with it great sadness. The 1989 volume was dedicated *inter alia* to my dear friend Professor Ken Hosking, who sadly departed this life in June 1991 at his home town of Camborne, set in the heart of his beloved tin-fields (see Haile and Hutchison, 1992). It was he who compiled the masterful syntheses of the tin deposits of South-East Asia. The great mining industry, that once dominated the SE Asian region and the world, has been decimated through lack of industrial demand and rapidly declining commodity price. Ken was here at its peak and dominated the geological analysis of tin mineralization.

I have assumed throughout the compilation of this volume that every reader should have ready access to my 1989 book. Accordingly, the regional geological picture will not be repainted. However, in describing the Cainozoic tectonic settings of the oil, gas and coal deposits, I have taken the opportunity to revise relevant sections and provide a reference list which can be used to update the earlier volume. Some sections in my 1989 volume have not favourably survived the test of time, but this is in the very nature of scientific writing.

The present volume is therefore intended as a supplement to Hutchison (1989*a*), specifically written for geologists and explorationists concerned with the search for and understanding of the major geological economic commodities of South-East Asia.

Exeter C.S.H.
1995

ACKNOWLEDGEMENTS

The compilation of this book has benefited from collaboration and discussion with many colleagues. In particular, I would like to mention K. R. Chakraborty, S. P. Sivam, R. B. Tate, C. R. Foss, G. H. Teh and Azhar Haji Hussin.

I am grateful to Simon Irwin and J.A. Eyer for help in obtaining production figures and to my son Timothy for teaching me the efficient use of word-processing programmes and for editing help. Ching Yu Hay, Roslin Ismail and Roshdy Abu Daud transferred my sketches into the final illustrations. I wish to thank the staff of Oxford University Press for their help in producing the book, the last in the earth sciences series to be published in Oxford.

Gratitude is expressed to Ungku Aziz and Syed Hussein Alatas, past vice-chancellors of the University of Malaya, for facilitating my research and writing throughout the years. I am also grateful to S. Kulvanich and G.R. Balce, erstwhile directors of the CCOP Secretariat, and W. H. Kanes, director of ESRI, for facilitating my inter-reaction with the worldwide community of geologists involved in the fascinating South-East Asian region.

Contents

1
Introduction

The geographical region covered in this book extends approximately from Taiwan in the north-east and Myanmar (formerly Burma) on the west to include the Philippines and Indonesia in the south and east (Fig. 1.1), but the boundaries are flexible. The region is covered extensively by seas, but not all are of similar character. The 200 m water depth is a useful demarkation line between shallow seas, which cover the continental shelves, and the deeper more oceanic seas. Generally the continental crust of the land masses extends with attenuated thickness as far seawards as the 200 m contour. South-East Asia has a number of marginal seas, where the water depth exceeds 3 km; these are floored by oceanic crust.

The marginal seas are separated by island or remnant arcs (Karig, 1971) made up of volcanic or accretionary wedge material. In addition, the arcs may be complicated by the addition of continental fragments which have been rifted from the larger continental masses of Asia and Australia, and carried away by sea-floor spreading.

One of the two main shelf sea regions is the Sunda Shelf, which is a submarine extension of continental South-East Asia, connecting together the peninsular landmass of Indochina and the Malay Peninsula with the islands of Sumatra, Borneo and western Sulawesi (Fig. 1.1). This large peninsular extension of Eurasia is commonly referred to as Sundaland. The other is the Sahul Shelf, which is a seawards continuation of continental Australia extending north to the Timor Trough. This shelf extends to link up with New Guinea via the Arafura Sea. Detached fragments of the Australian Platform occur in the islands of Sumba and eastern Sulawesi (Pigram and Panggabean, 1984; Hutchison, 1989*d*). Rifted continental fragments and attenuated continental lithosphere occupy much of the South China Sea and are assumed to have been detached from the stretched China shelf between Hainan and Taiwan. These fragments include north-east Palawan, and several submarine banks such as the Reed Bank and the Luconia Shoals (Taylor and Hayes, 1983).

Low sea levels during the Pleistocene Great Ice Age resulted in the landmass coastline of peninsular South-East Asia extending to the present day 200m contour line (Fig. 1.1). This landmass has been named Sundaland, but the term is used throughout this book to mean the peninsular extension of Eurasia which is constructed in major part of continental lithosphere. Southern Borneo (Kalimantan), eastern Java and western Sulawesi are part of Sundaland, but these areas are not strictly continental, and may represent accreted island arcs and uplifted marginal basin strata. They are not cratonic in the way that Sumatra, the Malay Peninsula and Indochina are, and lack coherent Palaeozoic stratigraphy.

Precambrian rocks outcrop along the eastern coastline of Vietnam in the Kontum Massif, and also form the basement to the Yangzi Platform of China (Hutchison, 1989*a*). There has been a long tradition of equating grade of metamorphism with age in South-East Asia. Thus, the amphibolite facies massifs of Thailand, west of the Nan-Uttaradit Suture, are shown as Precambrian on the geological map of Thailand (Chakkaphak and Veeraburus, 1982). Cobbing *et al.* (1992) mapped these amphibolite facies terrains as 'migmatitic granite'. Ahrendt *et al.* (1993) have shown from detailed U-Pb zircon dating that the metamorphism was definitely not Precambrian and occurred at 200 Ma during the Upper Triassic Indosinian Orogeny.

Palaeozoic and Mesozoic formations dominate eastern Myanmar and Thailand through Peninsular Malaysia and Sumatra. Upper Triassic granites, formed during the Indosinian Orogeny, also characterize the Malay Peninsula and Indonesia as far south as Billiton (Cobbing *et al.*, 1992).

The outer landmasses of Sundaland—Kalimantan, Java and Sulawesi—are dominated by Cretaceous and younger formations and they do not have an older Phanerozoic stratigraphy. A detailed account of the Phanerozoic geology of the region is beyond the scope of this book. For this the reader is referred to Hutchison (1989*a*).

1.1 ACTIVE PLATE MARGINS

A nearly continuous arc-trench system extends throughout the region, from the Ryukyu arc in the north-east, through the Philippines and Indonesia, to Myanmar in the north-west (Fig. 1.2). This active convergent plate margin is characterized by shallow seismicity and well developed Benioff Zones, trenches, connecting transform faults,

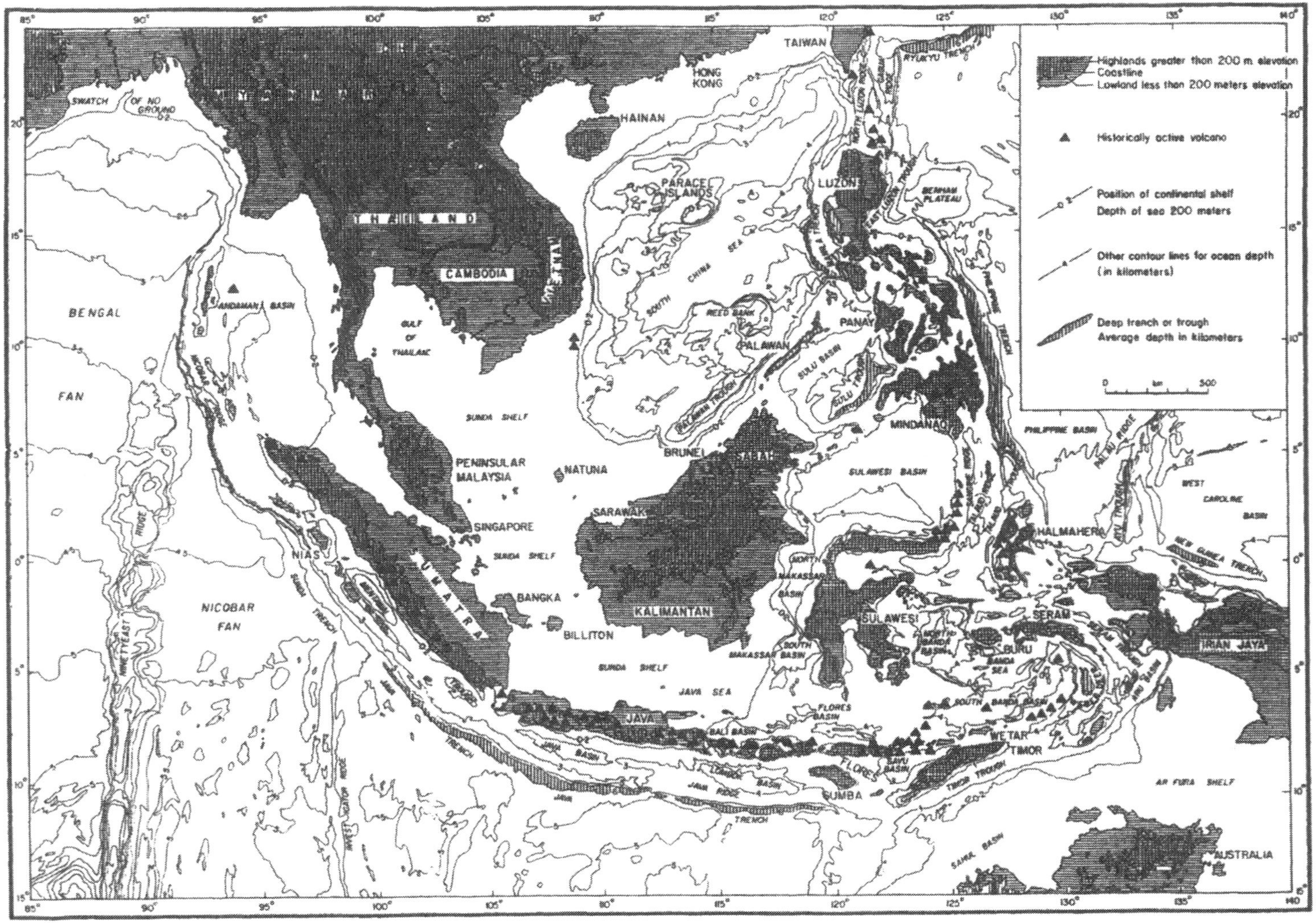

Fig. 1.1. Major physiographic features of South-East Asia, after Hutchison (1989*a*)

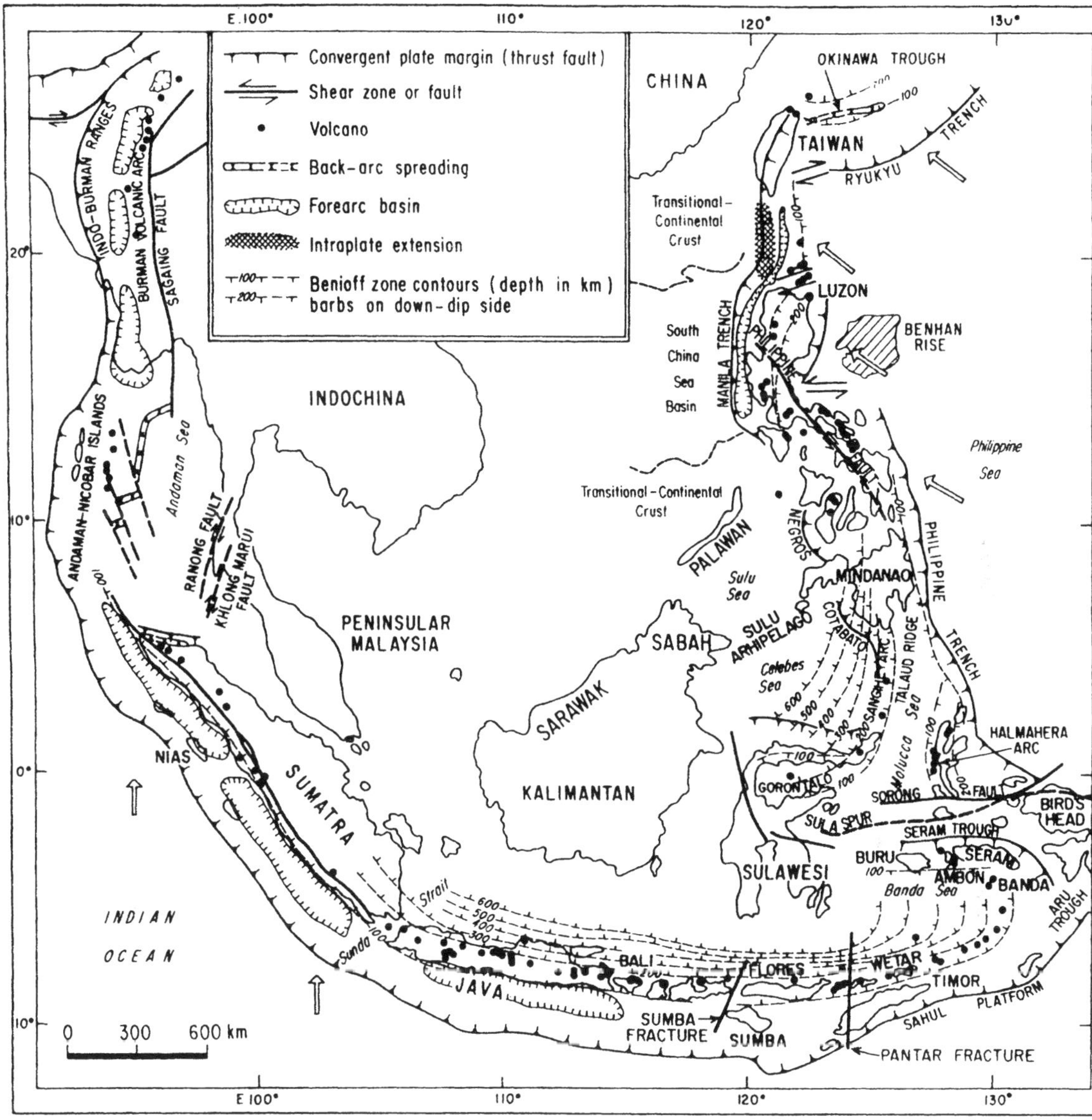

Fig. 1.2. The active plate margins of South-East Asia, showing the position of the trenches, arcs, main transform faults, and Benioff zones, from Hutchison (1989*a*).

accretionary wedges, and volcanic arcs. Some sectors are under collision, in contrast to the ongoing subduction processes, and plate reorganizations are inevitable as marginal basin lithosphere is consumed. Sundaland is outlined by the active arc-trench system, but the sector through Myanmar is under collision. The accretionary prism has been uplifted as the Indo-Burman Ranges and the volcanic arc is extinct. The arc-trench system related to the Manila Trench of the Philippines has collided with the continental shelf of China in Taiwan. There is also collision between Australia and the arc-trench system in the Timor region.

The well developed Benioff Zone of Java does not continue through Sumatra, which is characterized by only shallow seismicity and is dominated instead by right-lateral wrench faulting.

1.2 SEA-FLOOR SPREADING HISTORY OF SOME MARGINAL BASINS

The history of rifting of Sundaland is contained in the many Tertiary sedimentary basins of the region (Fig. 2.1), but there has also been a history of rifting events dating back to the late stages of the Upper Triassic Indosinian Orogeny. In addition to the sediment filled basins, which have attracted the attention of the petroleum industry, the marginal basins contain a good record of advanced rifting leading to sea-floor spreading which enables us to understand the Tertiary evolution of the region better. All of these complicated events took place within the broad back-arc region, behind the convergent arc-trench system. However, the details of the events show that there is no clear relationship between subduction and the rifting history, and models are discussed in Sections 2.1 and 2.2.

1.2.1 Banda Sea

Hartono (1990) has summarized the magnetic lineations which have been identified by Lapouille *et al.* (1985) and by Lee and McCabe (1986), who both agreed that the Banda Sea oceanic crust becomes younger northwards (Fig. 1.3). For the south Banda Sea, the anomalies range from M14 (131 Ma) to M6 (121 Ma), whereas for the north Banda Sea, they range from M3 (117 Ma) to M0 (112 Ma). However, the anomaly identification is not convincing, especially since it has not been confirmed by drilling and must remain tentative.

The central region, which coincides with the Banda Ridges, from which Silver *et al.* (1985) dredged samples of continental sedimentary and metamorphic rocks and volcanic arc andesites (radiometric dates given in Fig. 1.3), is characterized by an absence of magnetic lineations.

The identified magnetic lineations (Fig. 1.3) cannot simply be interpreted as defining the pattern of sea-floor spreading for the following reasons: they are not parallel to the basin floor geomorphology, and some have been drawn crossing the Banda Ridges. Accordingly, there is no unqualified acceptance of the pattern nor the given age of sea-floor spreading of the Banda Sea, and its composite nature has not been given due consideration by the compilers of magnetic anomalies.

The North Banda Sea has a general NW–SE structural pattern as shown by new bathymetric data, interpreted to have resulted from very large NW–SE strike-slip faults (Réhault *et al.*, 1991). The West Buru fracture zone can be traced for 300 km. Closer to Sulawesi lies the Tolo Thrust, which connects northwards with the Matano Fault Zone. These tectonic complexities render the magnetic anomaly interpretation of Lapouille *et al.* (1985) of doubtful value and revision will be necessary following details of the dredging programme (Réhault *et al.*, 1991, 1994), which show that the north and south Banda Sea basins are separated by a Triassic carbonate platform along the Sinta Ridge. South of Lucipara Ridge, Réhault *et al.* (1994) have dredged andesites and basalts which yielded K:Ar dates ranging from 4.8 to 6.0 Ma.

The heat flow is generally low, between 0.4 and 1.7 HFU, which together with the bathymetry, supports the interpreted Mesozoic age of the crust. However, higher values of 3.21 and 4.19 do not fit the picture of old extinct oceanic crust (Van Gool *et al.*, 1987). They indicate the heterogeneous nature of the sea, and infer more recent activity.

1.2.2 Celebes Sea

The magnetic anomalies 30 to 33 (65 to 72 Ma), which Lee and McCabe (1986) reported to have identified, have not been substantiated by subsequent drilling results. Ocean Drilling Program sites 767 (4°47.5′N, 123°30.2′E) and 770 (5°8.7′N, 123°40.1′E) were drilled in late 1988 (Silver *et al.*, 1989*a,b*) (Fig. 1.4).

The basement is of pillowed plagioclase-olivine phyric basalt, whose trace element geochemistry bears the signature of normal mid-ocean ridge basalt (MORB). The sea originated in an open ocean setting in the Middle Eocene (42 Ma). This age is constrained by radiolarian clays which directly overlie the basalt, and is consistent with the earlier identified anomalies 18 to 20 (42 to 47 Ma) of Weissel (1980), shown on Fig. 1.4.

By the Middle Miocene (14 Ma), the Sea received continentally derived turbidite deposits rich in quartz and plant debris (Fig. 1.4). These correlate with rapid uplift of the Crocker Formation of Sabah and with the fluvio-deltaic Tanjong and Sandakan Formations. By the late Miocene (6 Ma) nearby explosive arc volcanism produced thin ash layers in the basin, and continued thereafter to provide a major component to the sediments. No major carbonate units occur at site 767, but thin carbonate turbidites were deposited sporadically in the Middle to Upper Miocene and more commonly in the Pliocene.

The seismic refraction data reveal an apparent downwards flexure of the Celebes oceanic crust as it approaches the Sulu Ridge. This is interpreted by Rangin (1989) as the trace of a now extinct Upper Oligocene to Miocene trench paired to the Sulu Archipelago volcanic arc, which extends into the Dent and Semporna peninsulas of Sabah.

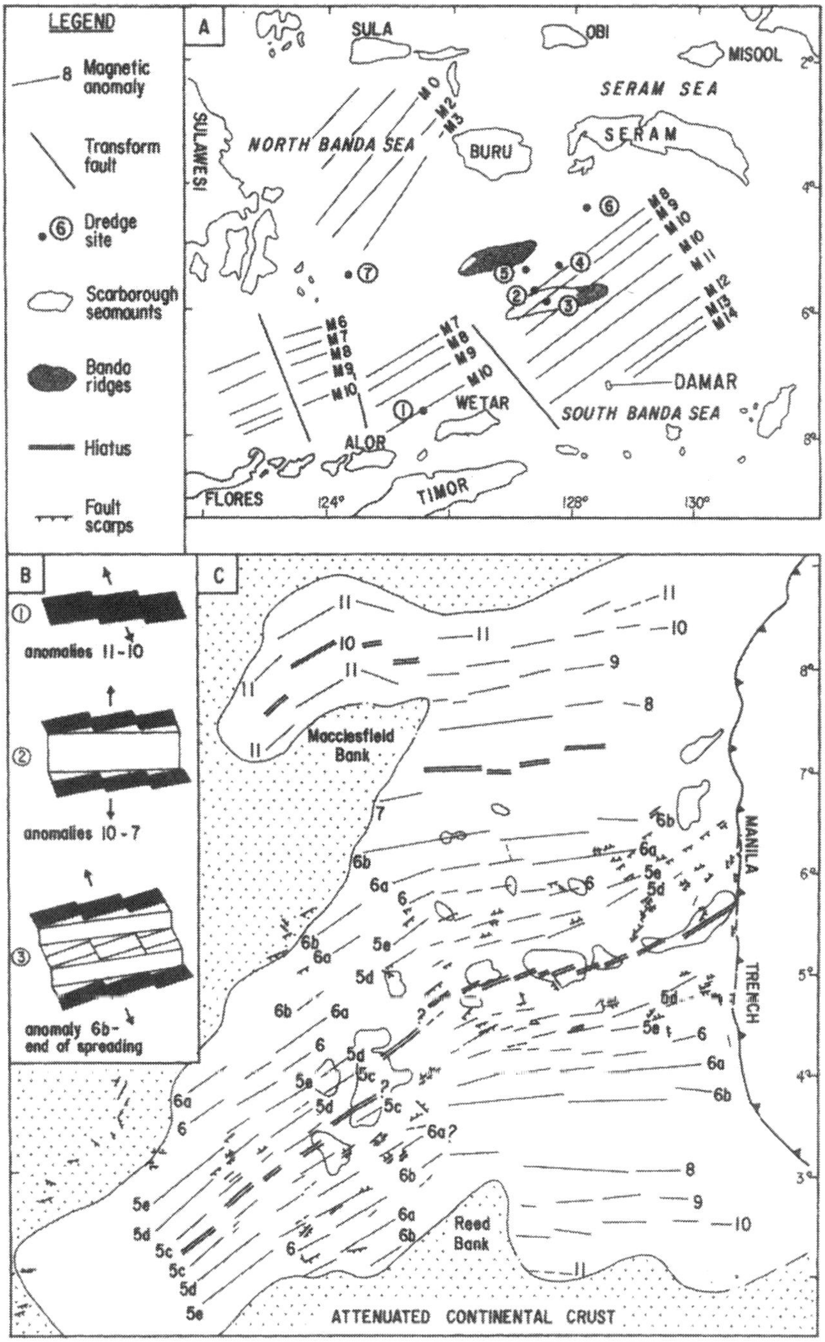

Fig. 1.3. **A**: Banda Sea magnetic lineations (after Lapouille *et al.*, 1985 and Hartono, 1990). Positions of the northern Banda micro-continents, dredge sites and K:Ar dates from Silver *et al.* (1985). Dredge details: 1 = basalt 0.40 ± 0.01 Ma; 2 = phyllite 22.50 ± 0.5 Ma; 3 = pyroxene andesite 6.16 ± 0.08 Ma; 4 = pyroxene andesite 7.03 ± 0.17 Ma; 5 = amphibolite 10.5 ± 0.1 Ma; 6 = pyroxene andesite 0.75 ± 0.02 Ma and 7 = meta-diabase 8.75 ± 1.48 Ma. **C**: Proposed magnetic lineations for the South China Sea Basin, after Briais *et al.* (1993). The Scarborough Seamounts delineate the relict spreading axis. **B**: The three distinct spreading episodes.

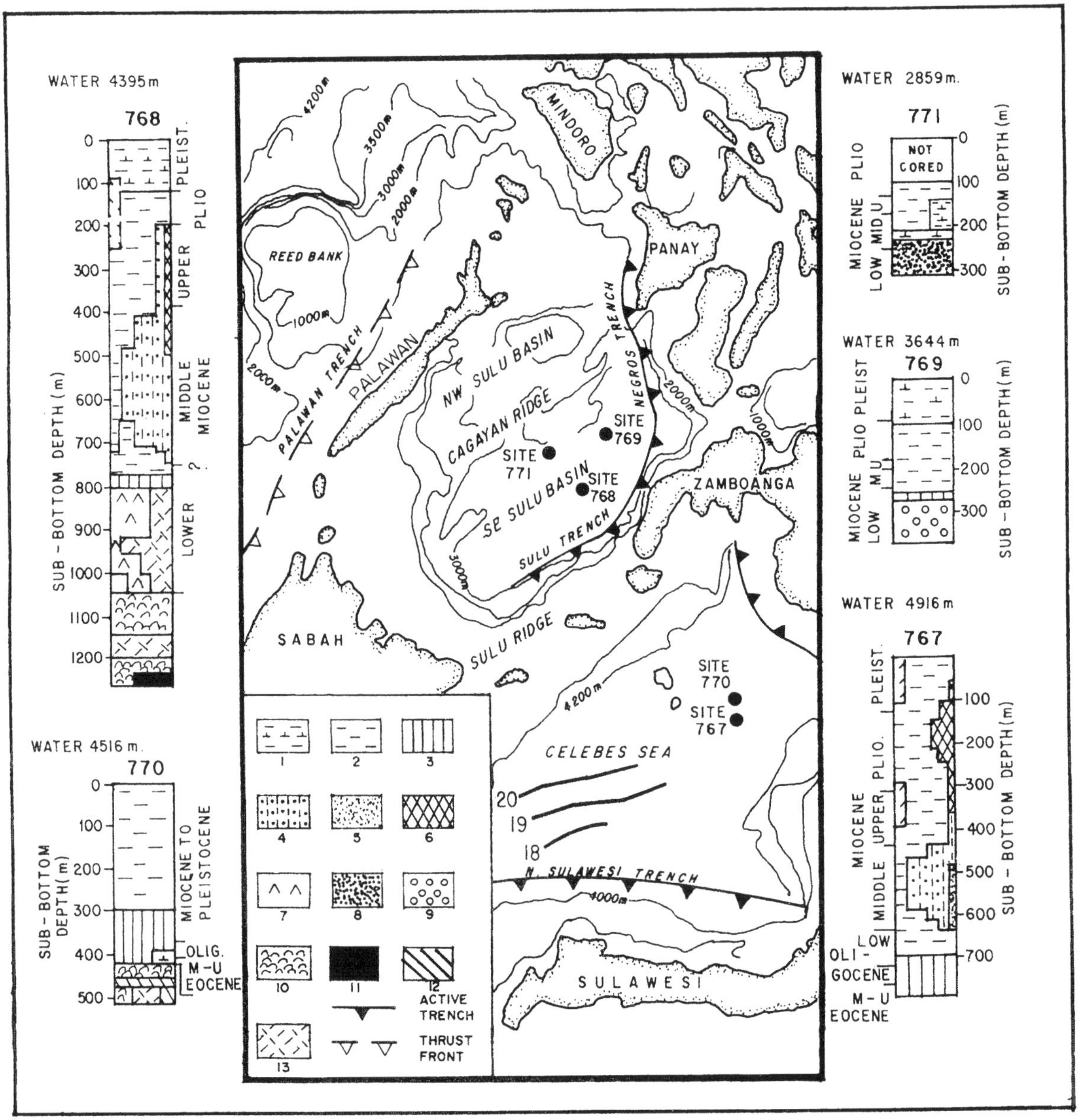

Fig. 1.4. ODP sites in the Celebes and south-east Sulu Sea (after Silver *et al.*, 1989*a*,*b*). Core lithologies are: 1 = nannofossil marl or nannofossil-foraminiferal marl; 2 = hemi-pelagic sediments including clay or silt(stone); 3 = pelagic brown claystone; 4 = terrigenous turbidite; 5 = quartz siltstone to sandstone; 6 = graded carbonate turbidite; 7 = fine ash or tuff; 8 = pumiceous rhyolitic to dacitic coarse tuff and lapilli-stone; 9 = andesitic to basaltic coarse tuff and lapilli-stone; 10 = pillow basalt; 11 = basalt sheet flow; 12 = brecciated massive basalt; 13 = diabase sill. Magnetic anomalies 18, 19 and 20 (42 to 47 Ma) are from Weissel (1980).

1.2.3 Sulu Sea

The geology of the Sulu Sea (Fig. 1.5) has been described by Hutchison (1992*a*).

1.2.3.1 The North-west Sulu Sea

The NW Sulu Sea has a thick crust. The seismic basement is of highly deformed layered rocks (Fig. 1.6) and the most attractive interpretation is that it represents a continuation of the Rajang Group of Sabah (Rangin, 1989; Rangin *et al.*, 1990). This includes a Lower Cretaceous ophiolite complete with Barremian chert (Basir *et al.*, 1985), overlain by the Oligocene to Lower Miocene Crocker Formation. Equivalents of the Sabah Rajang Group on Central Palawan have given nannofossils ranging in age from Maastrichtian to Upper Oligocene (Rangin *et al.*, 1990).

An obvious unconformity, shown to be early Middle Miocene in the drilled Balabac Basin (Beddoes, 1976), separates this basement from the overlying relatively undeformed Neogene strata, which commonly attain a thickness of 7000 m in the main basin depocentres (Fig. 1.6). It is therefore very likely that the deformed

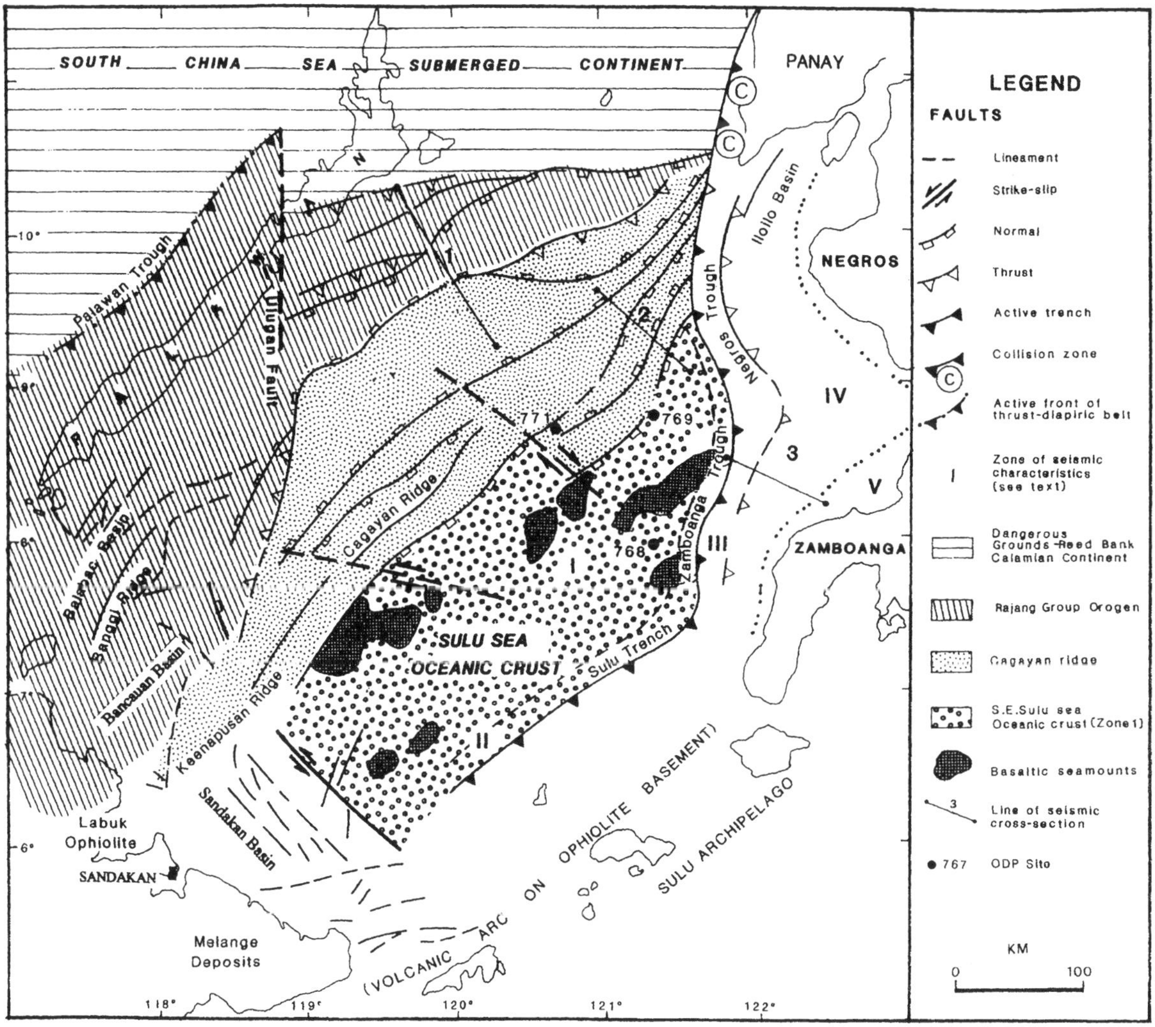

Fig. 1.5. The structural and geological features of the Sulu Sea (after Rangin, 1989; and Hinz and Block, 1990), showing also the location of the seismic lines of Fig. 1.6.

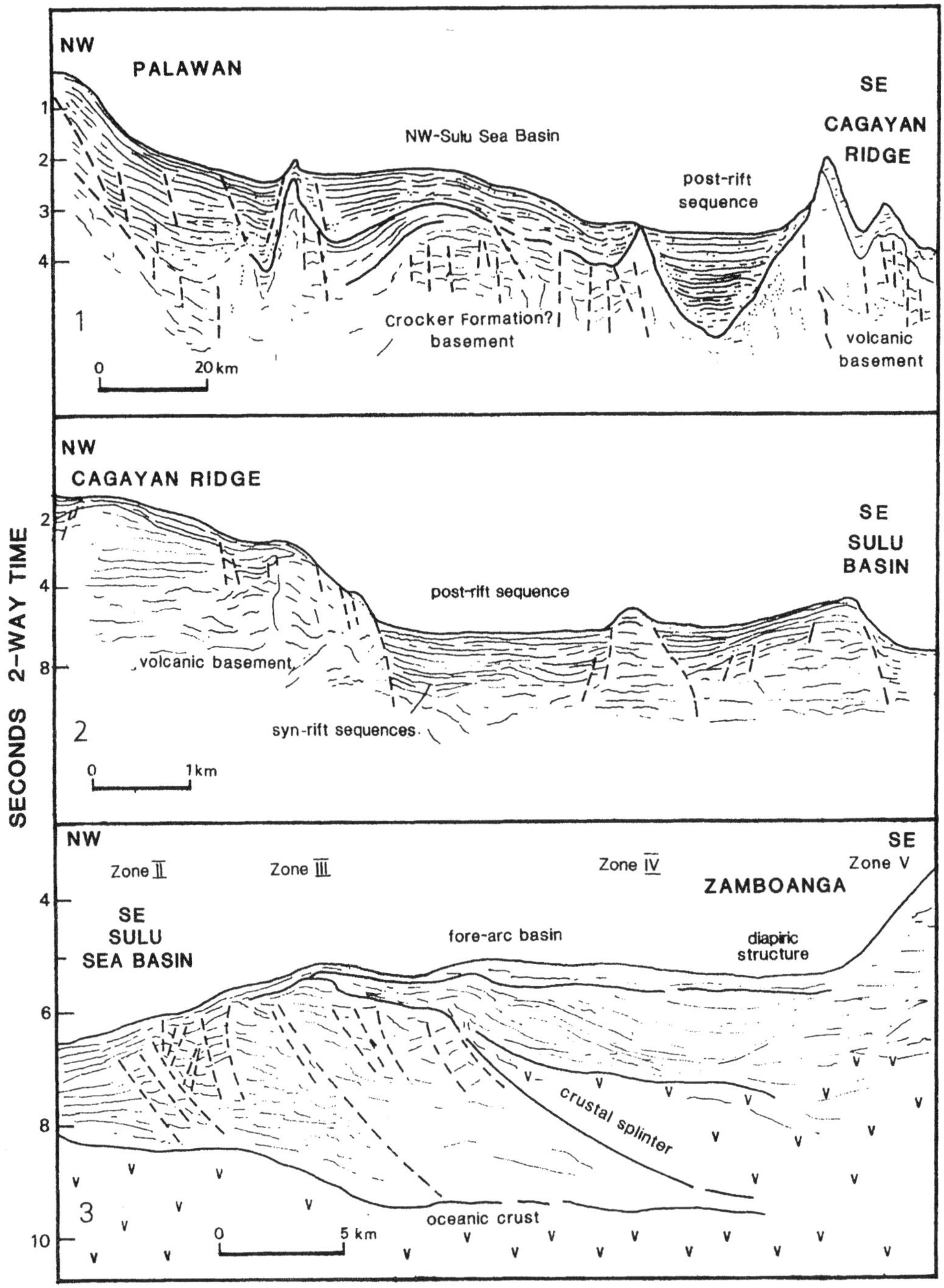

Fig. 1.6. Three NW–SE seismic reflection sections across the Sulu Sea, see Fig. 1.5 for locations. **1**: The North-west Sulu Sea (Hinz, 1983). A Middle Miocene age is assumed for the distinct unconformity between 2.5 and 3 s (2-way time). The thickness of sediments above it exceeds 7000 m. The chaotic and diapiric basement in the centre may be correlated with the pre-Middle Miocene Crocker fold-belt of Sabah. **2**: Profile across the south-eastern flank of the Cagayan Ridge showing the volcanic basement cut by SE-facing normal faults (after Rangin, 1989). **3**: Profile across the active Sulu Trench, showing structural zones II, III, IV and V (after Hinz and Block (1990).

Rajang Group, which now underlies the NW Sulu Sea, formed a landmass continuous with Sabah (the Sabah–Palawan Orogenic Belt) before the early Middle Miocene unconformity. This landmass was fringed along its SE cordilleran-type margin by the Cagayan volcanic arc. Early Middle Miocene rifting affected the whole Sulu Sea area and caused extensive subsidence, the greatest being centred in the SE Sulu Sea marginal basin.

1.2.3.2 The Cagayan Ridge

Andesite, containing phenocrysts of plagioclase, clinopyroxene and olivine, was dredged from the Meander Reef on the Cagayan Ridge (Kudrass *et al.*, 1986). The K:Ar age of 14.7 Ma is consistent with Lower to Middle Miocene carbonate sediments dredged from another seamount on the Cagayan Ridge. The basement at site 769 is of massive unstratified dark green coarse lapillistone of andesitic to basaltic composition, devoid of admixed sediment. ODP site 771 bottomed in late Lower to early Middle Miocene massive structureless volcaniclastic lapilli-stone underlain and capped by basaltic lava flows (Silver *et al.*, 1989*a*,*b*). The intercalations of flows and pyroclastic strata suggest close proximity to volcanic vents, which are likely to have extended above sea level and to have vented subaerially.

ODP Sites 769 and 771 on the flank of the Cagayan volcanic ridge, penetrated a thin sedimentary section devoid of turbidites (Fig. 1.4). The volcanic and pyroclastic rocks are immediately overlain by brown claystone containing late Lower to early Middle Miocene radiolaria. This age therefore represents the cessation of volcanism of the Cagayan Ridge as well as spreading of the SE Sulu Sea Basin. Rifting of the volcanic arc must have resulted in its rapid foundering. The basaltic to andesitic volcanic tuffs in 769 may correlate with the tuffs underlying the Sandakan Formation onshore Sabah, which contain Oligocene fossils (Rangin *et al.*, 1990). The andesite tuff, beneath the Sandaken Mosque, contains apatite crystals which have yielded a fission track age of 33.9 ± 7.7 Ma (Oligocene) (S. Bergman, pers. comm. Nov. 1994). Because the rock is a tuff, the apatite ages may indicate the age of the active volcanism which gave rise to the tuffs. This age is consistent with the fossil content.

1.2.3.3 The Sulu–Zamboanga–Negros subduction system

The South-east Sulu Sea has been shown by Hinz and Block (1990) to be divisible into five structural zones (Figs 1.5, 1.6). Zone I forms the major part of the deep SE Sulu Sea basin. Heat flow is relatively high, ranging up to 4.74 HFU. Its igneous basement is relatively smooth and has been drilled at ODP site 768. Its seismic signature is typical of oceanic crust, overlain by a sedimentary layer varying in thickness from 1 to 2 s (2-way time). It is dominated by a chain of basement highs trending 60°, from which basaltic rocks have been dredged. Hinz and Block (1990) report that there is no evidence for the existence of the magnetic lineations 17 to 20 of Lee and McCabe (1986), further disproved by the ODP drill data.

The basement at site 768 is of pillow basalt and sheet flows, intruded by dolerite sills (Fig. 1.4). The basalt has trace element geochemistry signatures transitional between MORB and island-arc tholeiite.

The SE Sulu Sea Basin seems to have originated as an intra-arc basin in the late Lower to early Middle Miocene as determined by radiolarian biostratigraphy in sediments overlying the volcanic basement (Fig. 1.4). Initial deposition of volcanic brown clay was followed shortly thereafter by rapidly deposited coarse mass-flow pyroclastic rhyolitic and dacitic tuffs composed of devitrified glass shards and pumice. Slow accumulation rates (9 m Ma^{-1}) marked the early part of the Middle Miocene, but in the late Middle Miocene (10.5 Ma) very rapidly deposited (300 m Ma^{-1}) continentally derived turbidites inundated the basin, and volcanic ash layers appeared at 6 Ma.

Zone II represents the 4550 m deep trench of the active subduction zone. The E to SE-descending oceanic basement dips at 10° and is overlain by 0.5 to 1.8 s (2-way time) of sediments. Heat flow lies between 1.87 and 2.37 HFU.

Zone III is of a sedimentary sequence characterized by imbricate thrust sheets. The faults dip SE and E and curve downwards to join a major décollement surface which lies just above the descending igneous oceanic crust.

An up-thrust slab of oceanic crust supports a sedimentary apron 2.4 s thick (two-way time; Fig. 1.6), which resembles a fore-arc basin; this forms Zone IV. The sediments have a gentle 8° dip towards the E. This fore-arc basin continues northwards to be represented by the Iloilo Basin of central Panay, which contains limestones, coarse clastics and basalt flows. Heat flow values range from 1.07 to 1.69 HFU.

Zone V forms the western slope of Zamboanga and Negros. A sedimentary apron, ranging from a few hundred metres to 2.8 s (2-way time), overlies the basement complex of the Sulu Archipelago and Zamboanga. The Sulu Archipelago islands consist of ophiolitic rocks (e.g. Tawi-Tawi) and Neogene volcanic rocks (e.g. Jolo).

The seismic profile across Zones II through V (Fig. 1.6) is remarkably like that of a normal arc-trench system, but on a miniature scale. The distance from the trench to the Zamboanga Peninsula and Jolo island is little more than 100 km and no Benioff Zone has yet been analysed for this region. The miniature dimensions of the subduction system make it virtually impossible for Sulu Sea oceanic lithosphere to have subducted to a depth adequate for creation of a volcanic arc. It is by no means certain that the volcanic forms on Jolo and Zamboanga are active, and they should not in any case be related to the immature Sulu Trench.

The borehole televiewer showed that the short axis of the elliptical drill hole at ODP site 768 is oriented north-east, representing the present maximum stress direction of the crust (Silver *et al.*, 1989*a*,*b*). This is consistent with north-easterly subduction beneath the active Sulu, Zamboanga and Negros trench system.

1.2.3.4 Related rift events on-land Sabah

Broken formations and mud-matrix mélanges are abundant, especially in the Dent Peninsula of eastern Sabah. They have been given a variety of names, such as the Garinono, Ayer, Kuamut and Kalumpang formations (Hutchison, 1992*a*). Their petrography and field geology have been described by Clennell (1991, 1992). They were formed in a marine environment during the late Lower Miocene and earliest Middle Miocene. They are not metamorphosed and do not form narrow linear belts, but are geographically extensive. Unfortunately outcrops of the mélanges have a short outcrop life in this climate of heavy rainfall. They must represent surficial formations. The clasts in the mud-matrix mélanges can be traced to contiguous Palaeogene formations, such as the Kulapis and Labang, and to the ophiolitic basement and associated Lower Cretaceous chert. It is concluded that the breaking of the formations and mélange formation are genetically related to the onset of extension and sea-floor spreading of the SE Sulu Sea Basin, which extrapolates south-westwards into Sabah (Hutchison, 1992*a*; Clennell, 1991). It should be noted that Hamilton (1979) was responsible for the 'law' that all South-East Asian mélanges are subduction-related. The eastern Sabah mélanges are the opposite—they are extension related, formed during rifting of a volcanic arc to form the SE Sulu Sea by intra-arc spreading (Hutchison, 1992*a*). The SE Sulu Sea is the only South-East Asian marginal basin to be demonstrated to fit and support the intra-arc rifting hypothesis of Karig (1971). This has come to be the most popular model for marginal basin formation, but it is supported only by the Sulu Sea and no other. The Cagayan Ridge is the remnant arc, which died off before deposition of the deltaic Sandakan Formation, and the Sulu Archipelago continued active as the arc-trench system migrated south-eastwards. The first sediments to be deposited into the SE Sulu Sea rift were volcanogenic (Fig. 1.4), in keeping with an intra-arc setting. Sea-floor spreading of the SE Sulu Sea would have been related to north-westwards subduction of the Eocene lithosphere of the Celebes Sea; the trench migrating south-eastwards, leaving the Cagayan Ridge behind (Hutchison, 1992*a*).

1.2.4 South China Sea

Most of the South China Sea is shallow water and underlain by attenuated continental crust. The deeper part, lying between Vietnam and the Philippines, is a marginal basin characterized by sea-floor spreading (Fig. 2.2).

1.2.4.1 The South China Sea marginal basin

The South China Sea Basin represents a part of Sundaland that rifted and subsided to the stage of marginal basin sea-floor spreading.

Sea beam surveys of the central axial region of the South China Sea Basin exhibit two major structural trends: scarps trending 50°, interpreted as normal faults, and scarps trending 140°, interpreted as fracture zones (Briais *et al.*, 1989). Magnetic and gravimetric surveys in the Scarborough Seamounts region show anomalies trending approximately 50°, disrupted by transform faults striking 140°, spaced at less than 20 to 30 km. Briais *et al.* (1993) have reported a rather detailed analysis of the data of Taylor and Hayes (1983) supplemented by a vast amount of Chinese data (Chen, 1987). Their analysis supports the general conclusions of Taylor and Hayes (1983) for the main basin, but is more comprehensive. The main spreading history remains from anomaly 11 (31 Ma, Oligocene) to anomaly 5c (16 Ma, Middle Miocene) (Fig. 1.3). The choice of magnetic anomaly designations for the main eastern basin was strongly influenced by the results of Hékinian *et al.* (1989), who dredged volcanics from the Scarborough Seamounts, radiometrically dated 10 to 15 Ma. These volcanics are interpreted to have been extruded along the still weak relict spreading axis.

Briais *et al.* (1993) have also proposed that the NW sector of the basin north of Macclesfield Bank has anomalies 10 and 11, and the SW sub-basin has anomalies 6b (23 Ma) to 5c (16 Ma); however, I have to admit that the identification of these anomalies is far from convincing (Briais *et al.* 1993, pages 6309–11). It has to be concluded that their analysis is the best that can be achieved with the present data, and a drilling programme is awaited. They analyse the spreading history of the South China Sea basin in three distinct episodes: 32–30 Ma (anomalies 11 to 10), characterized by NNW–SSE spreading; 30 to 26 Ma (anomalies 10 to 7), characterized by N–S spreading, then 23 to 16 Ma (anomalies 6b to 5c), characterized by NNW–SSE directed spreading (Fig. 1.3). The authors rightly admit that their interpretation and choice of anomalies are not unique.The spreading of the South China Sea terminates westwards against the major Red River wrench Fault.

1.3 OIL AND GAS PRODUCTION AND RESERVES

The petroleum industry is, with a few exceptions, almost exclusively concerned with Tertiary basins, and the pre-Tertiary is considered economic basement. However oil, sourced from the Tertiary Talang Akar Formation, has migrated down and is reservoired in fractured granodiorite and metaquartzite in the Kuang area basement to the South Sumatra Basin (Sardjito *et al.*, 1991). Similarly, Tertiary hydrocarbons in the Gulf of Thailand have been locally reservoired in Triassic limestone, and in fractured granite basement of the Mekong Basin. Indonesia dominates the industry, and its major oilfields are in Sumatra and Borneo (Kalimantan). Indonesia uses the term Kalimantan to refer to the whole island, a practice upsetting to the Malaysian and Brunei governments, who use it to refer strictly to Indonesian Borneo. Large areas of the South China Sea, such as the Spratly Islands and Dangerous Grounds (the Nansha Block of China), which are underlain by attenuated submerged continental lithosphere, generally await international agreement on ownership or production sharing before exploration.

The geology of the sedimentary basins, which contain the bulk of the hydrocarbons of the region, has been summarized in Chapter 3 of the 1989 volume. Chapters 2 to 6 of this book give a summary of the known oil and natural gas basins condensed and illustrated in Fig. 2.1.

The 1992 average daily production of oil in barrels (= 158.987 litres) and gas in cubic feet (28.32 litres or 0.02832 m^3) by individual countries is listed in Table 1.1, compiled from magazines such as *Petromin*, published in Singapore, and *World Oil* and *Oil and Gas*, in the U.S.A.. The reserves are also tabulated, but since these are estimated, there is commonly a range of values, depending upon the publication, its sources and viewpoint. In addition, there are analytical papers that provide a picture of the older country production and resources (Masters *et al.*, 1991).

An attempt has been made to calculate production from individual oil-bearing basins and to place the SE Asian basins in order of importance (Table 1.2). The raw data are predominantly from the *Oil and Gas Journal* (Vol. 90, No. 52, 1992). However, it is virtually impossible to calculate numerically accurate cumulate production for basins, because the published country data are by field and operator, and abandoned fields are not tabulated. Fields also may change name. Some countries do not even release data for individual oil fields. The analysis by Sumii *et al.* (1992) falls far short of acceptable accuracy in basin production. Table 1.2 represents the 'best' numbers which can be currently calculated. For Indonesia, I have been greatly assisted by reference to the Oilfield Atlases, published by the Indonesian Petroleum Association, which do tabulate abandoned fields. A further problem is that fields overlap, for example the Baram Delta and Sabah, West Natuna and Malay Basin, Malay Basin and Gulf of Thailand, NW Java and Sunda Basin. The allocation of a particular field to a basin therefore is a matter of personal opinion. The Bintuni Basin of eastern Indonesia has not been listed, because it is a disappointing basin and all fields are now abandoned.

Gas production is difficult to quantify since it is poorly documented, and gas tends to be maintained for future production according to local and regional demand. No

Table 1.1. Estimated country proven reserves and 1992 actual production

Country	Reserves (at Jan 1 1993)		1992 Production	
	Oil in 10^6 barrels	Gas in 10^{12} cubic feet	Oil in 10^3 barrels/day	Gas in 10^6 cubic feet/day
China	24 000 to 28 000	36.0 to 49.4	2840	1460
Indonesia	6581 to 10 900	64.4 to 104.2	1400	6500
India	5080 to 6125	25.9 to 42.4	583	1875
Malaysia	3600 to 3700	59.0 to 67.8	637	2142
Brunei	1350 to 1400	12.2 to 14.0	183	1010
Myanmar	50 to 1470	9.3 to 9.8	16	80
Vietnam	500 to 750	0.5 to 1.4	96	87
Bangladesh	1 to 500	12.7 to 25.4	0.5	558
Philippines	147 to 220	1.6 to 2.2	15	–
Thailand	150 to 242	6.6 to 8.4	49	777
Taiwan	4 to 50	0.7 to 2.4	1.3	86
Total world	997 041.7	4885.4	60 029	

Source: *Oil & Gas Journal* 1992 (Dec. 28), Vol. 90 (No. 52), and Petromin. (Singapore), Jan. 1993

Table 1.2. Cumulative production of the larger South-East Asian oil fields

Basin	Total Production[1]	Fields	Oil/ Gas	Oldest Field	Common reservoir	Top 3 Fields [total production[1], and year field began]
1. Central Sumatra	7577	108	O	1939	Miocene/Pliocene sandstone	Minas '44 [3 603], Duri '41 [655], Bekasap '55[426]
2. Baram Delta	3062	21	O	1910	Miocene/Pliocene sandstone	Seria '29 [1 003], SW Ampa '63[665], Champion '70 [346]
3. Kutei	2031	23	O	1897	Miocene/Pliocene sandstone	Handil '74[720], Attaka '70 [522], Sanga-Sanga '97[267]
4. South Sumatra	1705	83	O	1913	Oligocene/Miocene sandstone	Limau '28[216], Talang Jimar '37[173], Benakat '32 [140]
5. Malay	1073	12	O/G	1971	Miocene sandstone	Tapis '78 [252], Guntong '85 [113], Seligi '88 [71]*
6. North Sumatra	937	29	O/G	1926	Miocene/Pliocene limestone/sandstone	Arun '71[468 gas], Rantau '29[242], Tualang '73[31]
7. NW Java	820	32	O/G	1969	Oligocene/Miocene limestone/volcanics	B [122], Jatibarang '69 [90], E [161]
8. Sunda	614	18	O	1970	Oligocene/Miocene sandstone/limestone	Cinta '70 [191], Rama '73[98], Krisna '76[67]
9. S Burma Myanmar	509	5	O	1800	Miocene/Oligocene sandstone	Yenangyaung '02[192], Chauk-Lanywa '02 [128], Htaukshabin [88]
10. West Natuna	467	5	O	1974	Miocene sandstones	Kakap KH '80 [379], Udang '74 [63], Kakap KF '84[20]
11. Sabah	461	5	O	1971	Miocene sandstone	Samarang '72[308], St Joseph '75 [62], W Erb '71[52]
12. Salawati	352	19	O	1936	Miocene limestone	Walio '75[165], Kasim '73 [50], Klamono '36 [36]
13. Tarakan	323	9	O	1906	Miocene/Pliocene sandstone	Pamusian '06 [193], Bunyu '23 [85], Sembakung [23]
14. NE Java	165	33	O	1893	Miocene sandstones	Kawengan '26 [88], Ledok '93 [25], Lidah [16]
15. Barito	116	31	O	1938	Eocene sandstone	Tanjung '38[104], Warukin '66[11]
16. Sukothai	58	5	O	1981	Miocene sandstone	Sirikit '81 [57], Sirikit West '83[0.7], Thap Raet '83 [0.3]
17. Gulf of Thailand	52	5	G	1973	Oligocene/Miocene sandstone	Erawan '73 [26], Satun '80[13], Platong '76[8] gas
18. West Palawan	41	6	O	1977	Oligocene/Miocene limestone/sandstone	Nido [16], Cadlao [11], Matinloc [10]
19. Taiwan	27	4	O	1904	Miocene and Pliocene sandstones	Tiechengshan '62 [16]. Chingtsohu and CBK '67[4]
20. Bula (Seram)	19	6	O	1897	Mesozoic and Pleistocene	Bula '97[16], Lemun [3]

The oil basins are identified by the same numbers on Fig. 2.1.
Compiled from data in **Oil and Gas Special**: *Oil and Gas Journal* (Dec. 28, **1992, pp. 52–79**), with Indonesian control from the Field Atlases, Vols 1–6, published by the Indonesian Petroleum Association, Jakarta. ♣ By courtesy of Effendy Cheng Abdullah, Petronas exploration department.

1 barrel = 158.987 litres = 0.14 tonne (approx.)
[1] Production in millions of barrels oil (or equivalent) up to the end of 1991

attempt therefore has been made to assess the gas potential of the basins. The Gulf of Thailand, Luconia Province of Sarawak (Malaysia), Gulf of Martaban (from which gas is being piped to Bangkok) and the North Sumatra Basin are the most important established gas fields of the region.

Table 1.2 clearly shows the pre-eminence of the Central Sumatra Basin in SE Asia, followed by the Baram Delta. There are few giant fields in the region, mention should be made of the outstanding Minas in Central Sumatra and Seria in the Baram Delta (Brunei). Both actively produce oil.

The gas fields of Bangladesh are scattered SW and NW of Chittagong and its small oilfields near Dhaka and Agartala. No large fields have been discovered.

The Baram Delta Province extends westwards from Brunei into Sarawak and north-eastwards into offshore

Sabah. The onland Brunei Seria Field is the largest and has continued active since its discovery in 1929 to the present, but the onland Miri Field of nearby Sarawak has been abandoned.

Oil company activity in Myanmar has moved southwards into the Gulf of Martaban. In the Irrawaddy Delta Basin, the Prome Field has produced 7×10^6 barrels and Myanaung 21×10^6 since 1964 from Miocene sandstones. Generally Myanmar has proved disappointing and reserves continue to be re-estimated downwards.

In eastern India, there are a number of moderate sized fields in the Assam area. The Digboi Field has produced 106×10^6 barrels of oil since 1889, Nahorkatiya 376×10^6 barrels since 1953 and Rudrasagar 39×10^6 barrels since 1960. These fields continue to produce.

Eastern Indonesia has been generally disappointing and the Bintuni Basin fields have now been abandoned.

In northern Thailand, the Fang Basin contains two main oilfields, the Mae Suhn has produced more than 3×10^6 barrels of oil since 1963 from lacustrine Miocene–Pliocene sandstones. The fields are operated by the Royal Thai Army along the border of the 'Golden Triangle'.

There are no released production figures for Vietnam. Production is known from the Gulf of Bacbo (Tonkin), possibly up to a few tens of tonnes of oil per day. The

Table 1.3. Country and world 1990 production of mineral commodities (in metric tonnes)

	World Total	China	Myanmar	Thailand	Malaysia	Indonesia	Philippines	Vietnam
Antimony[1]	66 846	35 000	–	500	–	–	–	–
Barite[2]	5 594 657	1 750 000	9145	107 707	48 291	–	500	–
Bauxite[3]	109 601 000	2 400 000	–	–	398 000	1 206 000	–	–
Bentonite[4]	9 733 277	–	700	–	–	5914	16 484	–
Copper[5]	8 693 000	300 000	4400	–	24 300	178 000	182 000	–
Chromium[6]	13 541 837	25 000	–	416	–	7635	173 200	3500
♠ **Diamonds**[7]	101 566 000	1 000 000	–	–	–	30 000	–	–
Fluorite[8] **[1989]**	5 585 712	1 700 000	–	98 375	–	–	–	–
Gold[9] **[in kg]**	2 127 398	100 000	–	–	2594	11 158	24 591	–
Gypsum[10]	94 859 800	7 981 800	30 800	5 752 000	–	–	119 700	27 200
Ilmenite[11]	3 951 611	150 000	–	10 674	530 237	–	–	–
Iron ore[12]	548 494 000	58 000 000	–	71 000	210 000	84 000	–	–
Kaolin[13]	23 672 714	no report	–	347 690	152 972	160 098	–	750
Lead[14]	3 390 000	364 000	4 400	22 200	–	–	–	–
Manganese[15]	8 980 000	653 000	–	–	–	–	–	–
Mercury[16]	6263	700	–	–	–	–	–	–
Molybdenum[17]	122 232	15 700	–	–	–	–	–	–
Monazite[18]	21 841	–	–	377	3323	–	–	–
Nickel[19]	937 173	27 000	80	–	–	68 308	15 818	–
Phosphate[20]	48 628 000	6 200 000	–	3000	–	1000	1000	96 000
Salt[21]	182 303 000	20 000 000	263 000	218 600♠	–	598 000	490 700	340 000
Silver[22] **[1989]**	14 760	125	28	–	13	74	48	–
Sulphur[23]	58 154 000	5 370 000#	–	–	–	4000 N	319 000Ψ	–
Talc [T][24] **& Pyrophyllite [P]**	9 084 414	2 200 000	20 steatite	29 290 P 4360 T	–	–	–	
Tantalum[25] **& niobium [in kg]**	32 103 000	–	–	9000	2000	–	–	–
Tin[26]	218 518	42 000	634	14 635	28 468	30 200		900
Tungsten[27]	51 918	32 000	311	290				
Zinc[28]	7 319 600	619 000	2200	81 000	–	–	100	–
Zirconium[29]	852 512	15 000	–	490	4279	2500	–	–

Source: 1 = Llewellyn (1992); 2 = Searls (1992); 3 = Plunkert and Sehnke (1993); 4 = Virta (1993); 5 = Jolly (1993); 6 = Minerals Yearbook (1993); 7 = Austin (1992); 8 = Minerals Yearbook (1993); 9 = Lucas (1992); 10 = Davis(1992); 11 = Gambogi (1992); 12 = Kirk (1993); 13 = Virta (1993); 14 = Woodbury (1992); 15 = Jones (1992); 16 = Minerals Yearbook (1993); 17 = Blossom (1993); 18 = Hedrick (1992); 19 = Kuck (1993); 20 = Stowasser (1992); 21 = Kostick (1992); 22 = Minerals Yearbook (1993); 23 = Ober (1992); 24 = Virta (1992); 25 = Cunningham (1993); 26 = Carlin (1992); 27 = Smith (1992); 28 = Jolly (1992); 29 = Templeton (1993). **NOTES: 6** ≡ for 1989; **♠7** ≡ **Diamonds (in carats,** = 0.2g); **9** ≡ **Taiwan** 72; **10** ≡ **Laos** 113 400, **Taiwan** 1 700; **13** ≡ **Taiwan** 105 084, **Bangladesh** 7 000; **16** ≡ for 1989; **18** ≡ Monazite concentrate; **20** ≡ Rock phosphate; **21 ♠** ≡ Thailand [118 800 rock salt + 99 800 other]; **21** ≡ Bangladesh = 216 000; Taiwan = 82 500; Cambodia = 40 800; Laos = 8 160. **22** ≡ **Taiwan** 6; **23** ≡ P (from pyrite), N (native), B (by-product). # China = [4 400 000 P + 320 000 N + 650 000 B]; Ψ Philippines = [134 000 P + 185 000 B]; **23** ≡ Singapore 60 000 B; **24** ≡ **Taiwan** 22 123T; **26** ≡ **Laos** 500.

White Tiger oilfield of the Mekong Basin, discovered in 1973, produces about 50 000 barrels per day from Lower Miocene strata.

Most of China lies outside SE Asia. Its gas is mainly produced from the Sichuan Basin, containing the world's earliest producing oil- and gas field, and it is still producing on a world scale. The southern offshore region has been disappointing. The Pearl River Mouth Basin, SE of Hong Kong, produces oil, but China does not release individual basin or field data. Half of China's oil is produced from the giant Daqing field of NE China, and exploration will soon be focused on the remote NW Tibet basin of Tarim.

1.4 MINERAL COMMODITIES

General reviews of the mineral deposits and some production figures were given by Hutchison and Taylor (1978), Taylor and Hutchison (1979) and Hutchison (1987). A tabulation of the country production of mineral commodities (in metric tonnes) for 1990 is presented in Table 1.3, compiled from many sources listed at the foot of the table. Table 1.4 gives the production figures for the year 1991, taken exclusively from Mining (1992). There are discrepancies between the two tables, and I am not in a position to comment on which is the more accurate. The more one knows about official production figures, the more one becomes sceptical regarding their accuracy.

China has increased its share of the world output of antimony from 17% in 1979 (Hutchison, 1983*a*) to 52% in 1990. Thailand has fallen from 7% to 0.7% and Malaysian production (from Bau, Sarawak) has ceased.

China has increased its barite share from 6% in 1979 to 31% in 1990 and the SE Asian oil industry has given impetus to a local mining industry, especially in Thailand. Malaysian production will continue to decline because no new deposits have been found.

The largest chromite mines of SE Asia occur in the Philippines (Zambales), but the Philippine world share of chromium has decreased from 6% in 1979 to just over 1% in 1990.

The island arcs of the Philippines are characterized by a large number of porphyry copper and epithermal gold deposits. Gold, however, occurs fairly widespread throughout the more continental parts of South-East Asia, such as Kalimantan, Peninsular Malaysia and Thailand. Gold exploration and mining activity have dramatically increased in the past few years especially in Indonesia (notably Kalimantan and Sulawesi).

In the 1960s and 1970s Malaysia was a major producer of iron ore, but by the time this book is published all mines will have been worked out.

The island of Palawan (Philippines) used to produce about 4% of the world's mercury, but production has now ceased. The minor production from Bau (Sarawak) has long since gone.

Table 1.4. Country mineral production for 1991

	China	Myanmar	Thailand	Malaysia	Indonesia	Philippines	Units
Antimony	35.0	–	–	–	–	–	in 10^3 tonnes
Barite	1750	0.67	93.0	–	–	–	in 10^3 tonnes
Bauxite	2500	–	–	347.4	1240	–	in 10^3 tonnes
Chrome ore	–	–	–	–	–	172.8	in 10^3 tonnes
Clay (Kaolin)	not avail	–	255.0	186.7	–	–	in 10^3 tonnes
Coal[4]	1090	0.03	14.69	–	14.53	1.27	in 10^6 tonnes
Copper in concentrates	560.0[1]	14.2	–	98.3	656.5	144.9	in 10^3 tonnes
Fluorite	1700	–	60.6	–	–	–	in 10^3 tonnes
Gold	–	20.8	–	2747[2]	17,024	26 240	in kg
Gypsum[3]	8300	24.9	7196	–	–	–	in 10^3 tonnes
Iron ore	170.0	–	–	0.35	–	–	in 10^6 tonnes
Lead ore	300.0	1.4	39.2	–	–		in 10^3 tonnes
Nickel ore	29[1]	–	–	–	2300	563.2	in 10^3 tonnes
Silver	–	14.6	–	13 439	89 690	38 414	in kg
Tin	26.0	0.13	14.9	20.5	30.4	–	in 10^3 tonnes
Tungsten	30 000	6.7	440	–	–	–	in tonnes

Source: Mining 1992
[1] = refined metal [2] Mainly by product of Mamut porphyry copper
[3] = Laos: 100.0 [4] = Taiwan: 0.47

China has increased its output of molybdenum from 2% in 1979 to nearly 13% in 1990.

Nickel is mined as lateritized peridotite and serpentinite of ophiolite complexes in the Philippines and Indonesia, especially in eastern Sulawesi. Indonesia has grown in importance to more than 7% of the world nickel production.

The province extending southwards from eastern Myanmar and north Thailand, through Peninsular Malaysia to as far as the Indonesian island of Billiton constitutes the world's premier tin metallogenic province. This was once a region dominated by tin mining, with sea-going dredges at Phuket, Thailand, and Billiton, Indonesia, and dredges operating on man-made lakes in the great placer tin-fields of Kinta Valley (centred on Ipoh) and Kuala Lumpur in Malaysia. The world price was set daily at the Straits Trading tin smelter in Penang and the commodity price posted daily over the Ipoh Club bar. Overpricing by the tin cartel forced manufacturers to make tin plating thinner, and to seek alternative materials such as plastics. Rapid fall in demand led to plumetting tin prices and collapse of the cartel. What used to be the world's largest tin mining corporation (Malaysia Mining Corporation) no longer handles the commodity. The South-East Asian tin research centre (SEATRAD) in Ipoh, jointly financed by Indonesia, Thailand and Malaysia, has closed permanently because of declining interest in tin. In 1979 Malaysia claimed 25%, now its share is declining rapidly below 13% of the world total tin production. Indonesia, however, has increased its world production share from 10% to 14%.

Southern China continues to dominate the world scene not only in tungsten, but also in tin. It is a major metallogenic province of both these commodities, which occur together in the same lodes. This intimate co-existence is rare throughout the greater South-East Asian region. Malaysian and Indonesian tungsten production has reduced to zero and Thailand has fallen from 6% in 1979 (Hutchison, 1983*a*) to only 0.5% in 1990. Myanmar (formerly Burma) was in the first quarter of this century the world's foremost producer of tungsten, but mining activity has declined to insignificance.

2

Cainozoic and Mesozoic oil and gas-bearing basins: northern region

2.1 TECTONIC CLASSIFICATION

In my earlier volume (Hutchison, 1989*a*, p. 63) it seemed desirable to have a system of basin classification which incorporates the terminologies of plate tectonics. With the benefits of hindsight, I now believe that all attempts at classification of South-East Asian basins (Fig. 2.1), including my own (Hutchison, 1989*a*) have failed. The fitting of geological details into various favoured models has been contrived and attempts at relating the stress fields of the faulting patterns (e.g. Wood, 1985) and basin shapes to the active plate margins have been unsuccessful. The fore-arc category alone remains viable, but only where the plate margin has not been reorganized throughout the basin evolution, e.g. Sibolga, Bengkalis and South Java. All others, i.e. the 'back-arc' basins, remain genetically enigmatic and defy meaningful plate tectonic classification.

Accordingly, I will not again attempt a genetic classification of the basins, but in this book have grouped together those basins that share some geological similarities. My basin categories are descriptive and I hope, therefore, more enduring and useful. Genetic classifications are vulnerable, especially in South-East Asia, where the Cainozoic era has witnessed a continuously changing pageant of co-existing convergent and extensional tectonics of a complexity which so far has defied satisfactory analysis.

Genetic classifications also tend to be fitted to the current fashionable tectonic hypotheses. The indentation or extrusion hypothesis (see Hutchison 1989*a*, p. 60) is still alive and well amongst the geological community. It is a most appealing hypothesis, but requires a high degree of faith to believe that basins could be created as far as 4000 km from the point of impact of India with Eurasia. Support for the hypothesis has also led to a re-classification of most of the 'back-arc' basins as having been created by major strike-slip faults (shear or trans-extensional basins), and the folding of their sedimentary fill as trans-pressional. These concepts are topical, but may represent a passing phase. 'Flower structures' on seismic profiles have become synonymous with wrench tectonics, and these flowers are spreading like weeds. Do we know for certain that a flower structured fault system unequivocally means a basement wrench fault? Indeed it simply means there was compression across the fault plane, but compression should not be interpreted as trans-pression unless there is independent evidence for offset of geological features across the fault.

Daly *et al.* (1989) have presented a regional analysis of the origin of the South-East Asian Tertiary basins. Their model, like that of Tapponnier *et al.* (1986), fails to accommodate the anti-clockwise rotation of Borneo (Schmidtke *et al.*, 1990), and Peninsular Malaysia (?), and assumes clockwise rotations for all terrains. Daly *et al.* (1989) introduce a few changes to the extrusion tectonics model of Tapponnier *et al.* (1986), and also bring attention to some features that require special consideration.

The formation of the Sumatra basins has been suggested to be linked to oblique subduction of the Indian Ocean lithosphere at the Sunda Trench, initiated during the Middle Eocene (Packham, 1990; 1993), as also has the formation of the Malay and Gulf of Thailand basins (Daly *et al.*, 1989). The Middle Miocene onwards was an era of wrench tectonics and dextral shear, and trans-pression continued into the Pleistocene (Packham, 1993). However, the N–S morphology of the Gulf of Thailand is clearly inherited from the underlying and adjacent Upper Triassic Indosinian structures of the Malay Peninsula.

The widespread occurrence of structural inversions are considered by Daly *et al.* (1989). The Oligocene inversion of the Sumatra–Java fore-arc basins corresponds with the deceleration in the rate of subduction at the Sunda Trench. The Mid Miocene inversion in Kalimantan and Natuna was due to the collision of continental fragments in the South China Sea area, particularly to form the Rajang orogenic belt of Sarawak–Sabah. Pliocene inversion in the Barito and Kutei basins may have resulted from the collision of Australia with the Banda volcanic arc. However, many of these generalizations are speculative, and the chronology of the events also needs to be refined.

2.2 SUNDALAND

The Upper Cretaceous Sundaland continent formed a large SE peninsular protrusion of the Eurasian plate into

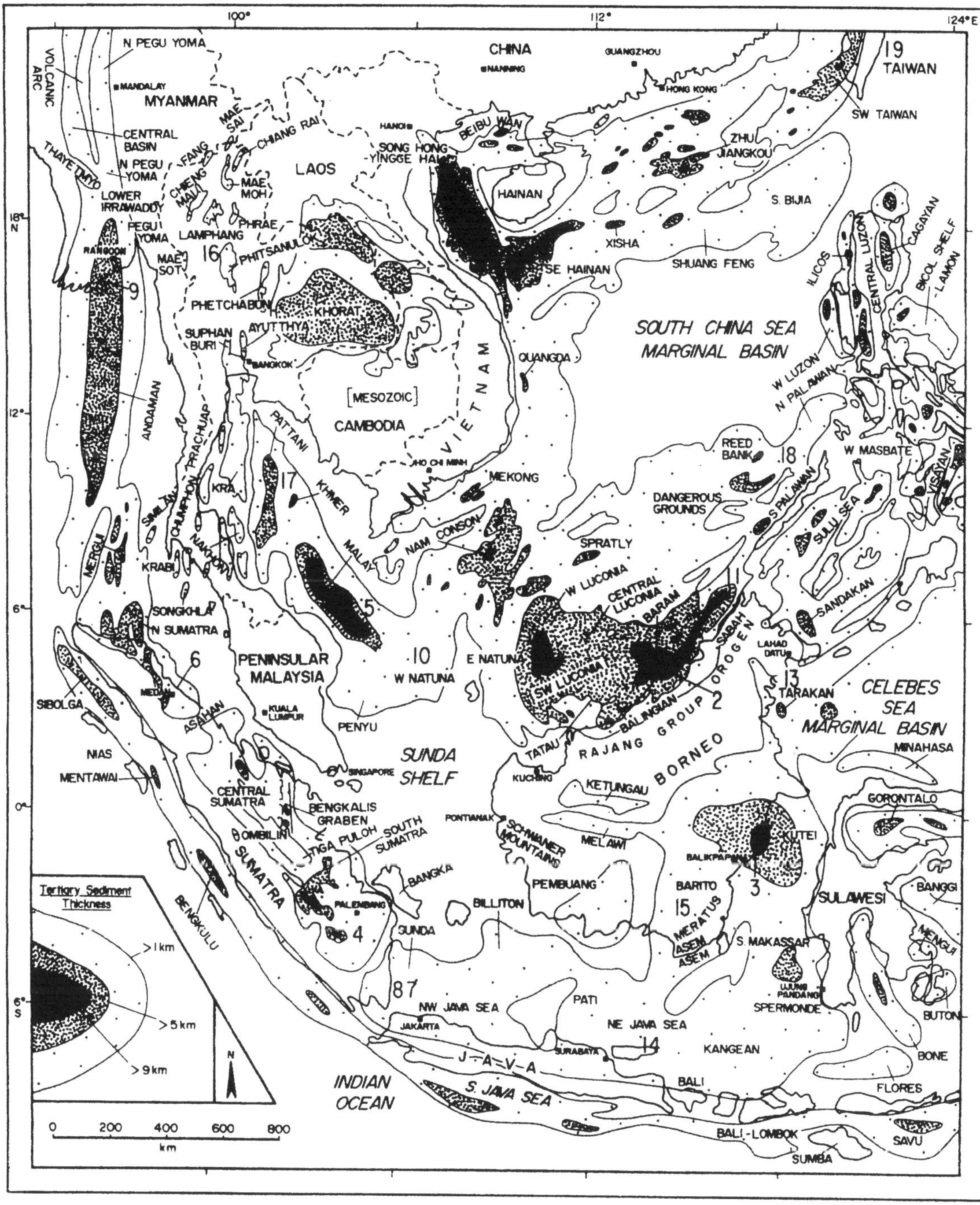

Fig. 2.1. Distribution map of oil and natural gas-bearing basins of South-East Asia showing Tertiary sedimentary thicknesses, simplified after CCOP (1991). The basins of Sumatra are modified after Davies (1984) and De Coster (1974). The onland basins of Thailand are taken from Chinbunchorn *et al.* (1989) and Burri (1989). The Khorat Basin is Mesozoic. The numbers are in reference to Table 1.2. Numbers 12 and 20 are off the map towards the SE.

South-East Asia, extending as far as eastern Java and western Sulawesi (Hutchison, 1992*b*) (Fig. 2.2). The continental slope along the southern margin of this landmass can be identified at Ciletuh Bay in SW Java, where the Eocene through Oligocene sequence is entirely bathyal marine, composed of well documented quartz-rich turbidite deposits, derived from land not too far to the north (Schiller *et al.*, 1991). The landmass was of moderate elevation, for the interpreted high eustatic sea levels of the Palaeocene (Haq *et al.*, 1987) did not result in marine ingressions into Sundaland, with the important exception of the Rajang Group Basin of Borneo. The rest of the landmass was devoid of marine influence (Fig. 2.2). An Eocene unconformity is widespread over the whole region, both throughout the north-east Indian Ocean, and on the extensive Lower Palaeogene landmass of South-East Asia, hereafter referred to as Sundaland.

East of the Sunda Shelf, the Sundaland continent was divided into two by an important ocean gulf, known as the Proto South China Sea (Ru and Pigott, 1986). However, this oceanic gulf had no relationship to the present day South China Sea, so that the term 'Proto South China Sea' is misleading. Haile (1994) has recommended a preferable name 'Danau Sea', after the suture ophiolitic rock suite. It most likely would have been a gulf of the Pacific Ocean, that penetrated deep into Sundaland as far west as the vicinity of Natuna. It was floored by oceanic crust and the first sediments are preserved as Lower Cretaceous (Barremian) ribbon cherts (Hutchison, 1992*b*). The sedimentary fill of the Danau Sea was mainly the Upper Cretaceous through Palaeogene predominantly turbiditic Rajang Group. The Rajang Group is generally known in Sarawak as the Belaga Formation and in Sabah as the Crocker Formation (Hutchison, 1989*a*). Misconceptions regarding the Rajang Group have led to flawed tectonic models of Borneo and the South China Sea. A clarification (based on Hutchison, 1996) is included in Chapter 3.

By late Eocene the Sarawak and central Kalimantan part of this large oceanic basin was compressed between the Schwaner Mountains continental block and the Nansha Block (Luconia province and Spratly–Dangerous Grounds). Most authors invoke southwards movement of the Nansha Block, requiring the existence of a Proto South China Sea. At this time the Nansha Block was not yet attenuated (this came only in the Oligocene) and formed a contiguous part of the marginal area of the Eurasian Plate. Southwards motion is unnecessary, for the Sunda Plate Margin (later to become the Sunda or Java Trench) was not yet an active subduction system, and the passive Sunda plate margin, together with the Schwaner Mountains Block, was pushed northwards as a result of sea-floor spreading in the South Indian Ocean.

The Crocker Formation basin of Sabah was not eliminated until the late Miocene, accomplished by Oligocene to Lower Miocene spreading of the South China Sea (*sensu stricto*) marginal basin (Briais *et al.*, 1993; Hutchison, 1989*a*, pp. 85 and 195). Elimination of the Danau Sea, and compression of its Rajang Group sedimentary fill, resulted in the Borneo Orogeny, of which there were two major episodes. An Upper Eocene event (Sarawak Orogeny) resulted in compressional uplift of the Rajang Group into a new mountain belt (Sibu Zone), which can be traced continuously eastwards through Sarawak (Hutchison, 1989*a*, p. 185). The Upper Miocene Sabah Orogeny resulted in folding and dramatic uplift of the Crocker Formation to form the Crocker Ranges of Sabah and Palawan (Hutchison, 1989*a*, p. 190). Erosion of the newly formed Sarawak and Sabah orogens (Sibu Zone and Crocker Range mountains) resulted in sand dominant fluvio-marine formations in the basins of west Sarawak and Kalimantan, and the Baram Delta Province, which will be described in Chapter 3.

Eocene grabens attracted the formation of lakes in the Sundaland fluvial system. Strongly subsiding grabens favoured the accumulation of oil prone source rock facies, if the grabens were relatively sediment-starved (Katz, 1991). On the other hand, slowly subsiding basins developed coal facies. Lacustrine source rocks account for over 80% of Indonesia's petroleum reserves—the Sumatran basins are predominantly of lacustrine origin. This proportion of lacustrine sourced oil is large on a world scale, exceeded only by China, which has more variable age lacustrine source rocks—Permian (Junggar) to Eocene (Pearl River), whereas Sundaland is exclusively Upper Eocene to Lower Oligocene (Katz, 1991).

The unconformity is interpreted to be due to the initial collision (at about 55 Ma) and indentation of India into flush collision with Eurasia, now well established at about 45 Ma, the time of magnetic anomalies 19 and 20 (Dewey *et al.*, 1989; Hutchison, 1993). This was also the time of separation of Australia from Antarctica (Veevers, 1984) and the creation of the new spreading system at the South-east Indian Ocean Ridge (Packham, 1990). The fossil spreading centre in the north-east Indian Ocean indicates that spreading ceased shortly after the formation of magnetic anomaly 20 (Liu *et al.*, 1983) and jumped to the newly created South-east Indian Ocean Ridge. This major reorganization of the spreading pattern of the Indian Plate may have been a consequence of the collision, and it is interesting that a major change of direction of the Hawaii–Emperor seamount and island chain occurred at 43 Ma, suggesting a worldwide tectonic event. However, Norton (1994) has shown that there was no 43 Ma regional tectonic event in the Pacific and the bend was probably due to a non-stationary hotspot. The plate reorganization in the Indian Ocean had a major

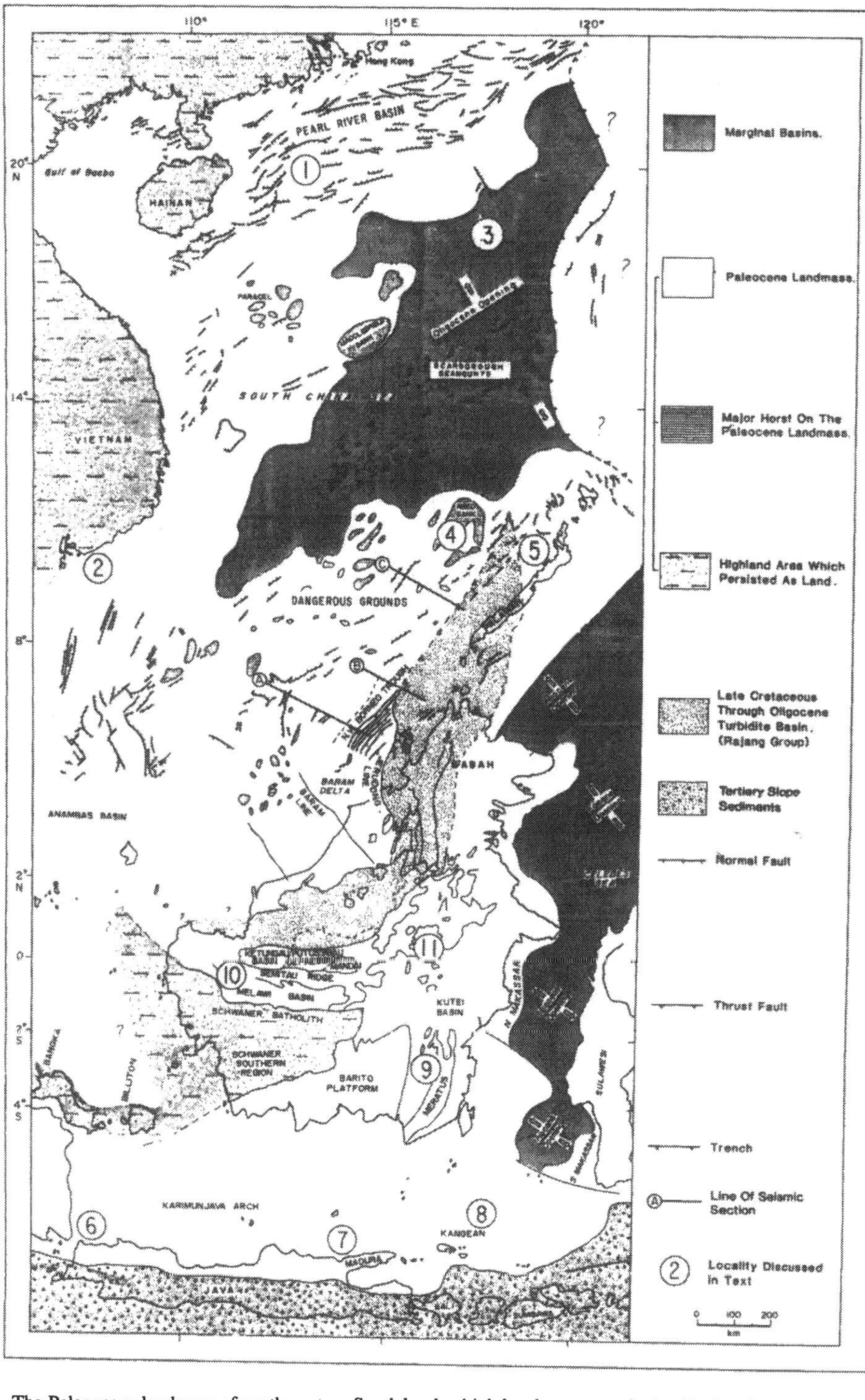

Fig. 2.2. The Palaeocene landmass of south-eastern Sundaland, which has been extensively rifted to form marginal basins and Eocene grabens filled by terrestrial alluvial and lacustrine deposits. Its eastern margins are ill-defined (after Hutchison, 1992*b*). Numbers 1 to 11 refer to stratigraphic columns of Figs 2.5 and 4.16. The seismic lines are shown on Figs 2.18 and 2.15.

impact on the tectonics and stratigraphy of the whole of South-East Asia. Clure (1991) speculated that subduction of the hot spreading centre (anomaly 20) during the late Eocene at the Sumatra Trench would have led to a higher potential for oil generation, and may have been responsible for the pre-eminent position of the Central Sumatra Basin.

In the Bay of Bengal, the Eocene unconformity subdivides the wholly marine sedimentary section into pre-collision continental rise and post-collision Bengal fan (Curray and Munasinghe, 1989; Curray, 1991). Before collision, northern India should have been topographically depressed as it was dragged down into the Tethyan Trench along its northern margin (similar to Australia at the present-day Timor Trough; Veevers, 1984), resulting in a sedimentary hiatus because India shed little or no detritus as far as the eastern Indian Ocean. The collision led to progressive uplift of Tibet and the Himalaya, erosion of which resulted in the post-unconformity Bengal fan. The unconformity therefore represents the time of the India–Eurasia collision (Curray and Munasinghe, 1989; Curray, 1991).

The Eocene unconformity on Sundaland (Fig. 2.2) requires a different explanation than that given for the submarine unconformity, but both need to be related to the collision of India.

2.2.1 Origin of the Eocene unconformity

The most elegant hypothesis for the Eocene initiation of faulting and subsidence is that of extrusion tectonics given by Tapponnier *et al.* (1982, 1986). These authors emphasized the propagation of faults radiating eastwards and south-eastwards from Yunnan. A somewhat modified model is presented here. The contrary unpopular view was expressed by England and Molnar (1990), who suggested that continental 'escape', extrusion' or 'expulsion' is an illusion, and that the northwards movement of India is taken up by N-striking right-lateral simple shear in eastern Tibet.

India began colliding with the Eurasian continental margin at 55 Ma, as shown by a distinct reduction in its northwards movement documented from work on the Ninety-east Ridge (Klootwijk *et al.*, 1992). Collisional events as old as 50 Ma are widely recorded along the Indus–Zangbo Suture zone (Crawford and Searle, 1992). The collision was flush by the time of magnetic anomaly 20 (45 Ma), when the spreading pattern of the Indian Ocean underwent a dramatic reorganization. From Eocene to the present, India rotated counter-clockwise (Dewey *et al.*, 1989) and made its greatest indentation into Eurasia at the Yunnan Syntaxis (Fig. 2.3). The northwards passage of the eastern margin of India was greatly facilitated by the prominent transform fault which lies immediately east of the Ninety-east Ridge. East of it was wholly oceanic lithosphere, but since it has a fossil spreading axis as young as anomaly 19 (Liu *et al.*, 1983), the hot and elevated lithosphere would have experienced difficulty in subducting at the Sunda Trench. Therefore there was also a kind of collision (or diminished subduction activity) along the Sumatra Trench, resulting in the north-easterly migration of Sundaland away from the NNE-moving Indian Ocean Plate. The present day plate margin (Sumatra Trench) also may not have become a subduction system until the Oligocene.

The continental margin of Eurasia at the time of initial collision is deduced to have been oriented WNW–ESE (Fig. 2.3) and lay near the equator (Metcalfe, 1991; Hutchison, 1989*b*). This is consistent with the palaeomagnetic data from Tibet, which indicate that the terrain around Lhasa has been pushed northwards by 17° of latitude (Achache *et al.*, 1983). Before indentation, the Lhasa terrain was contiguous with western Thailand, western Peninsular Malaysia and Sumatra (Fig. 2.3).

Sundaland was progressively oroclinally bent since 45 Ma into its present configuration. The consequences of such bending have been outlined by Tapponnier *et al.* (1982, 1986), who emphasized the need for clockwise rotation and faulting (escape or extrusion tectonics). Continental Sundaland should have been dominated by right-lateral shear parallel to the progressively bending tectonic fabric elements of the continental plate. Ongoing palaeomagnetic studies show that the Tertiary strata of several onland basins of Thailand and Upper Tertiary basalts indicate a clockwise rotation (Fuller *et al.*, 1991). However, after extensive palaeomagnetic research in South-East Asia, Richter *et al.* (1993) have concluded a complicated inconsistent pattern of rotations. The Tertiary basins of Thailand display inconsistent rotations, either clockwise or anti-clockwise. Richter *et al.* (1993) propose that the inconsistent rotations occur in small terrains within fault zones, outside of which the blocks will be translated by wrench faulting, but not rotated.

The tectonic response to oroclinal bending can be seen in a scaled-up model of stratal bending of an outcrop-scale fold. The outer western zone of the convex-westwards orocline should have undergone considerable stretching and pull-apart disruption (Fig. 2.4). The N–S opening of the Andaman Sea is a direct consequence of the bending resulting from maximum indentation at the Yunnan Syntaxis. Further south, the Sumatran Tertiary basins have resulted from the same bending. The Bengkalis Graben of Sumatra, whose sedimentation began with Eocene Pematang Formation continental redbeds and lacustrine deposits, is interpreted to have

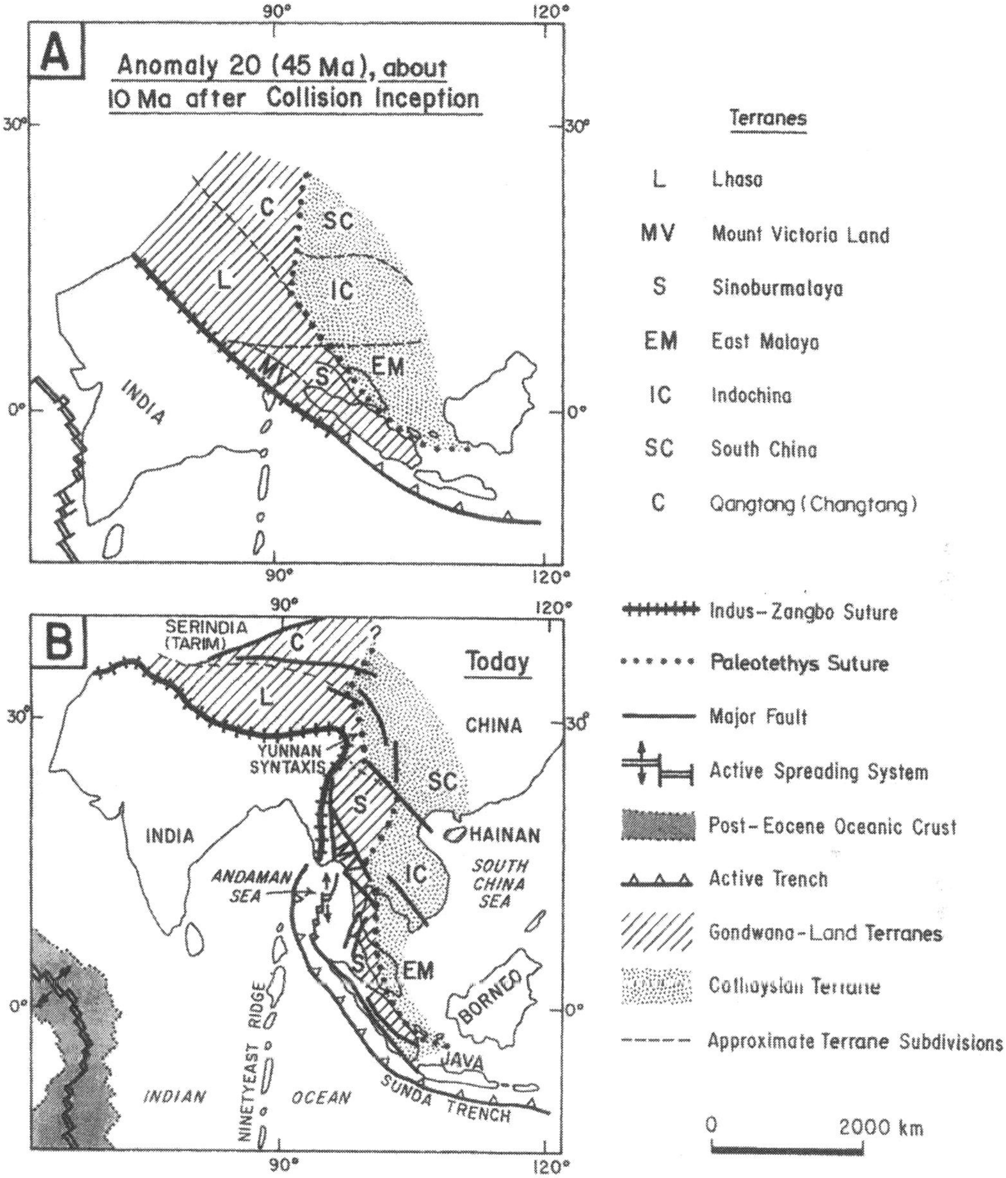

Fig. 2.3. Indentation of India into South-East Asia. A: inferred palaeogeography at 45 Ma (anomaly 20), about 10 Ma after initial collision and anti-clockwise rotation into flush alignment. B: present day geography showing maximum indentation at the Yunnan syntaxis and spreading from the SE Indian Ocean Ridge. Modified after Hutchison (1992*b*, 1993). Tectonic subdivisions of Tibet follow Sengör *et al.* (1988) and Görür and Sengör (1992).

resulted from right-lateral shear and pull-apart tectonics. The eastern inner zone of the orocline (East Malaya–Bangka–Billiton–Schwaner Mountains) was compressive during oroclinal bending and did not develop Tertiary basins. The compressive zone is displayed along the east coast of Peninsular Malaysia (Fig. 2.4), where Chakraborty and Metcalfe (1984) have demonstrated that the Upper Palaeozoic low grade metasediments (Mersing Beds) contain a record of three folding phases. The F_1 and F_2 fold axes are dispersed about steeply plunging F_3 fold

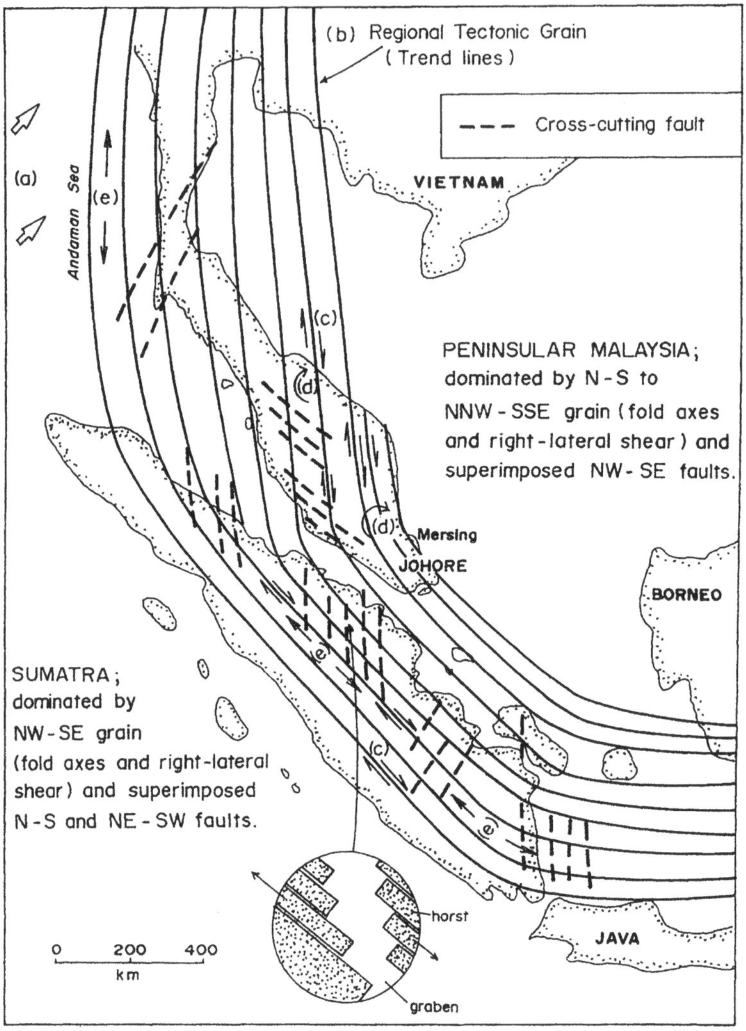

Fig. 2.4. Schematic plan of collision and oroclinal bending. (a): Indian indentation vector; (b): oroclinal bending of the regional trend lines; (c): right-lateral shear parallel to the orocline; (d): post-Eocene clockwise rotation of rigid blocks; (e): zone of maximum pull-apart giving Andaman Sea and the Sumatra Tertiary basins along the outer curvature of the orocline. The inner curvature (Peninsular Malaysia to Bangka and Billiton) is compressional and devoid of Tertiary basins. Inset: according to Moulds (1989), the N–S trending Bengkalis Graben has followed from pull-apart simultaneously utilizing NW–SE and NE–SW faults, resulting in a N–S trending graben (after Hutchison, 1992*b*, 1993)

axes, as a result of crustal bending on a sub-vertical axis. This east coast fold belt is the most deformed of the whole Malay Peninsula and represents a zone of considerable crustal shortening (Hutchison, 1993). Its highly deformed rocks extend northwards from the SE tip of Peninsular Malaysia to north of Kuantan.

Oroclinal bending (Fig. 2.4) can be accomplished by right-lateral shear on the pre-existing Indosinian fabric of the region. Good support for interpreting right-lateral shear is offered by the structure of the Mesozoic Tembeling Formation (Fig. 4.6). The orientation of fold axes in the uppermost Triassic to Jurassic Tembeling Formation is characteristically NW–SE (Koopmans, 1968; Harbury *et al.*, 1990; Loganathan, 1978; Gobbett and Tjia, 1973, pp. 319–24). In contrast, the Triassic Semantan and Gemas formations are folded along axes striking NNW (340°) to N–S (360°) (e.g. Gobbett and Tjia, 1973; Loganathan, 1977; Lum, 1979). The transverse fabric of the Tembeling Formation (Fig. 4.6) is most readily explained by right-lateral trans-pressional wrench motion along N–S faults in the basement (Fig. 4.8).

Oroclinal bending should also result in clockwise block rotation. Fuller *et al.* (1991) show that their palaeomagnetic work suggests some clockwise rotations. However their later work (Richter *et al.*, 1993) concluded a more complex picture in which only small fault-bounded blocks may have rotated. Enkin *et al.* (1992) interpret their palaeomagnetic data to conclude that the Chinese blocks have been extruded eastwards away from the Indian collision, resulting in internal deformation in larger blocks and small local rotations.

The oroclinal bending should have resulted in new topographic relief over Sundaland. The areas under compression were uplifted resulting in rivers with increased sediment load. The active right-lateral shear faulting resulted in river capture and diversion, the most spectacular example being the Mekong, and this is a common feature of most South-East Asian rivers (Hutchison, 1989*a*, p. 68). Pull-apart tectonics resulted in topographic depressions which attracted the rivers, resulting in lakes. The sediments which accumulated in these topographic depressions can be shown generally throughout Sundaland to date back only as far as the Eocene unconformity. Naturally, not all strata above the unconformity yield Eocene dates, some are as young as Oligocene, because the oroclinal bending was an ongoing tectonic process, initiated 45 Ma ago, but not all topographic depressions were created simultaneously, and the river network was constantly changing in response to the right-lateral shear.

Nevertheless, it is remarkable that most of the basal continental sediments overlying the Sundaland unconformity are of Middle to Upper Eocene age, as determined by regional palynological studies (Morley, 1991, and pers. comm. 1992). The considerably older age (Uppermost Cretaceous) for the basal Kayan Sandstone of Sarawak (Tan, 1986), which represents the western extremity of the Ketungau Basin of Kalimantan, is an outstanding anomaly which requires palynological re-investigation before acceptance.

The non-cratonic south-east peninsula of Sundaland, lying south of the Rajang Basin (Danau Sea) and east of Sumatra, reacted to the oroclinal bending in a less predictable manner. It is also important to bear in mind that a larger cratonic mass—Australia—is in collision with the Banda volcanic arc at the Timor Trough and extends much further north to Irian Jaya and the Sula–Buru–Seram region (Hutchison, 1989*b*). This eastern collisional indentation has had a major impact on the eastern extremity of Sundaland, causing its fragmentation and imposing an anti-clockwise rotation and left-lateral shear, notably along the Sorong Fault System. The well documented progressive anti-clockwise rotation of Sarawak and Kalimantan (Schmidtke *et al.*, 1990; Hutchison, 1991*b*) may be understood in terms of the indentation of Australia into eastern Sundaland, as opposed to that of India.

2.3 NORTHERN MARGINS OF THE SOUTH CHINA SEA

The sector lying between the northern extent of the South China Sea marginal basin and the coastline of China and Vietnam is of rifted attenuated continental crust now submerged beneath sea level (Fig. 2.2). The rifting history is well displayed by the Tertiary basins which are being increasingly investigated by petroleum companies. Authority for oil and gas exploration to the east of 113° rests with Nanhai East Corporation, Guangzhou, and west of 113° with Nanhai West Corporation, Guangzhou. Although independent, both are supported by the China National Offshore Oil Corporation (CNOOC), Beijing. The north-western parts of Sundaland had been land since the beginning of the Upper Triassic; and China, Indochina, Thailand and Peninsular Malaysia were widely covered by terrestrial Jurassic redbeds and Cretaceous sandstones (Hutchison, 1989*b*, 1992*b*).

The Jurassic-Cretaceous Yenshanian tectonic pulse resulted in the widespread emplacement of high-level granite and extrusion of rhyolite, especially along what is now the coastal zone of China and SE Vietnam. These igneous events weakened the crust and were the precursors of Cainozoic extension and subsidence, the stretching phase of which was entirely terrestrial.

The basins lying along the coast of China have a common stratigraphy and tectonic evolution, summarized by Su *et al.* (1989), who concluded that the lithosphere stretched by a factor of 1.8 from Upper Cretaceous to Upper Oligocene (Fig. 2.5). Lower Eocene red clastics are overlain by Lower to Middle Eocene dark source-rock lacustrine shales, overlain by Lower Oligocene coal-bearing coastal swamp deposits. These early deposits are preserved in grabens which extend NE–SW parallel to the coast. Spores and pollen are rare especially in the basal redbed alluvial fan deposits, so that the inferred Palaeocene beginnings are doubtful. The overlying lacustrine and swamp deposits are confidently dated Eocene. Mesozoic and Lower Tertiary strata outcrop widely along the coastal province of China towards the NE and have good petroleum source-rock potential. Most of the oilfields are in the medium to small range, 20–40 Mt oil-in-place, with one 200 Mt field.

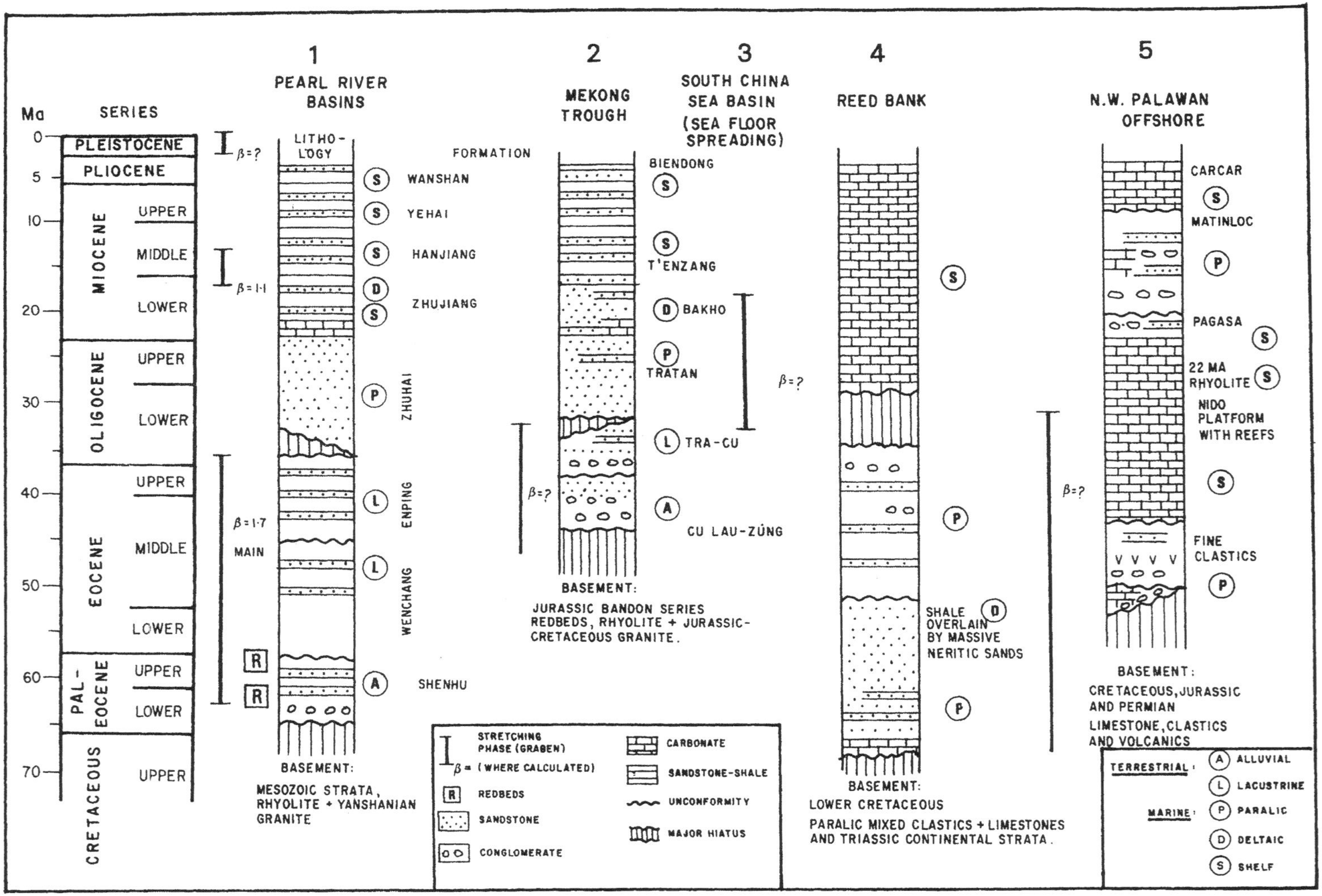

Fig. 2.5. Stratigraphic sections of northern South China Sea basins which have been drilled to the pre-Tertiary basement. 1 = after Su *et al.* (1989) and Li (1984). 2 = after Le (1986). 3 = from Taylor and Hayes (1983). 4 = after Taylor and Hayes (1980). 5 = from Saldivar Sali (in Hinz and Schlüter, 1985) and Holloway (1981). Locations are given in Fig. 2.2.

The Mekong Trough (Vung Tau) has a similar history (Le, 1986). The lower part of the sequence is of Eocene terrestrial alluvium and lacustrine deposits confined to major grabens (Fig. 2.5).

2.3.1 The intermontane basins of continental Thailand

There are 19 basins onshore and 11 offshore Thailand. The basins have a similar early history to many South-East Asian basins, which were initiated on the extensive Palaeocene landmass of Sundaland. The Thai basins started as N–S elongated narrow, rapidly subsiding rift grabens, initially filled by alluvium, but soon thereafter they became characterized by fault-controlled lacustrine sedimentation (Burri, 1989). The rifts were then aborted and flexural movements created wider depressions filled mainly by alluvial deposits. The marine incursions into the Malay, Sumatran and southern Gulf of Thailand basins were from the south and did not penetrate into basins of what is now on-land Thailand. The latter retained their landlocked intermontane character (Fig. 2. 6).

Polachan and Sattayarak (1989) and O'Leary and Hill (1989) have demonstrated that the formation of the basins has resulted from E–W extension resulting from trans-extensional dextral movement on the NW–SW trending Mae Ping and Three Pagodas Pass faults. The Nan-Uttaradit Papaep-Tethyan suture became reactivated in the Palaeogene as a left-lateral wrench fault, which played an important role in the development of the Sirikit Field in the Phitsanulok Basin. The trans-extensional wrench tectonics can be related to the major indentation of India into Eurasia at the Yunnan Syntaxis.The basins exhibit a range of characteristics, but may be grouped as follows (Burri, 1989):

- With a pronounced lacustrine phase, containing thick organic-rich clays. Most of the larger basins such as Phitsanulok and Chumphon belong to this category;
- With a dominant swamp environment and thick lignite deposits, e.g. Mae Moh, Krabi, Li;
- Small basins with predominantly alluvial deposits, e.g. Chiang Rai.

The truly intermontane basins, with rich lacustrine source rocks, have given rather indifferent exploration results. There are numerous oil-seeps, resulting from the very high fault density, leading to small hydrocarbon traps and highly compartmented and difficult oil fields (Burri, 1989). After some 70 exploration wells, only the minority of which were dry holes, only one medium sized accumulation, the Sirikit Field, three small deposits and the small Fang oilfield were discovered. Fang has 5 fields together producing some 1200 barrels of oil per day. Phitsanulok has 9 fields, totalling some 20 000 barrels/day oil and 30 million ft^3/day of associated gas. The Gulf of Thailand Pattani Basin has 13 fields, totalling approximately 20 000 barrels/day condensate and 555 million $feet^3$ gas per day.

2.3.1.1 Stratigraphy

The age of the early rift stages of the basins is generally held to be Oligocene, based on palynology, but Buffetaut *et al.* (1989) have shown that, for example the Krabi Basin of Peninsular Thailand, contains Eocene vertebrate fossils. They therefore cast doubt on the palynological age designations. The stratigraphy of the basins may be summarized into two sequences:

A. Syn-rift: The lower unit (Upper Oligocene to Lower Miocene) begins with alluvial and swamp deposits. In the Phitsanulok Basin, the Nong Bua Formation, up to 1000 m thickness, is of low energy alluvial red-brown claystone with minor coarse to fine lithic sandstone (Fig. 2.6). The deeper (western) side of the basins is occupied by coarse sandstone and conglomerate (Sarabop Formation, Fig. 2.6), formed by rapid sedimentation off the active bounding fault adjacent to the deep basin.

In the similar intermontane basin of Batu Arang, Peninsular Malaysia, the western fault-related facies is referred to as 'Boulder Beds' (Stauffer, 1973*b*). The middle unit (Lower to Middle Miocene) is of thick lacustrine deposits in the central basinal areas. At Batu Arang the deeper lacustrine facies shows turbiditic characteristics, whereas the eastern margin of the basin demonstrates shallow water to fluvial characteristics.

The middle unit is predominantly of mudstone of high organic content, with minor sandstones. In the Phitsanulok Basin, this lacustrine unit is 2 km thick (Lan Krabu and Chum Saeng formations). In the Pattani and Fang basins, this unit is 1000 m and 700 m thick respectively (Chinbunchorn *et al.*, 1989). The lacustrine shales are excellent source rocks containing types II and III kerogen and total organic carbon contents ranging from 1.1 to as high as 42% in the Phitsanulok Basin, 1.27 to 2.16% in the Chumphon Basin. The upper unit (Middle to Upper Miocene) is of fluvial deposits with ephemeral lacustrine deposits in the basin centre. In the Phitsanulok Basin, this unit is 2 km thick (Pratu Tao and Yom formations).

B. Post Rift: All basins clearly demonstrate an Upper Miocene unconformity (caused by the Sabah Orogeny ?). Listric normal faulting dies out at the unconformity. In the Phitsanulok Basin, the unconformity is clearly identifiable on the seismic section (Fig. 2.6), and it is characterized by

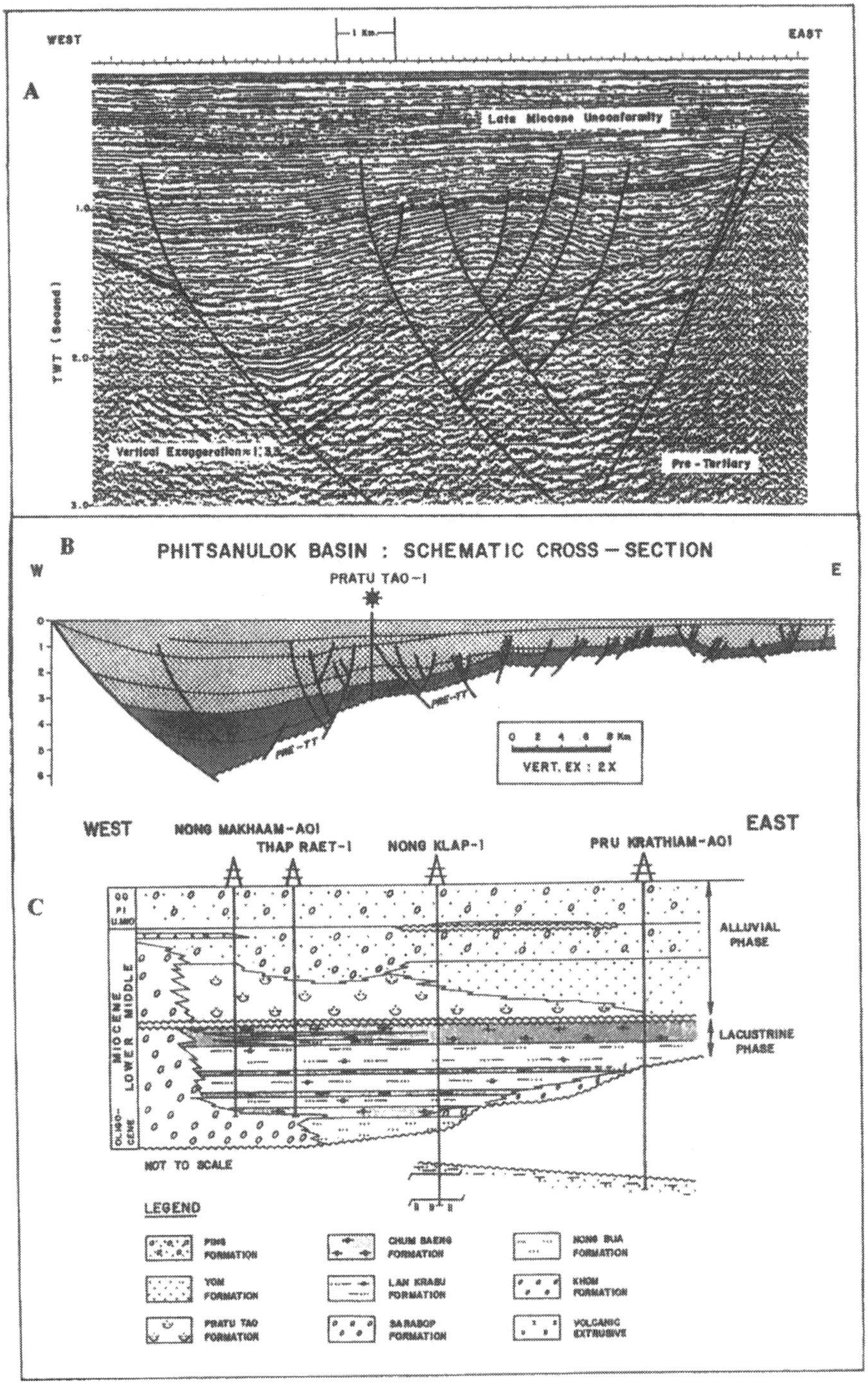

Fig. 2.6. The intermontane Tertiary basins of on-land Thailand. A: seismic line showing the half-graben and listric normal fault characteristics of Thai non-marine basins (after Chinbunchorn *et al.*, 1989). B: cross section through the Phitsanulok Basin, and C: stratigraphy of the Phitsanulok Basin (after Burri, 1989).

basaltic lava flows, dated 10.3 ± 0.2 Ma. The post-rift sequence is therefore unconformable on the underlying syn-rift sequence. The post-rift sequence is up to 1700 m, 1250 m and 700 m respectively in the Pattani, Phitsanulok and Fang basins. It consists of high-energy fluviatile coarse sand and gravel with some interbedded clays; these probably represent braided stream coalescing fans.

The basins have above average geothermal gradients. Pattani Basin has a gradient of 40 to 58° km^{-1} and generally the on-land basins have gradients around 30 to 40° C km^{-1}. However, the Fang basin has an exceptionally high gradient of 75° C km^{-1} (Chinbunchorn *et al.*, 1989). These basins would have reached maturity if they contain sediments in excess of 3 km thickness.

2.3.2 Mesozoic Khorat Basin

The Khorat Basin (Fig. 2.1) may be classified as interior cratonic. It occurs on a stable region built upon Precambrian basement and it experienced internal drainage leading to rather saline conditions by the Early Cretaceous. Seismic sections and drilling have confirmed that the uppermost Triassic to Cretaceous Khorat Group of NE Thailand is underlain by pre-Khorat Middle to Upper Triassic sequences which occupy half grabens (Sattayarak *et al.*, 1989). These Triassic half grabens have been developed unconformably upon a Permian and older basement (Fig. 2.7). From this information, the Indosinian Orogeny can be dated Lower Triassic. The orogen was then subjected to extensional tectonics in the Middle Triassic, giving rise to half grabens filled with the Middle–Upper Triassic pre-Khorat Formation. By uppermost Triassic or Liassic times, the Khorat basinal area then sagged and the half grabens were succeeded by more extensive flat lying to gently dipping Jurassic and Cretaceous Khorat Group (Cooper *et al.*, 1989). In Neogene time, the Khorat Basin experienced inversion in the region underlain by the thick pre-Khorat grabens, forming broad anticlines such as the Phu Phan anticlinorium (Fig. 2.7). The western margin of the Khorat basin was also tilted into a steeply dipping monocline.

The pre-Khorat half grabens show two major trends: ESE under the Phu Phan Uplift and N–S in the Loei-Petchabun foldbelt of the western margin of the Khorat basin (Cooper *et al.*, 1989). These orientations are controlled by the underlying fabric of the Permian and older basement (Fig. 2.7).

2.3.2.1 The pre-Khorat sequence

The sequence is variable, but generally started with basal conglomerate variable in thickness up to 200 m. Conglomerates are thickest near the major bounding listric faults. Fan conglomerates give way laterally to lacustrine black shales and limestone containing algae and fossil turtle remains. Above the basal conglomerates, fluvial sandstones occur with grey conglomerates. The sandstones are usually of redbed facies. These grade laterally into lacustrine limestones and carbonaceous shales. Some limestones are stromatolitic.

The upper sequence comprises yellow-brown fluvial sediments. The sandstones are lithic and hence of low porosity. Rhyolitic tuff and agglomerate occur locally (Sattayarak *et al.*, 1989), interbedded with the lower conglomerate.

The Huai Hin Lat Formation outcrops along the western escarpment of the Khorat Basin. It begins with 140 m thickness of basal polymict conglomerate, overlain by grey sandstones and shales which have yielded a good fauna of Carnian–Norian age (Hutchison, 1989*a*). It unconformably overlies Permian Saraburi Limestone in the Loei foldbelt. The Huai Hin Lat Formation was formerly taken as the base of the Khorat Group, but now is recognized to represent the lower part of the pre-Khorat sequence (Sattayarak *et al.*, 1989).

The Triassic pre-Khorat Formation has a thickness variable from 10 to 3000 m. The lacustrine limestones and shales have good source rock characteristics, but unfortunately the reflectivity measurements indicate over maturity.

2.3.2.2 The Khorat Group

In addition to lying on the Triassic half graben fill, the Khorat Group onlaps onto Permian and older basement rocks. The mesa-like geomorphology of the Khorat 'plateau' as seen from the south and west is due to the strongly up-tilted western and southern margins of the basin to as high as 1300 m above sea level, whereas the central surface of the basin is no more than 100 to 300 m above sea level. The stratigraphy has been summarized by Sattayarak (1983).

The grey beds of the Triassic pre-Khorat sequence are overlain by a lower redbed formation, called the Nam Phong and Phu Kradung Formations, which together attain a total thickness of 1456 m in the type locality. The basal part of the Phu Kradung Formation contains lacustrine micritic limestones. Freshwater fossils indicate an age of Lower Jurassic.

The middle sandstone sequence is composed of the Phra Wihan Formation, forming an escarpment north of the border with Cambodia. It is of white quartz sandstone of thickness up to 250 m. The informal upper redbed division includes salt beds at the top. The overlying Maha

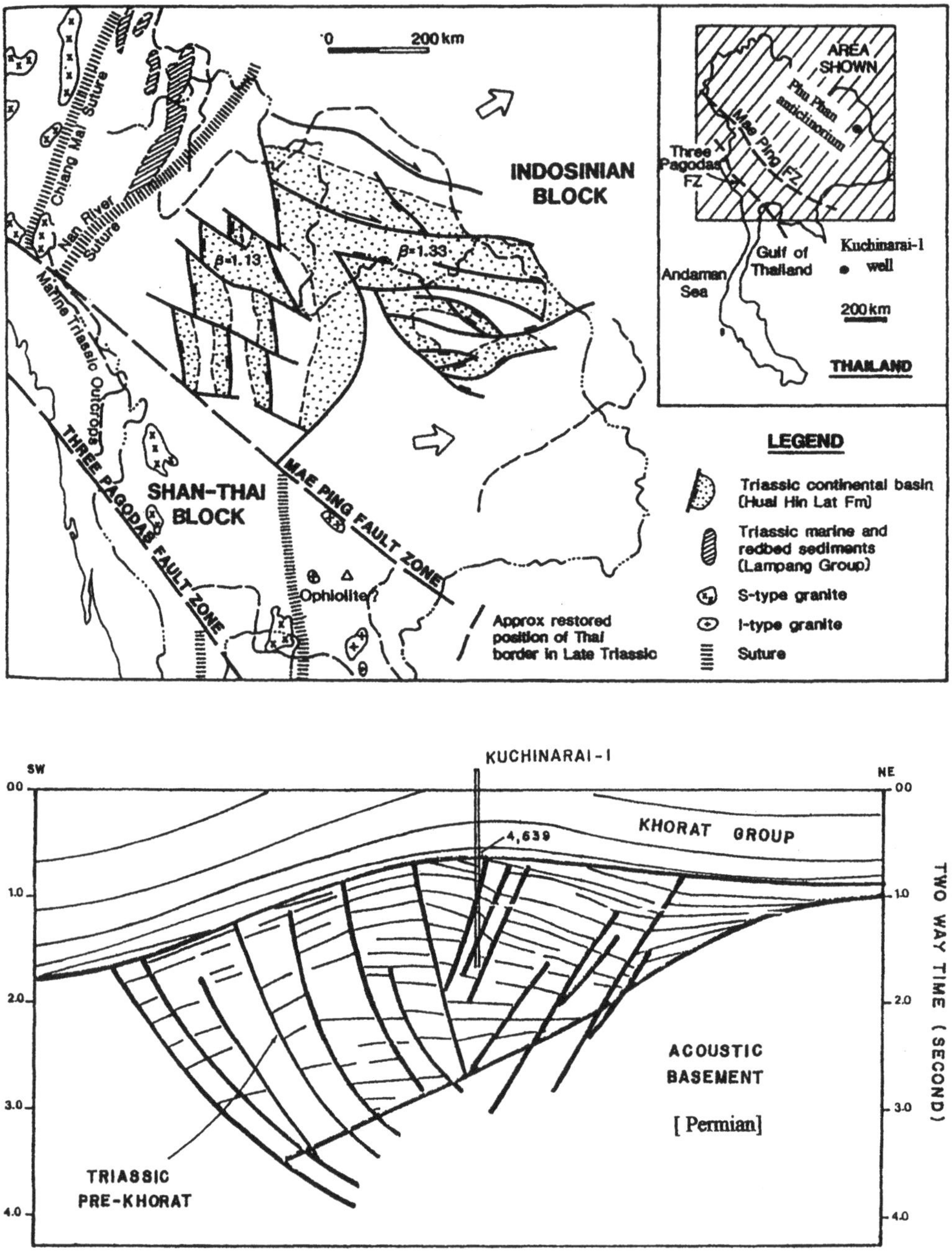

Fig. 2.7. **Top**: distribution of Middle–Upper Triassic half-grabens beneath the Jurassic Khorat Group (from Cooper *et al.*, 1989). The effects of Upper Tertiary extension, Upper Cretaceous to Lower Tertiary inversion and the clockwise rotation of the Khorat Basin have been removed. **Bottom**: seismic sketch cross section across the Phu Phan anticlinorium of NE Thailand. Width of the section is approximately 18 km. The Kuchinarai-1 well encountered the base of the Khorat Group at 1414 m (= 4639 ft). Beneath that, to a total depth of 3356 m, the well was in Triassic strata, which include the Huai Hin Lat Formation, formerly mapped in outcrop as Khorat Group (after Sattayarak *et al.*, 1989).

Sarakham Formation is of sandstone and mudstone. Its upper parts are mainly of halite, carnellite and sylvite, with subordinate anhydrite and gypsum. Its type section is in a borehole which penetrated 610 m of salt-bearing strata. The age is not well constrained, but is probably Upper Cretaceous, with extension into the Lower Tertiary.

2.3.3 The Gulf of Thailand

The morphology of the Gulf of Thailand is clearly a link between the N–S structural geology of Peninsular Malaysia with that of Central Thailand. Accordingly the fabric of the Gulf of Thailand is inherited from structures formed during the Triassic Indosinian Orogeny, accentuated by Tertiary faulting, usually parallel to directions of weakness in the pre-Tertiary basement. The Gulf is effectively divided into N–S horsts and grabens. The most prominent horst is the Ko Kra Ridge, which extends N–S for several hundred kilometres and separates the deep Pattani Trough on the east (Fig. 2.1) from a series of shallower grabens including the Hua Hin Basin, on the west. A discussion of subsidence rate analysis of the Gulf of Thailand is given by Pigott and Sattayarak (1993). The Gulf commenced its extension and basin subsidence in the Lower Oligocene and continued episodically. Crustal attenuation began first in the southern Pattani Basin at 36 Ma and propagated northwards. Similar development of the Chumphon and Hua Hin basins occurred at about 33 Ma, followed by the Kra Basin at 22.5 Ma. At 14 Ma the Gulf of Thailand temporarily experienced a time of quiet tectonic activity. Major strike-slip movements on the master faults at 5 and 2 Ma resulted in the dramatic deepening of some basins and caused the uplift of others (Pigott and Sattayarak, 1993).

The N–S border between the Pattani Trough and the Ko Kra Ridge appears to be the northwards extension of the Bentong–Raub Palaeo-Tethys Suture of Peninsular Malaysia. The prominent basement ridges (Ko Kra and Ko Pha Ngan) are therefore expected to be composed of Upper Triassic Main Range Granite. The geology of the Main Range, including Triassic to Jurassic basins, is expected to be present in the basement of the Gulf of Thailand. The basement of the Pattani Basin is likely to consist of rocks similar to the Central N–S Basin of Peninsular Malaysia, occupied by Permo-Triassic (Semantan Formation) as well as Jurassic (Tembeling Formation) sedimentary rocks. A widespread Middle to Upper Miocene unconformity can be traced throughout the Gulf of Thailand, throughout the Malay Basin and West Natuna, and must be related to the Sabah Orogeny, when the Crocker Formation was compressed and uplifted to form a mountain belt.

2.3.3.1 Pattani Basin

This is the largest basin of the Gulf of Thailand, covering an area of around 38 500 km^2. The deepest part of the basin contains more than 8 km of sediments, all thought to be Tertiary. Most of the basin fill is of fluvio-lacustrine sediments (Woollands and Haw, 1976). The lowermost known sequence TP-40 is of Upper Oligocene alluvial flood-plain and ephemeral lacustrine deposits (Fig. 2.8). However, it is likely that TP-40 is underlain by as yet unproved Eocene strata in the depocentre. Widespread lake conditions developed in the Lower to Middle Miocene. The richest source rocks are Lower Miocene (TP-30) and Upper Oligocene (TP-40) grey shales. Because of the very high geothermal gradients of 40 to 58° C km^{-1}, mature source rocks also occur in TP-20 in the deeper parts of the basin.

From Middle to Upper Miocene, a change in tectonic activity resulted in elimination of lacustrine conditions and alluvial flood-plain deposits became dominant. From Upper Miocene to Recent, subsidence of the Gulf of Thailand resulted in coastal plain marine sedimentation throughout TP-10 to TQR.

Eleven gasfields have been found since 1972. Five now produce at a daily rate of more than 600 M ft^3 of gas and over 20 000 barrels of condensate. Proven reserves of the basin are approximately 6×10^{12} ft^3 of gas and 170 M barrels of condensate. The gas reservoirs are associated with intensely faulted structures in Lower and Middle Miocene sandstones. The sandstone reservoirs are therefore of limited lateral extent, making the fields difficult to produce. The gas is usually produced from 1200 to 3000 m depth.

The high geothermal gradients of 40 to 58° C km^{-1} result in rapid decline in porosity and permeability of the Miocene sandstones with increasing burial depth (Trevena and Clark, 1986). The decline results from burial diagenesis. The best porosity (20 to 25%) and permeability (800 to 1300 mD) reservoirs are generally associated with large inter-granular pores in sandstones between 915 and 1980 m depth. At greater depths, most inter-granular pores have been eliminated and the porosity is then mostly secondary, related to dissolution of feldspars, but the secondary pores are commonly filled with kaolinite and illite.

2.3.3.2 Hua Hin Basin

The formation nomenclature is the same as for the Pattani Basin, except that the designations are TH instead of TP, and the cycles of Woollands and Haw (1976) are the same. The basin is small (c. 4500 km^2) with a maximum thickness of 4500 m of siliciclastic strata (Fig. 2.8). Non-marine limestones are common in the middle part of the sequence,

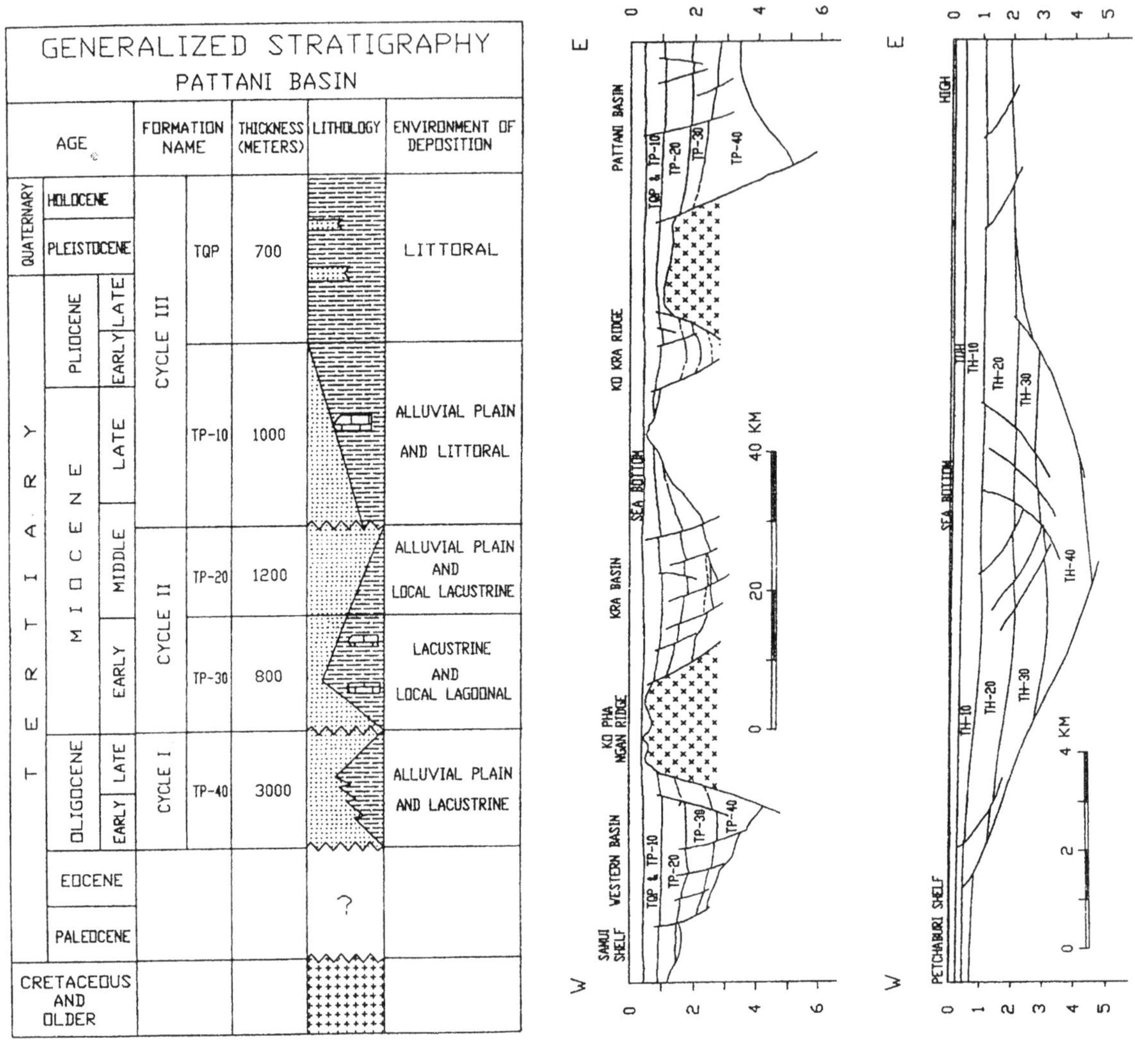

Fig. 2.8. The intermontane Tertiary basins of the Gulf of Thailand, which became marine by the end of the Miocene. Left: generalized stratigraphy of the Pattani Basin. The Hua Hin Basin shares the same terminology, but formation names substitute H for P, so that TP-40 in Pattani becomes TH-40 in Hua Hin and so on. There are more fresh water carbonates in TH-40 and TH-30 than in the Pattani Basin (TP-40 and TP-30). The Hua Hin lacks carbonate in TH-10. Middle: E–W cross section through the northern Pattani Basin. Right: E–W section through the northern Hua Hin Basin (after CCOP, 1991).

particularly in the north. In general TH-40 is lacustrine, while TH-30 and TH-20 are fluviatile with minor lacustrine intervals. There is a prominent late Middle Miocene unconformity (the Sabah Orogeny), above which the deposits are fluviatile to marginal marine in sequences TH-10 to TQH.

Good source rocks occur in TH-40 and the oil window lies between 2800 to 3900 m. However, the basin is not extensively drilled. The sandstone reservoirs are thin and, because of the high heat flow, porosity decreases rapidly with depth, but may be 15 to 20% in TH-40 and TH-30. The reservoir targets are Oligocene TH-40 and Miocene TH-30 sandstones as well as pre-Tertiary basement carbonate topographic highs.

2.3.4 The Zhu Jiang Kou (Pearl River Mouth) Basin

The basin is typical of the SE China Cainozoic epicontinental or intermontane basins (Li, 1984). The Palaeocene and Eocene formations are entirely terrestrial and deposited within grabens and half-grabens (Chen and Pei, 1993) (Figs 2.2; 2.5; 2.9). A detailed analysis of the basin geometry and its subsidence led to the conclusion that the evolution of the basin can be accounted for by one main phase of stretching from 60 Ma to 35 Ma ($\beta = 1.7$) (Su *et al.*, 1989).The Palaeocene through Eocene active rifting stage is represented by ENE–WSW trending normal faults (Fig. 2.2). This was followed by a Lower Oligocene through Middle Miocene downwarp. Since then it has continued to subside because of sediment load. The basin is sub-divided by major faults into six provinces: Zhu-1 to Zhu-3 depressions, North Faulted Terrace, Wheaten Local High, and the Shenhuansha Local High (Fig. 2.9).

The typical sedimentary sequence consists of red clastics of Palaeocene to Lower Eocene age deposited within grabens and half grabens, unconformably overlying a basement which contains Yenshanian granite plutons. The redbeds are overlain by dark argillaceous lacustrine shales of Middle to Lower Eocene age, and by Lower Oligocene

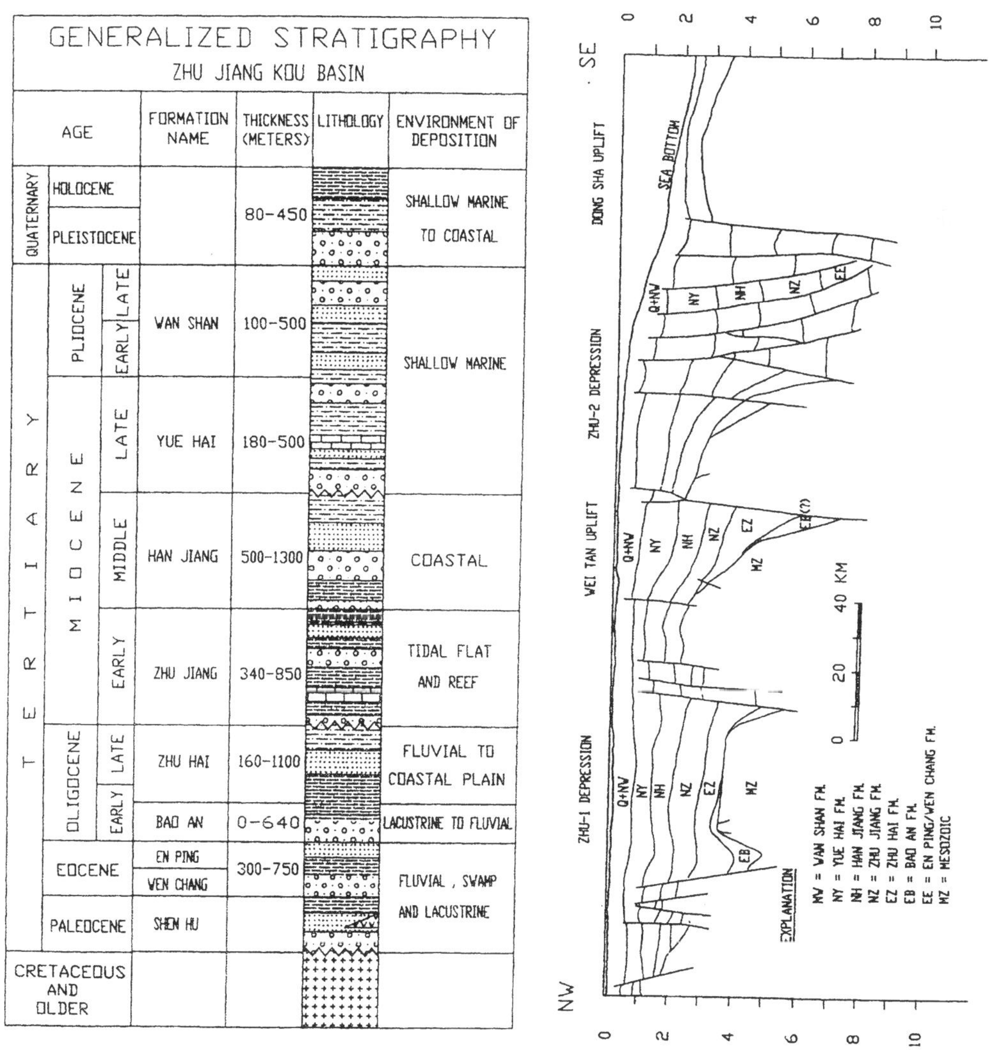

Fig. 2.9. The Zhu Jiang Kou (Pearl River Mouth) Basin. Left: generalized stratigraphy. Right: NW–SE cross section of the basin (after CCOP, 1991). The vertical scale is in km.

coal-bearing littoral plain and swamp deposits. Seismic stratigraphy indicates, however, that a marine sequence may be present in parts of the basin (CCOP, 1991). The first marine incursion was in the Lower Oligocene, considerably earlier than in the Malay Basin. The Shen Hu is dominated by red sandy conglomerate of fluvial fan facies, associated with igneous debris. The Wen Chang is of black lacustrine mudstones, with thin sandstones and siltstones (Fig. 2.9). The En Ping formation is dominated by floodplain and river sandstones with some swamp and lacustrine mudstones. The Upper Oligocene Zhu Hai Formation is of grey coarse sandstones and silty mudstones, interpreted to be paralic. The regional subsidence phase consists of Lower to Middle Miocene littoral clastics and shallow bank carbonates, and then by Upper Miocene to Quaternary neritic and bathyal clastics (Su *et al.*, 1989). A regional unconformity between the Middle Miocene Han Jiang and the Upper Miocene Yue Hai is known as the Dongsha Movement (Chen and Pei, 1993). Much more spectacular uplift events occurred at this time in West Natuna and in the Crocker Ranges of Borneo (Sabah Orogeny).

The main source rocks occur within the Eocene to Lower Oligocene lacustrine Wenchang and Enping Formations, with total organic carbon values of up to 7.62% (Chen and Pei, 1993). Oligocene and younger formations have total organic carbon contents less than 1.5%. Detailed geochemical analysis has shown that the source rocks are overwhelmingly of lacustrine origin, and the Eocene lacustrine rocks are mature (Chen and Pei, 1993; Fu J. *et al.*, 1993). The geothermal gradient of the basin is 42° C km^{-1} and the oil threshold depth is 2200 m, where the vitrinite reflectance has been measured at 0.8%. In the deeper grabens, the Enping Formation has been buried by as much as 5 km and reflectivity has been measured up to 1.25%, indicating maturity. Gas has been found in Oligocene and Miocene sandstones, conglomerates and limestones. Structures which may have trapped oil/gas are: compressional and roll-over anticlines, anticlines related to basement uplift, drape structures and half anticlines. There are also composite and stratigraphic traps related to buried hills, bio-reefs and facies changes. Presently there are 6 oilfields and 2 gas fields and 10 additional structures have been identified. Usually the fields are of 20–40 Mt of oil-in-place but one is of 200 Mt. Obvious reservoir targets are: Eocene, Oligocene and Lower Miocene sandstones, Lower to Middle Miocene biohermal limestone and reefs, and traps related to pre-Tertiary basement topographic highs (CCOP, 1991).

The largest oilfield discovered to date is that of Liuhua 11-1 on the northern part of the Dongsha High (Tyrrell and Christian, 1992), 220 km SE of Hong Kong. The giant in-place oil accumulation is in a 400 m thick Lower Miocene carbonate anticlinal structure, elongated NW–SE, the crest of which is in 300 m water depth. The limestone is underlain by water-bearing Oligocene to Lower Miocene Zhuhai sandstone, which unconformably overlies the Mesozoic igneous basement of the Dongsha High. The overlying oil seal is formed by upper Miocene shales.

The oil has migrated south-eastwards up-dip from the Huizhou Sag, sourced from Palaeogene lacustrine rocks (Fu J. *et al.*, 1993). All the oil within the carbonate became biodegraded when part of the carbonate complex was exposed to freshwater influx during lower sea levels of the Upper Miocene (Tyrrell and Christian, 1992).

The southern depressions of this basinal area are relatively unexplored, but seismic sections suggest that the Tertiary sequences may be underlain by bedded Mesozoic strata with source-rock potential.

2.3.5 Song Hong–Yinggehai Basin

This is a large (200 000 km^2) basin occupying the Gulf of Bacbo, between Hainan island and the coast of Vietnam (Fig. 2.1). The basin exists because of a seawards extension of the powerful NW–SE trending onland faults of Song Chay and Song Hong (Red River) (see Hutchison, 1989*a*, p. 144). The basin is 500 km long and 50 to 60 km wide, bounded on the NE by the No. 1 Fault; the west and south boundaries remain poorly known (Zhang and Zhang, 1991). The Yinggehai Sea contains at least 15 documented oil seepages.

2.3.5.1 Hanoi Rift

The onland Hanoi Rift is filled by the Red River Delta. The Neogene–Quaternary section of the delta is up to 3.5 km thick (Staritskiy *et al.*, 1973) in the narrow central graben, but shallows considerably towards the margins. Strong Neogene movements caused considerable deformation and angular unconformities. The Miocene to Pliocene deposits are mainly fluvio-lacustrine or littoral, resting unconformably on older basement (Tran *et al.*, 1979). Miocene sediments are mainly of conglomerate, sandstone, siltstone, and claystone, with lignite in the upper part. The thickness varies across the fault zone from 200 m to 3 km. The Pliocene includes siltstone and claystone with intercalations of conglomerate and sandstone, and thickness varies between 50 and 300 m.

2.3.5.2 Northwest part, offshore Vietnam

The fault system of the Hanoi Rift continues hundreds of kilometres offshore through the Gulf of Bacbo (Tran *et al.*, 1979). The basin is developed upon down-faulted

Palaeozoic basement and there is a strong but variable NW–SE magnetic lineation (Li, 1984). Ongoing wrench faulting is indicated by earthquake activity. The sedimentary fill varies from 2 to 10 km. The oldest known sediments are Palaeocene sandstones and siltstones of alluvial to fluvial origin, followed by Eocene lacustrine mudstones (Fig. 2.10). Continental sedimentation continues into the Oligocene, but by Lower Miocene the basin became shallow marine to deltaic (CCOP, 1991). Thin Miocene coal beds occur. The Song Hong-Yinggehai Basin shows the same style and intensity of Middle to Upper Miocene basin inversion (the Sabah Orogeny) comparable to the West Natuna Basin (Fig. 2.17), the evidence for which I have seen on unpublished seismic sections. This inversion resulted in many local unconformities and a major regional Upper Miocene unconformity.

Oil exploration is still immature. Only a few wells have been drilled and gas and condensate discovered. Miocene marine mudstones are the main source rocks. The average geothermal gradient is 40° C km^{-1} with the oil window between 3000 and 4100 m. Reservoir rocks are provided by Miocene deltaic sandstones, and roll-over anticlines and stratigraphic traps are common.

GENERALIZED STRATIGRAPHY
SONG HONG - YINGGEHAI BASIN (A)

AGE			FORMATION NAME	THICKNESS (METERS)	LITHOLOGY	ENVIRONMENT OF DEPOSITION
QUATERNARY	HOLOCENE		HAI HUNG	20 - 30		FLUVIAL
	PLEISTOCENE		VINH PHUC	30 - 50		DELTAIC
			HANOI	30		SHALLOW MARINE
			HAI DUONG	60		FLUVIAL
TERTIARY	PLIOCENE	LATE	VINH BAO	100		FLUVIAL TO DELTAIC
		EARLY		300		SHALLOW MARINE AND DELTAIC
	MIOCENE	LATE	DONG HOANG	200 - 300		FLUVIAL TO DELTAIC
		MIDDLE	TIEN HUNG	300 - 500		SHALLOW MARINE AND DELTAIC
			PHU CU	500 - 1000		
		EARLY	PHONG CHAU	500 - 1500		
	OLIGOCENE	LATE	THUY ANH	300 - 1500		DELTAIC TO DEEP MARINE FAN
		EARLY	DINH CAO	100 - 500		CONTINENTAL LACUSTRINE
	EOCENE		PHU TIEN	300 - 1000		LACUSTRINE FLUVIAL
	PALEOCENE					
CRETACEOUS AND OLDER					?	

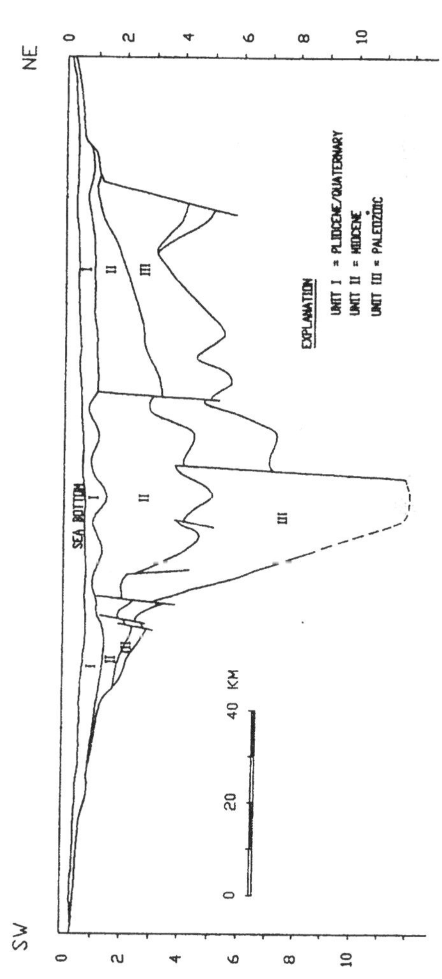

Fig. 2.10. Song Hong–Yinggehai Basin. Left: generalized stratigraphy of the nearshore Vietnam section. Right: SW–NE cross section parallel to the coastline (after CCOP, 1991). The vertical scale is in km.

2.3.5.3 South-east or Chinese part

The investigated part lying SW of Hainan island has a half graben structure (Fig. 2.1). The main boundary fault, known as the No. 1 Fault, delineates the NE side of the basin. The west and southern boundaries remain poorly known (Zhang and Zhang, 1991). Here the marine influence begins earlier than in the north—Upper Oligocene (Fig. 2.11)—and the succession includes shallow marine carbonates at several horizons. Possible reservoirs are Lower Miocene reefs.

There is a long hiatus between the Palaeogene and the Neogene, demonstrated as a major unconformity in the basin. Following basin inversion, the Miocene was deposited on an eroded surface (Zhang and Zhang, 1991). Mud diapirism is very common along the basin depocentre.

Source rocks occur in the Middle Miocene Meisham Formation, which contains organic carbon of 0.47 to 1.34% (average 0.69%). The average geothermal gradient is 45° C km^{-1}, and the source rocks lie within the oil window (Zhang and Zhang, 1991). Oil migration has been found to occur updip from the basin depocentre towards the east.

2.3.6 Beibu Wan Basin

The Beibu Gulf, between Hainan island and mainland China, appears to be a NE extension of the Song Hong

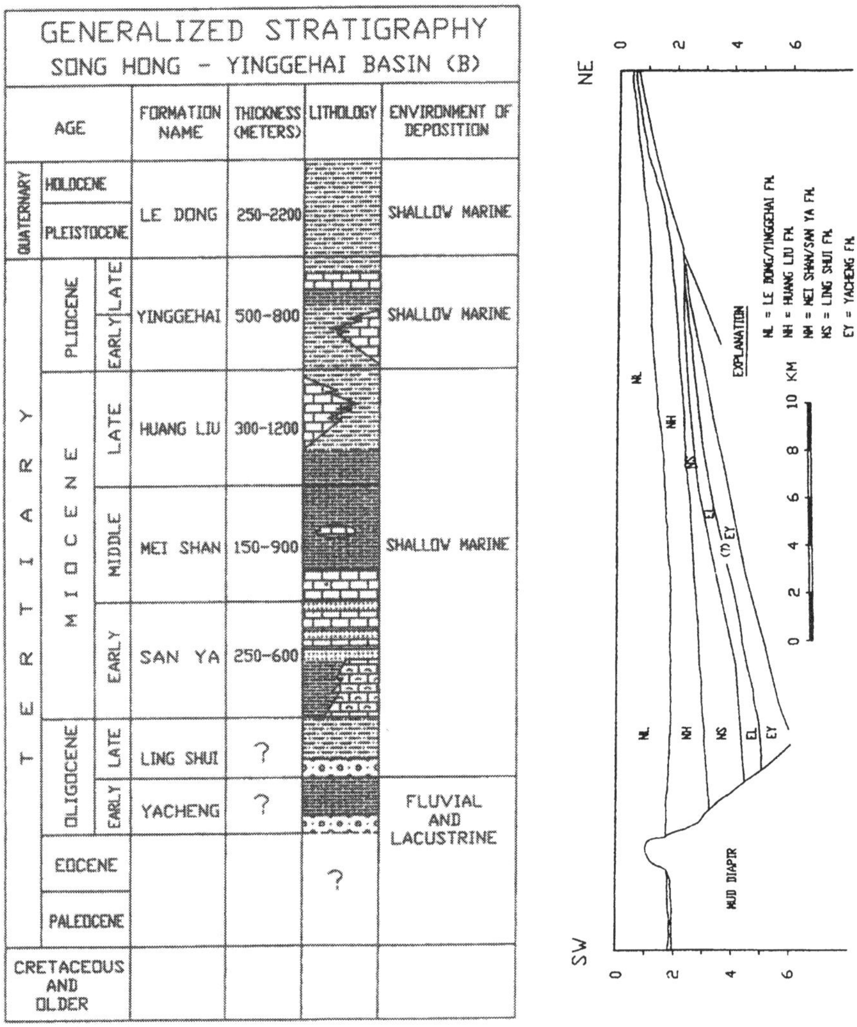

Fig. 2.11. Song Hong–Yinggehai Basin. Left: generalized stratigraphy of the Chinese portion adjacent to western Hainan island. Right: SW–NE cross section close to Hainan Island (after CCOP, 1991). The vertical scale is in km.

Basin. It is developed on down-faulted Palaeozoic basement, and has a sedimentary fill of 4–6 km (Li, 1984). The thick Palaeogene depocentres of Weiixinan, Haizhong, Wushi and Maichen form depressions between horsts, which are draped over by Neogene strata.

Some of the depressions have more than 6 km of Tertiary strata (Fig. 2.12). The Palaeocene Changliu Formation consists of continental mudstone, sandstone and conglomerate. The overlying Liu Shagang Formation contains organic rich lacustrine muds, which represent the main source rocks. Coastal to fluvial sand and mud were deposited in the Upper Oligocene. The Miocene started with a transgression and deposition of shallow marine limestone, conglomerate, sandstone and mudstone. Volcanic lenses occur in the Middle Miocene Jiaowei Formation (Fig. 2.12). Neritic Upper Miocene and Pliocene clastics represent slow regression. The basin is associated with Quaternary alkaline basalts, which are also common onland Hainan.

The average geothermal gradient is 40° C km^{-1}. Offshore drilling began in 1976 and exploration activity was high in the early 1980s. The Wei 10–3 oilfield was put on production in 1986. Since then new discoveries have been made. The main reservoirs are buried gas-bearing Carboniferous limestone topographic highs, Eocene sandstones, Eocene turbidites, and Oligocene to Lower Miocene sandstones within anticlinal structures (CCOP, 1991).

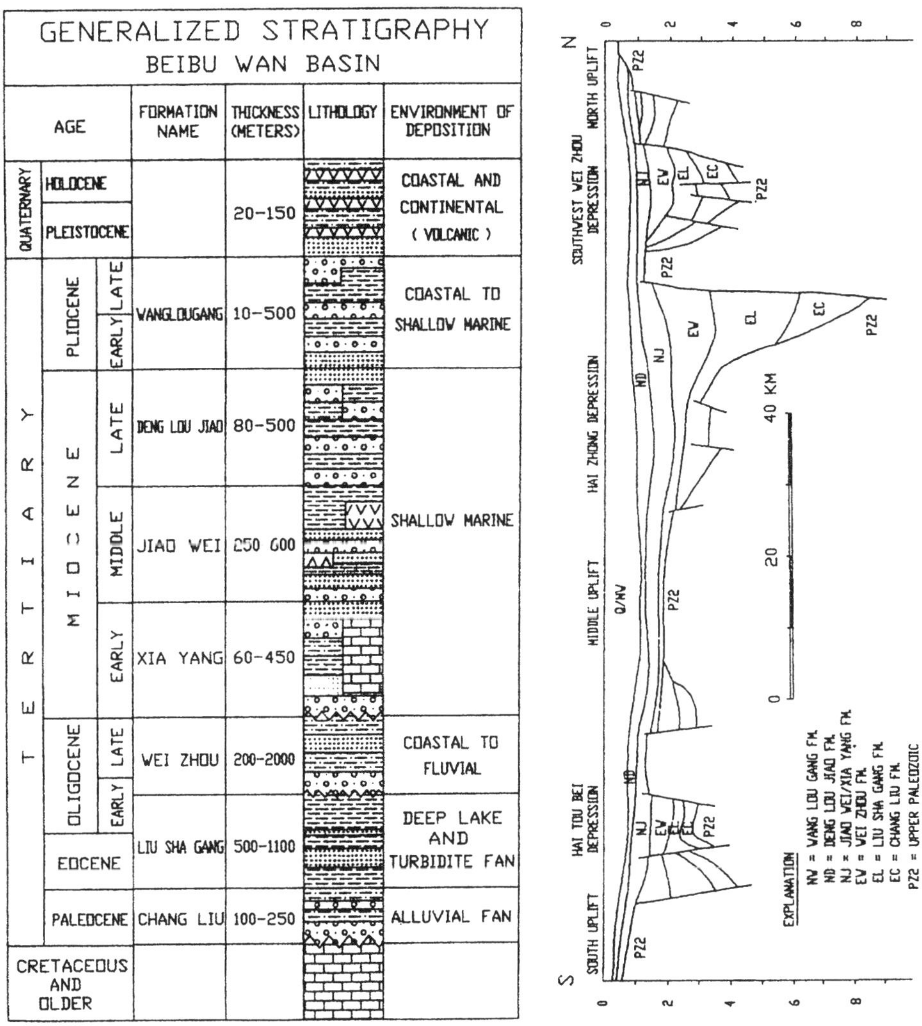

Fig. 2.12. The Beibu Wan Basin. Left: generalized stratigraphy. Right: N–S cross section (after CCOP, 1991). The vertical scale is in km.

2.3.7 The Mekong (Vung Tau) Basin

This is an E–W trending basin, connecting onshore with the Mekong delta. The thickness of Cainozoic strata in the delta varies erratically, and there are small bedrock outcrops within the delta area. The pre-Cainozoic basement is strongly fault-controlled. The delta passes into the narrow ENE-trending Vung Tau Basin, which has more than 5 km of Cainozoic strata in its deepest parts (Fig. 2.13). The trough extends 400 km NE from the mouth of the Mekong River and has a width of 100 km (Le, 1986). It is a typical epicontinental basin, with Upper Eocene graben formation followed by Upper Oligocene depression in the axial zone, down-warped some 2400 m.

The basement is of Liassic Bandon Series of 2 km thickness, folded and intruded by Dalat granite plutons, dated 60 to 150 Ma (Upper Jurassic to Cretaceous). Most of the faults trend E–W and some have a throw of the order of 200 to 1000 m.

The sedimentary sequence has a maximum thickness of 6 km. The lower part is of 3 km thick Upper Eocene Cu Lau-Zung Formation, which fills narrow half grabens. The continental formation is mainly of coarse grained conglomerate, gravel and sandstone. The overlying Lower Oligocene Tra-Cu Formation is of 250 m of conglomerate

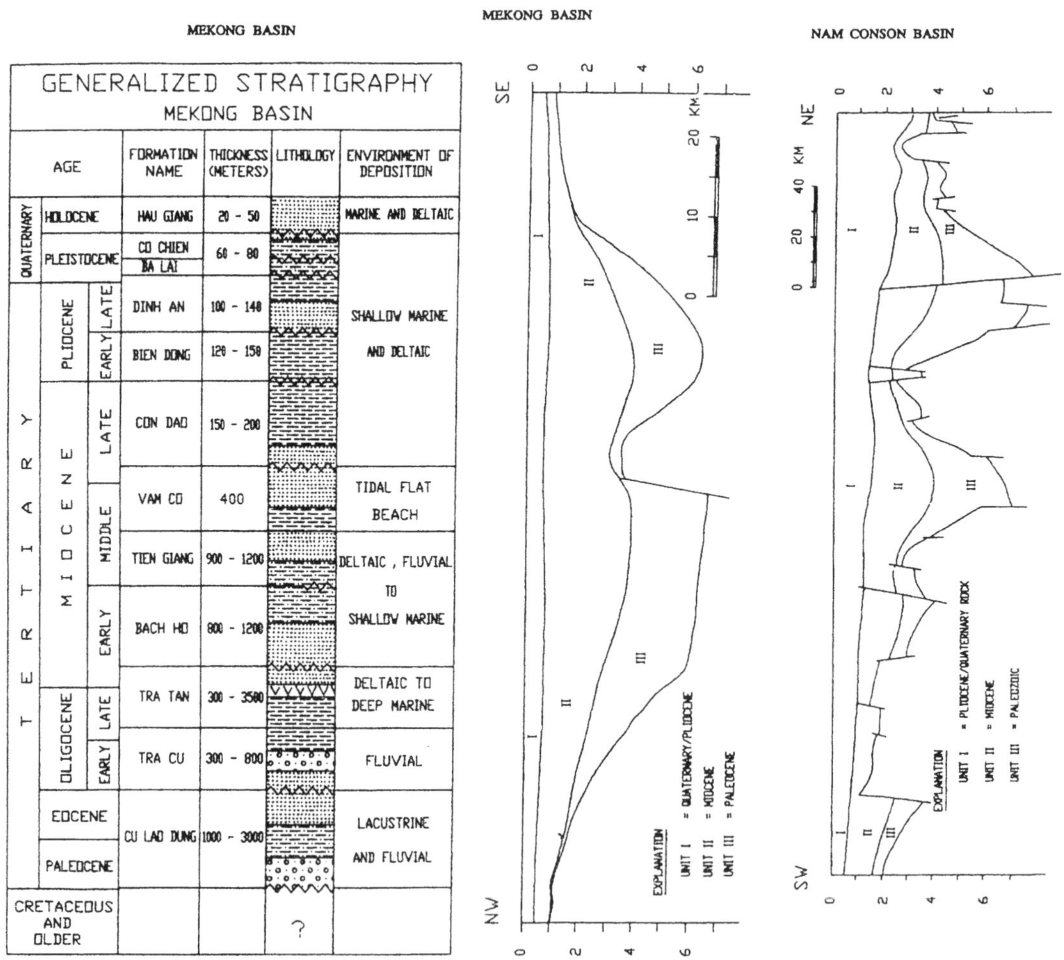

Fig. 2.13. The Mekong and Nam Conson basins of the South China Sea. Left: generalized stratigraphy of the Mekong Basin. The Nam Conson shares the same formation names, but has some lithological differences discussed in the text. Middle: NW–SE section through the Mekong Basin. Right: SW–NE cross section through the Nam Conson Basin (after CCOP, 1991). The vertical scales are in km.

and sandstone, followed by finer grained strata (Fig. 2.13). It unconformably overlies the Cu Lau-Zung Formation and has been interpreted as lacustrine. The Upper Oligocene Tratan and Lower Miocene Bakho formations occupy the central trough. The 2500 m thick Tratan Formation is of lagoonal and deltaic sandstone-shale and some limestone. The unconformably overlying 200 m thick Bakho Formation is of sandstone-shale delta facies. These last two formations have petroleum potential with good carbonaceous source rocks and reservoir sands on the basin margins (Le, 1986). Some basalt occurs in the Oligocene and Lower Miocene strata. These volcanic rocks are clearly seen on certain seismic sections as flat-lying horizons of thickness variable from 800 m to zero. The thicker parts show typical untectonized volcanic cones. There is a possibility of fractured volcanics possessing reservoir quality. Other seismic sections are devoid of volcanics. Marine facies are increasingly important in the Middle Miocene T'enzang and Lower Pliocene Biendong formations. They contain abundant seams of brown coal in their upper parts.

The basin is characterized by two regional unconformities (Le, 1986), dated Lower and Upper Miocene, and syn-depositional NE–SW, E–W, and NW–SE-trending fault systems decrease in intensity upwards and die out by late Miocene time. No major inversions have been seen in the seismic sections.

Plio-Pleistocene basic volcanics occur widely and volcanic cones characterize the sea floor in places near the east coast of Vietnam. The basalts contain mantle peridotite xenoliths and are probably rift related and alkaline, but details have not been published. Volcanism in this region is not extinct and in 1993 an onland eruption was observed but not publicized near the east coast (P.B. Woodroof, pers. comm., Aug. 1994).

Exploration for petroleum began in the late 1960s. The White Tiger Field, (see Hutchison, 1989*a*, pp. 179 and 183), discovered in 1973, now produces about 50 000 barrels of oil per day from Lower Miocene strata. Good production rates have been verified in the Lower Oligocene Tracu Formation and from fractured and weathered igneous basement.

2.3.8 Nam Conson Basin

This basin lies parallel to and south of the Mekong Basin, from which it is separated by the Conson Swell, and separated from the Malay Basin by the Khorat Swell. The NE–SW trending basin contains up to 15 km of sedimentary fill. There are major N–S trending faults indicating important basement control. The average throw of these faults is 800 m. The basin shares the same stratigraphic nomenclature as the Mekong Basin (Fig. 2.13). The Palaeocene to Eocene beginnings with fluvial to lacustrine deposits are similar to the Mekong Basin. The Palaeogene rift-fill sediments onlap onto NE–SW and N–S basement fault blocks. Wells drilled to-date have penetrated a transgressive mega-sequence in coastal plain, delta, estuarine and lagoonal environments (Matthews and Todd, 1993). Sedimentation broadly kept pace with extension and subsidence. A major difference between the Nam Con Son and the Mekong basins is that marine carbonates become important southwards in the Miocene. The first marine carbonate is dated late Lower Miocene. The carbonates formed on drowned basement highs. There are important limestones in the Vam Co and Con Dao formations During the Miocene, the basin was deltaic in the west with a gradual change to shallow marine and deep marine eastwards. Miocene land therefore lay to the west. A Middle Miocene rift-fill sequence is characterized by a prominent unconformity at its base (Matthews and Todd, 1993). A major late Middle Miocene truncational unconformity records a gentle compression and basin inversion, synchronous with the Sabah Orogeny. A general marine shelf environment with infill of sand, silt and mud, prevailed from Upper Miocene to Pleistocene. Platform-type carbonate build-ups, which developed on fault blocks during the flooding, were sequentially inundated by the easterly prograding palaeo-Mekong clastic wedge (Matthews and Todd, 1993).

Exploration for oil and gas is still in the early stages but is now accelerating, with emphasis on carbonate and sandstone reservoirs of Oligocene and Miocene age. Both oil and gas have been discovered in anticlinal traps, of which the Big Bear is the major oilfield now under developemnt (CCOP, 1991).

2.3.9 Malay Basin

The basin is approximately 500 km long in a NW–SE direction, and 200 km wide. The simple elliptical shape first appeared in Hamilton (1979), who did not give the source of his insight, and no subsequent publication has sought to modify it. The basin links up with the Pattani Basin on the NW and the W Natuna Basin on the SE. The change from N–S structures in the north and west, through NW–SE in the central region, to E–W in the Penyu and Natuna areas, mimics a corresponding regional change (oroclinal bending) in the onland pre-Jurassic structures from Thailand, through the Malay Peninsula to Singapore, South Sumatra to Bangka, Billiton and Borneo (Hutchison, 1993). Therefore this convex westwards oroclinal bend is largely basement controlled as illustrated in Fig. 2.4. However some rift structures, such as the Penyu Basin, continue beyond the region of basement control to discordantly transect the basement fabric. In the same way, the N–S Bengkalis Graben was discordantly superimposed upon the NW–SE pre-Jurassic fabric of Sumatra (Hutchison, 1993). Along the southern margin of the Malay Basin, a basement high, the Tenggol Arch, separates it from the Penyu Basin (Fig. 2.1).

2.3.9.1 *Structural evolution*

The structural history of the Malay Basin is generally held to comprise three stages (Fig. 2.14):

a. An Upper Eocene to Oligocene extensional stage, giving rise to basement grabens, which trend more E–W in the SE of the basin and swing to a NNW–SSE trend along the western side of the basin (e.g. Dungun Graben) (Liew, 1993). The E–W trending Oligocene half-grabens, down-faulted to the north, are well known in the West Natuna Basin (Wongsosantiko and Wirojudo, 1984). The main graben trends westwards to underlie the Bekok, Seligi and Pulai fields of the Malay Basin (Fig. 2.17). The Oligocene strata are thickest adjacent to the fault (up to 2.0 s 2-way time) and thin northwards towards a topographic high. By analogy with the Natuna area, the existence of Upper Eocene strata in the graben depocentres is strongly suspected. This E–W Oligocene half graben was tectonically inverted in Middle Miocene time to form the 'AR' High (Pollock *et al.*, 1984), which lies parallel to the more southerly Boundary High (Fig. 2.16). Over both these highs, Miocene sedimentation was thin. It is over these E–W highs (inverted Oligocene half-grabens) that the large Malay Basin oil fields are located (Fig. 2.17). Unfortunately these fundamental elements of the Natuna

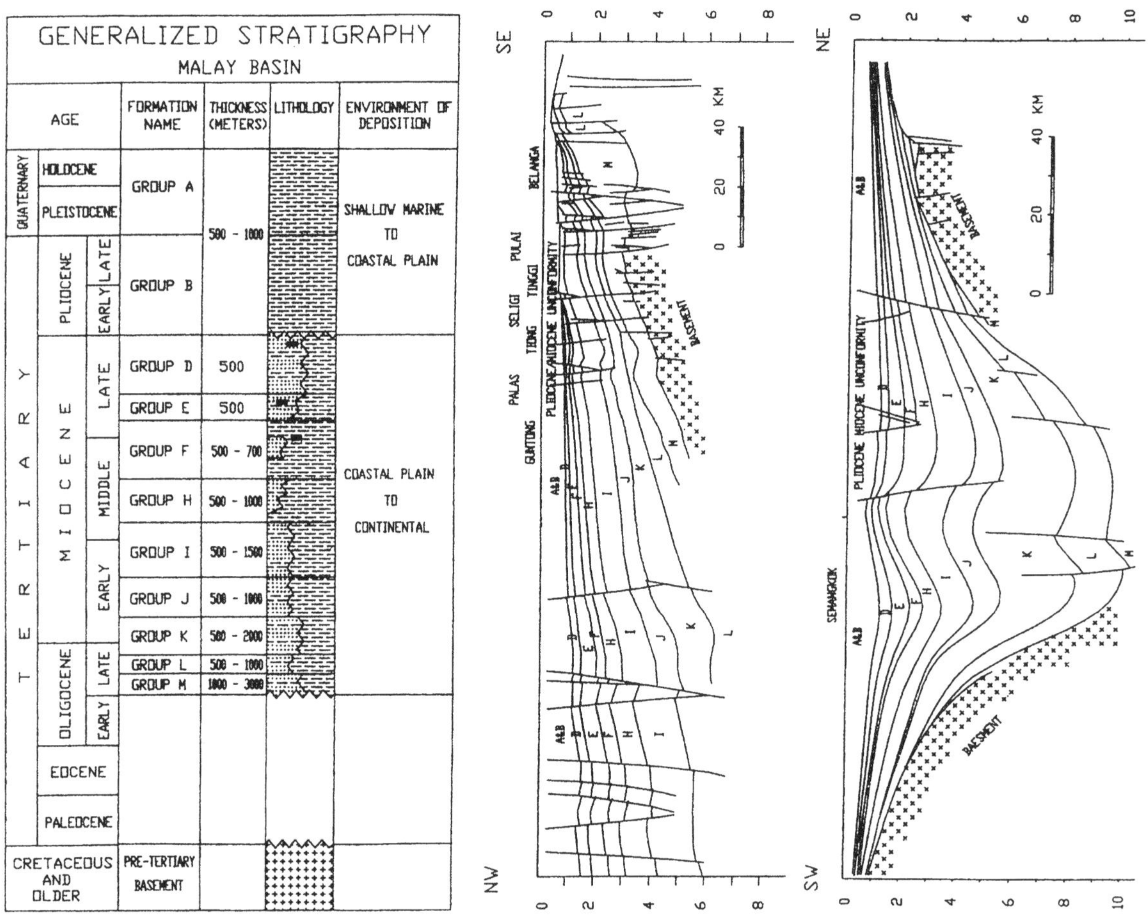

Fig. 2.14. The Malay Basin. Left: generalized stratigraphy. Centre: NW–SE longitudinal section of the south-eastern part of the basin. The Pliocene/Miocene unconformity becomes more pronounced SE-wards towards the West Natuna Basin. Right: SW–NE section across the basin (after CCOP, 1991).

Basin, which extrapolate into the adjacent Malay Basin, have never been described in publication by Exxon geologists, although Creaney *et al.* (1993) showed that we may expect some details in the future. The lack of study and publication reflects the rapidity with which EPMI (Exxon) found obvious trapping structures (e.g. Fig. 2.15) with excellent direct hydrocarbon indicators on their seismic, thereby the early exploration stages of the Malay Basin were obviated (Harun, 1987). The basement structure of the west-central Malay Basin is therefore identical on the western side to that of adjacent Peninsular Malaysia (N–S to NNW–SSE) (Figs 4.6 and 4.7) and on the eastern side to the West Natuna area of Indonesia (E–W). Middle Miocene inversion of the southern Malay Basin was less spectacular than in adjacent West Natuna, amounting to 1 km or less of uplift, making use of the bounding half graben faults. Because the inversion was less, the basin deeper and the Miocene section thicker, so that seismic resolution becomes lost at depth, the inversion has not been analysed in detail. Limestone is known from seismic data to form extensive basement; the geographic extent has not been published, but Fontaine *et al.* (1990) described Triassic limestone from a depth 2752 m in Sotong B-1 well, NE of the Tenggol Arch.

b. A Middle to Upper Miocene phase of N–S compression, giving en-echelon E–W trending anticlines, probably resulting from right lateral wrench movement, but the basement faults on which the wrench motions occurred have never been identified, nor their orientations known. The N–S compression becomes more pronounced towards the W Natuna Basin, and accordingly resulted from the Sabah Orogeny. The whole Neogene sequence is folded, usually overlying the inverted basement half grabens. The anticlines are the main oil exploration targets. During this compression, the SE part of the basin was strongly uplifted, and some Middle and all of the Upper Miocene sections have been eroded. The uplift may have been as much as 1300 m.

c. A Pliocene to Holocene extension. Pliocene formations drape over the whole basinal region, and towards the SE rest with an angular unconformity on Middle Miocene strata (Armitage and Viotti, 1977). In the south-east of the basin, the Upper Miocene–Pliocene Pilong Formation (Group B) rests with pronounced angular unconformity on the Bekok Formation (H) in the NW and progressively onlaps onto older formations, eventually on the Trengganu Group (J and K) sandstones in the SE (Nik Ramli, 1988). The generally E–W extension gave N–S oriented normal faults cutting the E–W anticlines. The sequence from Middle to Upper Miocene N–S compression to Pliocene E–W extension is a logical result of crustal relaxation after the orogenic peak. The young age of the E–W extension is consistent with the principal stress directions observed in borehole breakout studies (Mohd. Idrus, 1989).

Murphy (1989) pointed out that the Malay Basin exhibits typical inversion tectonics; its evolution being of 3 stages: 1) normal extensional faulting of the Cretaceous and older basement, 2) rapid sedimentation decreasing through time as activity along the normal faults decreased to zero, 3) a short lived and rapid compressive phase ending in quiescence. Murphy (1989) believes the Middle to Upper Miocene compressive phase was associated with trans-pression on through-going basement wrench faults.

Some compression continued after the main inversion, resulting in uplift and erosion of some 1200 m of section. The Lower Pliocene beds are arched upwards, but the compressive movement ceased and the youngest beds are perfectly flat-lying.

2.3.9.2 Stratigraphy

The geologists of EPMI (Exxon) collectively maintain that the oldest basinal strata are of Lower Oligocene age. These are continental alluvial sediments, difficult to date accurately, and I believe that the early grabens may contain Upper Eocene to Lower Oligocene continental sediments. The succession in the 6 to 10 km depocentre is unknown. By extrapolation from nearby West Natuna, the thick Oligocene strata are predominantly alluvial and lacustrine, but their environment of deposition has not been studied in the Malay Basin.

The presently used terminology of the Malay Basin is shown in Fig. 2.14. The alternatives are to use formation names (Armitage and Viotti, 1977), or a numerical system (Mohd. Nazri, 1988) in which stratigraphic unit I is equivalent to Group M (Upper Oligocene) and unit VIII equivalent to Groups (A+B) (Plio-Pleistocene).

The basin fill is predominantly siliciclastic. The Upper Oligocene to Lower Miocene strata are channel fill and alluvial fan deposits, but lacustrine deposits, though not described, do exist. These continental deposits grade upwards to fluvio-marine and coastal plain deposits. Land lay to the NW, W and E and the sea penetrated into Sundaland from a south-easterly direction (Mohd. Nazri, 1988). During the Middle to Upper Miocene, deltaic to shallow marine environments with dominant tidal influence prevailed (e.g. Mazlan, 1991). By this time, strongly elevated land was recently created by the Middle–Upper Miocene structural inversion in the Natuna area. Channels trending NW–SE have been identified on seismic profiles (Mohd. Nor Mansor and Rudolph, 1993). The erosional provenance for sediments was from the uplifted West Natuna province, with transport towards the NW. Coal beds are common at this time in the Malay

Basin. The Middle Miocene J reservoir sandstone is interpreted as a shoreface and shelf deposit, complete with offshore bars. The thick marine shales that overlie the offshore bars contain isolated sheet sandstones (Nik Ramli, 1986). Depositional texture, including degree of bioturbation, controls the J sandstone permeability. The non-bioturbated offshore bar and shoreface sandstones have permeabilities > 100 mD, whilst the bioturbated sandstones are < 30 mD (Nik Ramli, 1988).

2.3.9.3 *Tenggol Arch*

A major unconformity characterizes the southern part of the Malay Basin and the whole Middle Miocene section is missing (Mohd. Nazri, 1988). This part of the basin is also characterized by a prominent basement topographic high, the Tenggol Arch, which separates it from the Penyu Basin.

The basement rocks of the arch were drilled in Malong SG-17.2 well, 150 km offshore Trengganu (Mazlan, 1992). The arch is of highly fractured porphyritic rhyolite and granophyre, intruded by dolerite dykes. This basement formed a land surface which was deeply weathered to form a regolith, within which zircon from the rhyolite was concentrated. Unfortunately these zircons have not been dated.

Identical rocks characterize the eastern coastal outcrops of Johore and south Pahang and the offshore islands. By analogy with south Johore (Pengerang), the rhyolite and granophyre is currently thought to be Triassic, and by analogy with Kuantan the dolerite dykes may be Upper Cretaceous (Hutchison, 1989*a*; p. 166 and 290–291).

The arch is unconformably overlain by the Lower Miocene Tapis Sandstone (Fig. 2.15). The overlying Trengganu Shale represents a widespread transgression event. The Middle Miocene Malong Sands of the *Sand-Coal Formation* represent the main reservoirs of the Malong Field (Noor Azim and Mazlan, 1990). They were deposited in shoreface, inner shelf or fluvial channel environments. They have never been deeply buried (< 1500 m), and show few effects of diagenesis. The best reservoir characteristics are of 22 to 36% porosity and 100 to 800 mD permeability.

2.3.9.4 *Source rocks and hydrocarbon trap styles*

The source rocks are predominantly non-marine, and range from algal-dominated Lower Oligocene lacustrine strata of the M syn-rift group, to Miocene delta plain and coastal plain sources in Groups H and I. The Group K shale is mature and within the oil generating window in the Malay Basin centre and is over-mature along the SW flank of the basin. Oil and gas migrated along strata updip from the more mature SW region consistently towards the NE. Hydrocarbons in the over mature SW province contain CO_2 derived from older underlying basement limestones. In the K Group sandstone reservoirs the dominant process of hydrocarbon migration from SW towards the NE is of fill and spill (Creaney *et al.*, 1993). Since the hydrocarbon migration is overwhelmingly parallel to the dipping strata, and there are good seals, there are several independent hydrocarbon systems, derived from separate source rocks.

The Malay Basin is a well established hydrocarbon producing area (Table 1.2), which roughly can be divided into two provinces:

- The non-associated gas province. This is the northern and central parts of the basin where Middle to Upper Miocene strata provide the main reservoirs.
- The oil and associated-gas province. In the south-eastern part of the basin, oil is found in Upper Oligocene and Lower Miocene reservoirs (CCOP, 1991).

Geothermal gradients in the Malay Basin range from 45 to 52° C km^{-1} (Wan Ismail, 1990). The oil window should therefore lie between 1000 m and 3500 m (Hutchison, 1983*a*, p. 280). Therefore Oligocene and Lower Miocene formations in the basin centre should be overmature. Source rocks have been identified within K, L and M sequences (Creaney *et al.*, 1993). From publications on the adjacent West Natuna area (Pollock *et al.*, 1984), the source rocks are Barat Formation (= K and J) lacustrine shales.

There are six trap styles in the Malay Basin (Mohd. Tahir and Rudolph, 1992). Basement-supported traps related to basement highs occur on the basin flanks. Non basement-supported traps occur in the central portion of the basin and are related to compressional inversion and associated faulting. The larger oilfields are characterized by simple E–W trending elliptical domal structures, which probably owe their origin to the Middle to Upper Miocene inversion. In the Tinggi Field (Harun, 1987), the main reservoir is the J-18/20 sandstone. The structure is an E–W elongated doubly-plunging anticline cut by N–S trending normal faults, some downthrown to the W, others to the E. The structure forms a perfect hydrocarbon trap, enhanced for EPMI (Exxon) by excellent seismic direct hydrocarbon indicators (DHI) (Fig. 2.15).

Compressional anticlines also occur on the Tenggol Arch (Ng, 1987), where they occur *en echelon*, elongated either WNW–ESE or E–W. The *en echelon* nature and elliptical shape of these anticlinal traps suggest a transpressional origin, resulting from wrench faulting of the basement. Suitably oriented faults are the NNW–SSE Tenggol Fault Zone, and the compressional anticlines could be produced by right-lateral motion on these basement faults. However, neither the orientation nor the existence of basement faults beneath the Tinggi Field have

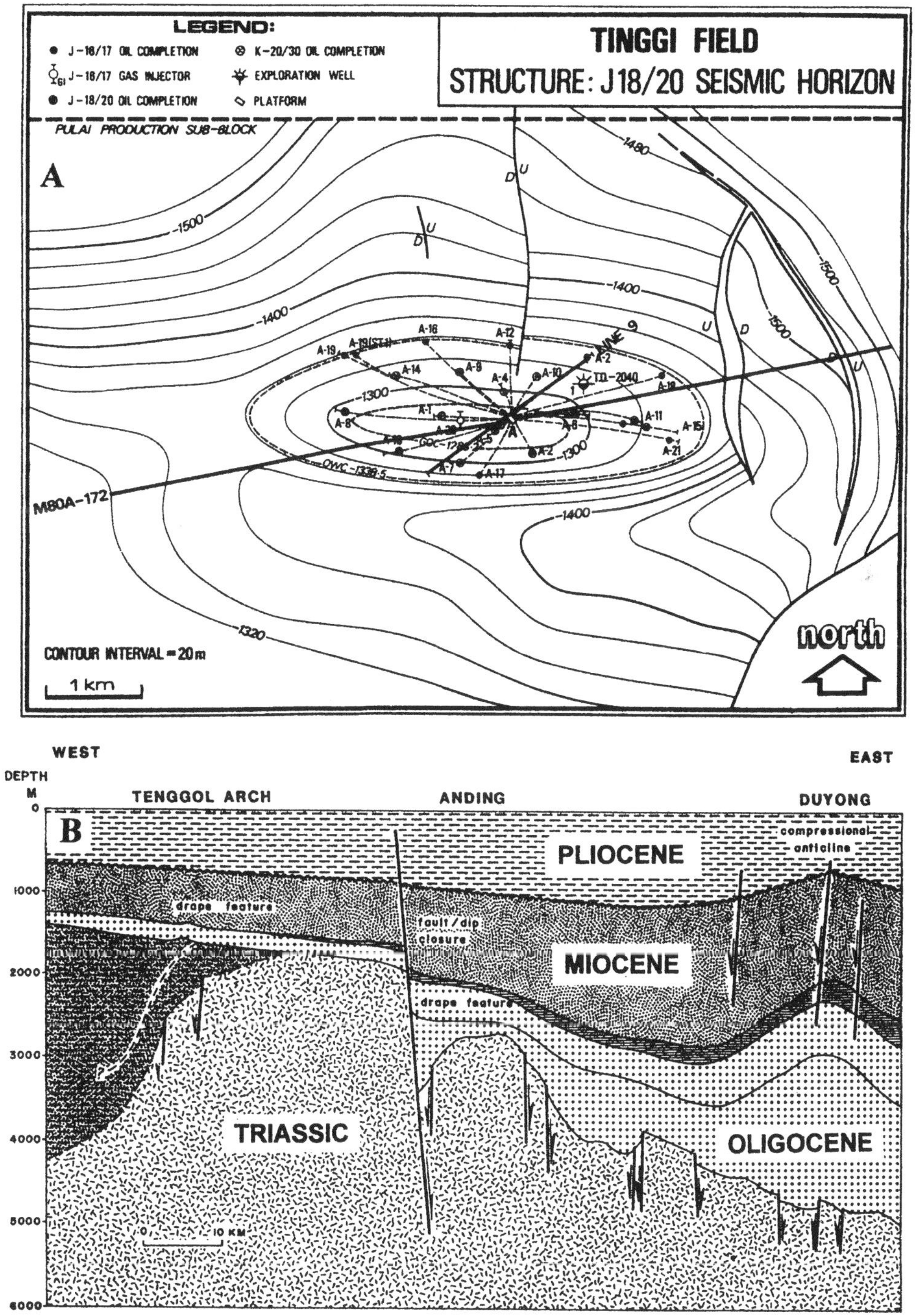

Fig. 2.15. Hydrocarbon trap styles in the southern Malay Basin. A: the E–W trending elliptical dome of the Tinggi Field (after Harun, 1987). B: E–W cross section over the Malong Field of the Tenggol Arch (after Ng, 1987). Irregularities in the faulted basement were filled by Oligocene continental sediments, then draped over by Miocene marine strata. The Upper Miocene regional unconformity is shown.

ever been discussed in the literature. This is surprising because at least two major NW–SE oriented right-lateral wrench faults are well described from the West Natuna Basin immediately to the SE of the Malay Basin (Fig. 2.16) and these basement faults lie in a suitable orientation to be used to give the N–S trans-pression and E–W trans-extension.

Another important trapping style is shown by Miocene strata unconformably draping over the topographic highs of the Tenggol Arch horsts (Fig. 2.15).

2.3.10 West Natuna Basin—Indonesia

The prominent Natuna Arch extends northwards at least 300 km past Natuna Island along longitude 108° E. The Arch is an extension of the Sunda Shelf, West Borneo Basement and Kuching Zone of Borneo (see Hutchison 1989*a*, p. 185). It stood above sea level and received no Oligocene sedimentation and accordingly it was a land barrier between the West Natuna and the quite different and unrelated East Natuna basin (White and Wing, 1978; Wongsosantiko and Wirojudo, 1984). The latter partly correlates with the Luconia Platform and Miri Zone (Greater Sarawak Basin), and accordingly is described in Chapter 3.

A major 200 km long N–S trending wrench (rotational) fault extends northwards from the Natuna Arch. It forms the eastern bounding fault (downthrown to the west), between 5 and 6° N, of a Miocene basin of thickness in excess of 2 s (2-way seismic time). The same fault forms the western margin (downthrown to the east) of a Miocene basin of at least 1.5 s (two-way seismic time) at about 7° N. The eastern margin of the Natuna Arch is interpreted as the northwards extrapolation of the Lupar Line of Sarawak. Accordingly, this major N–S wrench fault should correlate with the Lupar Fault, considerably downfaulted onland Borneo on its SW side to form the Ketungau Basin (Silantek Formation in Sarawak).

2.3.10.1 Depositional history

The West Natuna basinal area has been dominantly non-marine. Many wells have penetrated basement in this region (Hutchison, 1989*a*, pp. 179 and 183). A variety of Lower Cretaceous igneous rocks have been encountered. Palynology shows that from Oligocene to Lower Miocene the sedimentation was dominantly non-marine. There was no major marine incursion until the Middle Miocene. However, there was a short-lived marine incursion in the earliest Miocene (Pollock *et al.*, 1984). It is most convenient to subdivide the stratigraphy into four tectono-stratigraphic mega-sequences, following Ginger *et al.* (1993).

2.3.10.1.1 Syn-rift mega-sequence Initial extension created highly dissected grabens with many small fault blocks. Within the deeper grabens, the basement is unconformably overlain by either volcaniclastic rocks (Jonklaas, 1991; Sutoto, 1991) or by the 'Lama' Formation (Fig. 2.16), composed of thick dark grey mudstone with minor sandstones and siltstones. Spores are age-undiagnostic in this continental formation but an Upper Eocene age is preferred. The Benua Formation is predominantly of shale, representing deposition in a widespread lacustrine system. The Benua Formation is also known locally as the Belut (Pollock *et al.*, 1984). Unfortunately these lacustrine deposits, where drilled, are organically lean and do not represent good source-rocks. However, these Upper Eocene lacustrine shales must represent potential source rocks, because geochemical analysis shows that the oils, trapped in structures adjacent to the Anambas and other grabens, are from a lacustrine source (Ginger *et al.*, 1993). The overlying Lower Gabus Formation was also predominantly lacustrine, but the environment was oxidizing. This was due to deposition into half grabens and valleys of immature coarse clastic weathered products of eroding dominantly granitic outcrops. Sedimentation generally began in the Lower Oligocene. A minor angular unconformity at the top of the Lower Gabus Formation marks the end of rifting in the Anambas Graben. These lacustrine strata are characteristically of redbed facies, cored in several wells (Hutchison, 1989*a*, p. 183).

2.3.10.1.2 Post-rift mega-sequence Following the end of rifting, a period of tectonic quiescence ensued from

Fig. 2.16. The West Natuna Basin (after Daines, 1985). A: wrench fault orientations around the Udang Field, as traced on the pre-Tertiary basement. Inset: strain ellipse to relate the Neogene structures to the basement wrenching. The Boundary Graben is another name for the Anambas Graben, shown in Fig. 2.17. B: map of the Upper Palaeogene and Neogene compressional (anticlinal structures and reverse faults) and extensional (normal faults) structures. Normal faults strike 153° on average, reverse faults and fold axes trend 63°, and the regional activating wrench faults strike 122° on average. C: chronostratigraphy of the Cumi-Cumi PSC of the West Natuna Basin (after Ginger *et al.*, 1993). In Malay Basin terminology, A+B = Muda; H+I+J = Arang; K = Barat and Upper Gabus; Trengganu Shale = Barat; L + M = Keras Shale Member of the Gabus. ▶

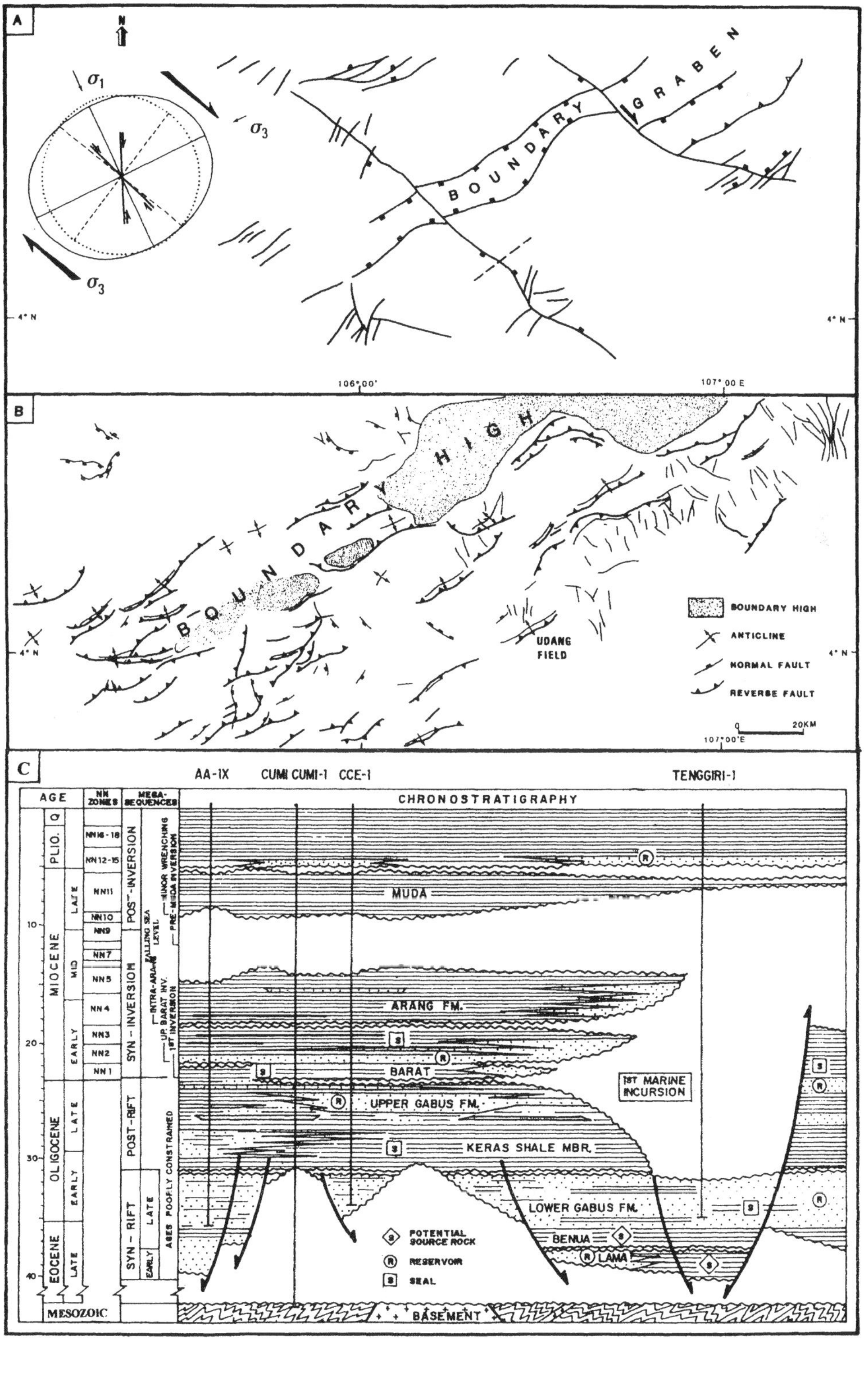
A
N
σ_1
σ_3
σ_3
BOUNDARY GRABEN
4° N
4° N
106°00'
107° 00 E
B
BOUNDARY HIGH
UDANG FIELD
BOUNDARY HIGH
ANTICLINE
NORMAL FAULT
REVERSE FAULT
4° N
4° N
0
20KM
107°00'E
C
AA-1X
CUMI CUMI-1
CCE-1
TENGGIRI-1
AGE
NN ZONES
MEGA-SEQUENCES
CHRONOSTRATIGRAPHY
Q
PLIO.
MIOCENE
LATE
MID
EARLY
OLIGOCENE
LATE
EARLY
EOCENE
LATE
MESOZOIC
NN16-18
NN12-15
NN11
NN10
NN9
NN7
NN5
NN4
NN3
NN2
NN1
POST-INVERSION
SYN-INVERSION
POST-RIFT
SYN-RIFT
LATE
EARLY
MINOR WRENCHING
PRE-MUDA INVERSION
FALLING SEA LEVEL
INTRA-ARANG
UP. BARAT INV.
1ST INVERSION
AGES POORLY CONSTRAINED
10
20
30
40
MUDA
ARANG FM.
BARAT
1ST MARINE INCURSION
UPPER GABUS FM.
KERAS SHALE MBR.
LOWER GABUS FM.
BENUA
LAMA
BASEMENT
POTENTIAL SOURCE ROCK
RESERVOIR
SEAL

Middle Oligocene to Lower Miocene, dominated also by lacustrine sedimentation. The extensive Keras Shale Member of the Gabus Formation was deposited in large shallow lakes, by which time most of the basement rift topography had been buried, resulting in low-lying topography. The Keras Shale has good source rock characteristics. Gradually the lake environments were replaced by prograding braided deltas and lacustrine plain deposits of the Upper Gabus Formation. Relatively coarse terrestrial clastics were brought in by fluvial systems, mainly of high energy braided streams. Silts and clays were deposited in distal and lacustrine environments. The Gabus Formation is composed of thick quartzose sandstones, up to 65 m thick, especially towards the top. They are interpreted to have been provenanced from the north. An Oligocene age is ascribed on the basis of palynology. The Upper Gabus sandstones have good reservoir characteristics. In some parts of the West Natuna basinal area, the uppermost section of the Upper Gabus Formation and the base of the Barat Formation contain marine foraminifera, indicating the very first, though short-lived marine incursion. The marine episode is not documented in all wells; elsewhere lacustrine conditions prevailed. The central part of the Malay Basin was probably a large lake at this time, fed by internal drainage.

2.3.10.1.3 Syn-inversion mega-sequence The first signs of inversion are seen at the base of the Barat Formation (Fig. 2.16). The Lower Miocene upper part of the Barat Formation (also known as Udang Formation; Sutoto, 1991) onlaps a surface within the Lower Barat Formation (Ginger *et al.*, 1993). During deposition of the syn-inversion formations, there were a number of transgression-regression cycles, so that the facies fluctuated from shallow marine to coal-swamp coastal plain. Some of these cycles may be tied to inversion events, related to the Sabah Orogeny, and others to eustatic sea level changes. The Barat Formation represents meandering streams within a low lying floodplain. Distal muds were deposited in quiet lacustrine settings. It is equivalent to the Pulai Formation (K sands) of the adjacent Malay Basin, which shares the same sedimentary-tectonic evolution with the West Natuna basinal area.

The extensive Lower to Middle Miocene Barat Formation is of lacustrine silts and shales, but locally contains a reservoir quality sandstone. It represents a period of tectonic instability, and in the adjacent Malay Basin, subsidence was more rapid, resulting in restricted water expulsion from the clays leading to abnormally high pore pressures.

The overlying thick Arang Formation is of marginal marine fluvio-deltaic fine sandstones and claystones with important coal beds. Provenance is deduced to be generally from the north, but much of the Arang Formation must have been locally provenanced from erosion of the newly inverted highs. Marine nanno-plankton indicate a Lower to Middle Miocene age. Fossils in the Mid to Upper Arang Formation indicate the first onset of marine conditions in the West Natuna area. An important and widespread Middle Miocene tectonic event (the Sabah Orogeny) caused widespread uplift and erosion. Drainage patterns were altered by the tectonic inversion. Former basement highs gradually became depositional lows, whilst former depocentres such as the Anambas Graben, were inverted to form the Boundary and other highs. Much of the Arang Formation sediments were derived from erosion of these inverted features.

2.3.10.1.4 Post-inversion mega-sequence An important regional basin-wide unconformity (equivalent to the Shallow Regional Unconformity of Sabah) marks the top of the syn-inversion mega-sequence. The unconformably overlying Muda Formation ranges from Upper Miocene to Recent. It is of shallow marine claystone, containing minor unconsolidated sands and thin limestones, but contains several unconformities representing periods of uplift, which allowed extensive river deltaic systems to develop (Pollock *et al.*, 1984). The present day water depth averages 90 m.

2.3.10.2 Structures

Daines (1985) has interpreted the West Natuna Basin as a failed intracontinental rift or aulacogen (the western extension of the South China Sea Basin ?), which was subsequently inverted in response to basement wrench faulting (Fig. 2.16). The majority of the anticlines (individual oil fields) are orientated NE–SW (average trend 63°), and are of consistent size (12 to 20 km long, 2.0 to 4.5 km wide). This suggests that the folds were formed by a single tectonic process. Most of the synclines are oriented NE–SW. The few anticlines that do not fit into the NE–SW pattern can be demonstrated to be due to pre-Tertiary basement topography and/or complex fault interactions.

The Boundary High (uplift), and the parallel 'AR' High to the north of it, represent inverted half-graben deeps; the Boundary High, before inversion, is known as the Anambas Graben. The southern flanks of the highs are composed of thrust eroded anticlines, whereas their northern flanks are relatively unfaulted. Uplift over the North Boundary area is estimated at 3 km, and the pre-Tertiary basement is now at 6 km. The Boundary High is a complex zone, demonstrating different rates of uplift. The anticlines are as a rule bounded on their SE flank by reverse faults. The reverse faults are oriented E–W to

NE–SW. Generally the throw decreases from the order of 350 m to about 10 m south-westwards away from the Boundary High. The faults typically extend from the pre-Tertiary basement up to the level of the mid Arang Formation. Local synclinal deeps occur on the downthrow of the reverse faults. The dominant strike of the normal faults is NNW–SSE (153°). Their throw is normally less than 100 m, although the old faults that cut basement may have a greater throw. Few normal faults penetrate the Muda and upper Arang formations. Wrench fault geometry is best displayed in the Udang and Sepat areas (Fig. 2.16), where rotated antithetic faults and Riedel shears demonstrate right-lateral movement. Seismic lines across the NW–SE oriented wrench faults (average strike 122°) indicate large thickness differences across the faults of the Gabus and Belut formations. However, most of the wrench faults show little thickness variation across them, and therefore they are difficult to map. *En echelon* folds give a clue to their existence. The wrench faults in the Udang area extend upwards from basement to the mid-upper Muda Formation, although movement since the Middle Miocene (Arang Formation) was probably confined to vertical readjustments.

The wrench faults are of regional significance, and Daines (1985) interprets them as defining an important shear direction extending through the Malay Basin, through the West Natuna Basin, across the Natuna Arch, continuing into Borneo by way of the Lupar Line, which provided a pre-existing line of weakness suitable for reactivation as a wrench fault.

Daines (1985) concludes that the original Anambas and AR grabens, which trend NE–SW, became compressive and inverted as a result of wrench motion on the Lupar–Natuna–Malay Basin NW–SE oriented right lateral shears. The inversion began at the beginning of the Miocene and extended up till the beginning of the Upper Miocene. It therefore coincides in time exactly with the orogenic compression and uplift of the Crocker Formation of Sabah. Accordingly the West Natuna inversion is an integral part of the Sabah Orogeny, which was a major mountain building event in eastern Borneo. The attempt by Daines (1985) to fit the tectonic evolution of the region into Molnar and Tapponnier (1975) escape tectonics is less than convincing, because clockwise rotation is required for the region, and this is in conflict with the anti-clockwise rotation of west Borneo, as defined by comprehensive palaeomagnetic studies (Schmidtke *et al.*, 1990).

2.3.10.2.1 Timing of inversion The earliest inversion event, seen on seismic sections (Ginger *et al.*, 1993), occurred within the lower part of the Barat Formation (Fig. 2.16). The next event is near the top of the Barat. Unfortunately chronostratigraphy is difficult in these continental strata, but dates of 23 and 22 Ma are interpreted. At least two seismically mappable unconformities are seen in the Arang Formation; the most significant being the Intra-Arang Unconformity, dated by nannofossils at 18.5 Ma. Further lesser unconformities occur before the major unconformity representing the base of the Muda Formation. It represents a regional unconformity, resulting from inversion and accentuated by a fall in sea level.

2.3.10.2.2 Analysis of inversion sequence Ginger *et al.* (1993) successfully analysed the sequence from half-graben formation through inversion by construction of balanced sections based on excellent seismic data. A summary is shown in Fig. 2.17. Restoration shows the half-graben geometry very well (Fig. 2.17, D). The half grabens have β extension factors of 1.28 to 1.31 in the Anambas Graben and 1.05 in the AA-1X graben.

The regional seismic data show clearly that the structures formed by the Miocene inversion events, including the oil-trapping anticlines, are directly related to the orientation, geometry and heave of the underlying graben boundary faults. There is also a correlation between the scale of inversion and the scale of the original Oligocene extension. The deepest grabens invert to give the largest magnitude inversion structures. The uplift of the rift fill sediments has amounted to as much as 3100 m. From the restored sections, the shortening across the Boundary High varies from 9 to 18%. During inversion, the half-graben fill and overlying cover sediments formed asymmetrical anticlines. Usually the steep limb of the anticline lies on the fault-bounded side of the half-graben. Similar inversion structures have been described alsewhere in Indonesia by Letouzey *et al.* (1990) and in the Alpine foreland by Roure *et al.* (1992). All of these studied examples show that the main half-graben boundary fault is of utmost importance during the inversion, and changes from a normal fault to a reverse fault at the inversion stage.

2.3.10.3 Source rocks

Generally the pre-Muda Formation kerogens are gas prone, but kerogen rich horizons in the lacustrine Barat Formation represent the main oil source rocks. They derived the kerogen from terrestrial plants together with algae. The kerogen type has an erratic unpredictable distribution, so that some fields are gas-prone and others oil-prone (Pollock *et al.*, 1984). The upper Gabus Formation also has potential.

The present-day geothermal gradients average 45.5° C km^{-1}, defining the theoretical oil window between 1100 and 3400 m depth. However the spectacular

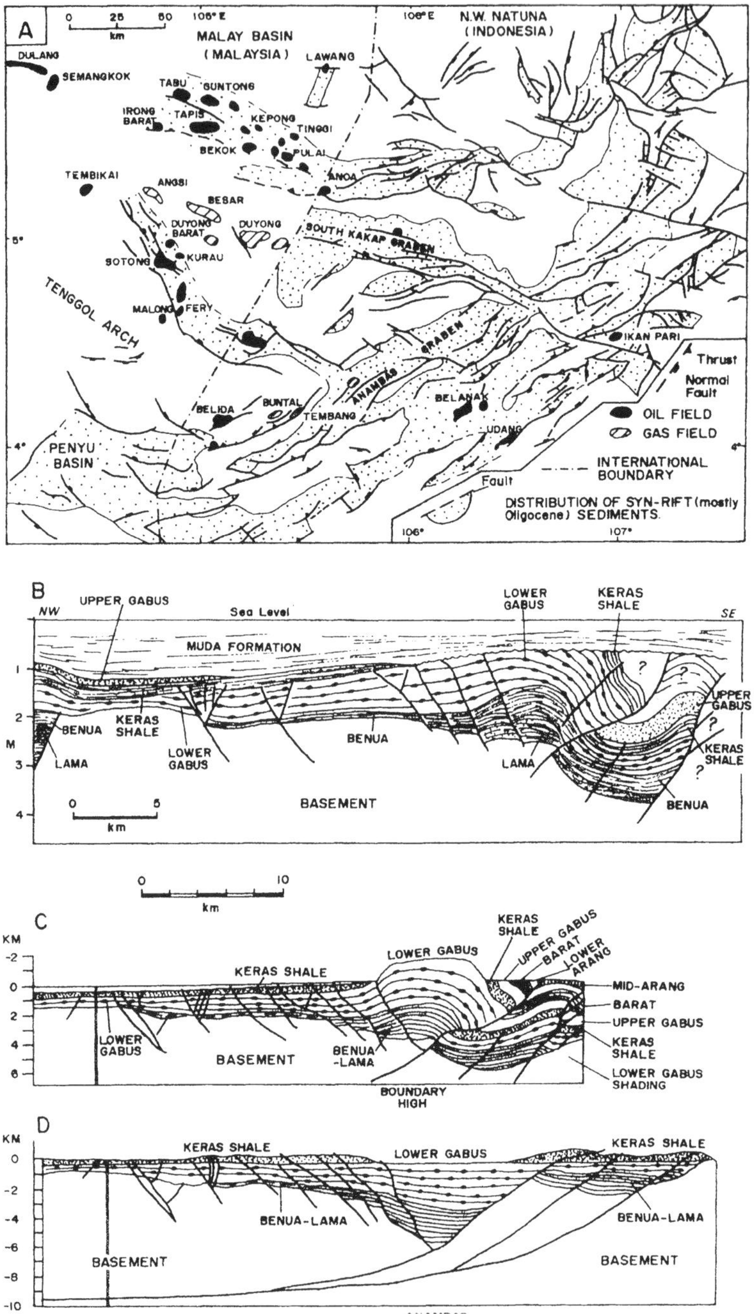

Fig. 2.17. The West Natuna–Southern Malay basinal area. A: main Palaeogene half-grabens and distribution of the syn-rift (Upper Eocene through Lower Oligocene) sediments. B: interpreted NW–SE seismic section across the Boundary High. C: computer generated cross section of the same seismic line at a time when the Boundary High was not yet eroded . D: computer generated cross section of the same line before inversion of the Anambas Graben to form the Boundary High. The β extension factor at the rift stage is 1.31. During inversion to form the Boundary High, shortening was 18% and inversion was accomplished using the bounding faults of the graben (after Ginger *et al.*, 1993).

Miocene inversion makes it impossible to define the actual oil window strictly. The Upper Eocene to Middle Oligocene sections (Barat and Gabus formations) are considered to be the mature source-rocks.

2.3.10.4 Reservoirs

Reservoir quality rocks occur widely in the Gabus and Arang Formations. In the KH Field, the Gabus sands have up to 28% porosity and 1500 mD permeability, with an average 20.5% and 500 mD. Secondary or dissolution porosity can reach 35% in places. The oil and gas traps are faulted domal anticlines and closures on the upthrown side of reverse faults.

2.4 THE SOUTHERN MARGINS OF THE SOUTH CHINA SEA

The Nansha Block sector lying between the southern extent of the South China Sea marginal basin and the Palawan–North-west Borneo Trough (Fig. 2.2) is interpreted as attenuated and rifted continental crust submerged beneath sea level. The results of a marine multi-channel reflection seismic array, of total length exceeding 13 000 km, covering the Dangerous Grounds, Palawan–North-west Borneo Trough, and slope of Palawan and Sabah, have been summarized by Hinz and Schlüter (1985) and Hinz *et al.* (1989). In addition, Sarawak Shell (Hazebroek and Tan, 1993) and Brunei Shell have acquired their own unpublished regional reconnaissance seismic lines, an unlocated example of which was given by James (1984). The South China Sea Institute of Oceanology, Guangzhou (Xia and Zhou, 1993), and the Ministry of Geology and Mineral Resources (MGMR), Guangzhou, have independently run a considerable number of high quality multi-channel regional seismic lines across the Reed Bank–Dangerous Grounds–Spratly Islands (Nansha Block). Financial support and processing of the seismic data came from the China National Offshore Oil Corporation (CNOOC), Beijing. The structure and stratigraphy have been analysed (Fig. 2.18) based on these seismic lines (Hutchison, 1992*b*), controlled by extensive critically located dredged samples along submarine horst margins (Kudrass *et al.*, 1986), and oil company data from the Reed Bank and offshore NW Palawan (Taylor and Hayes, 1980; Holloway, 1981). There is a remarkable similarity of geology, extending from the Reed Bank south-westerly to the region offshore Brunei and northern Sarawak (Figs 2.2, 2.18).

2.4.1 Calamian Island group

The oldest rocks of the Philippines occur in the Calamian Islands, north Palawan, Mindoro and Panay. The stratigraphy and structural relationships have not been sufficiently resolved, but the continental nature and antiquity are beyond doubt (Hutchison, 1989*a*; Fontaine, 1990; Mitchell and Leach, 1991).

The oldest palaeontologically dated rocks are Permian limestones (Calamian and Panay), Triassic (Calamian) and Jurassic (Calamian and Palawan), and black shales interbedded with sandstone, conglomerate and lignite beds on Mindoro contain Jurassic ammonites (Fontaine, 1990). The Mesozoic strata are underlain by a basement complex of unknown age.

2.4.2 Shelf off NW Palawan

The geology offshore the Calamian Islands has been summarized from drilling data by Saldivar-Sali *et al.* (in Hinz and Schlüter, 1985), Holloway (1981) and Hinz *et al.* (1985). The basement is of Upper Jurassic to Lower Cretaceous carbonate and shale. Upper Cretaceous to Lower Eocene strata are generally absent from the Palawan Shelf.

The major hiatus is followed by Upper Eocene transgressive sandstone overlain by a widely developed Upper Eocene to Oligocene carbonate platform (Nido Limestone), of 800 to more than 1200 m thickness. Local reef development characterizes the upper part. One well encountered a 22 Ma rhyolite.

The Nido Limestone is overlain by the Pagasa Formation, of claystone with conglomerate intervals, in turn unconformably overlain by the Mid to Upper Miocene Matinloc Formation, of conglomeratic sandstone, siltstone and mudstone, with local limestone, interpreted to be paralic.

The Matinloc is unconformably overlain by Upper Miocene to Pliocene Carcar Formation, of shelf and reefal limestone and mixed neritic clastics.

In NW Palawan the pre-Oligocene rocks may be a possible hydrocarbon source (Durkee, 1993). The oil is reservoired in the Nido Formation and found in a variety of structural trap types as well as in stratigraphic reservoirs and reefs. There are some sharp, high relief, wrench fault-related anticlines, such as West Linapacan; low relief undulating platform anticlines structurally higher on the Palawan Rise, such as the Calauit–Calauit South; pinnacle reefs along the outer edge of the Oligocene–Lower Miocene carbonate shelf, such as the Nido Complex; and turbidite fan deposits, such as the Galoc Field, overlying the carbonates (Durkee, 1993).

The NW Palawan Shelf terrain continues through northern Palawan island into the NE Sulu Sea, where it is known as the Cuyu Platform, extending almost to the islands of Panay and Mindoro. This region contains more than 7.5 km of sediments in places, and the geothermal gradient is low, at 22° C km^{-1}. The shelf area is modified

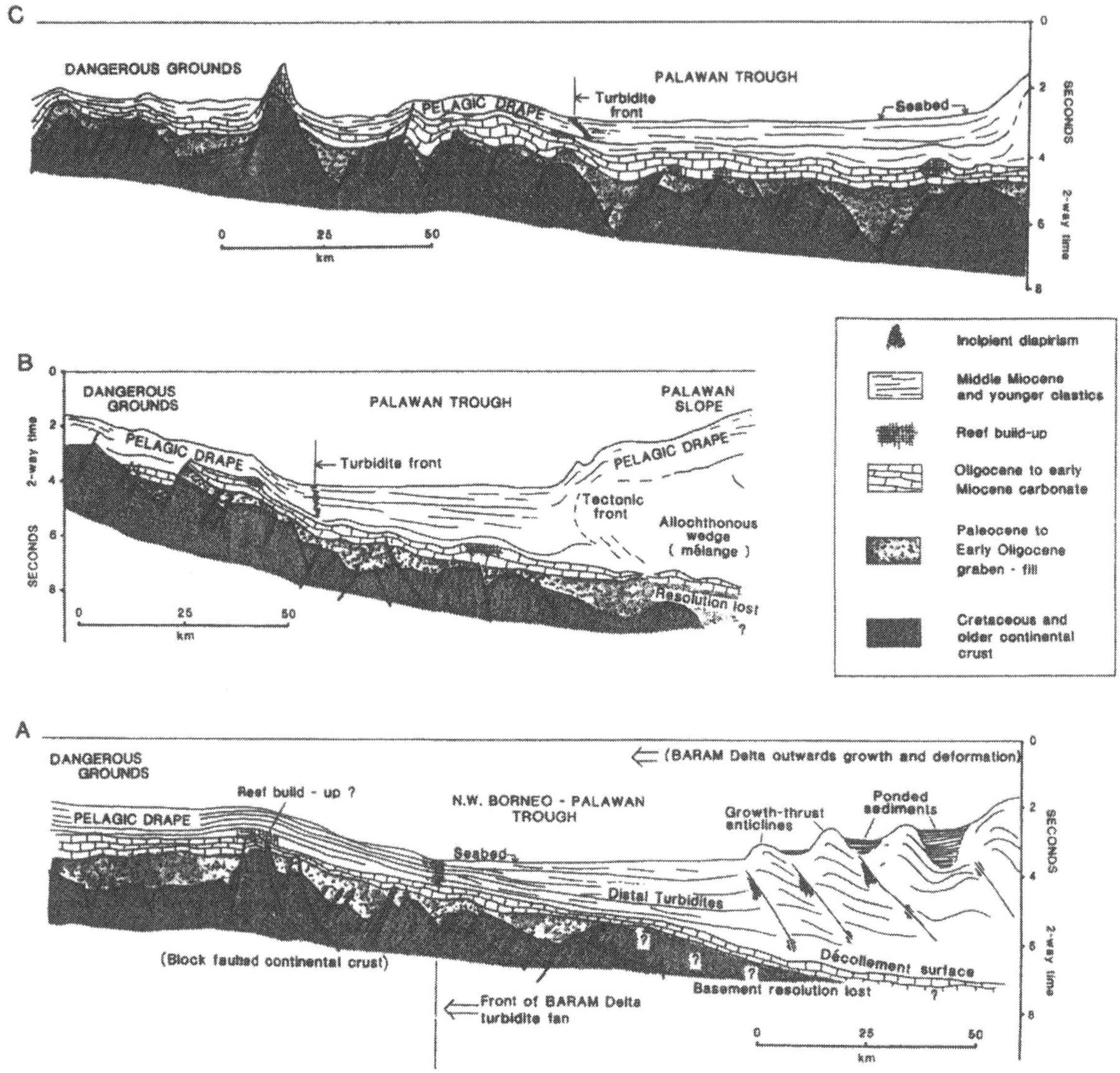

Fig. 2.18. Geological interpretations of three NW–SE oriented seismic profiles from the Nansha Block (Dangerous Grounds) across the Palawan–North Borneo Trough. A: composite section offshore Sarawak, based on Hinz *et al.* (1989). B: section adjacent to the Kudat Peninsula of Sabah, after Hinz and Schlüter (1985). C: section close to the Reed Bank and adjacent Palawan, after Hinz and Schlüter (1985). For location of the sections see Fig. 2.2.

by wrench faulting and shear zones trending NW–SE along the coast of Mindoro island. Reconnaissance seismic lines across the platform demonstrate the presence of large-scale folding of Lower Tertiary and Upper Mesozoic strata beneath the platform, which is composed of Tertiary limestones and Upper Tertiary volcanic rocks (Durkee, 1993).

2.4.3 The Reed Bank

The oldest dated sedimentary rock was dredged from the eastern side of a graben east of the Mischief Reef (Kudrass *et al.*, 1986). Light-brown to grey deltaic sandstone contains Upper Triassic to Lower Jurassic *Clathropteris* leaves. Dark grey marine claystone contains moulds resembling Triassic *Halobia* and *Daonella*. Gneisses and schists have given K:Ar metamorphic dates of Upper Jurassic to Lower Cretaceous. An Upper Oligocene to Lower Miocene carbonate platform was extensively developed during the inferred time of spreading of the South China Sea Basin. It was widely sampled and dated by Kudrass *et al.* (1986).

Drilling in the Reed Bank (Sampaguita-1) bottomed in a Lower Cretaceous complex of continental to paralic strata. They were described by Taylor and Hayes (1980) as: grey siltstone and shale with coal; sandstone and conglomerate composed of quartz and feldspar grains; intermediate lavas, quartz diorite and tuff; pale buff fine grained sandstone; and coal. At the end of the Cretaceous, the Reed Bank area was part of an emergent continent. Thick marginal marine sediments with coal of Lower Cretaceous age were exposed and eroded. The thickness of the Cretaceous and probably Jurassic strata appears to exceed 5 km in the south-eastern part of the Reed Bank (M. Rebora in Taylor and Hayes, 1980).

A major unconformity is overlain by a thin Palaeocene limestone, followed by neritic shale and massive deltaic pebbly quartz sandstone and siltstone. They contain clasts of quartzite, tuff, radiolarian mudstone and chert. The first Tertiary sea transgression occurred in the Palaeocene. A broad thin carbonate platform was deposited. Granitic basement, 'outcropping to the west', shed sands and coarse clastics to extensive deltas (M. Rebora, in Taylor and Hayes, 1980).

These were followed by Lower to Middle Eocene marine shale, calcareous sandstone and siltstone, persisting up to a major unconformity at the end of the Middle Eocene. The overlying coarse sandstone and mudstone sequence is interpreted as paralic to marginal marine. This sequence is unconformably overlain by massive shelf and reefal carbonates persisting from Upper Oligocene to the present day.

Kudrass *et al.* (1986) dredged Palaeocene siltstone from SW of Alicia Annie Reef, and Oligocene siltstone from the Sabina Shoal, 100 km south of the Reed Bank. Lower Miocene shallow water carbonates were dredged from 23 sites from the various submarine slopes, and used to define the seismic stratigraphy (Figs 2.18; 2.19).

Pliocene–Pleistocene basalt, yielding K:Ar dates ranging from 0.4 to 2.8 Ma have been dredged from localities north, east and south of the Reed Bank. These are directly comparable to Hainan island and the western Pearl River Basin, where there was a Plio-Pleistocene renewal of rifting and basalt eruption.

2.4.4 Dangerous Grounds (Nansha Block)

The Dangerous Grounds (Fig. 2.2) is an integral part of the large South China Sea submerged continent, which includes other shoal areas such as the Spratly Islands and the Reed Bank. Towards the east it stands above sea level as North Palawan and the Calamian Islands (Rangin, 1989).

The multi-channel seismic profiles of the South China Sea Institute of Oceanology (Xia and Zhou, 1993) and MGMR reveal a thick Palaeogene sequence overlying a variable basement which includes layered Mesozoic strata. The clearly stratified and widespread Palaeogene sequence occupies half grabens in the basement and is interpreted as neritic clastics (Xia and Zhou, 1993). This sequence was later strongly deformed into asymmetric folds and their anticlinal highs eroded. The unconformity above the Palaeogene sequence has been recognized in several seismic profiles in the reef-and-shoal areas (Spratly Islands–Dangerous Grounds). By long distance correlation from limited drilled areas, the unconformity is tentatively dated Upper Eocene by Xia and Zhou (1993), related to a major tectonic event in Borneo (the Sarawak Orogeny). Above this unconformity, the Neogene sequence is thin and occurs mainly as fillings of discrete half-grabens in the reef-and-shoal areas, but thick and continuous on the Reed Bank and Palawan Shelf and Luconia Province. The carbonate platform of the Reed Bank, characterized by a very distinctive regional seismic pattern, has been dated as Oligocene to Lower Miocene by dredged samples (Hinz *et al.*, 1989; Kudrass *et al.*, 1986). Xia and Zhou (1993) therefore conclude that the reef-and-shoals area (Dangerous Grounds–Spratly Islands) stayed mainly in an uplifted state since late Eocene time, in contrast to the shelf areas to the south and south-east. After acquiring the exploration rights of the Wan'an Bei Block 21 (8° N, 110° E), Crestone Energy Corporation (Denver) have acquired all the geophysical data and will carry out a seismic stratigraphical analysis and compare it with the contiguous Dai Hung (Big Bear) field of the Nan Conson Basin of Vietnam. Drilling, if it ever takes place, will depend upon a great amount of goodwill between the various countries that claim ownership to the region, especially between Vietnam and China.

2.4.5 The Palawan–North-west Borneo Trough

Many geologists repeat the view of Hamilton (1979) that the Palawan–North-west Borneo Trough is a NW-facing trench which became extinct in the Middle Miocene. When he published this view, the high quality reflection surveys (Figs 2.18; 2.19) had not yet become available. The seismic sections show that the Trough bears only a superficial though strong resemblance to a trench. The trough area is 60 km broad, of flat undeformed deep water Middle Miocene or younger sediments. The Dangerous Grounds terrain is not converging on Borneo, otherwise the sediments of the trough would be deformed. The sediments reaching the trough represent the turbidite front of the Baram Delta originating from erosion of the elevated Crocker Range of Borneo. These sediments are actively being deformed and redistributed away from the Borneo coastline, by thrusting, growing anticlines, ponding and spilling over of sediments oceanwards (Hazebroek and Tan, 1993). The delta front progressed rapidly oceanwards over under compacted pro-delta muds. Their subsequent de-watering resulted in active mud diapirism, which on some sesimic sections is seen as submarine mud volcanoes superimposed upon the flat-lying trough strata oceanward of the tectonic front.

The south-eastern margins of the South China Sea may now be resolved into several distinct structural units, based on the high resolution seismic work of Hinz *et al.* (1985, 1989), and Hinz and Schlüter (1985), supported by the dredged samples described by Kudrass *et al.* (1986).

2.4.5.1 *The North-west Borneo Trough*

The seismic data clearly display the Oligocene to Lower Miocene marker carbonate platform continuing from the Nansha Block to underlie the whole of Zone II (Figs 2.18, 2.19). There is no plate boundary as suggested by Hamilton (1979) and followed by many subsequent workers such as Tan and Lamy (1990). On the contrary, tilted fault blocks and syn-rift basins indicate the same structural style below Zones I and II. The thickness of the Miocene to Recent sub-parallel bedded sediments, which overlie the carbonate platform, increases from 1 s (two-way time) at the NW limit of Zone II to approximately 2.5 s at its south-eastern limit (Hinz *et al.*, 1989).

The flat sea floor of the North-west Borneo Trough lies beneath a water depth of 2800 to 2900m. The undeformed flat-lying Recent sediments of the Trough are locally penetrated by active sea-floor diapiric cones or ridges, unfortunately not included in the selection of Hinz *et al.* (1989). They occur within Zone II, only a short distance from its boundary with Zone III (Hutchison, 1992*b*).

2.4.5.2 *Ridge and trough belt*

Eight elongate imbricate thrust sheets trend 30 to 45° (Zone III of Fig. 2.19). Some continue over a length of 190 km. The distance between individual thrust planes varies from 3 to 15 km, and generally decreases towards the north-east. The individual thrust sheets overlap like roof tiles, all dipping towards the south-east. The bounding faults curve downwards to a sole fault or major décollement surface, defined by the top of the eastwards dipping Oligocene–Lower Miocene carbonate platform. Stratification within each individual thrust sheet is sub-parallel.

The deformation has resulted in a series of sea-floor ridges, which resemble anticlines, and have diapiric cores and stratal dips predominantly towards the south-east.

A distinct unconformity (B0, B1, B2) represents the top of the individual thrust sheets at the time of deformation (Fig. 2.19). This time-transgressive unconformity forms the base of the trough basins between the thrust sheet ridges. The thickness of the flat-lying infill decreases very systematically oceanward, indicating that thrusting was progressive in time and space, the active deformation front of the Baram Delta migrated oceanward from Middle Miocene to Recent. Zone III appears to have progressively widened oceanward at the expense of Zone II. The sediment source, whether from erosion of the land or by cannibalization of the older more elevated submarine ridges, was from the south-east, and landward troughs are filled up to the top of the ponding ridge, with progressively less fill in oceanward troughs (Fig. 2.18). The youngest troughs are still starved of sediment.

The inferred mechanism for producing the stratigraphy and tectonics of Zones I (the Nansha Block) to III is that the rapid sediment influx into the Baram Delta loaded the underlying crust much more closer to land than oceanwards at its turbiditic toe. This caused the underlying Oligocene–Lower Miocene carbonate platform basement to become progressively depressed landward beneath the neritic delta depocentre. Sediment loading squeezed out the pro-delta mud (Setap Shale), which flowed north-westwards up-dip above the shallowing carbonate platform, to reach the sea floor as mud diapiric ridges, which acted as dams to pond any new turbidite sediments and to divert them longitudinally parallel to the delta front. Mud diapirism was also active along the steep thrust faults resulting in the formation of diapiric ridge structures within the thrust sheets. Continuing mud withdrawal caused troughs to deepen landward of the diapiric fronts. As progressively more continentally derived sediment was deposited in the depocentre, the delta morphology attained an equilibrium geometry by progressively expanding oceanwards via an outwards migrating depositional/deformational front. The thrust faults of Zone III do not therefore result from regional NW–SE tectonic compression, but from north-westerly outwards growth expansion of the delta prism.

Farther east, imbricate thrusting also progressed oceanward from Pliocene to Recent and became more complex landward, where a basement of inferred Crocker Formation has been involved.

2.4.5.3 *Zone of thrust sheet basement*

This zone (IV of Fig. 2.19) varies between 25 to over 70 km width, and was interpreted by Bol and Van Hoorn

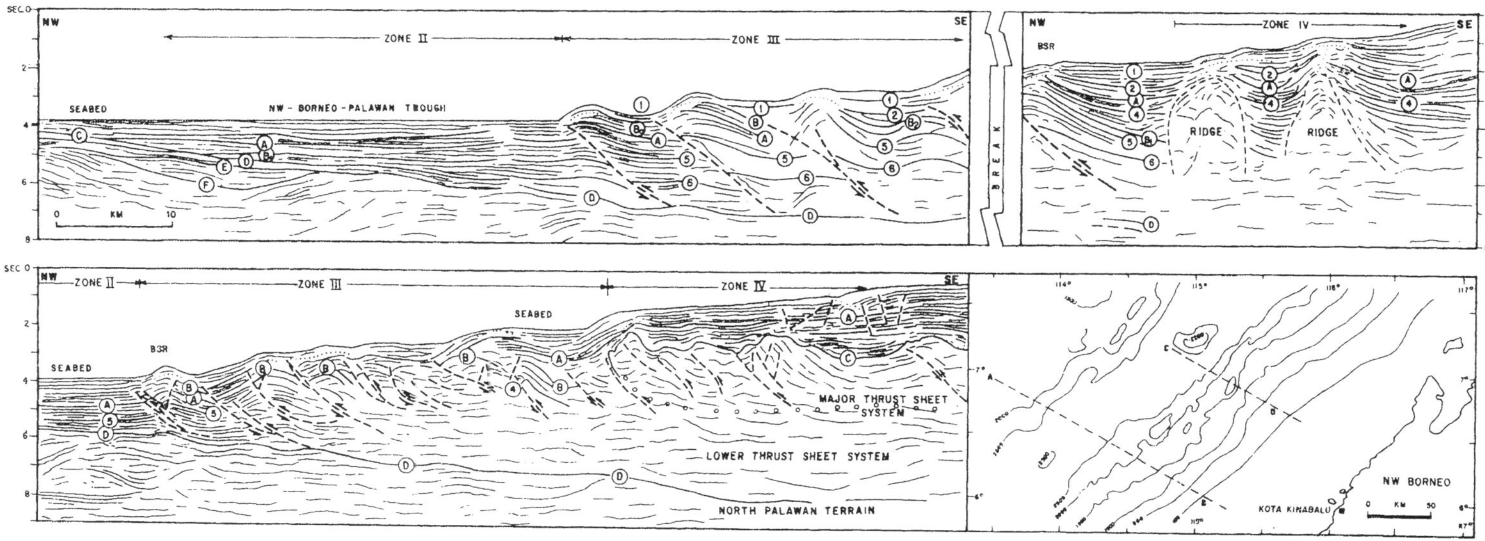

Fig. 2.19. Interpreted seismic sections from the Dangerous Grounds, across the North-west Borneo–Palawan Trough, onto the tectonically and sedimentationally active toe (front) of the Middle Miocene to Recent Baram Delta (after Hinz *et al.*, 1989). **Top**: section A-B of the location map; **Bottom (left)**: section C-D of the location map. A = base Upper Pliocene; B = late Middle Miocene; B_0, B_1, B_2 = time transgressive seismic unconformity; BSR = bottom simulating reflector; C = Lower Middle Miocene; D = Lower Miocene top of prominent carbonate platform; E = Oligocene base of the carbonate platform; F = Upper Cretaceous rift onset unconformity; 5 = Miocene-Pliocene boundary; 6 = Middle Miocene marker. Open dots represent the interpreted sole of a thrust sheet.

(1980) as a province of extensional and partly syn-sedimentary faulting of the Upper Miocene to Pliocene deltaic to shallow marine sediments.

The regional seismic profiles of Hinz *et al.* (1989) have been interpreted to consist of two superimposed duplex thrust sheets, buried by uniformly layered Middle Miocene to Upper Pliocene clastic sediments.

The lower thrust sheet system is interpreted to overlie the Oligocene–Lower Miocene carbonate décollement surface, but this is speculative.The boundary between the lower and upper thrust sheet is recognized as a change in seismic pattern (Fig. 2.19). The 2 to 3 s thick (2-way time) upper thrust sheet is interpreted to be of Oligocene-Lower Miocene Crocker Formation, which underlies the Deep Regional Unconformity (Bol and Van Hoorn, 1980), and is extensively uplifted in Sabah.

Diapirism localized along the steep SE dipping thrusts has upwardly penetrated and locally disturbed the overlying Miocene–Pliocene layered sediments, which remain flat-lying between the diapiric ridges.

The major thrust front, which separates Zone IV from III, bends southwards to extrapolate with the Morris Fault (Tan and Lamy, 1990) and extends onland as the Jerudong Line in Brunei (Figs 2.2 and 3.6) trending southwards to the margins of the Gunung Mulu massif (James, 1984). The Morris–Jerudong Line is a zone of extreme complication within the Baram Delta sediments, along which complex fault flower structures are abundant and interpreted to represent wrench motion within the basement. If the tectonic transport on the NE–SW oriented outer thrust boundary between Zones II and IV is towards the NW, then that part of the boundary which is oriented N–S (Jerudong Line) may represent left-lateral wrench faulting.

The boundary between Zones III and IV represents the northern and western extent of the Crocker Formation and Rajang Group, and there is no evidence for its existence as a basement to Zone III. Offshore, the Crocker Formation is overlain by Miocene–Pliocene deltaic-shallow marine sediments belonging to the same depositional scheme as the Baram Delta. The basement became locally uplifted to form complicated diapiric ridges on the sea floor during Miocene–Pliocene sedimentation, which was thickest in the intervening uncomplicated mud-withdrawal troughs. Examples of the complicated diapiric ridges outcrop on the Klias Peninsula and Labuan Island (Wilson, 1964). The offshore synclines (troughs) were the depocentres.

2.4.5.4 *Zone of advanced diapiric piercements from thrust basement*

This zone is a continuation landward of Zone VI, and contains the same overthrust basement of Oligocene–Lower Miocene Crocker Formation. There have been more advanced mud diapiric piercements into the unconformably overlying Neogene strata. Between the diapiric ridges, basinal troughs continued to subside. The zone is therefore structurally and stratigraphically complex.

2.4.5.5 *Pronounced diapiric piercement zone*

Diapiric ridges show a more mature development in the landward zones, and can occupy a vertical height of 4 s (2-way time). The intervening mud-withdrawal troughs are accordingly equally deep, and filled by flat-lying Neogene strata. The troughs have been called synclines. They are best interpreted as mud-withdrawal troughs, between intervening diapiric ridges. As the ridges grew, the strata of the troughs were marginally flexured upwards to form synclines (Fig. 2.19).

The diapiric ridges are interpreted to have continued growing as the intervening troughs subsided resulting in a balance between sedimentation, mud withdrawal, trough deepening and diapiric ridge growth. Therefore the Neogene sedimentary sequence on top of a ridge is greatly attenuated, and complicated, yet thick and simple within the troughs.

This zone is interpreted to be underlain by the same landward dipping platform of Oligocene to Lower Miocene carbonate platform as underlies the whole Baram Delta. This zone represents a more advanced stage of vertical diapirism and intervening subsidence because it lay closer to the sediment source than the more distal Zone III.

2.4.5.6 *Baram Delta*

The delta sediments continuously transgressed over the Dangerous Grounds continental lithosphere, loading and depressing it, resulting in its landward dip (Fig. 2.18). It is this landward dip that led to the erroneous interpretation that the Dangerous Grounds terrain (Nansha Block) is subducting SE-wards at the Palawan–NW-Borneo Trough. There are no scraped-up sediments and no deformation in the trough. These details have their exact analogy at the front of the Niger Delta, which is without doubt in a non-convergent setting (Hazebroek and Tan, 1993).

The Sarawak and Brunei sector of the Baram Delta (section A of Fig. 2.18) does not include Palaeogene Rajang Group. However, farther to the NE (sections B and C, Fig. 2.18), the equivalent of the Baram Delta has prograded oceanwards over a deformed basement of Crocker Formation. The north-westwards and westwards extent of the Crocker Formation basement is interpreted to be the tectonic front of sections B and C (Fig. 2.18), the offshore Morris Fault and the Jerudong Line of Brunei (Fig. 3.6).

Convergence between the Nansha Block, together with the onland Miri Zone of Sarawak, is required to have taken place and been completed (Hutchison, 1996) by the late Eocene (The Sarawak Orogeny). This convergence is held responsible for the orogenic folding and uplift of the Rajang Group (Belaga Formation) of the Sibu Zone. No known spreading of the present South China Sea can be held responsible for this, and hence we have to invoke a Danau Sea (Haile, 1994), which was completely subducted. Its ocean floor remnants are preserved on-land Borneo as ophiolitic sutures (Boyan, Lupar and Mersing lines), associated with Lower Cretaceous (Barremian) ribbon cherts (Basir *et al.*, 1985). Convergence to cause compression of the Crocker Formation Basin, causing the Upper Miocene Sabah Orogeny, may be related to the spreading history of the South China Sea Basin (*sensu stricto*), which ceased by the Middle Miocene. The tectonic front of sections B and C (Fig. 2.18) may represent this extinct convergent plate margin. It is clearly draped over by Middle Miocene and younger strata deposited after cessation of the convergence.

2.5 SUMMARY

The northern Sundaland region, reviewed in this chapter, was cratonized by the Lower Triassic Indosinian Orogeny, which resulted in extensive Upper Triassic granites and was responsible for establishing the structural fabric of the basement upon which the rift basins were formed.

The Khorat is the only pre-Tertiary basin to have been explored by the petroleum industry, unfortunately without success. Its rift history began with Middle to Upper Triassic half grabens containing lacustrine source rocks. The half grabens are overlain by a more extensive Jurassic and Cretaceous sag basin composed wholly of continental strata, dominated by redbeds in the lower part and salt deposits in the upper. It is a truly cratonic internal-drainage basin. The Khorat Basin has exact analogues, albeit smaller and more deformed, in Peninsular Malaysia, which have not been investigated by the petroleum industry. There are no continental Tertiary analogues of basins which became predominantly red and saline.

The Tertiary basins began their extensional history in the Upper Eocene or Lower Oligocene upon the geographically extensive Sundaland landmass which developed in the late Cretaceous. The first stage linear grabens were filled by continental alluvium followed by lacustrine deposits, which provided good source rocks. Many of the grabens were controlled in their orientation by the underlying basement fabric, for example in the Gulf of Thailand, but others seem to have developed in new directions independent of the basement. In western Thailand and the Malay Peninsula the basins continued continental throughout their history and were never connected to the sea. Yet they have produced oil in NW Thailand exclusively sourced from lacustrine strata. Elsewhere in the region, the basins became shallow marine by late Oligocene to early Miocene, and some of the shallow marine sequences also acted as source-rocks in addition to the underlying lacustrine strata.

Erosion of the extensive granite terrains of Sundaland, especially during Miocene times, resulted in good quality fluvio-marine quartz-rich reservoir sandstones. The rifting sequence from grabens to sag basins rarely continued to completion and the lower basin sequences have commonly been uplifted or inverted during the Middle to Upper Miocene to form topographic highs which became eroded and unconformably overlain by Upper Miocene to Pliocene post-unconformity marine strata. The inversions were not all equally spectacular, but the Middle to Upper Miocene inversion is universally recorded as a regional unconformity.

Oligocene through Miocene rifting resulted in sea-floor spreading in the eastern South China Sea, which has displaced southwards towards Borneo an extensive terrain of attenuated continental crust occupying shallow seas characterized by present day shoals and reefs. This terrain is known as the Dangerous Grounds, or Spratly Islands, and it has similar basinal histories to the terrain lying north of the South China Sea Basin. The southwards limit of this displaced terrain is the North-west Borneo Trough, towards which the Dangerous Grounds terrain dips southwards in a manner resembling subduction. The exact nature of the North-west Borneo Trough continues to be debated. Exploration of the extensive Dangerous Grounds–Spratly Islands terrain is hampered because it is claimed, wholly or in part, by China, Vietnam, the Philippines, Taiwan and Malaysia. Indonesia holds several inhabited islands in this region and is accordingly concerned about the dispute.

3

Cainozoic oil and gas-bearing basins related to the Sarawak and Sabah Orogenies

Many misconceptions are to be found in the literature regarding the 'Rajang Accretionary Prism' and orogenic events in Sarawak, Kalimantan and Sabah. Therefore it is necessary to review the data before describing the related Cainozoic basins. The following discussion is mostly based on Hutchison (1996). The basins occupy the northern coastal zone and offshore Borneo, the low elevation area between the Sarawak Border with Kalimantan and the Schwaner Mountains, and the eastern part of Sabah.

3.1 NEED FOR REVIEW

The models of tectonic evolution of the South China Sea region are flawed because they conflict with the documented geology of Borneo, and especially with the Rajang Group. Poorly constrained and imperfect models of the South China Sea have then been applied to analyse the tectonic evolution of Borneo. The model of Taylor and Hayes (1980, 1983) fares particularly badly for it shows active subduction of Mesozoic oceanic lithosphere continuously through the Palaeogene, becoming extinct by the Middle Miocene when the Luconia Shoals were wrongly interpreted to have collided with the North-west Borneo Trough. The model of Holloway (1981) is not significantly different. The mistake can be traced back to the authoritative Hamilton (1979), who described extensive tracts of Borneo as 'mélange', interpreted as always synonymous with subduction-related accretionary wedges.

The model of Ru and Pigott (1986) ignores the West Baram Line and wrongly extends the North-west Borneo (Palawan) Trough through the Luconia and Balingian provinces (their Fig. 1), and has active late Eocene south-eastwards subduction with impossible N–S oriented transform faults. These models wrongly concur that subduction continued until the early Miocene. The mistake of continuing the North-west Borneo (Palawan) Trough along the southern margin of the Luconia Province, ignoring the West Baram Line, is also made by Tongkul (1990), who paints a very abstract picture of the Upper Cretaceous to Middle Miocene tectonics, which cannot be related to the documented geology of Borneo.

Even the better models are flawed because they ignore the tectonics of the surrounding region. Thus, Ru and Pigott (1986) and Williams *et al.* (1988) accept the palaeomagnetic evidence that West Borneo has rotated anti-clockwise (Schmidtke *et al.*, 1990), but they ignore the evidence for clockwise rotation of Indochina (Achache *et al.*, 1983). Williams *et al.* (1988) bring Peninsular Malaysia and Sumatra together only in the Lower Tertiary; an impossibility because they share the same Palaeozoic and Mesozoic stratigraphy and structure (Hutchison, 1993).

The Mid Oligocene to Lower Miocene N–S extensional opening of the East Sub-Basin of the South China Sea, first detailed by Taylor and Hayes (1980) has been widely accepted, and forms the basis of most subsequent models of the region, including contiguous Borneo. However their identified E–W trending magnetic anomalies do not coincide with the geomorphology of the basin, and hence their orientation must be held in suspicion. There is also a lack of geological information to constrain their numbering of the magnetic anomalies. This practice led to serious misinterpretations in the nearby Sulu Sea (Lee and McCabe, 1986; Hutchison, 1992*a*).

The two main structural trends have been shown by Pautot *et al.* (1986) and Briais *et al.* (1989) to be 50° and 140°. The former may represent the true anomaly lineations, and the latter are likely to represent transform faults. However Briais *et al.* (1993) have supported the interpretation of Taylor and Hayes (1983) for the main basin, and described NE–SW trending anomalies 6b to 5c for the SW sub-basin (27 to 16 Ma). They related the Oligocene to Miocene spreading to left-lateral motion on the Red River Fault. Until there is direct age dating of the oceanic crust, it would be wise to hold the models as tentative and not apply them to Borneo.

A more carefully constructed evolutionary scenario of the Borneo margin of the South China Sea was painted by Tan and Lamy (1990), which benefits from comprehensive petroleum company data. However their inferred subduction history from Upper Cretaceous through Upper Eocene, using the Lupar and Mersing lines, and from Upper Eocene through Middle Miocene, using the North-west Borneo (Palawan) Trough, are not always matched by identified subduction-related volcanic arcs in Borneo.

3.2 THE RAJANG GROUP OF SARAWAK

The Rajang Group consists of the Belaga and Lupar formations, which are isoclinally folded, with dips usually between 80 and 90° (Fig. 3.1). It extends across the border into Kalimantan as the Embaluh Group (Tate, 1991). This monotonous flysch group is of interbedded sandstone and mudstone, now metamorphosed to lower greenschist facies phyllite and slate, commonly quartz veined (Kirk, 1957; Liechti *et al.*, 1960; Wolfenden, 1960). The group width perpendicular to the E–W strike from the Lupar Line to the Tatau Horst is 200 km. Thickness cannot be determined and has been guessed to be in the neighbourhood of 15 km (Haile, 1974). The Rajang Group constitutes the Sibu Zone of Sarawak (Fig. 3.1), outlined along the south by the Lupar Line ophiolitic mélange and along the north by the Bukit Mersing Line ophiolitic mélange (Hutchison, 1989*a*). The Rajang Group has been interpreted as a deep marine fan deposit, and outcrops have turbidite characteristics. Most outcrops have thin fine to medium grained sand layers which may be described as distal turbidite, but some outcrops show thicker sandstones. All are characterized by prominent sole marks.

In the south, adjacent to the Lubok Antu Mélange, the 5 km wide outcrop of the Lupar Formation is predominantly of a turbidite sequence similar to the contiguous Layar Member of the Belaga Formation (Liechti *et al.*, 1960; Tan, 1979). It was mapped separately because it incorporates large bodies of ophiolitic rocks and chert. Microfossils indicate an Upper Cretaceous age of Santonian to Maastrichtian. The Lupar Formation thus

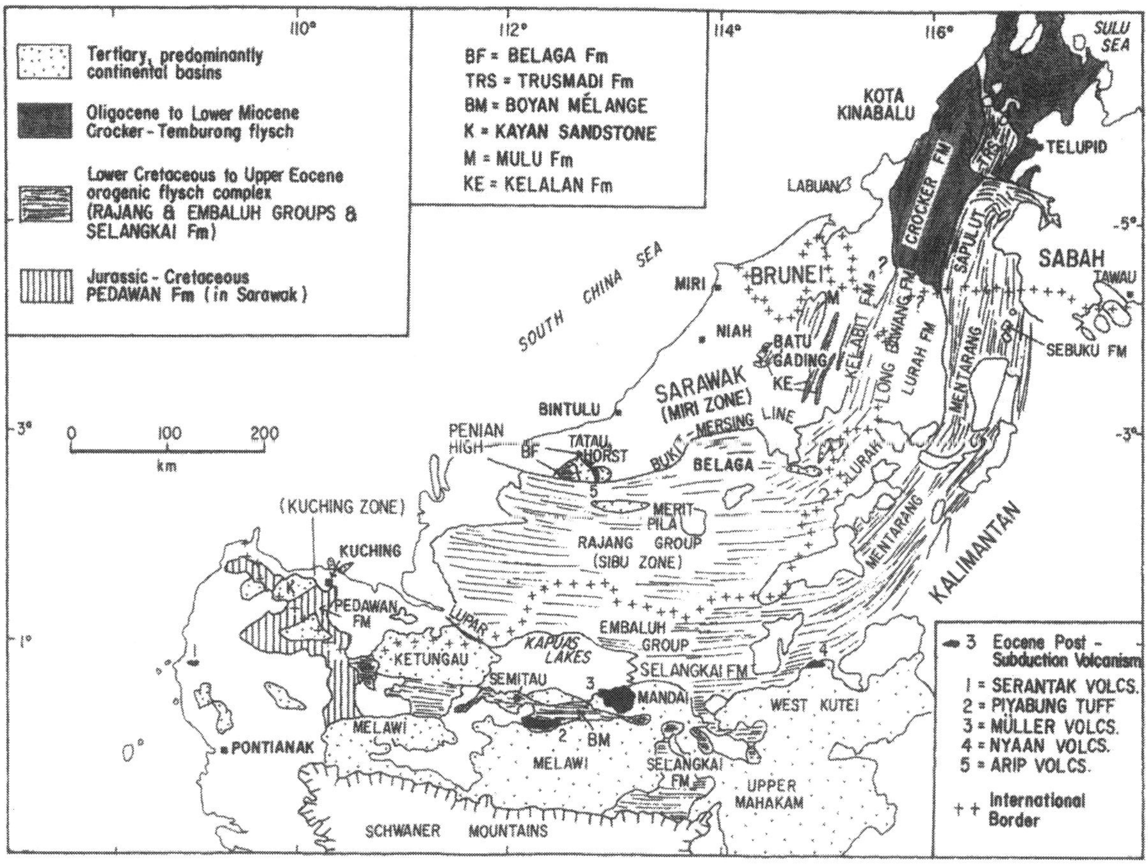

Fig. 3.1. Sketch geological map of Sarawak, Sabah and contiguous Kalimantan (Borneo) to emphasize the Rajang Group. Based on Tan (1982), Lim (1985), Pieters *et al.* (1987) and Pieters and Supriatna (1989, 1990).

represents the oldest part of the Rajang Group of Sarawak. From the south, adjacent to the Lupar Formation, to the Tatau Horst in the north, the Belaga Formation has been subdivided into the following members—Layar, Kapit, Pelagus, Metah and Bawang. Fossils indicate a progressive younging northwards from Upper Cretaceous Layar to Upper Eocene Bawang. In eastern Sarawak, the Kelalan Formation has been mapped by Haile (1962) as the correlative of the Belaga Formation. The steep dips, systematic northwards younging, and ophiolite and mélange association, have led to the very reasonable interpretation of a northwards-facing accretionary prism, related to late Cretaceous (85 Ma) to late Eocene (45 Ma) southwards subduction beneath the Kuching Zone of Sarawak.

Palaeo-currents in the Lupar Formation turbiditic sandstones indicate a major current direction towards 50 to 60°, with a subordinate maximum towards 80 to 90° (Tan 1979). The provenance of the earliest part of the Rajang Group turbiditic sandstones was from the Kuching Zone, lying to the SW and W. Most of this zone formed an integral part of the extensive Upper Cretaceous landmass known as Sundaland, upon which continental Eocene formations were widely deposited (Hutchison, 1992*b*). The fluvial system, which brought sediments from the interior of Sundaland, must have reached the coast in the neighbourhood of what is now known as the Lupar Fault. The geological record favours a narrow continental shelf as shown by outcrops of the Jurassic–Cretaceous Bau Limestone and Pedawan Format ion of the Kuching Zone (Ting, 1992), so it may be inferred that there was a rapid drop-off from land in the Kuching Zone to deep marine conditions in the Sibu Zone, with the fluvial system giving way rapidly to turbidite sedimentation. The lower Upper Cretaceous part of the continental Kayan Sandstone (Tan, 1986) may therefore be the on-land equivalent of the deep marine Lupar Formation and Layar Member of the Belaga Formation (Fig. 3.1).

3.2.1 Relationship to younger formations

One of the most spectacular Malaysian angular unconformities is exposed on the east end of the Tatau Horst (Wolfenden, 1960; Haile and Ho, 1991). The shallow dipping basal conglomerate of the Tatau Formation rests on steeply dipping and complexly folded Bawang Member turbidites of the Belaga Formation. The unconformity is palaeontologically dated Upper Eocene. The Tatau Formation is an Upper Eocene to Oligocene sequence of conglomerate, sandstone and shale. It is continental to fluvio-marine, becoming neritic upwards, and includes the subaerial Arip Volcanics. From this informative outcrop, it may be inferred that deformation, metamorphism, uplift and peneplanation of the Belaga Formation to form a landmass was completed before the end of the Eocene. I propose to name this the **Sarawak Orogeny** (Table 3.1). This information invalidates the tectonic model of James (1984), reproduced by Hutchison (1989*a*, p. 85).

An equally spectacular angular unconformity occurs at Bukit Besungai in the Batu Gading area in the upper reaches of the Baram River of eastern Sarawak, 85 km SSE of Miri. Here the steeply dipping Upper Cretaceous turbiditic Kelalan Formation, which is the eastern equivalent of the Belaga Formation, is unconformably overlain by flat lying highly fossiliferous Upper Eocene limestone (Adams and Haak, 1962). The tectonic significance of this quarry outcrop is identical to that of the Tatau area.

The steeply dipping Upper Eocene Basal Sandstone Member of the Silantek Formation occurs in fault contact (Lupar Fault) with the Upper Cretaceous Rajang Group (Lupar Formation) at the Marup Ridge (Tan, 1979). Palaeocurrents and conglomerate petrology indicate a provenance from the Lupar and Belaga formations, which formed an eroding landmass to the north-east. Hence the Silantek Formation rests unconformably upon the Rajang Group, but outcrops of the unconformity have been obscured by the later movements on the major Lupar Fault.

The Tatau Formation is overlain by a thin (550 m) shale-sandstone sequence known as the Buan Formation (Wolfenden, 1960). However, the major transgression over the landmass was accomplished by the Upper Oligocene to Lower Miocene Nyalau Formation, estimated to be about 5 km thick. It rests conformably on the Buan Formation and with marked unconformity on the Belaga Formation. The Nyalau Formation forms many outliers over the main Belaga Formation outcrops of the Sibu Zone (the largest at Merit Pila is shown on Fig. 3.1) and the unconformity is documented by Kirk (1957) and Wolfenden (1960). This is a very sandy formation in the west around Bintulu, becoming more muddy eastwards (Setap Shale). Detailed analysis by Shell indicates that a landmass on the west (Penian High) progressively gave way eastwards to a coastal plain rich in coal beds, then an inner neritic continental shelf (Fig. 3.2). The outer shelf was characterized by a narrow NNW–SSE trending carbonate platform (Subis Limestone), seen at the Niah Caves. The edge of the shelf was marked by a continental slope (known as the West Baram Line), east of which bathyal conditions prevailed in a region which became the Baram Delta in late Miocene time.

3.2.2 Southwards extension into Kalimantan

An understanding of the Rajang Group is impossible without knowledge of its southwards extension, perpen-

Table 3.1. Chronology of events related to the Sarawak and Sabah Orogenies

Ma	SERIES	*Schwaner Mountains* volcano-plutonic arc	*Post-subduction Volcanism & plutonism*	*Rajang Group deposition* Sarawak & *W. Kalimantan*	*Rajang Group deposition* Sabah & *E. Kalimantan*	*N.W. Sabah Crocker Ranges*	*South China Sea*
5	Pliocene		Metulang Volcs.				Post-inversion strata
10	*upper*		*Mt. Kinabalu*			Shallow Regional ∿∿∿	∿∿∿∿∿∿
15	*middle*					Deep Regional ∿∿	SABAH OROGENY ∿
20	Miocene		*Sintang*				
25	*lower*		intrusives	Deposition of the Melawi			Deposition of the
30	Oligocene			Ketungau and Mandai	Kelabit Fm.	[W. Crocker Fm]	South China Sea
35				*basins*		[*including the*]	*basins*
40			*Müller*	*Tatau Fm.*	*Batu Gading*	[Temburong Fm]	
45	*Eocene*			∿∿∿ SARAWAK OROGENY ∿∿∿	∿∿∿	∿∿∿	∿∿∿
50			*Nyaan, Piyabung*	↑	↑	[E. Crocker Fm.]	
55						? ... ?	
60	Paleocene		? Kayan Sstn. ?	[Belaga Fm]		Trusmadi Fm	
65		V		[Lupar Fm] ↑	[Sapulut Fm]		
70		V		[Embaluh Group]	[Mentarang Fm]		
75	Upper	V					
80	Cretaceous	P					
85		G [potassic granite]					
90		P					
95		P		↑ ↑			
100		P					
105	Lower	P					
110	Cretaceous	P					
115		P		[Selangkai Fm]			
120		P		[Pedawan Fm]			
125	Valanginian						
130							
135							

Key: ↑ end — time duration of turbidite sedimentation; ∿∿∿ unconformity; V = volcanism; P = plutonism; G = granite

dicular to strike, into Kalimantan. The Lupar Formation and the Layar and Kapit Members of the Belaga Formation are known in northern Kalimantan as the Upper Cretaceous to Eocene Embaluh Group (Fig. 3.1; Table 3.1), which is not sub-divided (Tate, 1991). This flysch sequence is strongly deformed. Bedding is in places overturned, and two fold phases have been identified (Williams *et al.*, 1988). The main fold axis orientation is horizontal, trending 280°. The folding is asymmetric, with southerly dips predominant. However, overall stratigraphic younging is towards the north, as in Sarawak. South-dipping thrust faults have been identified.

The Embaluh Group continues southwards as the Selangkai Formation, generally mapped as flysch, and shallow marine (shelf) southwards towards the Schwaner Mountains. Fossils range from Valanginian (130 Ma) to Cenomanian (Zeijlmans van Emmichoven, 1939). The formation youngs towards the north. The formation is variable, but large areas are of turbidite, less metamorphosed than the Embulah Group to the north. Tate (1991) has remarked on the great similarity between the Selangkai Formation and the Lupar Formation. However, it seems appropriate to interpret the Selangkai Formation as partly accretionary prism and partly fore-arc basin sequence. It also seems reasonable to conclude that the flysch belt is of Valanginian age in the south (Selangkai Formation), younging northwards through the Embaluh Group (= Lupar Formation and lower Belaga Formation), and eventually reaching its youngest (Upper Eocene) age near the Mersing Line at the Tatau Horst. The Selangkai Formation is in large part similar both in turbiditic character and age to the Pedawan Formation, which lies west across the Lupar Fault in the Kuching Zone.

The Lupar Fault appears from the Sarawak viewpoint to represent the most southerly termination of turbiditic rocks. However, the flysch belt extends southwards into the Sintang region of Kalimantan. The Lupar Line has therefore been given undue importance. It is the prominent fault margin between the Upper Eocene basal Silantek Formation and the flysch belt. Other such faults occur farther south on the margins of the Semitau Ridge. Indeed, within the Kuching Zone of Sarawak W and SW of the Lupar Line, there are abundant Lower to Upper Cretaceous turbidites known as the Pedawan Formation.

Ophiolitic mélange zones are not confined to the Lupar Line. They also occur along the Bukit Mersing Line (Hutchison, 1975), and along the Semitau Ridge, known as the Boyan Mélange (Williams *et al.*, 1988). All of these mélange belts lie close to and parallel to major faults which bound superimposed Tertiary basins. In this they have a similarity to the Meratus Mountains (Figs 4.17 and 4.18).

3.3 THE PAIRED VOLCANO-PLUTONIC ARC

Most authors (e.g. Hamilton, 1979; James, 1984) presume that the Upper Cretaceous through Upper Eocene Rajang Group of Sarawak (now modified to begin in the Lower Cretaceous by including Kalimantan) represents a northwards-facing accretionary prism, but there has been no clear identification of an appropriately positioned contemporaneous subduction-related volcano-plutonic arc. Gower (1990) concluded that the data from Borneo 'support the postulate of southeasterly-directed subduction of the South China Sea beneath Borneo during the early Tertiary'. At the same time he admitted that 'there are no convincing reports of early Tertiary magmatism, coeval with the postulated subduction'. He fell into the common trap of placing theory before mapped geology. The following discussion may clarify the situation of the appropriately distanced terrain lying 200 to 250 km to the south.

3.3.1 Convexity of the arcs

The island of Borneo is dominated by a framework of arcuate structures, all convex towards the south and south-west (Fig. 3.1). This feature was emphasized by the earlier Dutch workers (Van Bemmelen, 1949), but has never been discussed since the advent of plate tectonics. Hamilton (1979) noted the importance of the Schwaner Mountains volcano-plutonic arc thus: 'Presumably these igneous rocks are paired to the Cretaceous mélange complexes that occur in both Northwestern and Southeastern Borneo'. He did not refer to the convexity, but favoured subduction from the north, even though the convexity would favour subduction from the south.

The ophiolitic rocks of the Meratus Mountains have been dated Lower Cretaceous and are associated with calcareous mudstone, sandstone and limestone. These would seem to be ideal candidates to be related to the Lower Cretaceous volcano-plutonic arc of the Schwaner Mountains (Fig. 3.1). However, Sikumbang (1986) considered that there was no subduction relationship between them, and he concluded that the Meratus Mountains contain the record of a compressed marginal basin and paired volcanic arc.

The Meratus ophiolite formed part of the Upper Cretaceous landmass upon which diamond-bearing serpentinite conglomerate (the Pamali Breccia) was fluvially deposited (Bergman *et al.*, 1987). However, the Meratus were not uplifted to form the present day mountains until the late Miocene (Hutchison, 1992*b*). Before the uplift, the now separate Barito, Asem-Asem and Pasir basins were once one and the same, and the present day Meratus Mountains floored the Tertiary basin depocentre.

3.3.2 The Schwaner Mountains

The rigid 'core' of Borneo is formed by the Schwaner Mountains. Its northern boundary forms the southern margin of the Tertiary Melawi Basin (Fig. 3.1), and extends inland along an E to ESE trend, which cuts the equator east of Pontianak. The Schwaner Mountains extend eastwards more than 4° of longitude to disappear beneath the Cainozoic NW Kutei Basin, and they extend over 2° of latitude southwards to disappear beneath the Quaternary sediments of the coastal plains, which are underlain by the Billiton and Barito basins of S. Kalimantan. The region was divided by Van Bemmelen (1949) by an approximately E–W line along latitude 1° S into the 'Schwaner Zone' and a 'Southern Region'.

The Schwaner Zone, east of Pontianak, is mostly of I-type (subduction-related) calc-alkaline plutonic rocks ranging from granodiorite to gabbro, associated with a variety of metamorphic rocks mapped by Pieters and Supriatna (1989) as Palaeozoic, probably Carbo-Permian, though their age is indefinite. The 'Southern Region', lying east of Ketapang, is composed of Jurassic–Cretaceous flysch and contact metamorphosed limestone (the Ketapang Complex of Van Bemmelen, 1949), and volcanic rocks (the Matan Complex of Van Bemmelen), intruded by granitic batholiths (Pieters and Supriatna, 1989). The Bunga basalt and Kerabai volcanics, formerly included in the 'Matan Complex', range from basalt, through andesite, to dacite and rhyolite, and represent a mature volcanic arc, which Tate (1991) paired with the Upper Cretaceous part of the Rajang Group, representing southwards subduction.

Since the 'Schwaner Zone' is predominantly of plutonic and metamorphic rocks, it should be interpreted as a more uplifted region from which the Jurassic–Cretaceous sediments have been eroded. Conversely the 'Southern Region' is dominated by Upper Cretaceous volcanic and plutonic rocks, with only local occurrences of Lower Cretaceous granitoids. Accordingly it has not been uplifted.

The plutonic complex of the Schwaner Zone has been dated by the whole-rock K:Ar method by Williams *et al.* (1988) at 104 (Albian) to 123 Ma (Upper Valanginian). Earlier, Haile *et al.* (1977), using biotite and hornblende mineral separates, determined the K:Ar age range from 75 (Campanian) to 115 Ma (Barremian). The two data sets overlap and extend over the Lower and Upper Cretaceous.

Williams *et al.* (1988) report only four K:Ar dates from the potassic granites of the 'Southern Region', lying within the range 86 to 91 Ma. However Haile *et al.* (1977) report ten mineral separate K:Ar dates from this province, eight of which lie within the range 78 to 86 Ma, but two granites gave considerably older ages of 127 and 153 Ma. Therefore this province contains important Upper Cretaceous plutonic activity, dominated by alkaline granite, but there also occur some Lower Cretaceous to Upper Jurassic granitoids.

The volcanic rocks of the 'Southern Region' were considered too weathered by Haile *et al.* (1977) for accurate dating. However, Pieters and Supriatna (1989) reported Upper Cretaceous K:Ar ages of 65 to 75 Ma (Maastrichtian–Campanian).

In summary, most of the plutonic rocks of the Schwaner Mountains may be paired to the Selangkai Formation flysch belt. Volcanism and high-level plutonism continued at least into the Maastrichtian, when the Embaluh Group, Lupar Formation and Layar Member of the Belaga Formation were being deposited. The Cretaceous part of the Rajang Group may therefore be reasonably interpreted as a northwards-facing accretionary prism, paired to the Cretaceous Schwaner Mountains volcano-plutonic arc (Table 3.1, Fig. 3.2).

3.3.3 North of the Schwaner Mountains

A narrow E–W trending zone through the Upper Kapuas region of Central Kalimantan, of sporadic occurrences of Lower Eocene acid volcanic and pyroclastic rocks (Fig. 3.1), of restricted 48 to 50 Ma age, has been interpreted by several authors to be post-subduction and related to the early rift stages of formation of the Ketungau, Mandai, Melawi and West Kutei basins (Doutch, 1992; Tate, 1991; Van de Weerd and Armin, 1992). They can be demonstrated to be post subduction, because in several localities the Eocene volcanics unconformably overlie either the Boyan Mélange or the Embaluh Group (= Rajang Group).

The Piyabung Volcanics, of gently dipping subaerial lithic and vitric tuff and agglomerate up to 200 m thick, extend 40 km E–W from 112° to 112° 30′ within the Semitau Ridge at 0° 18′ N (Pieters *et al.*, 1987). A K:Ar whole rock age of 49.9 was obtained (Tate, 1991).

The Nyaan Volcanics, of 200 to 900 m of acid to intermediate tuff, agglomerate, ignimbrite and dacite, occur much farther east in the Mahakam area of the West Kutei Basin, at 114° 50′ E and 0° 50′ N. They gave a K:Ar date of 48.6 Ma (Tate, 1991) and 50 ± 2.5 Ma (Pieters *et al.*, 1987). In the Kelian district of the Upper Kutei, Van Leeuwen *et al.* (1990) describe Upper Eocene rhyolitic volcanic and pyroclastic rocks in a region dominated by Miocene volcanism. The Eocene igneous rocks are exclusively silicic, and cannot be interpreted as subduction-related. In the same general region, Wain and Berod (1989) show Upper Cretaceous to Eocene volcanic rocks on their maps, but do not describe them.

Near Singkawang, far to the west near the west coast at 1° N, the Bawang Dacite gave a K:Ar date of 51.3 Ma. It is part of the Serantak Volcanics, which are less than 300 m thick.

A much more extensive body of intermediate to basic volcanics lies entirely within the Mandai Basin in the Müller Mountains at 113° 20′E, 0° 30′N, K:Ar dated 40.9 Ma (Pieters *et al.*, 1987). These intermediate to basic tuffs and flows, many hundreds of metres thick, are interbedded with the basal sandstones along the Mandai River. The volcanic rocks were first documented by Molengraff (1902). Zeijlmans van Emmichoven (1939) confirmed the existence of volcanic rocks, predominantly of basic-intermediate to basic composition in the region of the Mandai Palaeogene basin east of the Lake District (Kapuas), equivalent to the Müller Mountains of Molengraff (1902).

On the north side of the Rajang Group, the acidic Arip Volcanics and associated Piring Granodiorite of the Tatau Formation have not been dated but post-date the metamorphism of the Rajang Group. They are comparable to volcanics described above from the southern side of the orogenic flysch belt.

It may therefore be concluded that there is no Eocene subduction-related arc to the north of the Schwaner Mountains, and that the geographically scattered predominantly silicic and pyroclastic eruptions are an integral part of the basin forming process. It is therefore doubtful if the term accretionary prism should be applied to the whole of the Rajang Group, for only its Cretaceous part can be paired to a volcanic arc, within the Schwaner Mountains. The Palaeocene–Eocene part should not be classified as an accretionary prism, for subduction-related volcano-plutonic activity had ceased, and a collision situation became operational.

3.3.4 Tectonic scenario

The sequence of tectonic events is portrayed schematically in Fig. 3.2., intended to replace the often quoted

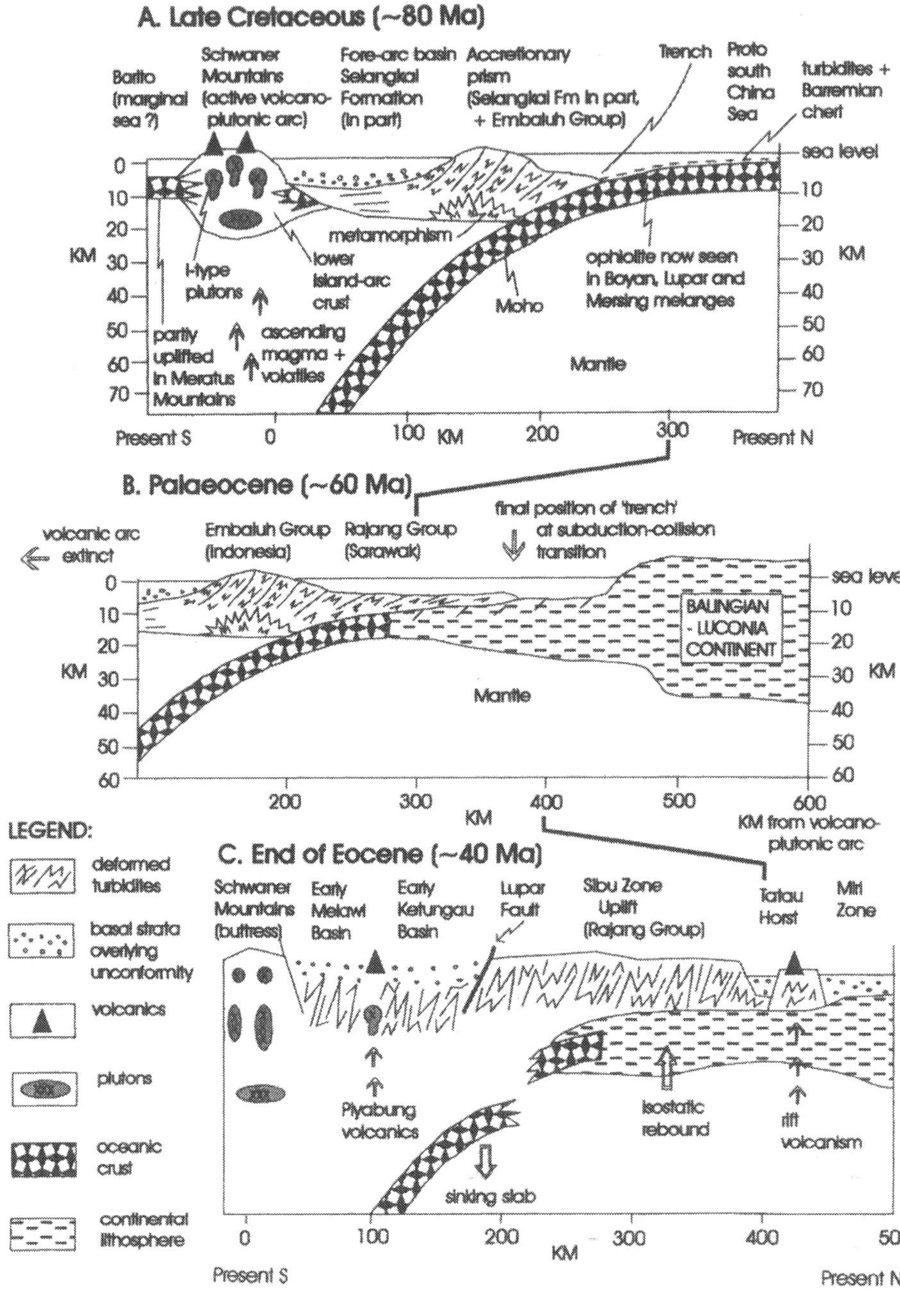

Fig. 3.2. Schematic N–S tectonic cross sections through the Schwaner Mountains and Rajang Group (based on Hutchison, 1996). Vertical and horizontal scales are very approximate. A: Upper Cretaceous north-facing arc-trench system. B: Palaeocene cessation of subduction, forwards migration of the 'trench' and incipient collision. C: Upper Eocene collision, compression and uplift of the Rajang Group between the Schwaner Mountains and the Balingian–Luconia Continent, followed by normal faulting and deposition of Upper Eocene strata unconformably upon peneplaned Rajang Group. The scenario of this figure is in strong contrast to the interpretation of James (1984), reproduced by Hutchison (1989*a*, p. 85).

diagram of James (1984), which I reproduced, thereby adding to the confusion (Hutchison, 1989*a*, page 85). The diagram (Fig. 3.2) shows that the Tertiary basins were unconformably superimposed on the already uplifted and metamorphosed Rajang Group. Uplift of the Rajang Group of Sarawak (Sibu Zone) may be modelled by southwards underthrust Balingian–Luconia continent, followed by isostatic rebound.

Subsidence of the Ketungau–Melawi basinal area may be related to sinking of the oceanic lithosphere slab after it became de-coupled from the Balingian–Luconia continental margin. The Ketungau and Melawi basins were one and the same at their earlier stages, and became two only when the Semitau Ridge (which contains the Boyan Mélange) rose up at the same time as the Meratus Mountains and other Sabah Orogeny inversions. The diagram by James (1984) has the wrong time frame, and gives the false impression that the high level Sintang Intrusives (Table 3.1) may have been an Oligocene volcanic arc. They are Miocene and related to the Sabah Orogeny and intrude through the Ketungau Basin strata. The real subduction-related volcanic arc is in the Schwaner Mountains (Fig. 3.1) considerably farther south at an appropriate distance.

3.4 THE RAJANG GROUP OF SABAH AND E KALIMANTAN

The Rajang Group of Sabah, as presently defined, had a much longer depositional history (Table 3.1) and its stratigraphy and structure have not yet been satisfactorily resolved. However, it can be clearly subdivided into a **Lower Part**, whose sedimentation history extends from Upper Cretaceous to Upper Eocene, and an **Upper Part**, extending from Eocene to Lower Miocene (Lim, 1985). It is the Lower Part only which extrapolates into Kalimantan and Sarawak. The underlying basement is known as the Chert-Spilite Formation, whose cherts are now dated Lower Cretaceous (Barremian) by Basir *et al.* (1985).

The Belaga Formation appears to continue east of 115° as the Upper Cretaceous to Eocene Mentarang Formation, extensively exposed and of distal turbidite, strongly folded and displaying slaty cleavage (BRGM, 1982). Northwards, along strike in SW Sabah, it has been mapped as the Sapulut Formation (Collenette, 1965). It is surprising that no slaty cleavage has been mentioned by Collenette, although previous workers noted the slaty characteristic. The farthest extrapolation into Sabah is known as the Trusmadi Formation, well exposed in accessible outcrops from Ranau (near Mount Kinabalu) along the main road towards Telupid. The predominantly Eocene Trusmadi Formation is of tightly folded, generally steeply dipping, phyllitic distal turbidite, characterized by quartz veining, of identical appearance to large parts of the Lupar and Belaga formations of Sarawak.

The upper parts of the Mentarang and Sapulut formations show indications of shallowing of the sea, and limestones and jasper conglomerates are recorded in both (Hutchison, 1989*a*). In Kalimantan, the Mentarang Formation is unconformably overlain by the Upper Eocene Sebuku Formation basal conglomerate, followed by reefal limestone (BRGM, 1982). Turbiditic sedimentation was therefore eliminated by tectonic uplift by late Eocene, a situation identical to that in Sarawak (Table 3.1).

A folding and metamorphic event (the Sarawak Orogeny) clearly must separate the Trusmadi from the unmetamorphosed Crocker Formation, but only fault contacts have been observed between the two (Collenette, 1965; Jacobson, 1970). The very sandy Crocker Formation turbidite, containing several extremely thick sandstones, is usually taken as Palaeocene to Lower Miocene, but it overlies the Trusmadi Formation unconformably, and hence must be Lower Oligocene to Lower Miocene (Lim, 1985).

The upper western part of the Crocker Formation has been mapped as the turbiditic Temburong Formation by Wilson (1964). It is very fossiliferous and gives an age of Oligocene to Lower Miocene.

The Oligocene to Lower Miocene Crocker and Temburong Formations have no equivalent in Sarawak, where the Rajang Basin ceased to exist by late Eocene time. The unconformity at the top of the Trusmadi and Mentarang formations indicates that the Crocker and Temburong formations should not be classified as part of the Rajang Group of Sarawak.

Palaeocurrent analysis of the Crocker and Temburong formations (Stauffer, 1968) indicate northwards directed currents. The provenance is most likely to be the Trusmadi, Sapulut and Mentarang formations, which were uplifted before the end of the Eocene. If the Mentarang, Sepulut and Trusmadi formation turbidites represent an accretionary prism, then the Upper Cretaceous to Eocene paired volcanic arc needs to be sought in the northern arm of Sulawesi, for it cannot be seen in Sabah. The northern arm of Sulawesi was contiguous with eastern Sabah before the late Eocene opening of the Celebes Sea (Hutchison, 1989*a*). However, the volcanic arc of north Sulawesi began in the Lower Miocene. Older rocks have been interpreted as ophiolite, with pre-Tertiary metamorphic and intrusive rocks (Carlile *et al.*, 1990).

An outstanding problem remains, and that is the relationship between the shallow marine Upper Cretaceous to Eocene Laruh Formation (BRGM, 1982) and the Crocker-

Temburong Formation. The distribution indicates that while the Crocker Formation was being sourced from the south, there lay a landmass, predominantly made of the Laruh Formation, in that direction. Just across the border in Sarawak, the Upper Cretaceous to Eocene Kelalan Formation has been mapped by Haile (1962). Outcrops at Batu Gading indicate that the Kelalan Formation is turbiditic and indistinguishable from the Belaga Formation. Unfortunately little is known about the region south of the Crocker formation and north of the Laruh Formation. The relationship should be in southern Sabah, near the border with Kalimantan, a remote region, where the geology is poorly known (Collenette, 1965).

3.4.1 Uplift of the Crocker Formation

Deformation resulting in spectacular episodic uplift of the Crocker Formation is well documented in the offshore oilfields by several prominent unconformity surfaces (Levell, 1987; Tan and Lamy, 1990). The first is the Deep Regional Unconformity (15 Ma, Middle Miocene) and the latest is the Shallow Regional Unconformity (9 Ma, Upper Miocene). The tectonic front moved progressively oceanwards (Fig. 3.6). The unconformities extended quite close to the continental slope (tectonic front), so that the fluvial system brought sands to the edge of the very narrow continental shelf to spill over the slope into the bathyal area as slumps, such as the one described by Hutchison (1994*a*), and to feed turbidite fans. The persistence of outer neritic to bathyal marine conditions west and north-west of the tectonic front has been documented by Rice-Oxley (1991). The Upper Miocene tectonic front of the Crocker–Temburong Orogen approximately coincides with the Jerudong Line, Morris Fault, and Transition Zone between the Inboard and Outboard belts of Shell (Fig. 3.6).

The Upper Miocene inversion of the Crocker–Temburong basin to form the Crocker Ranges of Sabah has never been given a name, but the term **Sabah Orogeny** seems appropriate. The spectacular Upper Miocene Mount Kinabalu (4101 m) is a product of this orogeny, with whole rock K:Ar dates commonly as old as 10 Ma (Jacobson, 1970).

3.4.2 The Sabah Orogeny regionally

This orogenic event was widespread throughout the South China Sea, but only in the Zhu Jiang Kou (Pearl River Mouth) Basin has it been given a name—Dongsha Movement (Su *et al.*, 1989). However, this is not a good regional name, since the inversion there is not very spactacular. The nearby Song Hong–Yingge Hai Basin contains very impressive Middle to Upper Miocene inversion structures, well displayed in regional seismic records, but they have not been published.

The best published account of spectacular basin inversion is by Ginger *et al.* (1993) in the West Natuna area, where the Anambas Graben became the Boundary High during Middle to Upper Miocene inversion (Fig. 2.16). The eroded anticlines are unconformably overlain by flat lying uppermost Miocene–Pliocene Muda Formation. In the contiguous Penyu and Malay basins, inversion structures have not yet been analysed and described because of a lack of seismic resolution at depth. However, the Upper Miocene unconformity has been well documented in the southern end of the Malay Basin (Armitage and Viotti, 1977; Nik Ramli, 1988). Even though the Middle to Upper Miocene inversion structures may not have been identified and described, the Upper Miocene unconformity is widely documented throughout South-East Asia.

All of the South China Sea basins rapidly returned to post-inversion subsidence after erosion of the inversion structures, so that the basins remain today beneath sea level.

Much farther south, the Upper Miocene inversion is well displayed in the Meratus Mountains, which formed the basement to the Eocene through Miocene sequence, then rose up only in the late Miocene to form a ridge which subdivided a formerly much more extensive basin (Hutchison, 1992*b*). The post-inversion unconformity is also seen in Central Sumatra (Hutchison, 1993). Both in the Meratus and Central Sumatra the uplifted basins have not subsequently subsided beneath sea level.

3.5 EAST NATUNA PROVINCE

The fault orientation through the Paus to Sokang area is NW–SE, but becomes more N–S northwards towards Nam Conson (White and Wing, 1978). This trend reflects the underlying basement grain, interpreted to be a seawards extrapolation of the Sibu and Embulah Zone of Sarawak and Kalimantan (Hutchison, 1996), which curves from E–W in east Sarawak and central Kalimantan, to NW–SE along the coastline and towards Sokang, then swinging N–S northwards to Nam Conson. White and Wing (1978) extrapolate the Lupar Line northwestwards from the Sarawak coastline, swinging due north between Natuna Island and the Paus High. The Sibu Zone, predominantly of Upper Cretaceous to Upper Eocene steeply dipping phyllite floored by Cretaceous oceanic crust, lies east of this line, and west of it lies the Sundaland continental crust.

The N–S to NNW–SSE trending arcuate ridge of the Paus High is a pre-Oligocene feature that had a strong influence on Upper Tertiary deposition patterns. This ridge, which is the northwards extrapolation of the Sibu Zone of Sarawak, is of phyllite, drilled into in several wells (see Hutchison, 1989*a*, pp 179 and 183–4—wells 24, 25, 26, 28, 29)

To the east of the Paus High is a N–S trending Miocene carbonate platform, with a carbonate shelf lying along its western margin in shallower water (May and Eyles, 1985; Rudolph and Lehmann, 1989). The carbonate is known as the Terumbu Formation, which overlies Oligocene to Lower Miocene clastics of the Arang and Gabus Formations (Fig. 2.16). The deltaic Arang Formation [equivalent to Cycle I and II of the Balingian Province (Nyalau Formation)], is considered the most likely source of gas in the Terumbu Limestone. At least 1600 m of the limestone facies has porosities commonly up to 25%, but unfortunately the gas is 72% CO_2, 27% methane and 0.5% H_2S, the remaining 0.5% being other hydrocarbon gases (Rudolph and Lehmann, 1989).

3.6 BALINGIAN PROVINCE

The Balingian Province grades northwards into the more stable Central Luconia Province, while to the south lies the Anau–Nyalau Fault Zone, which is an offshore continuation of the Tatau Horst. The eastwards onland extrapolation is the Bukit Mersing Line (Hutchison, 1989*a*, p. 185 and 267), which forms the boundary between the Miri Zone (which includes the Balingian Province) and the Sibu Zone. The Balingian Province is bounded by a major fault, the NNW–SSE trending West Balingian Line, similar to the Lupar Line. The West Balingian Fault extends seawards from the Tatau Horst region. It is a major tectonic feature, with the Penian High on its west and a vertical downthrow of up to 3.0 s (2-way time) of sediments to its east (Swinburn, 1993). On the east, the Balingian Province is bounded by the parallel West Baram Line.

Palaeogeographically, the Penian High represents the landmass of uplifted Rajang Group formed by the Upper Eocene Sarawak Orogeny, whose continental shelf extended NE-wards over the Balingian Province, extending as far as the West Baram Line. The latter represents the shelf edge, immediately to the NE of which lay the continental slope. The West Baram Line in all probability represents the NE-wards limit of continental crust forming an Atlantic-type margin. This inference leads to the conclusion that the Baram Delta depocentre may be underlain by oceanic crust (Hutchison, 1994*b*). There should be a mirror-image NW–SE trending continental slope and shelf somewhere towards the NE, perhaps in the Reed Bank–Calamian area, but it has never been specifically demonstrated. The stratigraphic nomenclature of the Balingian, Central Luconia and Baram Delta, as developed by Shell, is based on a number of sedimentary cycles from I (27 Ma) through VIII (1 Ma), described by Ho (1978).

3.6.1 Sedimentation history

Seismic stratigraphy, reinforced by a large number of wells and biostratigraphic studies, have led Shell to a rather complete understanding of the palaeofacies distribution of the Balingian Province (Doust, 1981), and this has been supplemented by subsequent Shell data and by Agostinelli *et al.* (1990), whose work has reinforced the previous interpretation of Doust (1981).

3.6.1.1 Cycle I (Upper Oligocene: 27 Ma)

The palaeofacies zones, as they are found in their present day orientation, trend perpendicular to the present day coastline, with a broad zone of lower coastal plain deposits trending NW–SE and outcropping extensively around and west of Bintulu, known as the Oligocene to Lower Miocene Nyalau Formation. It is a sandstone-dominated coastal formation, characterized by numerous thin coal beds, and showing tidal features and numerous fluviatile channels. Ripple marks are dominantly oriented NNW–SSE and the onland facies distribution indicates marine conditions to the east and land to the west. The same formation continues offshore, of several km thickness of fluvial and estuarine channel sands with overbank clays and coals, grading into marine shales with thin interbedded limestones (ASCOPE, 1981). The Cycle I sea progressively deepened in a NE direction, through coastal, holomarine inner neritic, through middle neritic to outer neritic and bathyal beneath Miri and what became the depocentre of the Middle Miocene to younger Baram Delta. Onland outcrops show that the sandy Nyalau Formation grades north-easterly into holomarine Setap Shale, formed of monotonous mudstone containing sideritic concretions.

3.6.1.2 Cycle I/II (early Lower Miocene: 23 Ma)

A carbonate platform developed, extending NW-wards from the onland outcrops of the Subis Limestone at Batu Niah. The width of the carbonate platform terminates NE-wards at the West Baram Line, beyond which the sea floor sloped downwards to the bathyal future Baram Dalta depocentre (Agostinelli *et al.*, 1990). Onland, the Subis Limestone is succeeded eastwards by extensive outcrops of Setap Shale, extending as far as the western foothills of the Lambir Hills.

Lower coastal plain to coastal environments are widespread in the SW and the palaeo coastline was at right angles to the present coastline. The landmass west of the coastline was known as the Penian High (Doust, 1981, Swinburn, 1993), and more than 30 wells in the West Balingian Basin have a stratigraphic hiatus at this time (23 Ma).

At least two fluvio-deltaic systems have been identified, with provenance in the Penian High, and sediment transport towards the NE (Fig. 3.3). By 18 Ma (top of cycle II) the Penian High became more extensive (regression) and the coastline trended NW-wards from Bintulu. More abundant fluvio-deltaic sedimentary input in a NE-direction put an end to the carbonate platform of Fig. 3.3 (A). From detailed analysis of the D18 Field (Almond *et al.*, 1990), the sequence is interpreted as a N to NE progradation and abandonment pattern of small delta lobes in a river-dominated lower delta plain setting.

3.6.1.3 Cycle III (late Lower Miocene: 16.5 Ma)

The palaeo coastline is now generally E–W, with fluvio-deltaic deposition growing northwards. The West Baram Line, however, continues to be a continental slope into the bathyal zone to the NE, which will become the future Baram Delta (Fig. 3.3. (A)).

3.6.1.4 Cycle IV (Middle Miocene: 13 Ma)

By this time present day Sarawak was generally land, but sea still deepened in a NE direction beneath Miri. A major delta began building NE-wards in the Lambir Hills area where the sandy Lambir Formation sits abruptly on muddy Setap Shale, while to the NW, along strike, a second less widespread episode of carbonate deposition was established adjacent to the West Baram Line at the edge of the continental shelf (Fig. 3.3 (B)).

3.6.1.5 Cycle V (Upper Miocene: 9 Ma)

The palaeogeography is now approximately similar to present day conditions, with the two coastlines parallel.

Spectacular uplift of the Crocker Formation during the Sabah Orogeny, and creation of the Tinjar Fault, resulted in the major Baram Delta being established by 9Ma, building outward in a north-westerly direction, its western margin being formed by the continental slope (West Baram Line). The water was bathyal so that initial sedimentation must have been by turbidity currents. Fluvio-deltaic sediments covered over and caused extinction of the carbonate zone of Cycle IV. At this time, little sedimentation reached the Central Luconia Province, which became established as a province of major carbonate build-ups.

The details of the Balingian Province demonstrate that the Sibu Zone (Belaga Formation) was uplifted first on the west (Natuna–Sunda Shelf) in the late Eocene and the locus of major uplift and erosion progressively migrated eastwards through Cycles II to V, when rejuvenation of the eastern part by the Sabah Orogeny resulted in the Baram Delta.

3.6.2 Structure

Tertiary deformation was related to periodic movement along the major bounding fault lines of the West Balingian and West Baram. The latter is thought to have transformed from a continental slope to a fault (Tinjar Fault). Movements may be related to the opening of the South China Sea Basin. During the late Oligocene to early Miocene (Cycles I and II), the interpreted bathymetry of the Central Luconia, Balingian and Baram Delta region indicates a strong control by important tectonic events (Hageman, 1987). By contrast, the bathymetry of the Middle Miocene to Pliocene was largely controlled by eustacy.

The Balingian Province may be divided into three sub-provinces (Swinburn, 1993), each with different structural trends and timing. The nature of the pre-Tertiary basement has a strong control on the intensity of Tertiary deformation. Deformation is less over strong coherent basement horsts. By contrast strong Tertiary deformation zones overlie a basement which shows weak seismic reflectors. Many Balingian anticlines appear to have diapiric cores, as seen on seismic sections (Du Bois, 1985).

3.6.2.1 East Balingian

This sub-province is an area of strong Upper Miocene to Pliocene wrench-related deformation, with structural axes oriented NE–SW. Individual structures are typically large asymmetric high amplitude folds, bounded by reverse faults which converge at depth and can be related to oblique strike-slip movements.

3.6.2.2 SW Balingian

This sub-province is an area of Oligocene to Lower Miocene wrench-related deformation of varying intensity. Structural axes trend NW–SE. The dominant fault trend is NE–SW and the province is characterized by basement faulting in the west and growth faults on the flanks of the Balingian Basin in the SE.

3.6.2.3 NW Balingian

This sub province was subjected to several phases of strong deformation from Oligocene to Pliocene and is

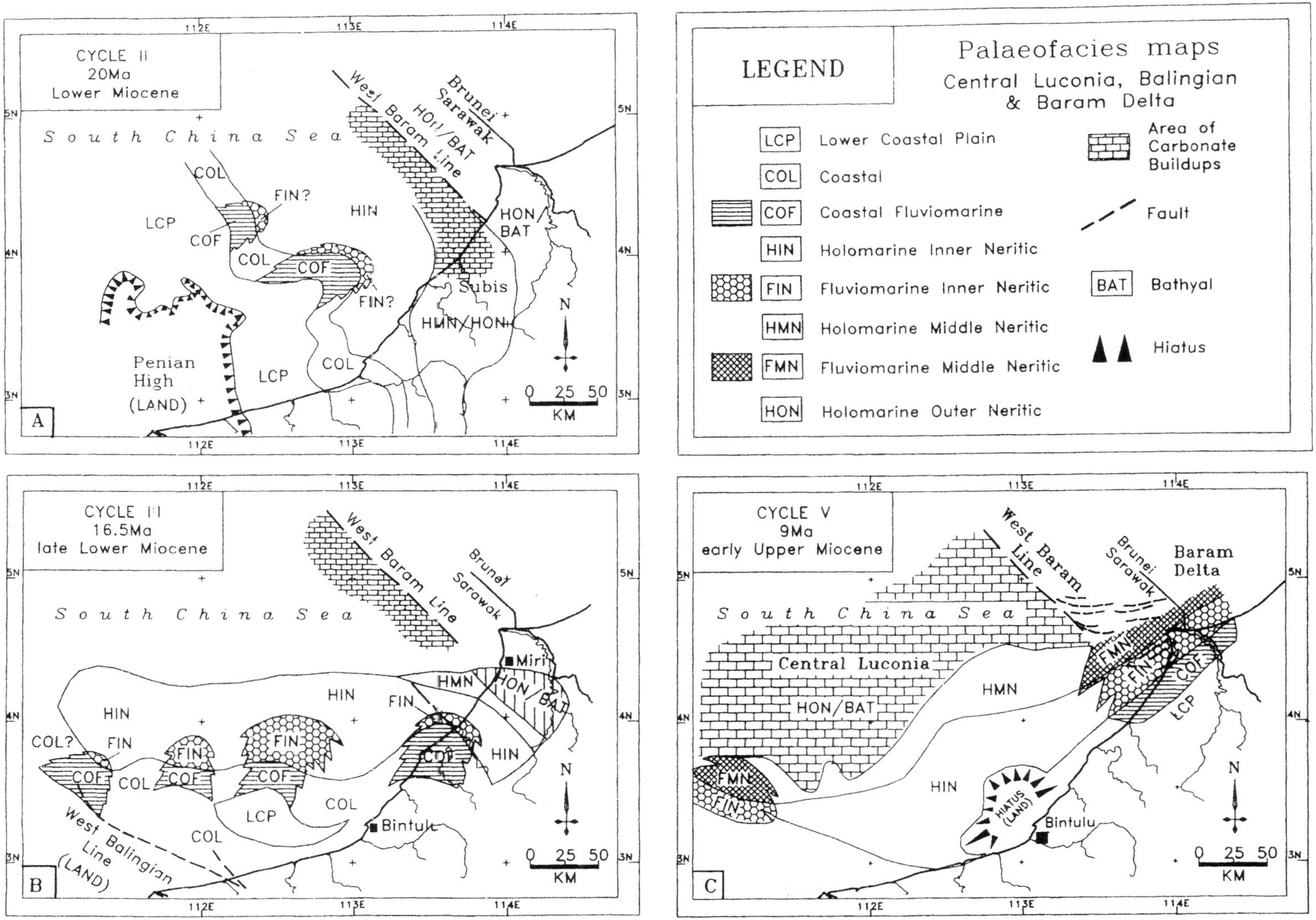

Fig. 3.3. Present day actual palaeofacies distribution for NW Sarawak, based on drill data and seismic analysis, modified after Doust (1981) and Agostinelli *et al.* (1990). A: for Cycle II (20Ma); B: for Cycle III (16.5 Ma). The shaded carbonate area adjacent to the West Baram Line is shown here, but did not develop until Cycle IV; C: for Cycle V (9 Ma).

characterized by *en echelon* NW–SE trending folds with complex fault patterns. This has created highly faulted ridges, depressions and deep narrow half grabens (Du Bois, 1985).

3.6.3 Source rocks and reservoirs

The average geothermal gradient is 42° C km^{-1}, consistent with the contiguous Central Luconia Province. Source rocks are considered to be coals and organic-rich clays of the Cycle I and II Oligocene to Lower Miocene Nyalau Formation. Kerogen compounds within the same beds constitute a secondary source of gas and oil (Du Bois, 1985).

The principal reservoirs are sands of coastal plain and nearshore origin within the Oligocene and Lower Miocene Cycles I and II. Individual sands are commonly less than 20 m thick and may have limited lateral extent (Du Bois, 1985). The D18 Field has productive reservoirs of Lower Miocene (Cycle II) lower coastal plain and delta deposits (Almond *et al.*, 1990). The field is a structurally complex highly faulted anticline. The sands exhibit a wide range of reservoir quality, with 25% average porosity.

3.7 LUCONIA PROVINCE

The Central Luconia province comprises a large area on the Sarawak Shelf, characterized by a Middle Miocene carbonate platform from which rise carbonate mounds and pinnacle reefs. The province has been relatively stable since the Lower Middle Miocene and its stratigraphy has been closely controlled by eustatic sea level changes (Hutchison, 1986). To the NE the province crosses the NNW–SSE trending continental slope (West Baram Line) into the Baram Delta Province (Fig. 3.5). The stability of the Luconia Province must result from an underlying continental basement (Fig. 3.4), whose NE margin is the continental slope (West Baram Line), The same continental basement probably underlies the Balingian Province, as far inland as the Tatau Horst. Pre-Miocene faulting produced a province of horsts and basins, which controlled the distribution of carbonate growth (Mohammad and Amita, 1993).

3.7.1 Stratigraphy

Carbonate deposition began during widespread marine transgression in the Middle Miocene (Cycle IV of Ho, 1978) at a time when this part of the continental shelf lay well away from the coastline and was in an open marine setting, rather like the Reed Bank is today. Platform type bank deposits were developed to a thickness of 200–300 m on the horsts. Pinnacle-type build-ups are distributed within the basinal areas (Mohammad and Amita, 1993). Maximum marine transgression took place during the Middle to Upper Miocene (Cycle V) and widespread reefs and reef associated carbonates were developed (Fig. 3.4). Some carbonate build-ups exceed 20 km in length and 1.5 km in thickness (Fig. 3.4). During this time, uplift of the Crocker Formation and rejuvenation of the Rajang Group landsurface of Sarawak by the Sabah Orogeny resulted in rapid erosion and progradation of the Baram Delta, which was directed into the deeper water oceanwards of the West Baram Line continental slope. By the end of Cycle V the Baram Delta had extended over much of the Balingian and southern parts of the Luconia Province, burying several of the reefs and build-ups.

Carbonate deposition was succeeded by open marine and coastal plain environments and deposition of more than 1 km of clays and sands, smothering the reefs and providing a seal for the hydrocarbons (Epting, 1980).

3.7.2 Structure

The province is characterized by a network of NNE–SSW trending faults which divide it into horsts and grabens. These faults are basement controlled, and they have affected the sedimentation and carbonate build-up pattern. To the east and west the province floor descends below younger deltaic sediments associated with the Baram Delta and East Natuna areas respectively.

3.7.3 Hydrocarbons

The high geothermal gradient of 43.6° C km^{-1} contrasts very strongly with an average 25.8 across the West Baram Line. The interpretation of the West Baram Line as a continental slope is in keeping with this difference, indicating an absence of a continental crustal basement to the Baram Delta (Hutchison, 1994*b*). Wells located off the shelf edge in the West Natuna area also have a much lower geothermal gradient.

The upper parts of the carbonates contain impressive concentrations of gas, now being piped ashore at Bintulu and liquified for export. This is a good place to pose the question why carbonates in South-East Asia usually reservoir gas and not oil. The answer is not in geothermal gradients, but must be in the source rocks themselves, which are low in sulphur. However, the Central Luconia province is not exclusively gas-bearing, and the carbonates do contain small accumulations of oil, but no publications have been forthcoming. Source beds are occasionally noted near the base of the carbonate build-ups and may be present within the time-equivalent off-reef facies. The carbonates themselves may also be source rocks (Du Bois, 1985).

The West Luconia Province is poorly known. Unlike the carbonate province to the east, it represents a depocentre filled by Neogene siliciclastic sediments of great thickness showing growth faults and large anticlinal structures.

3.8 BARAM DELTA

The Baram is a great and prolific Neogene delta. Oil was first discovered in 1910 in the Miri Field (Sarawak), long since abandoned, and in 1929 in the Seria Field (Brunei), which continues to be an active giant productive field. With the exception of Seria, oil company activity has moved offshore, with a trend towards progressively deeper water exploration.

On its SW side, the Baram Delta is bounded by the NW-trending West Baram Line, the pre-existing continental shelf edge (slope) of the Central Luconia–Balingian Province (Fig. 3.3). Based on interpretation of the memoir maps of Haile (1962) and Kirk (1957), the same (Cycles I and II) shelf edge may be continued SE-wards from the coast inland 120 km to the boundary between the Miri and Sibu zones (Bukit Mersing suture line), approximately following the NW-trending sector of the Tinjar River, to the eastern margin of the Usun Apau Mountains. The sand-dominant Nyalau Formation predominates west of this line and rapidly gives way eastwards to the time equivalent muddy and silty Setap Shale. Dykes of fresh olivine basalt outcrop along the straight Tinjar River (Haile, 1962), taken to represent a major fault line. The Tinjar Fault should be interpreted as having developed on the structurally weak continental slope (West Baram Line) as a result of subsequent compression which has folded the Nyalau Formation and Setap Shale into ENE-trending anticlines and synclines.

The SE margin of the Baram Delta is formed by the north-western tectonic front of the Crocker fold-thrust belt, which extends seawards as far as the Morris Fault, curving southwards along the Jerudong Line (Figs 2.2, 3.6). In late Miocene time the Morris Fault–Jerudong Line was a tectonically active (normal fault) shelf edge, downthrown to the west into the deep water Baram depocentre (Levell and Kasumajaya, 1985). The deep water re-entrant into the Miri Zone (bounded by the West Baram and the Morris Fault–Jerudong lines) was extensively sedimented by Miocene Setap Shale, but also includes major basement uplifts of Temburong (and 'Old Setap Shale') and Mulu formations.

The Baram Delta has built out into this re-entrant to form a fan shaped delta, draping over the Crocker tectonic front (Morris–Jerudong Line) (Fig. 3.6) and, to a lesser extent, the Balingian–Luconia continental shelf (West Baram Line). The bathyal re-entrant could not receive delta sediments until the Sabah Orogeny uplifted the Crocker Formation and hinterland, resulting in high erosion rates.

A major river system which fed the Baram Delta came from a south-westerly provenance, probably along the direction of the Jelalong and Grabit synclines, where Haile (1962) has mapped thick Belait Formation overlying Nyalau Formation, thence towards the Lambir Hills, and only thereafter bending into a more northerly direction (Fig. 3.3). One cannot therefore avoid the conclusion that the Middle Miocene to younger Baram Delta deposited sediments that had been eroded from thick outcrops of Upper Oligocene to Lower Miocene Cycle I and II Nyalau Formation. Many authors have stated that the Baram Delta resulted from erosion of the Rajang Group (Belaga Formation). This is by no means obvious and remains unproven. What is obvious is cannibalization of the nearby Nyalau Formation. It is also obvious that the recently uplifted Crocker Formation was extensively cannibalized in Mid to late Miocene time. The rivers only had a short distance to flow to the coastline and the continental shelf of the Morris Fault–Jerudong Line (Fig. 3.6). An interesting provenance study, based on heavy mineral analysis, remains to be carried out and published on the Nyalau and Crocker formations and the various derived Baram Delta formations, such as the Miri and Belait.

The outermost margin of the Baram Delta is clearly defined as the North-west Borneo Trough (Figs 2.2 and 2.18), which lies parallel to the Crocker tectonic front and the structural elements of the Crocker fold-thrust belt (North-west Borneo Trend). The tectonic significance of this trough will continue to be disputed (see Hutchison, 1992*b*; Hazebroek *et al.*, 1992)

3.8.1 Stratigraphy

The stratigraphic terminology is extrapolated from the Balingian Province, described by Ho (1978). There are three major eustatically-controlled cycles of regression and transgression (Rijks, 1981). The Delta began with Cycle V (Middle to Upper Miocene) and extended up to Cycle VIII (Upper Pliocene) (Figs 3.4 and 3.5). It continues today to prograde over the North-west Borneo Trough (Figs 2.2 and 2.19). Each of the cycles has a coastal plain environment in the south (sands, silts and clays) and grades in a northerly direction into holomarine neritic sediments (clays, silts and minor sands). The regressive sands of Cycles V and VI provide excellent oil reservoirs and may be up to 1 km cumulative thickness (Du Bois, 1985). The Baram Delta toe thrust zone (Fig. 2.18) shows

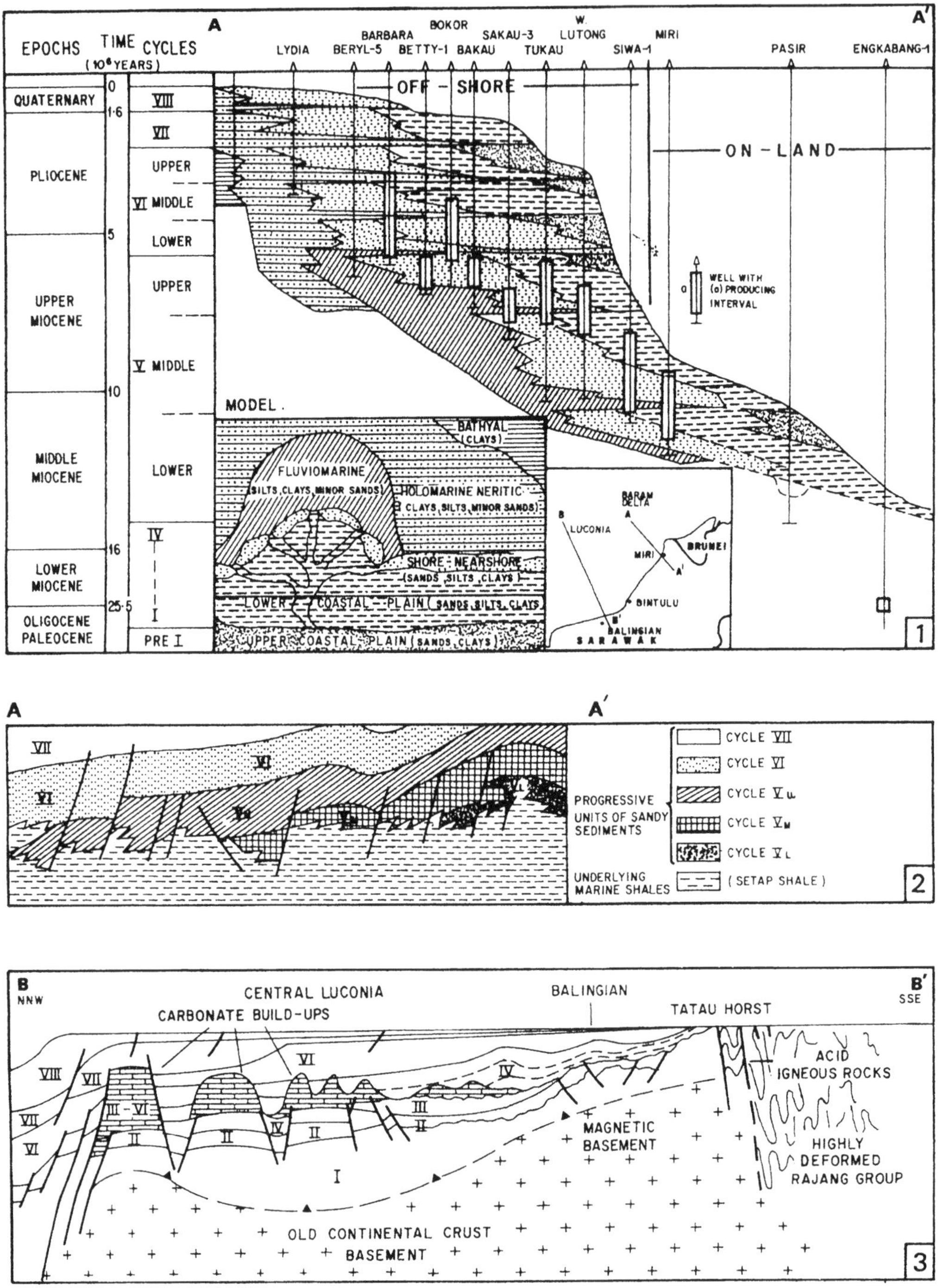

Fig. 3.4. 1: Time–stratigraphic profile of the Baram Delta (after Ho, 1978; ASCOPE, 1981). 2: Structural profile of the same section (from Rijks, 1981). 3: Structural cross section across the Tatau Horst, Balingian and Central Luconia provinces (after Epting, 1980).

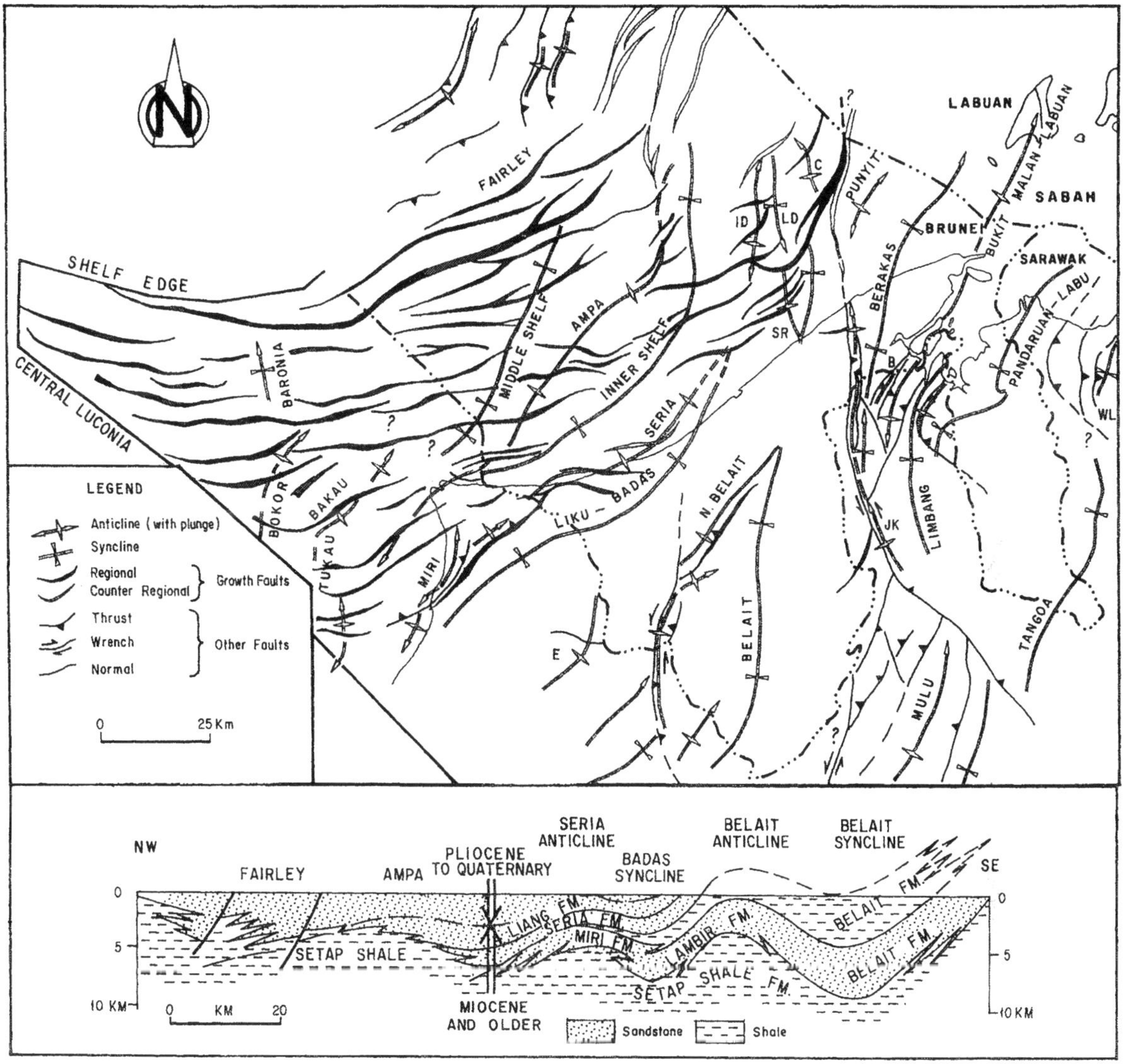

Fig. 3.5. Simplified structural map and cross section of the Baram Delta province of Miri and Brunei (after James, 1984, and Rijks, 1981). Unnamed anticlines: B = Bergapar, C = Champion, E = Engkabang, ID = Iron Duke, JK = Jerudong-Keramit, L = Linei, SR = Scout Rock. Unamed synclines: LD = Littledale, WL = West Lawas.

features, such as active submarine diapirism, that prove present day activity as a result of gravitational tectonics and shows many features in common with the Niger Delta, which is devoid of convergent tectonics. This zone, forming the NE-trending boundary between the Baram Delta and the North-west Borneo Trough, had previously been thought related to subduction, but is now ascribed to delta front (toe) tectonics (Hazebroek and Tan, 1993). Sedimentary draping over the deep and flat North-west Borneo Trough (Fig. 2.19) must be effected by turbidite sedimentation, so that the turbidite front of the delta actually extends as far as the NW margin of the North-west Borneo Trough and its contact with the Dangerous Grounds (Nansha Block) terrain (Fig. 2.18).

The Baram Delta province shows a close association of sedimentation to faulting. The Morris Fault trends southwards between the Samarang and Glayzer fields to continue onland as the Jerudong Line (Fig. 3.6). It is generally interpreted as a major left-lateral wrench fault (Bol and van Hoorn, 1980; Tan and Lamy, 1990). However, during the late Miocene, this fault was an active west-dipping normal fault with a downthrow to the west of up to 1.1 km, as shown on seismic sections, and it defined the shelf edge (Levell and Kasumajaya, 1985). The Pliocene sediments show a growth index of 2 to the west across the Morris fault.

The fault plane is characterized by extensive slump scars by which typically 1 to 5 km^3 of sediment were redeposited downslope westwards towards the deeper basin. These slumps would have fed turbidite sedimentation into the deeper basin. The time of slumping correlates with eustatic low sea level. The slump scars have been drilled into and are mud based. Such major slumping had never been seen onland until Hutchison (1994*a*) described the phenomenon at Masin Hill, south of Pangkalan Batu along the Jerudong Line in Brunei Darussalam. He described a large mud based (Setap Shale) slump containing numerous convoluted and rolled blocks up to 2 or 3 m diameter of sandstone. Deformation of the sand indicates soft sediment slumping as an olistostrome down slope into a deeper water environment.

The beginning of the Baram Delta is recorded in good onshore outcrops at the southern foot of the Lambir Hills, along the road from Miri to Batu Niah and Bintulu. The pro-delta Setap Shale, which forms extensive outcrops all the way to Batu Niah, is abruptly overlain by the very sandy Middle Miocene Lambir Formation. Progressing north-eastwards along the road to Miri, thence into Brunei Darussalam, the sand dominated Baram Delta formations young north-eastwards and are variously known as the Lambir, Tukau, Miri, Seria, Belait and Liang formations (Fig. 3.5).

The Baram Delta province extends NE into Brunei and along the coast of Sabah by way of Labuan island. There is no sharp demarkation between the Baram Delta and the structurally more complex Sabah Basin. However, from Labuan NE-wards, the provenance of the Neogene sediments was the Crocker Formation, and fluvial transport was by rivers other than the main Baram responsible for the Baram Delta.

The geology of Brunei Darassulam has been summarized by James (1984), Wilford (1961) and Tate (1976). Following the Middle Miocene Sabah Orogeny, uplift and erosion of the Baram Delta hinterland resulted in deposition of a thick pile of Neogene clastic sediments, which gradually prograded seawards. Sedimentation was strongly influenced by tectonic events, and up to 6 km of strata accumulated in basins such as the Belait Syncline (Fig. 3.5). Much of the progradational wedge consists of regressive mega-sequences separated by thin regionally extensive shales, which were deposited during eustatic sea level rises (Fig. 3.4). The depositional environment of the Neogene strata has been interpreted by James (1984) as predominantly coastal plain and delta front facies.

The Baram Delta prograded over the thick pro-delta Setap Shale Formation, which exceeds 3 km thickness in places. It is mainly of silty mudstone with occasional marly clay and thin sandstones. Bedding is poorly defined and indicated by discontinuous layers of sideritic concretions. In Brunei it has been dated Lower to Middle Miocene and is conformably overlain by the Belait Formation. However, there is an older 'Setap Shale', which outcrops south of Bangar (Temburong District of eastern Brunei). It is somewhat metamorphic and is unconformably overlain by the Belait Formation. It has appropriately been assigned by Brondijk (1963) to the Temburong Formation, equivalent to the West Crocker Formation.

The Middle to Upper Miocene Belait Formation reaches a total thickness in excess of 6 km. It is predominantly of interbedded thick sandstones and thin clays. Thick coal seams, point bar sands and alluvial flood plain silts form the lower part. The coal has been mined in Brunei and on Labuan. Peat dies out upwards as the formation becomes characterized by mangrove swamp silts, channel sands and flaser-bedded clays and silts. The formation conformably overlies and is partly time equivalent to the Setap Shale and is time equivalent also to the Miri and Lambir formations (Fig. 3.5).

The Upper Miocene Miri Formation is similar to the Belait and is near-shore marine, varying from coastal to fluvial. The lower part is argillaceous, whereas the upper part is dominated by sandstone, except in the Belait anticline, where the reverse is true, and the upper part passes into the Belait Clay.

The Middle to Upper Miocene Lambir Formation is predominantly of sandstone and shale, and equivalent in time to the Belait Formation. In the Belait Anticline, the Lambir is overlain by shaly facies Miri Formation.

The Upper Miocene Seria Formation attains a thickness of 2 km. The lower part is sandstone-rich, and ironstone nodules are common. It conformably overlies the Miri Formation, and laterally changes to shale northwards into the Setap Shale Formation. It is overlain by the Liang Formation.

The Upper Miocene–Pliocene Liang Formation attains a thickness of 600 m. Its main lithologies are sands with occasional conglomerates and lignite beds. It is unconformable on the Belait Formation near Lambang.

3.8.2 Structure

As is characteristic of most well developed deltas, the Baram Delta has many gravity induced listric growth faults, generally E–W trending and downthrown to the north. They are associated with normal faults. The listric growth faults curve to the NW, sub-parallel to the West Baram Line and SE they are directed NE–SW, sub parallel to the Mulu Shear Zone and the NW Borneo structural fabric of the Crocker Formation. The timing of faulting is older in the south (lower Cycle V), becoming progressively younger to the north (Cycles VI, VII to VIII). The strata are folded into major synclines, which are structurally simple, and anticlines, which are structurally complex (Fig. 3.5). These NE–SW trending compressional folds originated in the late Miocene (Rijks, 1981).

3.8.3 Source rocks and reservoirs

The geothermal gradient is low, in keeping with the inference from the Balingian Province (Fig. 3.3) that the West Baram Line and from Sabah and Brunei that the Morris Fault–Jerudong Line represent the continental slope, suggesting that the intervening Baram Delta re-entrant may be underlain by old cold oceanic crust, or extremely thin continental crust (Hutchison, 1994*b*). The geothermal gradient ranges from 18.2 to 32.5° C km^{-1}, with an average of 28° C km^{-1}, which is in strong contrast to the contiguous Balingian and Central Luconia province.

Source rocks are carbonaceous shales of lower coastal plain and fluvial environments. They contain land plant derived materials discharged at the delta fronts. All outcropping formations, especially the Miri and Tukau formations, contain abundant nodules of amber. The shales are organic rich and the palaeo-environment must have been closely similar to the present coastal zone of Sarawak. All marine inlets (e.g. Baram River mouth) are characterized by unpolluted black water, enriched in abundant fine grained organic material. There is also abundant reworking of brown coal/lignite into coal balls, seen at the coarse base of thick sandstones in Belait Formation coastal outcrops in Brunei. The underlying pro-delta Setap Shale bathyal muds must have contributed to the source of oil in this very rich delta.

The oilfields are generally located at the intersections of growth faults and anticlinal folds. Regressive sands, mostly associated with Cycles V and VI, form the principal reservoirs. In Baronia, for example, the sand bodies range from 3.05 to 73.15 m thickness. Although the sand bodies are characteristically medium to fine grained, they are well sorted and therefore have permeabilities up to 1000 mD, and porosities up to 30%. Some parts of the Baram Delta are abnormally pressurized. Abnormal fluid pressures are due either to under-compaction or to inflation, whereby normally pressurized reservoirs become over-pressurized by injections (leakage) via joints and faults from underlying under-compacted sequences (Schaar, 1976). The under-compacted sequences are unfavourable for hydrocarbon accumulation, whereas the inflated sequences, directly overlying under-compacted sequences, contain abundant hydrocarbons.

3.9 EAST BARAM DELTA AND NW SABAH

The Baram Delta extends east as far as the Jerudong Line and its seaward continuation the Morris Fault (Fig. 3.6). The Morris Fault is shown in publications as a left-lateral wrench fault (Levell and Kasumajaya, 1985; Hazebroek and Tan, 1993), but such an interpretation obscures its real character and regional significance.

3.9.1 Structure

NW Sabah is divided into an Inboard and an Outboard Belt by the transition zone. The Inboard Belt is characterized by intense compressional features, interpreted to have been caused by wrench movements in the basement (Wee, 1993). The dominant feature is the NE–SW trending Hankin–Tulak High, continuing northwards as the Saracen High. This high area represents a tight crestally-collapsed anticline. By contrast, the Outboard Belt is characterized by extensional deformation, marked by NE–SW down-to-basin normal faults at the shelf slope breaks (Wee, 1993). Sabah is characterized by several significant unconformities. They result from episodic uplift of the Sabah Orogen (Crocker Formation fold-thrust belt) and tilting oceanwards towards the NW and W. Typically each unconformity passes from an erosion surface to an on-lap surface towards the NW. There is an overall successive oceanwards shift with time of the uplifted and deformed blocks (Fig. 3.6). However, this regional pattern of progressive regional tilting is disrupted by separate movement of individual structural provinces (Levell, 1987). The style of deformation varied. Erosion of the Upper Intermediate Unconformity and Horizon II was preceded by open flexure folding and tilting, whereas the Lower Intermediate Unconformity and the Shallow Regional Unconformity were associated with tight folds and reverse or strike-slip faulting (Levell, 1987).

3.9.1.1 Deep Regional Unconformity

This unconformity (DRU) marks a major event. It angularly separates the uplifted Crocker Formation turbidites from an overlying Middle Miocene and later sequence of NW prograding clastic wedges, most of which were deposited in shallow water. The unconformity is dated

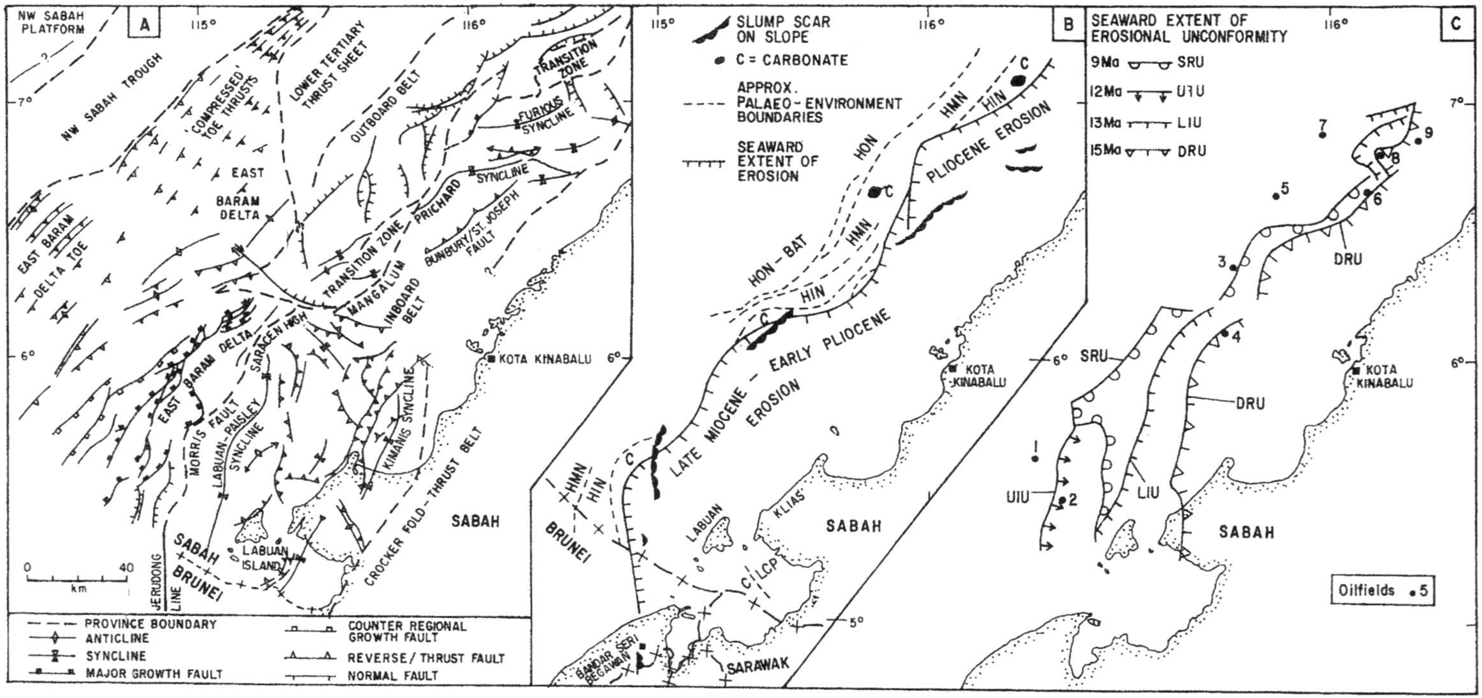

Fig 3.6. The East Baram Delta and NW Sabah. A: Main structural elements (after Rice-Oxley, 1991; and Tan and Lamy, 1990). B: Upper Miocene palaeo-environments for stage IV E/F (after Rice-Oxley, 1991), showing also the positions of identified slump scars along the Upper Miocene continental slope (shelf edge), from Levell and Kasumajaya (1991) and Hutchison (1994*a*). C: Seaward limit of the important regional unconformities, which successively migrated away from the uplifting Crocker Formation fold-thrust belt, based on Levell (1987). Palaeo-environments – LCP = lower coastal plain, C = coastal, HIN = holomarine inner neritic, HMN = holomarine middle neritic, HON = holomarine outer neritic, BAT = bathyal. Unconformities – DRU = Deep Regional Unconformity, LIU = Lower Intermediate, UIU = Upper Intermediate, SRU = Shallow Regional Unconformity. Oilfields: 1 = Samarang, 2 = Glayzer, 3 = Erb West, 4 = Ketam, 5 = Tembungo, 6 = St Joseph, 7 = SW Emerald, 8 = Furious, 9 = Barton.

early Middle Miocene (15 Ma), coincident with the end of spreading of the South China Sea Basin (anomaly 17). The unconformity is a planar surface without palaeo relief, so that uplift of the Crocker Formation was followed by a period of peneplanation.

3.9.1.2 *Lower Intermediate Unconformity*

This unconformity (LIU) resulted from a submarine deformation in which structures rarely reached above sea level and hence avoided erosion. The result was to form an onlap surface even over the crests of growing anticlines, seen, for example, at the Glayzer anticline and from Labuan to Ketam. In all these areas, the onlap surface is dated late Middle Miocene (13 Ma),

3.9.1.3 *Upper Intermediate Unconformity*

This unconformity (UIU) is present throughout the West Sabah Basin, but it has commonly been removed by the overlying more important Shallow Regional Unconformity. From its presence in the Glayzer and South Sabah Ridges, the unconformity has been dated at latest Middle Miocene (12 Ma). Slump scars occurred along the western edge of the unconformity area (Levell and Kasumajaya, 1985).

3.9.1.4 *Shallow Regional Unconformity*

Prior to the unconformity, the offshore Sabah region clearly was subdivided into independently moving basement blocks (Levell, 1987). The unconformity is overlain by marine deposits, of middle Upper Miocene age (9 Ma) representing a shallow marine erosion plane, on which no palaeo relief has been found.

3.9.1.5 *Horizon II*

Pliocene deformation occurs farther to the west and northwest in comparison with Miocene deformation (Bol and Van Hoorn, 1980). The earliest of the Pliocene deformations was close to the Lower/Upper Pliocene boundary. The deformation style was of gentle open anticlines and synclines with a general NW–SE orientation (Levell, 1987).

3.9.1.6 *Horizon I*

Major uplift, commonly of the order of 600 to 1000 m occurred over several structures in the Pleistocene, associated with tensional faulting over the crests of many structures, but the characteristic deformation is of gentle open folding.

3.9.1.7 *Tectonic controls on sedimentation*

In late Miocene time (Stages $IV_{D\ \&\ E}$ of Sabah = late Cycle V to early Cycle VI of the Baram Delta), Rice-Oxley (1991) showed that the Morris Fault and the boundary between the 'inboard' and 'outboard' belts represented a very steep continental slope, with land on the south-east side rapidly giving way north-westwards (and westwards in Brunei) into holomarine outer neritic to bathyal conditions. This time (Stage $IV_{D\&E}$) coincides with the 'shallow regional unconformity' (SRU) of Bol and Van Hoorn (1980) and Tan and Lamy (1990). It was a time of maximum uplift and seaward extension of the Sabah Orogen (Crocker Ranges), a time coincident with low eustatic sea level, which accentuated the steep continental slope off the Crocker Formation basement block into the deeper ocean. Levell and Kasumajaya (1985) have documented that in the Morris Fault area, the continental slope (normal fault) had a westwards drop (downthrow) of as much as 1.1 km from the land (Upper Miocene Unconformity). Voluminous erosion products off the Sabah Orogen were fluvially brought to the coastline and slid westwards down the unstable continental slope as slump scars into the outer neritic to bathyal Baram Delta. A growth index of about 2.0 characterizes the Lower Pliocene section westwards across the slope. A coalescing series of elongate spoon or scoop-shaped unconformities has been mapped along approximately 150 km of the 250 km long Upper Miocene shelf edge (Fig. 3.6).

The scars are interpreted as submarine slumps. Most retain a smooth slump scar morphology, but some may have been modified by erosional down-slope turbidity currents, for example to be a source for the Tembungo oil-bearing basin. One of these scars has been described as olistostrome in Setap Shale along the Jerudong Line onland south of Bandar Sri Begawan, Brunei (Hutchison, 1994*a*).

At Samarang, the slump scars were cut into a 500 m thick sequence of seismically-recognized fore-setted slope clays and overlying shallow water topset beds on the continental shelf (Levell and Kasumajaya, 1985). Earthquake activity or slope instability along the submarine fault scarp were responsible for the downslope slumping into the bathyal basin (Proto Baram Delta). There was either one downslope movement, or if there was more than one, they were closely spaced in time. This continental slope became a rather effective eastern margin to the Baram Delta and good reservoir sands slid down the slope into a deeper water environment. The dumping of good reservoir sands into deeper water, where source rocks are abundant, made for a perfect oilfield situation and explains why all the main oil-fields, such as

Samarang, lie west and north-west of the Jerudong Line–Morris Fault.

This configuration is clear from Brunei where east of the Jerudong Line, the coastal Belait Formation rests unconformably upon an older landsurface composed of 'Older Setap Shale' and Temburong Formation in the Temburong District of eastern Brunei. By contrast, west of the Jerudong Line, deeper water prevailed and there is a conformable transition from Setap Shale into deeper water (neritic) Miri Formation. Hazebroek and Tan (1993) have sought to prove about 30 km of left-lateral wrench motion across the Jerudong Line based on displacement of the palaeo-environment facies (coastal to inner neritic). This argument is invalid, because water depth was up to 1.1 km deeper west of the Jerudong Line compared to the east, and hence the palaeo-coastline west of the Jerudong Line naturally lay some 30 to 40 km farther inland, south of where it was in the Berakis Syncline to the east of the Jerudong Line. This is not a displacement by wrench faulting but results from the steep westwards sloping continental margin represented by the Jerudong Line. The argument for wrench motion in the East Baram Delta and NW Sabah has not been proven, and there is no onland evidence for it around the Jerudong Line (Hutchison, 1994*a*). The evidence seems to be entirely based on flower structures (local compression) rather than solid geological evidence.

The slumping of sands down the slope between the 'Inboard' and 'Outboard' zones has resulted in Upper Miocene turbidite deposition in the bathyal environment of the Outboard Zone in the St Joseph and Tembungo areas of NW Sabah. In the St Joseph area, the slump scars are cut into a slope and shallow water sediments along a steepening continental slope. Slumping occurred repeatedly during deposition of a 1.5 to 2.0 km thick sedimentary sequence over a long period of time. The proximity of the slump scars to the palaeo-coastline (there was virtually no continental shelf) resulted in extensive deposition of sandy turbidites offshore the continental slope.

Typically 1 to 5 km^3 of sediments were re-deposited down-slope in any one slump scar episode (Levell and Kasumajaya, 1985). Drilling into the scars shows that they are composed typically of deep water claystone. The erosional destruction of a slump scarred slope can be correlated with deposition of sand-bearing turbidite in the offshore bathyal basin.

The Sabah slump scars bear witness to active uplift of the Sabah landmass along a very steep tectonic fault (continental slope) and the proximity of this steep slope to the coastline. The dramatic nature of the continental slope and narrowness of the shelf were emphasized by Upper Miocene low eustatic sea levels.

3.9.2 Hydrocarbon occurrences

All commercial accumulations discovered to date are in the Middle to Upper Miocene clastic reservoirs of the Inboard Belt, Outboard Belt and Baram Delta provinces. The main hydrocarbon accumulations in the Inboard Belt are in Stages IVA and IVC lower coastal plain to fluvio-marine sands in complex reverse-faulted anticlines (e.g. Barton, S. Furious, Ketam). Hydrocarbon generation in the Inboard Belt is generally believed to have occurred prior to the Upper Miocene phase of deformation, particularly in the southern region (Tan and Lamy, 1990). Subsequent deformation resulted in complicated structure and secondary migration, resulting in small and complicated accumulations (e.g. Ketam and SE Collins).

The Erb West Field (Fig. 3.6) is a fairly simple NE–SW trending anticline, interpreted to have formed by wrench faulting of the basement (Johnson *et al.*, 1987). Oil and gas are contained in a 250 m thick Upper Miocene sequence of Stage IVD shallow marine shales and sandstones of variable reservoir quality. The sands were provenanced from the nearby emergent Erb High. Reservoir quality depends on the depositional environments. The best quality is found in massive sandstones, with porosity of 26% and permeability up to 2200 mD. Other lithologies are of generally poor quality; bioturbated sands have only 15% porosity and 20 mD permeability, heterolithic sands have 24% and 150 mD, and hummocky cross bedded parallel laminated sands have 25% porosity and 350 mD permeability (Johnson *et al.*, 1987).

In the Outboard Belt the main reservoirs are Stage IVD deeper marine sands in simple faulted anticlines (e.g. Tembungo), whereas the East Baram Delta hydrocarbons are found in Stages IVC, IVD and IVE shallow marine sands in relatively simple fault and dip-closed structures (e.g. Samarang). Here the hydrocarbon generation occurred before, during and after the main Upper Pliocene trap-forming deformation (Tan and Lamy, 1990).

All oils in this region belong to one family. Source rocks are typically lean and are dominated by terrestrial kerogens (Woodroof and Carr, 1992). In the Outboard Belt the Middle Miocene-derived oil and gas have migrated, often vertically, up faults and fractures to charge the Upper Miocene reservoirs. In the Inboard Belt, migration has been primarily along Middle Miocene reservoir conduits. Inboard traps are generally shallow and faulted and gas has been leaked to the sea-floor (Woodroof and Carr, 1992).

3.9.3 Tembungo Field

The Tembungo Field is the only oilfield in Malaysia to produce oil from turbidite sandstones. They are Upper

Miocene bathyal sandstones of porosity 20 to 25% and permeabilities in the order of hundreds of milliDarcies (Whittle and Short, 1978). The turbidites lie with a regional unconformity upon shallow inner littoral claystone and silts containing poorly developed sandstones. The discontinuous Upper Miocene proximal turbidite sands have been sourced by slump scars on the steep Upper Miocene continental slope (Fig. 3.6), as a lobate fan complex. Individual sands are up to 30 m thick and are generally massive. There are three types of sand bodies:

(a) widespread shaly sands at the base of the fan complex, formed as the turbidity currents spread their load over the relatively flat abyssal plain;
(b) local lenticular sand bodies deposited as channels within the upper part of the lobate fan complex; and
(c) thin discontinuous overbank crevasse splay sand deposits, formed when the currents broke out over the channel levees.

The turbidite sequence is folded into NNE-trending elliptical 4-way dipping anticlinal closures, which are strongly cross faulted by several normal NNW–SSE faults (Izman and Idrus, 1993). The morphology suggests that the trap structures are inversion related, although Whittle and Short (1978) invoke wrench tectonics. The uplift and structuring is of Recent age and all faults are sealing. Wrench faulting is interpreted from the occurrences of flower structures on the faults.

3.10 KETUNGAU, MELAWI AND MANDAI BASINS

Folding, metamorphism and uplift of the predominantly Cretaceous flysch belt (Rajang Group, *sensu lato)*, during the Sarawak Orogeny, was followed by mid Eocene rhyolitic subaerial volcanism. Between early Upper Eocene and mid Oligocene time, continuing episodic compression of this orogenic belt produced structural highs and a shifting pattern of complementary flanking basins and depocentres (Doutch, 1992). In places, the borders between the structural highs and the flanking basins became prominent normal faults, the best known example being the Lupar Fault, but others fringe the Semitau High. The uplifted orogenic flysch belt became a northern provenance for the Tertiary basins. In the south, the Schwaner Mountains provided a bulwark to compression, although its northern margin was upwarped as a consequence. The interpretation (Williams *et al.*, 1988; James, 1984) that the basins formed in a fore-arc setting is wrong. The Cretaceous part of the Rajang Group may have been an accretionary prism, but it was deformed, metamorphosed and accreted to the Sundaland landmass by the Sarawak Orogeny before the late Eocene beginnings of the unconformably superimposed basins. Pieters *et al.* (1987) and Van de Weerd and Armin (1992) concur that the basins were initially sedimented simultaneously in the Upper Eocene by a similar 'basal sandstone unit'. Doutch (1992) casts doubt on this conclusion and queried some of the correlations (Table 3.2). Upper Eocene larger foraminifera have been recovered from rare shallow marine interbeds in this predominantly alluvial sequence (Zeijlmans van Emmichoven, 1935). Reworked Cretaceous fossils collected from some outcrops resulted in wrong age assignments (Williams *et al.*, 1988) and caused considerable confusion.

3.10.1 Basal Sandstone Unit

Typically the Basal Sandstone Unit is of white quartz sandstone with minor conglomerate and mudstone, and rare thin coal seams. The sandstone is dominantly composed of milky quartz with minor plagioclase and alkali feldspar, chert, quartzite, acid volcanic and other rock fragments (Pieters *et al.*, 1987). Volcanic rocks commonly form the base, but at the Müller Mountains, 41 Ma old volcanics overlie the sandstone several hundred metres above the basal unconformity. The thickness of

Table 3.2. Correlation of formations and events between the basins of west-central Kalimantan and contiguous Sarawak, modified after Doutch (1992).

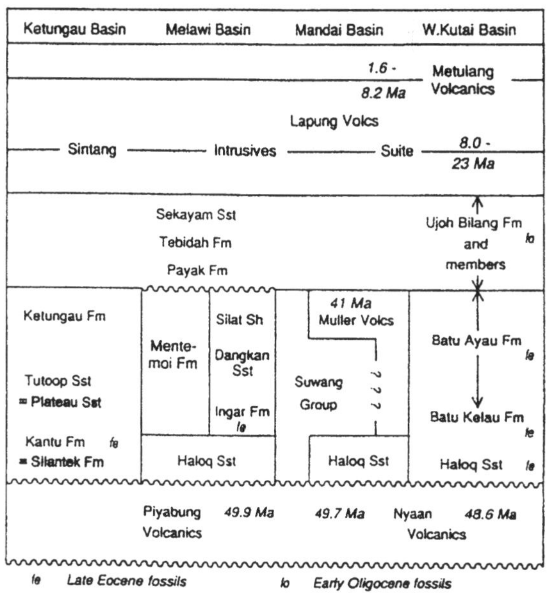

the Basal Unit varies considerably. The Kiam Haloq Formation along the Upper Mahakam River is only a few hundred metres thick, but increases westwards to about 500 m in the Barito River headwaters. The escarpment of the Behaba Sandstone rises to 800 m above the unconformity (Zeijlmans van Emmichoven, 1935). A minimum thickness of 1 km is assessed for the divide between the Keriau and Mandai and Melawi rivers (Pieters *et al.*, 1987).

3.10.1.1 Silantek Formation and Lupar Fault

In Sarawak, the Ketungau Basin strata are known as the Silantek Formation. Only the lower part, which outcrops adjacent to the Lupar Line, can be dated by fossil content. The Upper Eocene Basal Sandstone Member outcrops on the Marup Ridge and is steeply dipping (Tan, 1979). Its thickness is 1600 m. The strata flatten out southwards away from the Lupar Fault. It is obvious that the Basal Sandstone Member must overlie the Rajang Group with an angular unconformity at depth, but the only outcropping contact between the two is the Lupar Fault.

The average down-dip palaeocurrent direction is 228° from ripple marks and 194° from cross bedding (Tan, 1979) indicating that the provenance was from the northeast. This is consistent with conglomerate and sandstone petrology. They contain clasts of quartzite, schist, metagreywacke and mafic volcanics, indicating provenance from uplifted and eroding Lupar Formation and Layer Member of the Belaga Formation. The base of the lower member of the Silantek Formation is nearshore to shallow marine, but it rapidly becomes continental upwards and has redbeds in the upper part, followed by the thick continental cross bedded Plateau Sandstone. The Sibu Zone was therefore inverted by orogenesis to form a landmass before the Silantek Formation was deposited. The steep dips of the lower Silantek Formation adjacent to the Lupar Line imply that the Rajang Group orogenic belt continued to be uplifted, and the Ketungau Basin to subside after deposition of the basal Silantek Formation, resulting in the boundary becoming the Lupar Fault. The rapid elimination of marine conditions from the Silantek Formation indicates continuing inversion.

Accordingly it is wrong to refer to the Silantek Formation (Ketungau Basin) as fore-arc, since the Sibu Zone ocean had been eliminated and the Rajang Group transformed from an accretionary prism into an orogenic belt welding together the Schwaner Mountains Zone of Sundaland onto the Balingian–Luconia Province (Miri Zone).

The Lupar Fault is the margin of the Ketungau Basin, which developed by extensional tectonics affecting the Rajang Group orogenic belt. Studies to the north in the East Natuna area have shown that a similar situation persists in which Tertiary basins are unconformably superimposed on phyllitic basement (White and Wing, 1978), which can be equated to the Rajang Group. The basin margins likewise commonly form prominent faults, similar to the Lupar Fault, but the trend has been bent into a N–S alignment northwards from the coast of Sarawak.

3.10.1.2 Upper Silantek Redbed Member

The Silantek Formation continues upwards as red micaceous shale and mudstone interbedded with red siltstone and red sandstone, collectively named the Upper Kantu Beds by Haile (1957), and the Upper Silantek Redbed Member by Tan (1979). Their truly continental nature has precluded dating, and they are conformably overlain by the Plateau Sandstone, which also cannot be dated.

3.10.2 Plateau Sandstone

The Plateau Sandstone forms extensive and continuous mesa outcrops along the Klingkang Range which borders Sarawak and Kalimantan. It is a sequence of thick massive and cross bedded polymict and quartz sandstone with thin shale intercalations (Tan, 1979).

Zeijmans van Emmichoven (1935) has shown that the Plateau Sandstone gently dips southwards to form the lower unit of the Ketungau Basin, where its thickness has been estimated around 1800 m. It has subsequently been named the Tutoop Sandstone (Table 3.2), and it is conformably overlain by the Ketungau Formation, which outcrops along the synclinal axis of the Ketungau Basin. It is of interbedded sandstone, siltstone, carbonaceous mudstone and coal seams. The fossils indicate a shallow marine to brackish environment with freshwater influence, but they are not age indicative.

The Plateau Sandstone has also been mapped within the Melawi Basin (Zeijmans van Emmichoven, 1935), now called the Dangkan Sandstone, succeeded by thick black Silat Shale, thought to be a lake deposit, equivalent to the Ketungau Formation (Pieters *et al.*, 1987).

In Sarawak, Tan (1979; 1982) thought that the Plateau Sandstone may be Miocene, but the evidence from Kalimantan (Table 3.2) favours an Upper Eocene to lowermost Oligocene age.

3.10.3 Oligocene Formations

A major angular unconformity has been described in the Melawi Basin between the Eocene Dangkan Sandstone (including the overlying Silat Shale) and the overlying Melawi Group (Table 3.2). The Group comprises fluviatile sandstone of the Payak Formation and sandstone and mudstone of the Tebidah Formation. The Melawi Group unconformably overlaps the basement across the lower Kapuas River (Pieters *et al.*, 1987).

There is no direct evidence for the age of the Melawi Group, but Pieters *et al.* (1987) have tentatively correlated it with fine grained marine clastics and carbonates to the east in the Upper Kutei Basin.

3.10.4 Oil and gas potential

The presence of lacustrine shales in these predominantly continental basins has attracted reconnaissance surveys by international oil companies. Currently a large area is under concession to Canadian Oxy, having previously been leased to a French company. Rare oil seeps have been reported, but Pieters *et al.* (1987) conclude that the economic potential for coal mining is greater than that for oil and gas.

3.10.5 Upper Kutei Basin–Mahakam River

The ages of the Ketungau, Melawi and Mandai Basin stratigraphy have been based on tentative correlations with the more marine strata of the Upper Kutei Basin, where the geology has been described by Wain and Berod (1989).

Eocene strata rest unconformably on a landmass of Cretaceous turbidites and tectonic mélanges. Up to 1500 m of conglomerates and coarse pebbly sands at Kiham Haloq form the base of the sequence. These coarse clastics are of limited areal extent and are interpreted as braided stream deposits, possibly in alluvial fans, suggesting the existence of high-relief synsedimentary fault scarps. The Kiham Haloq Formation is overlain by deltaic and shallow marine sediments of the Batu Kelau and Batu Ayau formations, estimated to be up to 1500 m thick. Eocene pyroclastics, of minimum thickness 300 m, are present in the Kelian River (Van Leeuwen *et al.*, 1990). The Batu Ayau is overlain by up to 700 m of Lower Oligocene platform carbonates of the Batu Belaq Formation (Table 3.2). In more deeply subsiding parts of the Kutei Basin, the Batu Ayau grades upwards and laterally into latest Middle Eocene to Oligocene deep marine shales and turbidites of the Ujoh Bilang Formation, including volcaniclastic turbidites of the Len Muring Sandstone Member.

3.10.6 Sintang high level intrusive suite

Mid Oligocene to Miocene stocks, sills and dykes occur widely in the region stretching from the Bau district of Sarawak, across the basins and intervening highs, to the Upper Kutei basin. Twelve K:Ar dates from near Sintang range from 30 to 23 Ma in the south, and 18 to 16 Ma in the north (Williams and Harahap, 1987).

They intrude the Bau Limestone, the Silantek Formation, the Plateau Sandstone, and seem to prefer the basinal areas of the Ketungau, Melawi and Mandai, but also intrude the intervening basement highs. They range from pyroxene andesite and dolerite to microgranite. Chemically they are similar to I-type granitoids and granodiorite compositions predominate (Williams and Harahap, 1987). The chemical characteristics favour a mantle or deep crustal origin. The intrusions are classified as post-subduction. They are post-orogenic, for the outcrops are undeformed, but their tectonic significance remains enigmatic. Locally they have domed up the country rocks and have caused contact metamorphism, for example of Silantek Formation coals to anthracite. They have also been held responsible for mobilizing gold, for example in the Bau region of Sarawak (see Chapter 10). The basins of this region are said to be characterized by a high geothermal gradient, manifested by the Sintang Intrusive Suite.

3.10.7 Metulang volcanics

Miocene and Pliocene volcanics were widely extruded in central Borneo, subaerially upon a peneplained surface formed on the Embaluh Group and Rajang Group flyschbelt. The large volcanic plateaux form a major feature of Borneo and occur commonly in remote poorly studied areas (e.g. Kirk, 1957, 1968). Their petrogenesis is virtually unknown, so that their tectonic significance remains enigmatic. K:Ar dates range from 1.6 to 8.2 Ma (Tate, 1991).

3.11 SUMMARY

The region described in this chapter is characterized by a complicated tectonic history in which two deep marine flysch basins have been uplifted to form a land surface. The geology of the largely inaccessible hinterland of Borneo is still incompletely known, but the two major orogenic events are now well established. The oil basins owe their existence to these orogenic events: the Upper Eocene Sarawak Orogeny and the Middle to Upper Miocene Sabah Orogeny.

In the Sarawak Orogeny, the Rajang Group flysch was compressed between the Luconia–Balingian–Miri microcontinent and the Schwaner Mountains volcano-plutonic block and uplifted to form a newly added part to the Upper Eocene landmass of Sundaland, upon which the continental Ketungau, Melawi and Mandai basins were formed. These elongate depressions were fed by fluvial systems from the north-western interior of Sundaland and they found the sea at the western part of the deltaic Kutei Basin. The Lupar fault, which is parallel and adjacent to the Lupar Line ophiolitic mélange, represents the NE bounding fault margin of the Ketungau Basin.

The Middle to Upper Miocene Sabah Orogeny resulted in deformation and spectacular uplift of the Crocker Ranges of Sabah. Its cause is obscure, but it resulted in a new mountain belt, formed of sand-rich Oligocene to Lower Miocene West Crocker Formation turbidites, erosion of which resulted in the major Middle to Upper Miocene Baram Delta. Bathyal seas had pre-existed east of the West Baram Line continental slope of the Balingian Province and offshore Brunei and Sabah, but they were starved of sand until the Crocker Ranges were uplifted.

Coals formed an important source of hydrocarbons in the Ketungau Basin, where they are of Upper Eocene to Lower Oligocene age, and in the Balingian Province of Sarawak, where they are Upper Oligocene to Lower Miocene. The Setap Shale pro-delta bathyal muds of the Baram Delta must be an important source of hydrocarbons. However the Middle to Upper Miocene delta sequences are all rich in plant amber and allochthonous coal balls, so that coaly material continued to be an important source. The reservoir sands have been cannibalized from the well sorted West Crocker Formation turbidites. Baram Delta sands therefore have good reservoir quality. The trapping mechanisms are those of a classic delta.

The offshore Luconia province, which was responsible for the Upper Eocene compression of the Rajang Group flysch-belt, is characterized by numerous carbonate buildups upon a stable horst and graben basement. These carbonates have reservoired impressive gas deposits, piped ashore and liquified at Bintulu. The source-rocks may have been the carbonates themselves.

4

Cainozoic basins: western and southern Indonesia

4.1 NORTH SUMATRA BASIN

The North Sumatra Basin occupies a roughly triangular area of approximately 55 000 km^2, mostly onshore, and extending into the Straits of Malacca and Andaman Sea. It is separated from the Central Sumatra Basin by the Asahan Horst (Fig. 2.1).

4.1.1 Sedimentation history

Tertiary sedimentation began in isolated N–S elongated sub-basins (grabens). Most authors show the earliest graben fill formed of Tampur dolomite and limestone, which has been encountered in wells and also outcrops in hills along the SW margin of the basin. The palaeontology is ambiguous and ages as diverse as Lower Carboniferous to Oligocene have been published (see Caughey and Wahyudi, 1993). A thick sequence of Lower Oligocene Bruksah conglomerate (basal Bampo Formation) unconformably overlies the Tampur and older rocks. Therefore it is best to lump the Tampur in with the basement. The basement of the Cunda area is of Triassic phyllite. The basal sediments are locally derived sandstone and conglomerate of the Oligocene Bruksa Formation (previously known as the Parapat Formation). It is thick (up to 30 m) in the basins, and thin or absent over palaeo-topographic highs. The basal clastic sequence is succeeded by shales of the Oligocene to Lower Miocene Bampo Formation (Fig. 4.1). This formation is up to 1500 m thick on the flanks of the Jawa Deep (Fig. 4.1), decreasing to 150 m over the N–S oriented Arun basement high, and is completely absent over the crest of the Cunda High (Burnaman *et al.*, 1985). Carbonates occur on the flanks and crests of the basement highs.

Pre-Tertiary basement is overlain on the 'B' Block by predominantly fluvial Oligocene Parapat Formation sandstone, but elsewhere sedimentation may have begun earlier in the deeper grabens (Maliki and Soenarwi, 1991).

By early Miocene time, the sub-basins had coalesced and subsidence was less active, resulting in widespread carbonates on the shelf areas and Lower Miocene highs. These carbonates are widespread on the Malacca Shelf, but confined elsewhere to the horsts (topographic highs) such as the N–S Ibu Horst (Anderson *et al.*, 1993). The Peutu Formation is of marine calcareous shale containing reefs over depositional highs (the Arun Member). The thickness of the Peutu Formation in the 'B' Block ranges from 30 m over the highs, to more than 630 m in the basinal deeps. Drilled sections of the Arun Limestone reef facies range from less than 6 m to over 300 m (Burnaman *et al.*, 1985). Heavy mineral analyses of the Lower Miocene Peutu Formation sandstones indicate provenance from a predominantly granitic terrain lying to the E and SE, probably the Main Range of Peninsular Malaysia (Morton *et al.*, 1994).

A major regional tectonic event (= Sabah Orogeny) resulted in a distinct basin-wide unconformity at the end of the Middle Miocene (Anderson *et al.*, 1993). A phase of non-deposition and erosion caused the widespread absence of the Peutu Formation and reef facies. The Peutu Formation was buried by thick Middle Miocene shale (the Baong shales), representing the maximum marine transgression, and a period of maximum subsidence. Upper Miocene and Pliocene regressive sands of the overlying Keutapang, Seurula and Julu Rayeu formations were provenanced from the uplifting Barisan Mountains (Morton *et al.*, 1994).

4.1.2 Structure

The basement architecture is dominated by a N–S striking sequence of horsts and grabens of Eocene age. From west (97° E) to east (99° E), the major horsts are the Peusangan–Arun High (Fig. 4.1), the Idi Shelf–Alur Siwah High–Ikan Horst, the Ibu Horst, the Yang Besar Horst, the Pakol Horst, and finally the Glagah–Asahan Horst (Fig. 2.1), which separates the North Sumatra from the Central Sumatra Basin (Anderson *et al.*, 1993). Deeps in excess of 5 s two-way seismic time occur only on either side of the Peusangan–Arun High—the Jawa Deep on the west and the Lhok Sukon Deep on the east. East of 97.5° E the deeps lie at 3.0 to 5.0 s two-way time. Faults oriented NW–SE are restricted to near the Malaysia-Indonesia border in the Straits of Malacca. N–S striking right-lateral transcurrent fault motion on the N–S striking Eocene normal faults is considered to be fundamentally important in the development of the North Sumatra Basin (Situmorang and Yulihanto, 1985; Davies, 1984). The

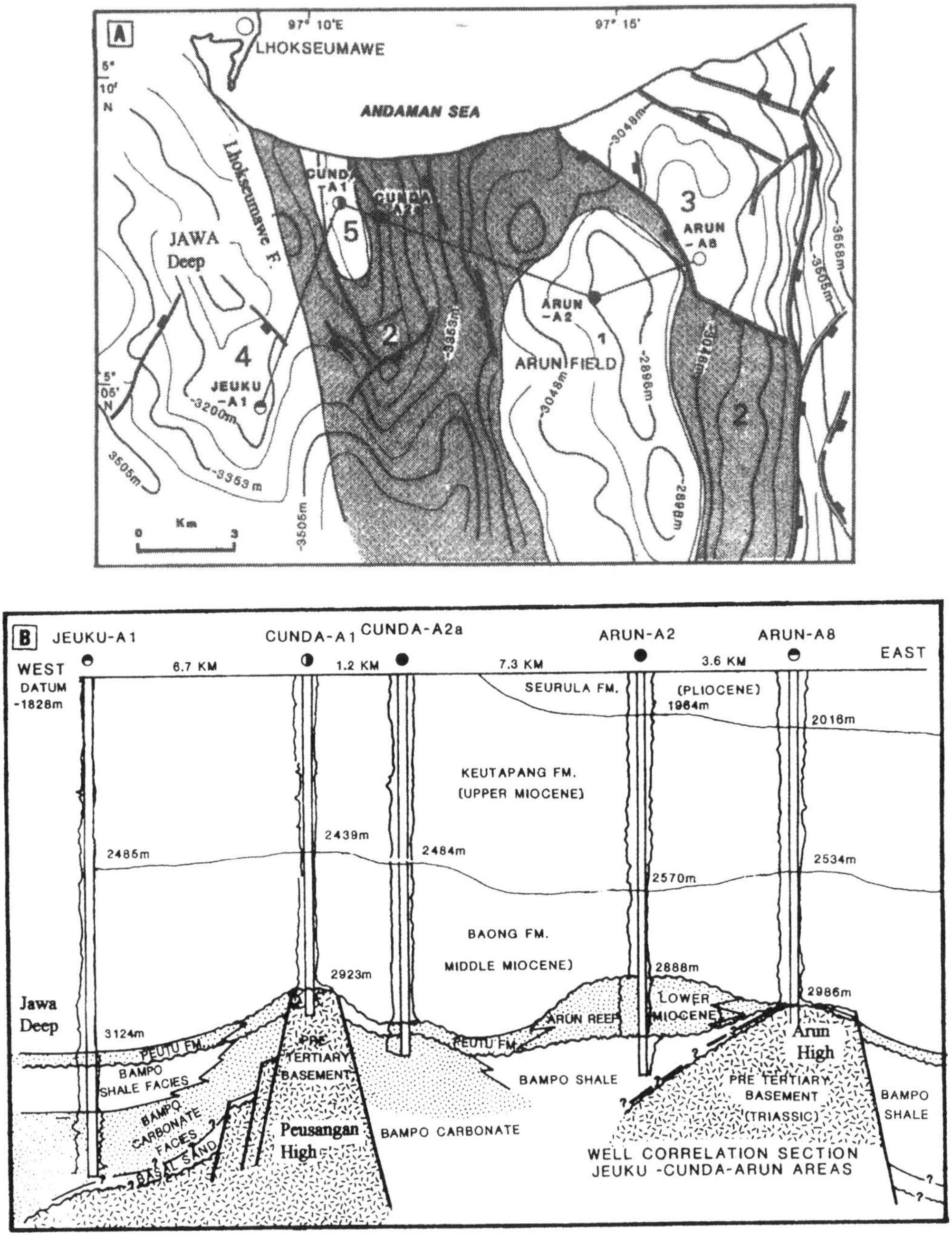

Fig. 4.1. The Arun–Cunda field of the North Sumatra Basin (after Burnaman *et al.*, 1985). **A**: Depth below sea level to the base of the Middle Miocene Baong Formation. Areas determined by seismic stratigraphy are: 1 = Thick carbonate (> 105m) on Upper Oligocene Bampo shale. 2 = Thinner carbonate (30 to 105m) on Bampo shale. 3 = Thin carbonate (< 15m) on pre-Tertiary basement. 4 = Multiple carbonates interbedded with Lower Miocene Peutu or Upper Oligocene Bampo shales. 5 = Baong Shale on Basement. **B**: E–W cross section of the field based on well control. Depths in metres below sea level.

same may be said of the Central and South Sumatra basins (Eubank and Makki, 1981; De Coster, 1974) where the interaction between the N–S striking faults and the NW–SE striking faults may also be of major importance in the localization of reservoir structures.

The strike of the faults in the acoustic basement (pre-Tertiary) is overwhelmingly N–S. The major N–S faults which have been named are the Arun, 98, Tanjung Pura and Tanjung Morawa fault zones (Situmorang and Yulihanto, 1985); the latter occurring about 30 km east of Medan (Fig. 4.2).

The fold axes in the Keutapang Formation and in the surface outcrops trend predominantly NW–SE. The faults also strike NW–SE. The evolution of the basins is elegantly described by Situmorang and Yulihanto (1985). Right lateral shear along the 98, Tanjung Pura and Tanjung Morawa faults caused pull-apart basins by imparting a NW–SE extensional stress pattern (Fig. 4.2). The Tertiary sediments, which filled these basins, were folded along NW–SE trending axes as a result of NE–SW compression resulting from the on-going right-lateral wrench faulting on the N–S basement faults (Fig. 4.2).

The Arun–Cunda basement platform high is thought to be bounded by N–S striking wrench fault systems (Fig. 4.1). It is flanked on the west by the Jawa and on the east by the Lho Sukon Deep.

The regional analysis by Davies (1984) suggests that the right-lateral Khlong Marui wrench fault zone of Thailand (Fig. 1.2) extends into Sumatra as the Lhokseumawe Fault (Fig. 4.1). This right-lateral faulting can accomodate the anti-clockwise rotation of the Sunda Plate, which has been documented by palaeomagnetic studies mainly on western Sarawak, but also with some support from the Malay Peninsula. Upper Oligocene horst and graben structures formed within the reactivated right-lateral wrench fault zones.

Lower Miocene uplift took place along older reactivated structures, resulting in widespread erosion of the Cunda–Arun high. Considerable amounts of Lower Miocene Peutu Formation and Oligocene formations were removed at this time. Uplift in Lower to Middle Miocene resulted in further erosion, and the Arun and other reefs were exposed at this time enabling subaerial processes to enhance porosity.

4.1.3 Source rocks

The primary source rocks are postulated to be lacustrine and marine shales of the Oligocene Bampo Formation (Anderson *et al.*, 1993). Geochemical modelling in the basin deeps suggests the onset of oil generation was as early as 22 Ma and the onset of gas as early as 19 Ma. Gas contents of the Peutu limestones is variable and may contain variable amounts of CO_2, posing an important obstacle to development in this basin. Contents vary from 7% to as high as 82%. Arun has only 15% CO_2. The substrata of Tampur Carbonate rather than shale suggests that the source of the CO_2 results from thermal metamorphism of the limestone (Caughey and Wahyudi, 1993), resulting from the relatively high geothermal gradient of 47° C km^{-1}.

4.2 MERGUI BASIN

This basin, referred to as South Andaman in Hutchison (1989*a*), is actually a northwards extension of the North Sumatra Basin, and links it to the Andaman Sea marginal basin. Accordingly it is a region dominated by powerful right-lateral wrench faulting (see Hutchison, 1989*a*, pp.17–18). It is inferred that the Mergui Basin is underlain by similar Palaeozoic and Mesozoic rocks that form north Sumatra and the Malay Peninsula (Hutchison, 1993), that have been down-faulted. The basin is bounded on the west by the Mergui Ridge and on the east by the Ranong Ridge.

There may be Eocene strata, but they have not been proven. The basin is thought to have started in the late Oligocene as a series of N–S trending half grabens. The stratigraphy of the Mergui Basin (Polachan and Racey, 1993; CCOP, 1991) can be divided into 4 units. TM-40 Oligocene sediments were transported southwards from Peninsular Thailand and deposited as fluvial-deltaic sequences in the north and east, known as the Ranong Formation, and in deeper water as marine shales and turbidites in the south, known as the Yala Formation (Fig. 4.3). During the Lower Miocene TM-30, local sub-basins coalesced to form a coherent Mergui Basin. The fluvial to shallow marine Payang Formation in the north gave way southwards to deep marine shales and occasional turbidites, known as the Katang Formation, deposited in most parts of the basin. Reefs, known as the Tai Formation, developed on structural basement highs mainly in the east on the Central High area. The basin continued to subside and deep marine shales were deposited from Mid Miocene to the present. The Middle Miocene is represented by the Trang Formation, with a facies change northwards to the shallower water Surin Formation. The Upper Miocene to Recent is represented by the Takua Pa Formation (Fig. 4.3).

4.2.1 Tai Formation

The Tai Limestone, equivalent to the Peutu of the North Sumatra Basin, was an obvious target for oil and gas exploration. The type section is in Tai well, between

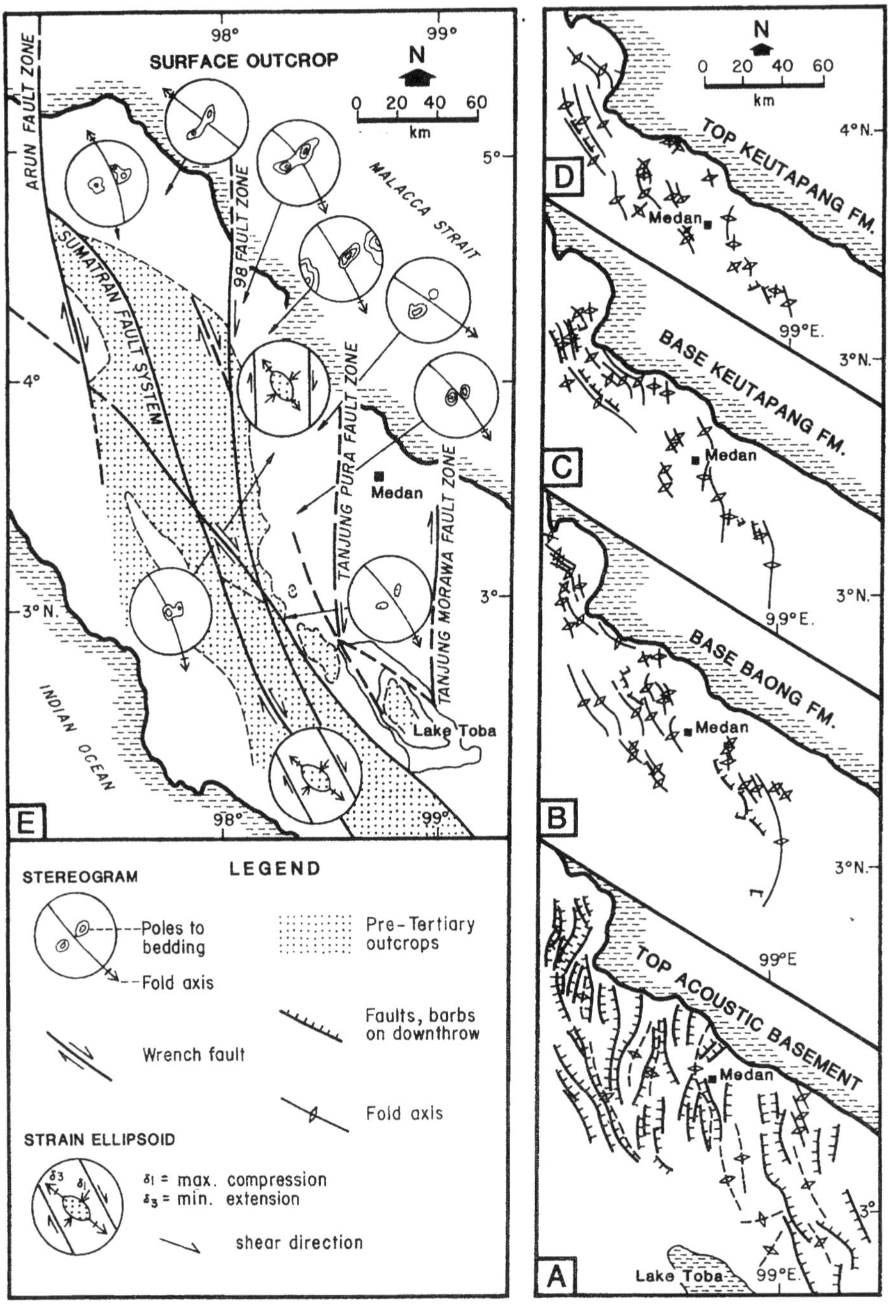

Fig. 4.2. **A**: The N–S striking right-lateral basement fault system of the North Sumatra wrench basin (after Situmorang and Yulihanto, 1985). **B,C,D**: By contrast, the majority of the faults and fold axes in the Keutapang and Baong formations trend NW–SE. **E**: surface outcrop fold axes also trend NW–SE. The basin origin and Tertiary sedimentary fill deformation is therefore appropriately related to right-lateral wrench motion on the basement faults, as shown by the strain ellipsoid (for mechanism, see Fig. 4.8).

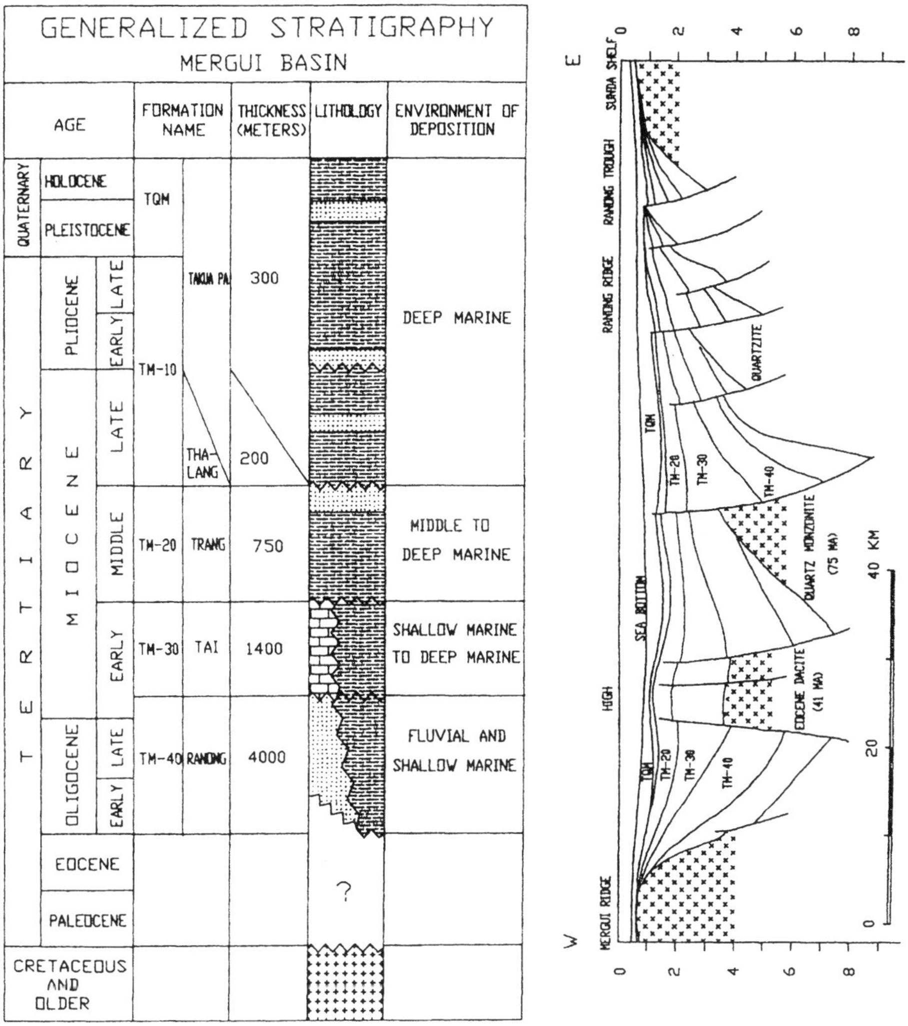

Fig. 4.3. Stratigraphy and E W cross section across the Mergui Basin, after CCOP (1991).

depths of 1128 and 1737 m. The Lower Miocene Tai Formation unconformably overlies a basement of quartz-chlorite schist on the Central High and has been penetrated only in the three wells shown in Fig. 4.4 (Polachan and Racey, 1993). The formation can be subdivided into three units: a basal unit of interbedded anhydrite, dolomite, shale and sandstone; a middle unit of coral-algal reefal limestone; and an upper unit of calcarenite interbedded with silty shale and sandstone (Polachan and Racey, 1993). It is everywhere unconformably overlain by shales of the Middle Miocene Trang Formation.

4.2.2 Hydrocarbon potential

Twelve wells have been drilled, but results have been disappointing. Marine shales of Oligocene and Lower Miocene are proven source rocks. Compared with the basins of Sundaland and Sumatra, this basin has a lower geothermal gradient of 33° C km^{-1}. Middle Miocene and younger shales are the units richest in oil-prone kerogen, but they seem to be immature throughout the basin. The potential traps are Oligocene deltaic sandstones, mid-fan turbidites and Lower Miocene reefs, which have good reservoir characteristics.

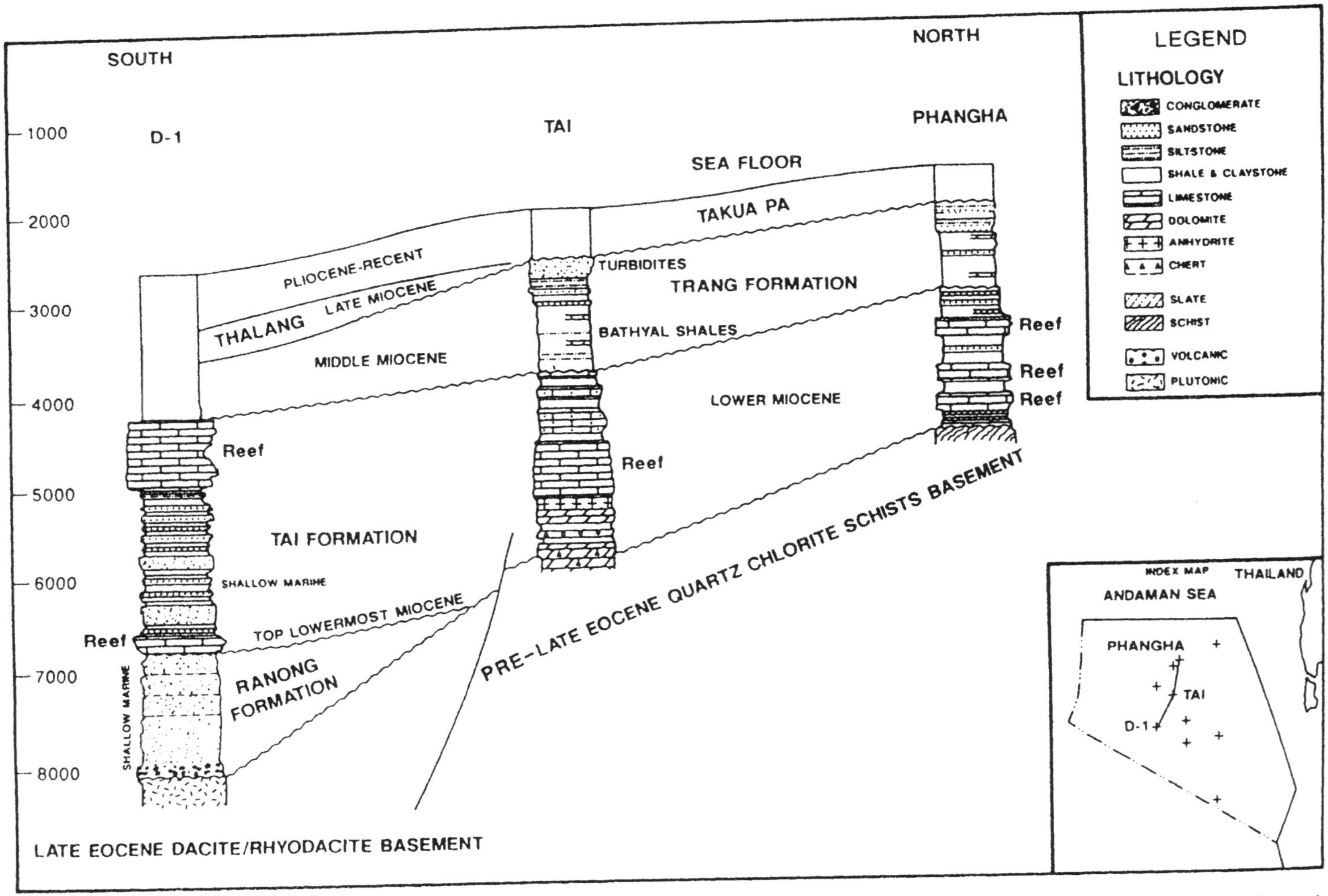

Fig. 4.4. N–S cross section across the Mergui Basin, after Polachan and Racey (1993). These are the only three wells that have penetrated the Tai Formation. The vertical scale is in feet.

It was hoped that the Tai Limestone would have potential similar to the Arun Field of North Sumatra. However in the Arun Field, the carbonate build-ups are on the flanks of the basement highs. In Mergui, they are directly on top of the highs, farther from potential source rocks. There are smaller carbonates to be seen on seismic sections in the deeper parts of the basin on the flanks of major structures, nearer to the source rocks. They may in future offer better potential for hydrocarbon exploration.

4.3 PRE-TERTIARY BASEMENT OF SUMATRA

Before describing the Tertiary basin of Central Sumatra, it is useful to consider the basement to the basinal area and its correlation with nearby Peninsular Malaysia, based on Hutchison (1993).

Peninsular Malaysia, east of a N–S line through Bentong and Raub, has definite Upper Permian Cathaysian affinities and is referred to as East Malaya (Metcalfe, 1990*a*) and Eastmal (Hutchison, 1989*a*). Metcalfe (1991) continues to show an extension of East Malaya into Sumatra. A West Sumatra terrain, formerly referred to in part as the 'Djambi (Jambi) Nappe', lies between Sinoburmalaya and the Semangko Fault of Sumatra. It has distinct Cathaysian characteristics, but of unique Lower Permian age. Tobler (1922), Zwierzycki (1930) and Van Bemmelen (1970) proposed that the 'Djambi Nappe' was allochthonously thrust SW over what is now called Sinoburmalaya and was rooted in the autochthonous Eastmal.

The continental fragments of Eastmal and West Sumatra were separated from Sinoburmalaya by the Palaeotethys Ocean, which began narrowing when the Gondwana blocks of South-East Asia started drifting northwards (Ridd, 1980; Sengör, 1984). Collision and suturing of the markedly different blocks resulted in the Lower Mesozoic Indosinian Orogeny (Fromaget, 1927; Belov *et al.*, 1986). The timing of suturing continues to be a matter for debate. Harbury *et al.* (1990, 1991) suggested an Upper Palaeozoic closure, but the traditional Upper Triassic age of the Indosinian Orogeny has been maintained by Hutchison and Sivam (1992) and Görür and Sengör (1992) in response to their unacceptable arguments.

The Palaeo-Tethys Suture is well defined in Peninsular Malaysia as the Bentong–Raub Line (Hutchison, 1975), but it has been extrapolated through Sumatra only in very generalized terms (Tjia, 1989; Metcalfe, 1990*a,b*; 1991).

Through northern Thailand, it has been mapped as the Uttaradit–Luang Prabang–Dien Bien Phu suture (Hutchison, 1989*a*). A Lower Permian (Kungurian) K:Ar age of 269 Ma has been determined from actinolite in mafic schists of the Pha Som Group, which consists mainly of fine grained chlorite of greenschist facies and blueschists (Barr and MacDonald, 1987). In a later paper, Barr and MacDonald (1991) proposed that the central Sukhothai terrain amalgamated with the Indosinian terrain during the Permian. These amalgamated terrains then collided with the Shan-Thai terrain during the Triassic. The boundary between the Shan-Thai and Sukhothai–Indosinian terrains is interpreted by Barr and MacDonald (1991) to be a cryptic suture, which they correlate with the Bentong–Raub suture of Peninsular Malaysia.

No continuous single suture zone can be delineated to separate Sinoburmalaya from both Eastmal and the West Sumatra terrain. The Cathaysian terrain of West Sumatra appears to be in contact with Gondwanaland Sinoburmalaya along a zone parallel to the active Semangko right-lateral wrench fault: the Cainozoic motions by which the disparate terrains have been juxtaposed remain to be resolved.

4.3.1 The Bentong–Raub Suture location

The Bentong–Raub Suture separates the western Gondwana affinity terrain of Peninsular Malaysia, which Gatinsky and Hutchison (1986) named *Sinoburmalaya*, from the eastern Cathaysian affinity terrain of Eastmal. Metcalfe (1986; 1990*a*) used the term *Sibumasu* for the Gondwana affinity terrain, which specifically included Sumatra. Hutchison (1993) has tried to make the two terrain definitions synonymous. As defined by Hutchison (1975), the Bentong–Raub Suture was shown trending SE from the vicinity of Gunung Ledang (Mount Ophir), to transect the gabbroic rocks of south Johore and Singapore, extending SE through the Riau Archipelago (Fig. 4.5). This interpretation was largely based on the distribution of granite types (Hutchison, 1977), and the misinterpretation of the gabbroic rocks as ophiolitic. The Main Range biotite granite west of the suture is distinctively megacrystic and of ilmenite series, whereas the granitoids of East Malaya are more equigranular, contain both hornblende and biotite, and many are of the magnetite series (Hutchison, 1989*c*; Ishihara *et al.*, 1979). Hutchison (1983*b*, 1989*a,b,c*) later concluded that the gabbroic rocks belong to the extended calc-alkaline series of the eastern volcano-plutonic belt, and that the Bentong–Raub Suture extends into the Straits between Malacca and Muar (Fig. 4.5), where serpentinites and amphibolites have been reported near the coast about 11 km SE of Malacca (Khoo, 1978; Hutchison, 1973).

The proposed suture extrapolation through the Riau Archipelago towards Bangka and Billiton has never been located in the field. Rocks which can be ascribed with

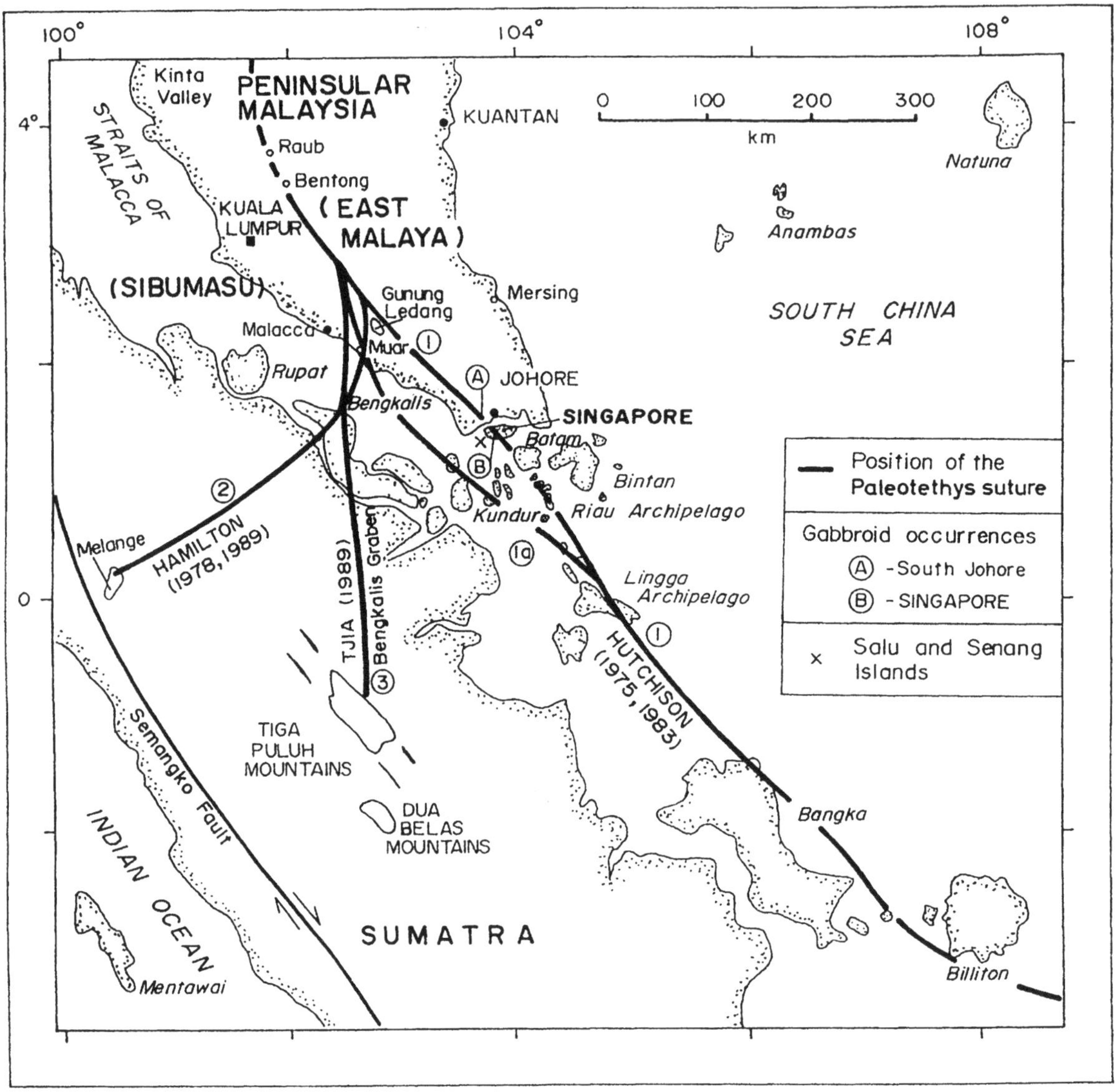

Fig. 4.5. Positions of the Bentong–Raub Suture as proposed by various authors. It was also named the Medial Malaya Suture Zone by Sengör (1984), who followed the interpretation of Hutchison (1983*b*). None of these can separate all Gondwanaland from Cathaysian affinity blocks (see Hutchison, 1993).

certainty to a suture assemblage do not outcrop. However, the islands to the SW of Singapore (Salu and Senang) contain chert and spilite (Singapore, 1976) (Fig. 4.5). A case can be made for extending the suture SE between Kundur, where the granitoids are ilmenite series, and Batam, where they are magnetite series (Wikarno *et al.*, 1988). Further extrapolation to separate Bangka from Billiton is not well founded, for the granitoids of both islands are predominantly of the ilmenite series. Furthermore Cobbing *et al.* (1986, 1992) were unable to discover any petrological distinction between granitoid types throughout these 'tin islands'.

Hamilton (1979) showed the suture swinging W from Malacca (Fig. 4.5), through a rock assemblage near 0°10′ S and 100°10′ E, interpreted as mélange. He has not revised its position (Hamilton, 1989), despite much new significant geological data. His extrapolation has to be rejected because it cuts across the stratigraphic distributions and the structural geology.

Undoubtedly impressed by their geographical similarity of N–S alignment at 102.3° E (Fig. 4.5), Tjia (1985, 1989) linked both the Palaeo-Tethys Bentong–Raub Suture of Peninsular Malaysia and the Palaeogene Bengkalis graben of Sumatra and coined the term Bentong–Bengkalis Suture. The interpretation that the suture continues south along the Bengkalis Graben is an interesting but probably wrong hypothesis (Hutchison, 1993), followed by Metcalfe (1990*a*,*b*; 1991).

The Bentong–Raub Suture defines the western margin of the Central Basin of Peninsular Malaysia, dominated by deformed predominantly Carnian–Norian strata (Hutchison, 1975, 1989*a*,*b*). This Triassic belt continues through the Riau and Lingga archipelagos to Bangka and Billiton (Fig. 4.6). The suture therefore should logically be expected to continue parallel to the 'tin islands' , and to lie somewhere in the neighbourhood of Bangka and Billiton. Unfortunately south and south-east of Malacca it is buried beneath Cainozoic formations and sea, and geophysical evidence has not yet been produced to define its likely location.

The Indosinian Orogeny, which resulted in the Palaeo-Tethys Suture, gave rise to an impressive array of granites in the Peninsular region of South-East Asia. Several major publications have detailed the petrology, geochemistry and radiometric dating of the granite belts since the publication of Hutchison (1989*a*). However, the overall picture remains intact. The major work is that of Cobbing *et al.* (1992). The new addition to our knowledge is the inclusion of Rb:Sr dating of granitoids from the Tavoy–Mergui–Davis Island Andaman Sea coast of Myanmar, named the Central Granitoid Belt by Khin Zaw (1990). Granitoids of this belt consistently have ages within the range 35 to 94 Ma. A considerable amount of new petrological, Rb:Sr, K:Ar and fission track dating of the Main Range of Peninsular Malaysia has been given by Krähenbuhl (1991). The fission track and K:Ar dating indicate a slow post-orogenic cooling within the range 1–5° C Ma^{-1}.

4.3.2 Structural similarity between Peninsular Malaysia and Sumatra

Peninsular Malaysia and Sumatra share many structural similarities, not yet fully documented. The shallow Straits of Malacca should not be held responsible for concealing a major discontinuity since none has been reported from petroleum company seismic programmes (Liew, 1994).

Both Sumatra and Peninsular Malaysia are dominated by major structures which strike SE and S-to-SSE (Fig. 4.6). The dominant strike and fold axis trends of Peninsular Malaysia, especially of the mainly Carnian-Norian Semantan and Gemas Formations of the Central Basin, are S to SSE, controlling the shape of the peninsula (Gobbett and Tjia, 1973; Tjia, 1989). A possible NE-trending hinge line may be imagined approximately through Johore Baharu (Fig. 4.6) [through Padang and Linggiu (Fig. 4.7)]. This is shown by the distribution of the Carboniferous–Permian formations (Fig. 4.7) and of the Triassic Central Basin (Fig. 4.6). SE of the hinge line the fold axes swing SE through the Jurong Formation of south Johore and Singapore (Singapore, 1976; Burton, 1973). The SE structural trend dominates the whole Riau and Lingga archipelagos (Tjia, 1964), and continues to Bangka (Katili, 1967), whence it swings eastwards through Billiton (Fig. 4.7). The dominant direction of strike and fold axes in Sumatra is NW–SE, controlling the shape of the island (Pulunggono and Cameron, 1984). The NW–SE Cainozoic faults of Sumatra, mostly known from oil company investigations (e.g. Moulds, 1989), are controlled by the anisotropy of the underlying basement (regional lithology and fold axes trend).

The NW–SE structural grain direction of Sumatra is paralleled in Peninsular Malaysia by a set of prominent extensional faults, commonly filled by major multi-phase quartz dykes such as the Klang Gates, which may have had subsequent wrench motion (Gobbett and Tjia, 1973).

The N–S structural grain direction of Peninsular Malaysia (strike and fold axes) is paralleled in Central Sumatra by major Lower Cainozoic horst and graben faults that subsequently experienced right-lateral wrench motion. They include the spectacular Bengkalis Graben, aligned at 102.3° E, extending over a distance of 265 km (Figs 4.6; 4.13). N–S faults also dominate much of the wider region, for example the North Sumatra Basin (Davies, 1984), the Sunda Basin (Molina, 1985), NW. Java (Reminton and Pranyoto, 1985), and Bangka (U Ko Ko, 1986). However, there are also some major faults oriented NE–SW (Fig. 4.6).

4.3.3 Evidence for right-lateral shear

The Semangko or Sumatran Fault is the most spectacular right-lateral active system of South-East Asia (Katili, 1974*a*; Hamilton, 1979; Hutchison, 1989*a*; Curray, 1989). It extends along the length of Sumatra and has splays extending westwards to the coast, and beyond to the trench (Karig *et al.*, 1980; Curray, 1989). The fault zone is characterized by bodies of highly sheared serpentinite. Its fundamental nature is also illustrated by the fact that the Benioff Zone from the Sunda Trench is completely absent north-east of the fault. Dextral motion

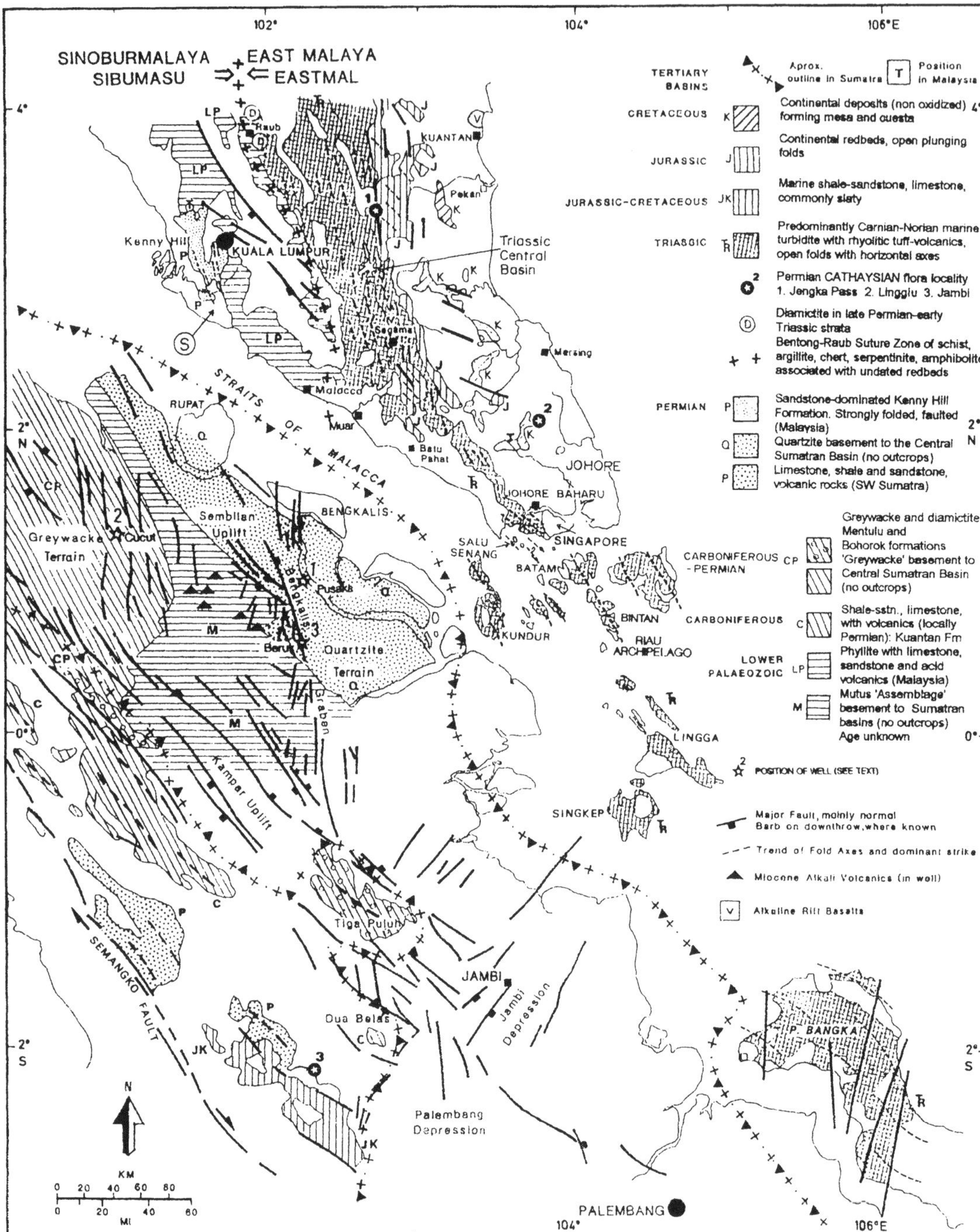

Fig. 4.6. Simplified geological map emphasizing the Upper Permian to Triassic Central Basin of Peninsular Malaysia and its extension into adjacent Indonesia, and the Palaeozoic correlatives between Malaysia and Sumatra. Based mainly on Malaysia (1985), Fontaine and Gafoer (1989), Eubank and Makki (1981), Pulunggono and Cameron (1984), Katili (1967) and Moulds (1989). Well names: 1 = Pusaka No 1; 2 = Cucut No 1; 3 = Beruk NE No 1. After Hutchison (1993).

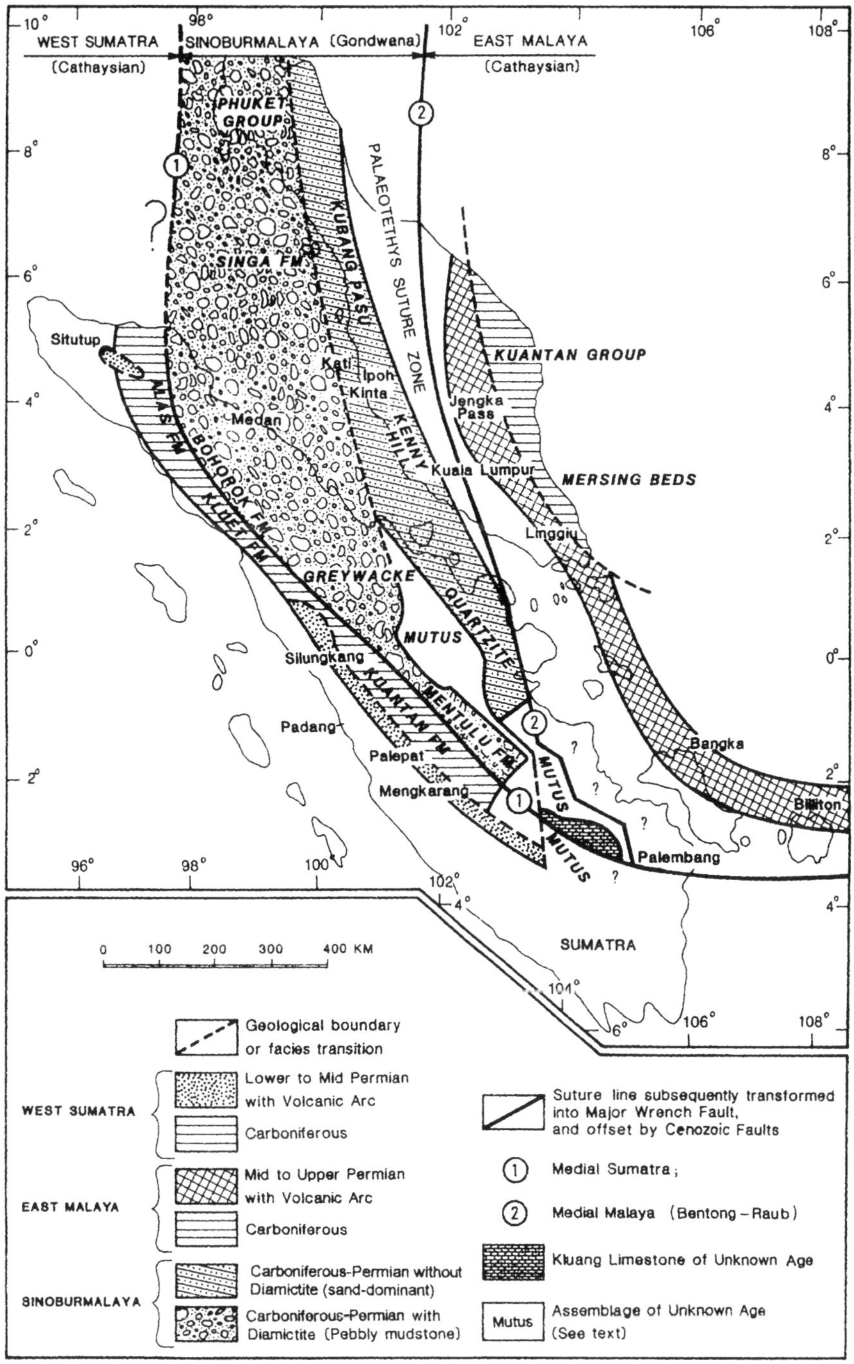

Fig. 4.7. Proposed subdivisions of Peninsular Malaysia and Sumatra into Gondwanaland and Cathaysia Carboniferous–Permian entities. The major formations are named: 'Greywacke', 'Quartzite', 'Mutus', and Kluang Limestone, and are confined to the pre-Cainozoic basement, accessible only by drilling. After Hutchison (1993).

on the fault has been estimated by various authors, up to 400 km, but usually around 45 km (Wajzer *et al.*, 1991) and attributed to oblique convergence at the Sunda Trench (Fitch, 1972; Hutchison, 1989*a*; Curray, 1989; Moore *et al.*, 1980).

Folding of the Tertiary strata of the Bengkalis Graben, and elsewhere in all Sumatran basins, is dominated by a style of fold, called the 'Sunda Fold' by Eubank and Makki (1981), a Cainozoic analogue of the transpressional folding of the Jurassic Tembeling Formation of Peninsular Malaysia, illustrated in Fig. 4.8 (Hutchison, 1993). The folding is interpreted to be caused by transpression resulting from Cainozoic right-lateral motion on the dominant N–S striking faults.

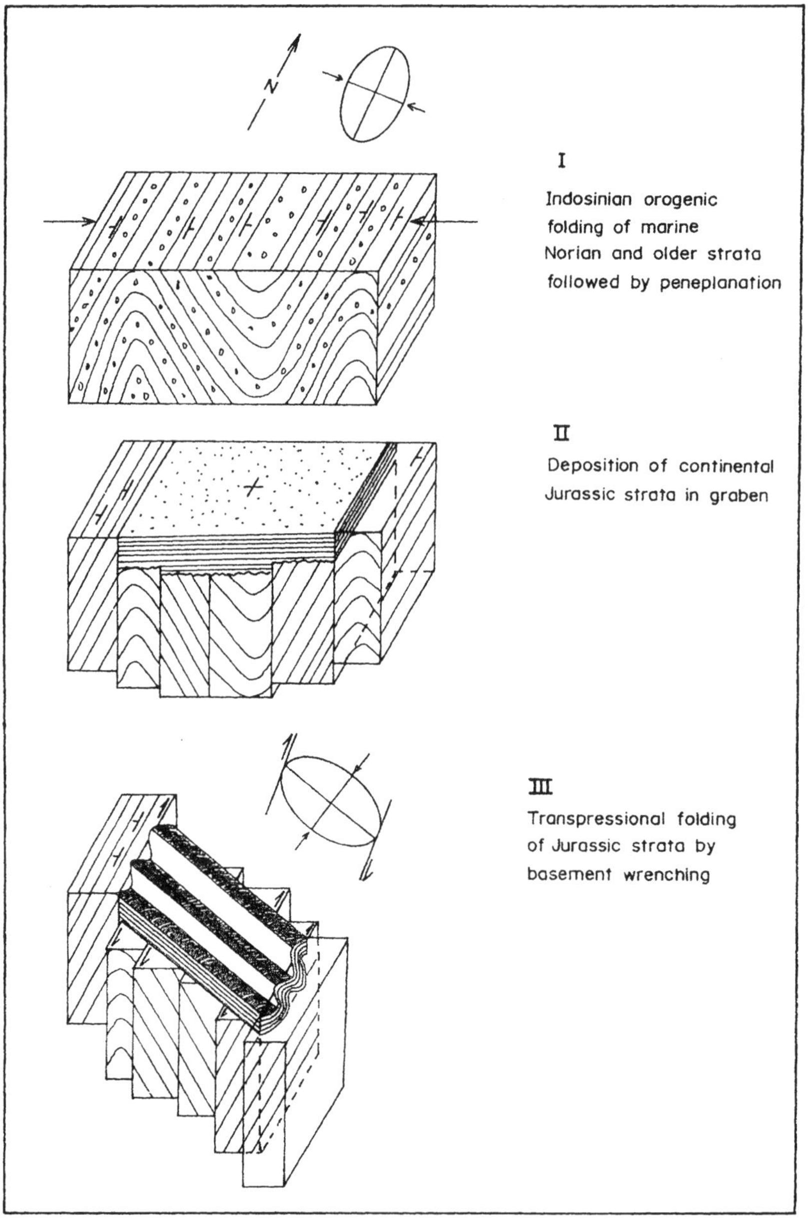

Fig. 4.8. Simplified schematic block diagrams to show how the structures of the Tembeling Formation of eastern Peninsular Malaysia may have become transverse to the main Indosinian orogenic fabric. Such a trans-pressional transverse folding style has been named the 'Sunda Fold' in Sumatra.

4.3.4 Stratigraphic correlation between Peninsular Malaysia and Sumatra

The older basement of Sumatra is extensively covered by Cainozoic rocks and is too complicated to be resolved by conventional seismic interpretation. It is known from several core descriptions and from generalizations made by oil company geologists (Eubank and Makki, 1981). The outcropping pre-Tertiary formations have been summarized by Cameron *et al.* (1980) and Pulunggono and Cameron (1984). In the following section, I comment on the basement of the north-eastern coastal plains of Sumatra and its possible correlation with Peninsular Malaysia. However, correlation between scattered core data and outcropping terrain represents a difficult task.

4.3.4.1 The 'Greywacke Terrain'

The Central Sumatran Cainozoic Basin is extensively underlain by a 'greywacke terrain' basement (Fig. 4.6) known from several drill cores (Eubank and Makki, 1981; Koning and Darmono, 1984). The 'greywacke' is described as predominantly pebbly mudstone, similar to the outcropping Carboniferous–Permian Bohorok Formation of the Tapanuli Group of Cameron *et al.* (1980) and the Mentulu Formation of the Tigapuluh Group (Fontaine and Gafoer, 1989). These formations are the Sumatran equivalent of the glacio-marine Phuket Group of south Thailand (Stauffer and Mantajit, 1981) and the Singa Formation of Langkawi Island of NW Peninsular Malaysia (Stauffer and Lee, 1986).

Cucut Well No. 1 revealed the most significant data (Koning and Darmono, 1984). The pebbly mudstone contains abundant granitic, volcanic and metamorphic clasts in an unmetamorphosed mudstone matrix, containing Lower to Middle Carboniferous continental palynomorphs. One granite clast was K:Ar dated at Lower Carboniferous (348±10 Ma), setting an older limit for the formation age. Some of the clasts display striations interpreted to be glacial (R.T. Eubank, pers. comm., 1989).

The interpretation of Carboniferous tilloids beneath the Central Sumatran Basin implies that a belt extends southwards from Langkawi (Singa Formation) under the Straits of Malacca and Central Sumatra Basin to reappear on Sumatra as the Bohorok Formation and Tigapuluh Group (Fig. 4.7).

4.3.4.2 The 'Quartzite Terrain'

The north-eastern half of the coastal plains of the Central Sumatran Basin is underlain by a basement described by Eubank and Makki (1981) as 'quartzite'. The only lithological descriptions have been given by Koning and Darmono (1984) who described brecciated metaquartzite. Pusaka No. 1 well cored into dark grey silty fractured shale, which yielded palynomorphs suggesting an age near the Devonian/Carboniferous boundary. A similar age was determined for quartzite with shale intercalations drilled near Rupat Island (Fontaine and Gafoer, 1989). The brecciated metaquartzite of Beruk Northeast No. 1 well contains mica along shear zones, K:Ar dated at 276±10 Ma (Lower Permian). Presumably this is a metamorphic age, but no data are available on the grade of metamorphism of the interbedded shales, except that it must be very low since the oil company descriptions are 'shale' and 'silty shale'. Lee (1982) assumed the basement rocks to the south-east of Rupat Island to be Mesozoic, but I agree with Fontaine and Gafoer (1989) that the 'quartzite terrain' is most likely to represent a southwards continuity of the sandstone-dominated Kenny Hill Formation (Stauffer, 1973*a*) of Peninsular Malaysia. The dominant strike directions of the southern outcrops of the Kenny Hill Formation are NNW–SSE (Abdullah, 1983). The trend is therefore towards the 'Quartzite Terrain' and the formation disappears southwards beneath the coastal alluvial plains (Fig. 4.6).

The sandstones of the Kenny Hill Formation are characterized by an interlocking mosaic of quartz grains indicating considerable post depositional changes such as pressure solution resulting in quartz overgrowths (Stauffer, 1973*a*). No detailed studies have been carried out to define the metamorphic grade of the inter-bedded shales, which in places have been described as phyllitic. Most of the sandstones have accordingly been named metaquartzite by Rosly (1979). Cores of such rocks would be named 'quartzite', the term used for the basement of Central Sumatra.

The Kenny Hill Formation is poorly fossiliferous. Rosly (1979) found palynomorphs and radiolaria in chert near Dengkil, but his work was inconclusive. Abdullah (1983, 1985) described a Lower to Middle Permian ammonoid and non-diagnostic crinoids from rocks assumed to be Kenny Hill Formation in the Sepang area on the borders between Selangor and Negri Sembilan (S on Fig. 4.6). The present data therefore suggest a broad Carboniferous to Middle Permian age for both the 'Quartzite' basement of Sumatra and the Kenny Hill Formation of Peninsular Malaysia.

4.3.4.3 The 'Mutus Assemblage'

This is an unresolved 'assemblage' of different rock types (Fig. 4.6), informally named by Eubank and Makki (1981), and has subsequently been extrapolated beneath parts of the South Sumatra Basin (Pulunggono and Cameron, 1984). In my view it does not represent a single formation but a structurally complex terrain of rock lithologies of different ages. The lithologies included are: radiolarian chert, meta-argillite, red-mauve shale, thinly bedded limestone, and a thinly bedded sandstone-shale sequence, interpreted as deep water. Pre-Cainozoic basalt (age not published) was recorded in one well, and tuff along its western margin with the greywacke terrain was K:Ar dated 222±3 Ma. There are many occurrences of Upper Triassic granites within the 'Mutus Assemblage' (Koning and Darmono, 1984). It was tentatively suggested by Eubank and Makki (1981) that the Mutus Assemblage may correlate with the lithologically similar Middle to Upper Triassic Kuala Formation of NW Sumatra (Cameron *et al.*, 1980). From this suggestion, the boundary between this assemblage and the 'Quartzite Terrain', the Kerumutan Line, was inferred to have ophiolitic characteristics, but an ophiolite assemblage has not been documented. Without supporting evidence, Pulunggono and Cameron (1984) correlated the Mutus Assemblage with the Triassic Semanggol Formation of Peninsular Malaysia. They also suggested that the Mutus Assemblage basement may have played an important role as a hydrocarbon source, because the richest oilfields of Central and South Sumatra are closely associated with a basement identified as Mutus Assemblage.

I propose that all the sedimentary lithologies of the Mutus Assemblage can be found within the Lower Palaeozoic envelope of the Main Range Granite of Peninsular Malaysia, and that a Triassic age may be unwarranted. The radiometrically dated igneous rocks should not have been included within the Mutus Assemblage for they have Triassic analogues in the Main Range Batholith complex.

Let us scrutinize each of the lithologies, and find its possible Malaysian correlative. Radiolarian chert and meta-argillite are important parts of the Bentong–Raub suture zone, but meta-argillite dominates the whole envelope of the Main Range Batholith, present in the Kuala Lumpur area as the Lower Palaeozoic Hawthornden Schist (Hutchison, 1989*a*). Radiolarian chert has also been recorded in the Kenny Hill Formation south of Kuala Lumpur (Rosly, 1979). Red-mauve shale occurs widely around the southern end of the Main Range Batholith at Port Dickson and Malacca, where the Lower Palaeozoic schist has been deeply lateritized. Thinly bedded Ordovician limestone occurs as intercalations in the Hawthornden Schist, and Silurian limestone is abundant in the Kuala Lumpur district (Gobbett and Hutchison, 1973). The rhythmite sequence of the Mutus Assemblage could in part correlate with some outcrops of the Kenny Hill Formation, and would resemble the Kubang Pasu Formation. The pre-Tertiary basalt has its equivalents within the Lower Palaeozoic schist of the Bentong–Raub suture zone, where they are usually converted to amphibolite. The most southerly occurrence of meta-basalt in Malaysia is at Bukit Larang, SE of Malacca (Gobbett and Hutchison, 1973).

It is likely that the 'Mutus Assemblage' (Fig. 4.6) represents a Palaeozoic subduction-accretion complex by correlation with the similar envelope lithologies of the Main Range Batholith of Peninsular Malaysia. Its apparent Triassic age results from the dating of Triassic granitoids which are well known in Peninsular Malaysia to have intruded into the subduction-accretion complex (Hutchison, 1989*a*).

4.3.4.4 Mid Miocene Basalts

Although sedimentation was continuous in the deepest parts of the Central Sumatra Basin, there is a major unconformity characterized by erosion of Lower Miocene and early Middle Miocene rocks. This unconformity is marked by an important phase of alkaline basalt dykes and pyroclastic extrusions, K:Ar dated within the range 17 to 12 Ma. They are covered by post-unconformity strata. These rocks have been drilled at seven localities (Fig. 4.6) (Eubank and Makki, 1981). Similar alkaline basalts form sills intruding the Gumai and Talang Akar Formations of the South Sumatra Basin, where they have been K:Ar dated 11 to 16 Ma (Pulunggono and Cameron, 1984).

The basaltic dykes and rift-related alkaline lava flows of contiguous Peninsular Malaysia range from Jurassic to Lower Pleistocene (Bignell and Snelling, 1977; Haile *et al.*, 1983). Tholeiitic basaltic dykes at Kuala Brang (150 km NNW of Kuantan) have been K:Ar dated 189 Ma (Lower Jurassic), and at Kuantan (Fig. 4.6) a dyke swarm has been dated 129 to 97 Ma (Lower Cretaceous). A single K:Ar date of 62 ±2 Ma (Lower Palaeocene) was obtained from the rift-related Segamat alkaline basalt, and the Kuantan alkaline basalt gave Upper Pliocene to Lower Pleistocene K:Ar ages of 2.5 to 1.2 Ma. More radiometric work in the Segamat area is warranted in the search for the yet undocumented Middle Miocene episode in Peninsular Malaysia.

4.3.4.5 Kluang Limestone

Hard massive limestone forms an extensive pre-Cainozoic basement to the South Sumatra Basin in the Palembang district (Figs 4.6 and 4.7) (Adiwidjaja and De Coster, 1973). No fossils have been found in cores. Pulunggono and Cameron (1984) refer to it as the Kluang Limestone and tentatively ascribe to it a Jurassic to Cretaceous age, presumably because the Gumai Mountains to the south have outcrops of Cretaceous and Jurassic Sepingtiang Limestone (Fontaine and Gafoer, 1989).

An older age for the Kluang Limestone should not be ruled out. If the Mutus Assemblage is in part a correlative of the Hawthornden Schist, then the Kluang Limestone is in an appropriate geographical location to be a correlative of the Silurian Kuala Lumpur Limestone (Gobbett and Hutchison, 1973).

4.3.5 Carboniferous–Permian distribution in Sinoburmalaya

4.3.5.1 Tilloid-bearing formations

This distinctive facies occurs throughout a belt extending more than 2000 km from western Yunnan (Wopfner and Jin, 1993; Jin, 1994), through central Myanmar (Burma) southwards to Sumatra. It is known by various names: the Mergui Group of Myanmar (Mitchell, 1992), the Khlong Kaphon Formation of the Phuket Group of Thailand (Burton, 1986), Singa Formation of NW Peninsular Malaysia (Jones, 1978; Ahmad bin Jantan, 1973; Gobbett, 1973), Bohorok Formation of north Sumatra (Cameron *et al.*, 1980), Mentulu Formation of the Tigapuloh Group of central Sumatra (Fontaine and Gafoer, 1989), and the 'greywacke' basement terrain of the Central Sumatra Basin (Eubank and Makki, 1981; Koning and Darmono, 1984). The rocks are typically laminated, dark grey, poorly sorted mudstones with scattered megaclasts, few fossils, and common soft sediment deformation.

Ice-rafted dropstones were documented in Langkawi by Stauffer and Lee (1986), west Yunnan by Wopfner and Jin (1993) and R.T. Eubank (pers. comm, 1989) has seen striated megaclasts in the greywacke cored from the basement of the Central Sumatra Basin. It should now be beyond doubt that these diamictites are of glacial marine origin, and Stauffer and Lee (1986) have shown that they are of shallow water origin.

The paucity of fossils is characteristic of the whole tilloid-bearing facies, but a Carboniferous–Permian age is generally accepted (Jones, 1978; Gobbett, 1973). A Lower Permian cool-water fossil assemblage has been described from south Thailand (Waterhouse, 1982). The most reasonable source for the glacially-derived clasts is the northern margin of Gondwanaland, and tilloids of the Australian basins are confined to the range Namurian to Artinskian (Veevers, 1984). Hence it is reasonable to conclude this age range for the SE Asian tilloid-bearing formations, and it is not in conflict with the palaeontology and stratigraphy.

The relations of the Bohorok Formation, Tigapuloh Group, and the greywacke terrain to older and younger formations are undocumented in Sumatra. However, the Phuket Group and Singa Formation are underlain by Devonian Bannang Sata Group in Thailand, of black slates intercalated with arenites and lenticular limestones (Burton, 1986) and the base of the Singa Formation is of latest Devonian to earliest Carboniferous red mudstone and sandstone unconformably overlying Lower Palaeozoic formations in contiguous Peninsular Malaysia (Gobbett, 1973). The Phuket Group and Singa Formation are conformably overlain by Chuping and Ratburi Formation carbonates, whose lower horizons have yielded late Lower Permian fossils (Gobbett, 1973; Fontaine, 1989).

Mitchell (1992) points out that 'no stratigraphic contact between the Mergui Group and Carboniferous or older rocks has been found in Myanmar and Thailand'. This gave him freedom to propose that the Mergui Group, which includes the pebbly mudstone-bearing Phuket Group and the Singa Formation, was not part of the Shan-Thai Block (= Sinoburmalaya or Sibumasu) 'until the late Triassic or early Jurassic, when they were emplaced as a nappe from the present west' (over thrust eastwards over Sinoburmalaya). The inferred W-dipping thrust contact is shown by Mitchell (1992) extrapolating southwards through the Langkawi islands of Malaysia. The well exposed geology of the Langkawi islands and NW Peninsular Malaysia (Jones, 1978), which I know in great detail, not only clearly demonstrate the unconformable base and conformable transitional top of the Singa Formation, but also render impossible the 'nappe theory' of Mitchell (1992).

4.3.5.1.1 Singa Formation base The Singa Formation can be shown in the field to overlie with angular unconformity both the Ordovician to Lower Silurian predominantly calcareous Setul Formation and the Upper Cambrian sandy Machinchang Formation (Jones, 1978; Gobbett, 1973). The basal sequence is very distinctive and outcrops widely. It is a condensed sequence (< 35 m) of Upper Devonian to Lower Carboniferous fossiliferous marine redbeds, commonly shaly, locally sandy and conglomeratic. The best exposures are on the Langkawi islands of Pulau Langgon and Pulau Rebak. Only a short distance ESE of the basal redbeds, typical Singa Formation mudstones, with thin slumped fine grained sandstone beds, contain spreads of glacial dropstones, the largest of which is of exotic rounded pegmatite surrounded by locally provenanced smaller angular sandstone clasts, outcropping on a small tourist-frequented headland on Pantai Cenang. From Pulau Rebak Besar east to Pulau Singa Kechil and Pulau Jong (base of the Chuping Limestone), the Singa Formation dips constantly ESE at 20°, allowing the total Singa Formation thickness to be estimated at 2300 m (Ahmad bin Jantan, 1973).

It is therefore necessary to conclude that both the Upper Cambrian Machinchang and the Ordovician to Lower Silurian Setul formations formed an extensive

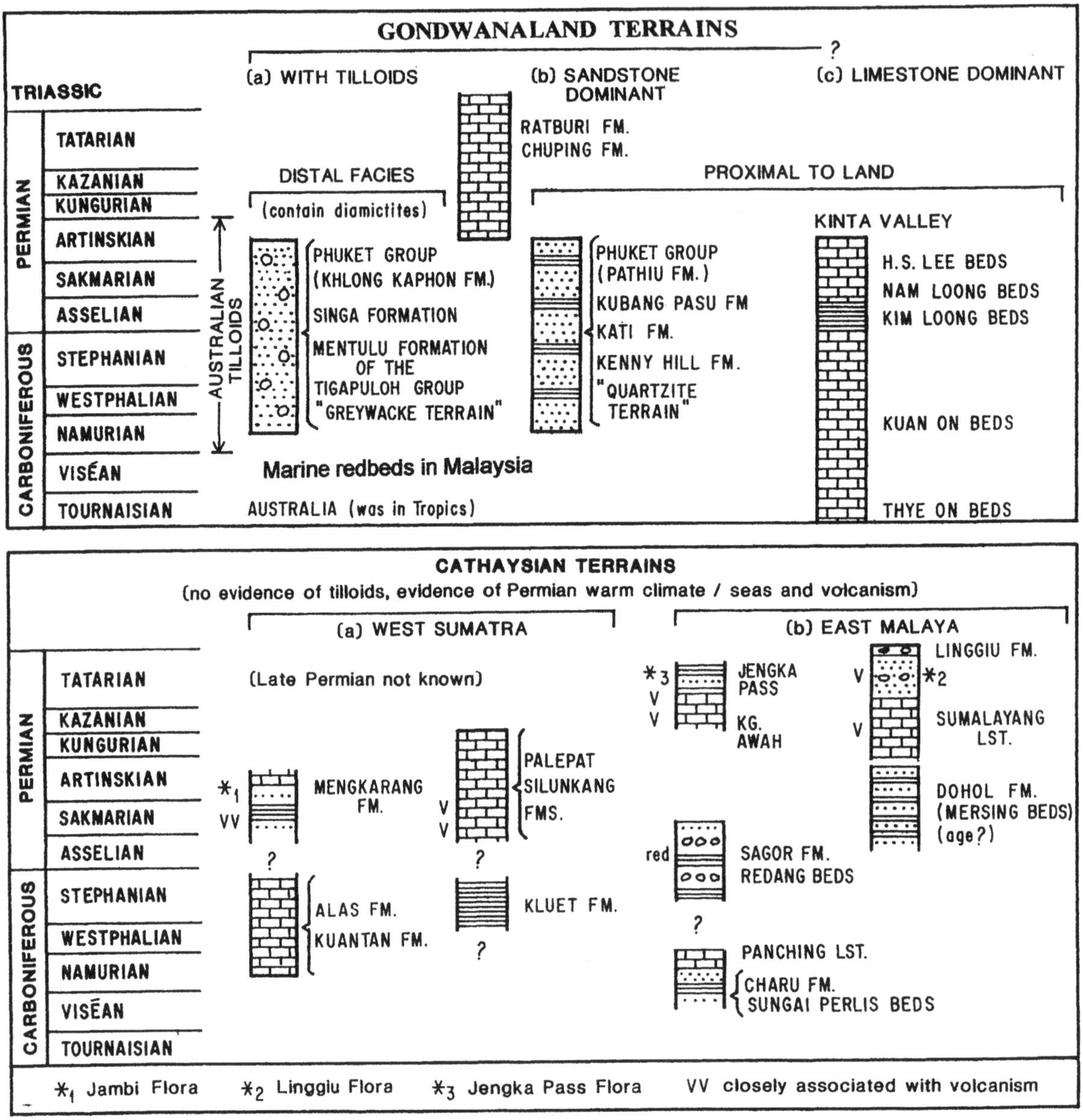

Fig. 4.9. Summary Carboniferous–Permian stratigraphy of the Gondwanaland and Cathaysia blocks of Peninsular Malaysia and Sumatra. The information source is discussed in the text. After Hutchison (1993).

peneplained oxidized land surface, the weathered products of which were deposited offshore as the Upper Devonian to Lower Carboniferous basal Singa marine redbeds, which were rapidly followed by typical Singa Formation containing iceberg-melt dropstone spreads. It is very probable, therefore, that the underlying surface of Machinchang and Setul formations may have experienced continental glaciation, but no such geological evidence has yet been found.

4.3.5.1.2 Singa Formation top The top of the Singa Formation is well exposed at the base of Pulau Jong and nearby Pulau Singa Kechil, where the conformable passage beds dip ESE at 12° (Jones, 1978, p. 71). The transitional top is also exposed on nearby Pulau Singa Besar. The conformable transition from the Singa Formation to the calcareous Chuping Formation is completely conformable and palaeontologically dated late Lower Permian (Fig. 4.9). In contiguous Thailand, the latter is known as the Ratburi Limestone.

The Chuping and Ratburi Formation limestones are characteristically impoverished in fauna so that the environment is considered to have been unsuitable for diverse and prolific life. Fontaine (1989) has recorded *Iranophyllum* at Phangnga near Phuket in south Thailand. This solitary rugose coral occurs also in the tilloid-bearing Lhasa–Gandise terrain of Tibet, but of course not in the tilloids (Chang and Pan, 1984). For reasons never given, *Iranophyllum* is interpreted as a warm water coral (Chang and Pan, 1984; Fontaine, 1989), so that the Lhasa–Gandise and Sinoburmalaya terrains might be considered of mixed-affinity. This is unnecessary. If Sinoburmalaya and the Lhasa–Gandise terrains rifted away from Gondwanaland during the early Permian, as deduced by Metcalfe (1990*a*), they would be expected to have the marine glacial deposits, but their late Lower Permian to Triassic limestones need not have any Gondwanaland signature, and Upper Permian *Glossopteris* flora need not have colonizedSinoburmalaya and the Lhasa–Gandise terrains. This interpretation is strongly supported by the Lower Permian brachiopods of the Kinta Valley (Shi and Waterhouse, 1991). They deduced that the brachiopod provinciality of Sinoburmalaya changed rapidly in Sakmarian time from southern cold temperate, showing a strong affinity with eastern Australia, to subtropical showing an affinity with Tethyan and Uralian faunas. This implies a rapid shift of the climatic belts in the Lower Permian due to the end of the Ice Age (Shi and Waterhouse, 1991). Görür and Sengör (1992) argue that the Lhasa–Central Burma Block was rifted off Gondwanaland only in the Upper Triassic and that its rapid northward drift took place in the Upper Triassic–Lower Jurassic.

4.3.5.2 *Non-glacial formations*

The diamictite-bearing belt is flanked on the east by a sandstone-dominant facies (Fig. 4.7). It is also characteristically impoverished in fossils. No tectonic dislocation separates the two facies. In NW Peninsular Malaysia, the base of the sandy facies (Kubang Pasu Formation) is represented in the same way as the Singa Formation by the same fossiliferous Upper Devonian to Lower Carboniferous marine redbeds, well exposed at Gunung Utan Haji (Jones, 1978) and at Wang Kelian (Lee and Azhar, 1991). In Perlis, the Kubang Pasu Formation is conformably overlain by the same Upper Permian to Triassic Chuping Limestone which overlies the Singa Formation on the Langkawi islands. The conformity is marked by a sequence of very fossiliferous passage beds of late Lower Permian age, well exposed on the western hillside of Bukit Temiang, where the strata dip 30 to 40° E and at the base of Bukit Tengku Lembu (Jones, 1978). In contiguous Thailand, the Chuping Limestone is known as the Ratburi Limestone.

The two facies (Singa and Kubang Pasu formations) are therefore time-equivalent lateral variations. I infer that the sand-dominant facies was deposited closer to land than the tilloid-bearing facies, following an identical demonstration by Redfern (1991) for the Grant Group of the Australian Canning Basin, which is inferred to have been contiguous with this part of Sinoburmalaya at that time (Hutchison, 1989*a*). The main Carboniferous–Permian land therefore lay to the present-day east, but part of the Langkawi islands also formed part of the land mass, as argued above. This has been suggested by several authors and the western margin of the Phuket Group represents the line of separation from Gondwanaland (Stauffer and Lee, 1986, p. 92). Palaeocurrent analysis of the Kubang Pasu, Kenny Hill and Singa Formations and the Phuket Group is either lacking or has yielded conflicting results (Burton, 1986, p. 349), but Ahmad bin Jantan (1973) inferred a palaeo-slope for the Singa Formation towards the east, with land on the west.

Mitchell (1992) proposes that Lower Permian diamictites were not part of the Shan-Thai (= Sinoburmalaya) until the late Triassic or early Jurassic, when they were emplaced as a nappe from the present west, probably as part of an arc system rifted from Gondwanaland. The regional analysis of Görür and Sengör (1992) concludes that Lower Permian glacial diamictites occur on the Lhasa and Central Burma Block, whereas continental clastic rocks (sand-dominated) characterize the Quangtang and Sinoburmalaya blocks, and that the major Palaeo-Tethys Ocean separated them at that time. These two blocks are now separated by the Lower Cretaceous Banggong Co-Nu Jiang Suture in Tibet. Görür and Sengör (1992) therefore would like to establish a similar suture (Mandalay Suture) between the Central Burma and the Sinoburmalaya blocks, and the nappe hypothesis of Mitchell (1992) satisfies their preconception perfectly! However, the Upper Permian basal Chuping and Ratburi limestones stratigraphically and conformably overlie both the diamictite-bearing and the non-diamictite-bearing sandstone-dominated formations, presenting a severe problem for Mitchell's (1992) nappe theory. The base of both facies is an Upper Devonian to Lower Carboniferous condensed redbed sequence. The nappe theory of Mitchell

(1992) is accordingly not viable, and the field evidence disproves a suture in NW Peninsular Malaysia and SW Thailand.

The non-glacial formations are briefly discussed from north to south. The Pathiu Formation of the Phuket Group of South Thailand (Burton, 1986) is mainly a marine submature arenite, with significant amounts of greywacke and mudstone. The fossils confirm an Upper Carboniferous to Lower Permian age. The Kubang Pasu Formation of north Peninsular Malaysia is largely composed of thick-bedded quartz and feldspathic sandstone inter-bedded with mudstone (Gobbett, 1973). The formation is sparsely fossiliferous, but definitely Carboniferous to Lower Permian and passes conformably up into the Chuping Limestone. The Kati Formation consists of inter-bedded sandstone, mudstone and siltstone, and in part is metamorphosed to metaquartzite and phyllite (Foo, 1983). Unfortunately it has not yet yielded useful fossils.

The Kenny Hill Formation of the Kuala Lumpur district and the 'Quartzite' basement of Sumatra form the southernmost extent of this sandstone-dominated belt. The Kenny Hill Formation was described by Stauffer (1973*a*) as a monotonous sequence of inter-bedded sandstone and mudstone. Low-grade metamorphism is ubiquitous, and many outcrops are appropriately named metaquartzite and phyllite (Abdullah, 1983). The environment of deposition has been interpreted as an unstable shelf under moderate water depth, not far from a low-elevation eroding land mass consisting mainly of sedimentary rocks (Stauffer, 1973*a*). The extremely sparse fauna indicates a Carboniferous–Permian age (Abdullah, 1985).

4.3.5.3 The Kinta Valley sequence

This sequence, lying on strike between the Kubang Pasu, Kati and Kenny Hill Formations, is unusual but very important within the regional picture ('Kinta' on Figs 4.7 and 4.9). There are 700 m of Carboniferous through Lower Permian carbonates with 100 m of shale and argillaceous sandstone (Gobbett, 1973; Suntharalingam, 1968).

The Nam Loong beds contain early Lower Permian (Sakmarian) brachiopods, which Shi and Waterhouse (1991) have shown shared a southern cold temperate provinciality with southern Thailand (Waterhouse, 1982). This fauna shows a strong affinity with eastern Australian faunas of similar age and has been interpreted to mean that during Asselian to Sakmarian times Sinoburmalaya was attached to eastern Gondwanaland (Shi and Waterhouse, 1991).

By contrast, the overlying H.S. Lee beds (Suntharalingam, 1968) contain fusulinids, ammonoids and abundant gastropods of late Lower Permian (Sakmarian to Artinskian) age, interpreted as a warm water fauna. By contrast with the Nam Loong beds, the fauna of the H.S. Lee beds have Uralian and Tethyan warm water provinciality and share no affinities with Gondwanaland. The Australian Gondwanaland basins of this age have a characteristic impoverishment in carbonate and an absence of fusulinids (Veevers, 1984; Fontaine, 1989). The abrupt change of faunal affinity, within a period of less than 5 Ma, between the Nam Loong and H.S. Lee beds, caused Shi and Waterhouse (1991) to infer a rapid northern movement of Sinoburmalaya crossing the Palaeo-Tethys after rifting off the northern margin of Gondwanaland in early Permian (Sakmarian) time. The breakup of the continental margin may have caused a reorganization of warm currents within the Palaeo-Tethys and change of climate along its margins, thus accelerating the faunal change. Alternatively the change may have been caused by a migration of climatic belts following the end of the Ice Age.

4.3.6 Eastmal Cathaysian Block

Lying to the east of the separating Bentong–Raub Line, this block has virtually no Carboniferous–Permian similarities with Sinoburmalaya (Figs 4.7, 4.9). Instead, its stratigraphy is linked to Indochina and South China (Cathaysia), with which it is likely to have stratigraphic and structural continuity beneath the Cainozoic Malay Basin (Fontaine *et al.*, 1990).

The oldest exposed dated strata are the Viséan to Namurian Charu Formation and Panching Limestone (Metcalfe *et al.*, 1980). The former is a near shore sandstone and shale formation, containing abundant fauna and flora, while the limestone contains Namurian fusulinids and conodonts. In Namurian time the Palaeo-Tethys was expanding as the northern Australian part of Gondwanaland rapidly drifted from 30° to 60° S (Veevers, 1984)· *Lepidodendron* and *Stigmaria* are common to Eastmal (Hutchison, 1989*a*) and Australia (Veevers, 1984), but fusulinids are absent from Australia, indicating the prevailing cold water conditions.

Upper Carboniferous to Lower Permian shallow marine to continental conditions are indicated by the Sagor Formation and Redang Beds (Hutchison, 1989*a*). The Dohol Formation and Mersing Beds are undated. Late Lower to early Upper Permian fusulinids are abundant in the Sumalayang Limestone, at Jengka Pass and at Kampong Awah, where the limestone is intimately associated with andesitic volcanic rocks (Gobbett and Hutchison, 1973).

Eastmal is confidently classified as Cathaysian *sensu stricto* because of Upper Permian *Gigantopteris* flora at Jengka Pass (Kon'no and Asama, 1970) and at Linggiu

(Kon'no *et al.*, 1971) (Fig. 4.6). By contrast with Sinoburmalaya, it is also characterized by Upper Permian andesitic volcanism, and by diverse and abundant fauna, including fusulinids.

4.3.7 West Sumatra Block

Previous writers have assumed that the Carboniferous–Permian formations across most of Sumatra represent lateral facies variations (Pulunggono and Cameron, 1984), despite the lack of demonstrable relationships, and Cameron *et al.* (1980) included them together within the Tapanuli Group. Such a grouping results in an unlikely juxtaposition of cold and warm water fauna and equatorial climate flora, notably in the Lower Permian strata. A possible solution is to give greater significance to the major NW–SE striking faults which define the SW limit of the tilloid-bearing formations (Fig. 4.7). I suggest that the Carboniferous–Permian formations to the SW of this Medial Sumatra Line have been juxtaposed against the tilloid-bearing Sinoburmalaya terrain by Cainozoic wrench faulting. The Medial Sumatra Line of Fig. 4.7 is not the Semangko Fault (Katili, 1974*a*), but is parallel to it, and should be considered part of the great right-lateral Sumatran Fault System. A strongly faulted terrain, composed of the Woyla Group (Cameron *et al.*, 1980) lies on the SW side of the Semangko Fault. It is now interpreted (Wajzer *et al.*, 1991) to be composed of Triassic to Cretaceous ocean floor sediments, oceanic tholeiite and arc volcanics, which have been accreted onto Sumatra and is not a micro-continent like the Gondwanaland and Cathaysian terrains discussed in this section.

The Kuantan Formation (Figs 4.7, 4.9) is the oldest of West Sumatra, composed of shale and sandstone, with thick limestone intercalations. The lower part is mainly of sandstone and shale with volcanic rocks, and may be correlated with the Kluet Formation of North Sumatra. The upper part is dominated by limestone, which may correlate with the Alas Formation to the north. The Kuantan Formation limestones have provided a good Viséan fauna, showing close affinities with Eastmal, Laos, Vietnam and East Thailand (Fontaine and Gafoer, 1989). This fauna has not been found in the Alas Formation, thus a possibility exists that the Medial Sumatra Line, which separates the Kuantan Formation from the tilloid-bearing formations, may continue to the north-west to lie on the Indian Ocean side of the Alas and Kluet formations. In that interpretation, the Alas and Kluet formations (Tapanuli Group) will be considered lateral facies variations of the Bohorok Formation, and accordingly be included within Sinoburmalaya.

The Alas Formation consists of limestone, shale, sandstone and conglomerate (Cameron *et al.*, 1980). The fauna includes Upper Viséan conodonts. It is said to be conformable on the poorly fossiliferous Kluet Formation, which consists mainly of turbiditic quartz arenite sandstone inter-bedded with slaty mudstone and siltstone (Cameron *et al.*, 1980). Fontaine and Gafoer (1989) maintain that the Alas and Kluet formations show a palaeontological similarity with the formations of the eastern part of the Malay Peninsula. However, they point out that the Alas Formation limestone is much thicker than any other Carboniferous limestones in NW Peninsular Malaysia and West Thailand. Accordingly, it remains uncertain whether the Alas and Kluet formations should be included within Sinoburmalaya, or within the West Sumatran Block.

The Lower Permian Mengkarang Formation consists of alternating sandstone, siltstone, shale, conglomerate and tuff, with limestone and coal intercalations (Fontaine and Gafoer, 1989). The limestones are closely associated with andesitic volcanic rocks. The Mengkarang Formation is the recent name for what Zwierzycki (1930, 1935) and Van Bemmelen (1970) called the Djambi (Jambi) Series. The Jambi Series, divided into three parts (top to base): Air Kuning Beds, Salamuku Beds and Karing Beds (see Hutchison, 1989*a*, p. 198), contains the famous Jambi Flora described by Jongmans and Gothan (1935). The flora are from a horizon lying above a Sakmarian fusulinid limestone and are considered to be Artinskian (Fig. 4.9). The more recent studies by Asama (1976, 1984) indicate that the flora are not the same as the Upper Permian *Gigantopteris* of SE Asia, but may represent an early evolutionary form of Cathaysian flora. Fossil wood from the Mengkarang Formation is characterized by an absence of growth-rings, interpreted to indicate growth under tropical conditions (Vozenin-Serra, 1989). Similar wood has been described from Cathaysia.

The fossils of the Mengkarang Formation are abundant and varied, remarkably unlike Lower Permian Gondwanaland fauna and flora, and interpreted as a product of a tropical climate. In particular, algae, fusulinids, compound corals and the Jambi Flora are unknown in Australia (Fontaine and Gafoer, 1989). A unique rich Lower Permian algal micro-flora have been described from the Mengkarang Formation (Vachard, 1989). The Lower Permian micro-fossils also include abundant small foraminifera and fusulinids (Nguyen, 1989). The West Sumatra Block (Fig. 4.7) could not have been contiguous with Sinoburmalaya during the late Carboniferous to early Permian.

Volcanic rocks are widely distributed along a NW–SE belt from Jambi Province, through Silungkang in Central Sumatra, to Gunung Situtup in the north. They are closely associated with fusulinid limestones of late Lower

Permian age. The Silungkang Formation is of limestone, shale and volcanic rocks. It borders the Kuantan Formation along its NE flank, but the exact boundary is ill defined. The Pelepat Formation is composed of andesite, dacite, tuff, volcanic breccia, conglomerate, sandstone, siltstone, shale and limestone. These formations have yielded abundant late Lower Permian (Kungurian) algae, fusulinids and corals, which show distinct affinities with Asia, but none with Australia (Fontaine and Gafoer, 1989).

The presence of Upper Permian in Sumatra has not been established, and in our present state of knowledge the West Sumatra Carboniferous–Permian Block is not equivalent to the East Malaya Block. Both represent a volcanic arc, but of different age. Both are rich in fusulinid limestones, and have Cathaysian-affinity flora, but even the flora are different.

4.3.7.1 The Jambi 'Nappe'

The Jambi Series terrain (called the West Sumatra Block in this section) was interpreted as allochthonous by Zwierzycki (1930) and Van Bemmelen (1970), and to have great similarities with Bintan Island and East Malaya. The Jambi Series (Mengkarang Formation) is separated along its SW margin by the Lematang Line from slaty Jurassic rocks (Pulunggono and Cameron, 1984). In their nappe theory, the Lematang Line was interpreted as a surface of shallow north-eastward dip, and to be the sole of a thrust (see Hutchison, 1989*a*, p. 198). However, Katili (1970) showed that the Lematang Line represents a nearly vertical transcurrent fault, and that shallow dips are only local features, perhaps resulting from local flower-structure trans-pression. A Jambi Nappe, rooted in the complexly deformed Bentong–Raub zone of the Eastmal Block, is neither feasible nor necessary. Both the Jambi and Eastmal terrains have similar volcanic arc characteristics, are rich in fusulinid limestones and contain Cathaysian flora, but all these features are of different age (Fig. 4.9). Hence I interpret the Western Sumatra Carboniferous–Permian block to be independent from Eastmal, and to have slid into its present position by major right-lateral Cainozoic transcurrent faulting. Likewise, the Woyla terrain was accreted to Sumatra and has slid into its present position by at least 45 km of right-lateral wrench motion on the complex Semangko Fault System (Wajzer *et al.*, 1991).

The Medial Sumatra Line, which separates the Western Sumatra Carboniferous–Permian block from Sinoburmalaya, disappears south-eastwards beneath Cainozoic formations, and probably links up with the Medial Malaya Line (Bentong–Raub Suture) beneath a thick cover of Cainozoic strata. Since major transcurrent motion is inferred for the Medial Sumatra Line, such motion is also most likely for the Medial Malaya Line. It has been interpreted as a Lower Mesozoic suture (Hutchison, 1975; 1989*a*,*b*). However the present juxtaposition of Eastmal and Sinoburmalaya is very likely to have resulted from Cainozoic wrench reactivation of the suture.

4.4 CENTRAL SUMATRA BASIN

This is the most outstanding oil-producing basin of South-East Asia (see Table 1.2). Its pre-eminence is illustrated by the fact that in 1973 its production accounted for no less than 74% of the total Indonesian production. Its basic framework has been summarized by Mertosono and Nayoan (1974), De Coster (1974), and Eubank and Makki (1981). There are many subsequent papers, which have given details of specific areas within the basin.

The basin is bounded on the SW by faults, south-west of which the pre-Tertiary basement rocks form outcrops along the Mountain Front of the Barisan Mountains (Fig. 4.10). On the NE the Central Sumatra Basin continues beneath the Straits of Malacca (Liew, 1994), but shoals rapidly towards the coastline of Peninsular Malaysia. Separation from the North Sumatra Basin is clearly defined by the Asahan Arch. However, there is no clear structural separation from the South Sumatra Basin (De Coster, 1974). The division is geographical rather than geological. Conventionally the artibrary division may be drawn northwards from the Tiga Puloh Mountains along the Bengkalis Depression (Fig. 4.6).

4.4.1 Stratigraphy

The stratigraphic succession has been summarized by Mertosono and Nayoan (1974) and De Coster (1974). The latter used certain terminologies which have not persisted in subsequent literature; that of Mertosono and Nayoan (1974) has prevailed, and is used here (Fig. 4.10):

Formation	Age
Minas Formation, (or Alluvium)	–Pleistocene
Petani Formation	–Middle Miocene to Pliocene
Telisa Formation	–Lower to Middle Miocene
Sihapas Group	–Lower Miocene
Pematang Group	–Upper Eocene to Oligocene

The Pematang Group was deposited within grabens on the Sundaland continent. The continental strata are difficult to date, and include excellent source-rock lacustrine deposits, which are responsible for the pre-eminent position of the Central Sumatra basin in South-East Asia. The Sihapas Group and Telisa Formation resulted from marine transgression as the basin sagged (Soeparjadi *et al.*, 1975: reproduced in Hutchison, 1989*a* p. 74). Subsequent uplift caused regression and deposition of the Petani Formation, and Sumatra has since remained above sea level, unlike the offshore basins of South-East Asia.

4.4.1.1 The Pematang Group

The Pematang Group may be analysed in terms of seven lithological facies. A zone 0.5 to 2 km wide along the bounding fault margin of the basins may be analysed as a series of coalescing alluvial fans. Fluvial facies includes meandering and braided stream deposits. Inter-distributary mudstones separate the meander channel sandstones and provide seals to a variety of stratigraphic traps (Williams *et al.*, 1985). Distributary and flood plain facies are characteristically red in colour. Delta front facies provide the best reservoir characteristics. Lacustrine facies are both of shallow and deep lakes. The organic rich shales of the deep lacustrine facies contain from 1 to more than 10% organic matter. Minor sandstones may represent turbidites. The shallow lacustrine facies contains abundant humic material and coals. Marsh and bog facies contain coals and black shales. This facies occurs throughout the Kiri Basin in a zone 15 to 30 m thick above the shallow lacustrine facies. Platform facies is recognized only in the Balam Basin. It consists of very shallow fresh water gastropod-rich mixed carbonate and terrigenous muds with little coarse clastic material.

The Pematang Basin region was dominated by a humid tropical climate throughout the Palaeogene as indicated by the palynological data. The Pematang Group may be subdivided into the following:

4.4.1.1.1 Lower Red Bed Formation It consists of variegated grey, green, mauve and red mottled mudstone, siltstone, sandstone and minor conglomerate. Shallow lacustrine or marsh-bog conditions existed in the deeper basinal areas (Fig. 4.12).

4.4.1.1.2 Brown Shale Formation It conformably overlies or is the lateral equivalent of the Lower Red Bed Formation. It is of organic-rich dark brown or black well laminated shale with siltstone stringers generally deposited in quiet waters, but convoluted bedding may have resulted from tectonic slumping. Thin sandstones occur in the deeper basins and they may be of turbidite origin. Transverse delta sandstones interfinger with lacustrine shales along the gentle hinge margin. Shallow water mixed carbonate-terrigenous platform mud flats developed locally.

The sediment input was predominantly axial to the basins, with minor transverse components. The maximum drilled thickness is 580 m.

4.4.1.1.3 Coal Zone Formation It is in part the lateral equivalent of the Brown Shale Formation, and in part slightly younger, and occurs mainly in the Kiri Basin. It is of interbedded shallow lacustrine shale, coal and minor sandstone. It attains a thickness of more than 610 m, but is generally less.

4.4.1.1.4 Lake Fill Formation It is of fluvial and deltaic sandstone, conglomerate and shallow lacustrine shale. It was rapidly deposited predominantly by axially prograding fluvio-lacustrine delta systems, with variable transverse input. The thickness exceeds 610 m locally.

In the deeper basinal areas, this formation is conformable on the Brown Shale or Coal Zone formations, but it may be transgressive with angular discordance.

4.4.1.1.5 Fanglomerate Formation It consists of sandstone, conglomerate and minor red-green mudstone. It was deposited predominantly along the troughs bounding fault scarps as a series of coalescing alluvial fans (Figs 4.11, 4.12). The maximum thickness attained is in excess of 1830 m. It is laterally and vertically transitional into the Lower Red Beds, Brown Shale, Coal Zone and Lake Fill formations. It grades laterally into the units of the axial parts of the basin.

4.4.1.2 Sihapas Group

The Sihapas Group is mostly of sandstone with minor shale intercalations and local limestone at the base. The sands are medium to coarse grained and have good reservoir characteristics. The oils of the Central Sumatra Basin are reservoired in this formation. The main provenance of these quartz-rich sands is the Main Range Batholith of Peninsular Malaysia, transported by such rivers as the Muar (proto Pahang River) (Fig. 4.13). The Sihapas Group is transgressive and fines upwards. The environment is fluvial to coastal plain.

4.4.1.3 Telisa Formation

The Formation is mostly of shale with minor siltstone, forming a good seal over the underlying reservoirs. It is the most widespread of all the Tertiary formations, formed during a time of maximum marine transgression. It is rich in foraminifera and was deposited in a shallow to open marine environment. Limestone is locally present.

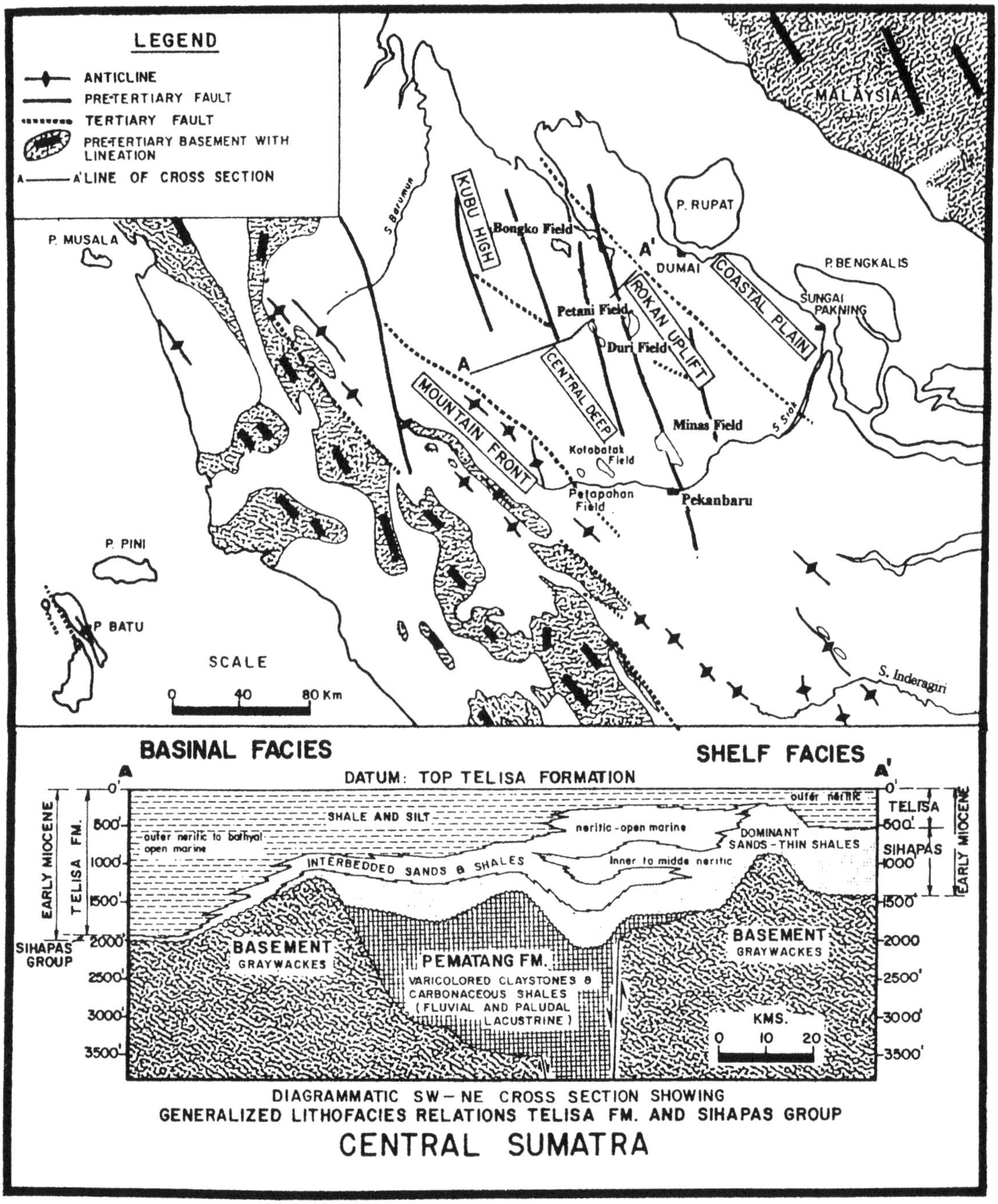

Fig. 4.10. Major features of the Central Sumatra Basin, after Mertosono and Nayoan (1974). The main pre-Tertiary tectonic fabric trends NW–SE (see Figs 4.6 and 4.7). These directions have been re-activated as faults. There are also prominent N–S faults parallel to the fabric of Peninsular Malaysia. The stratigraphy is shown in the cross section A–A'. The depth scale is in feet. The Telisa Formation is overlain by the Middle Miocene to Pliocene Petani Formation.

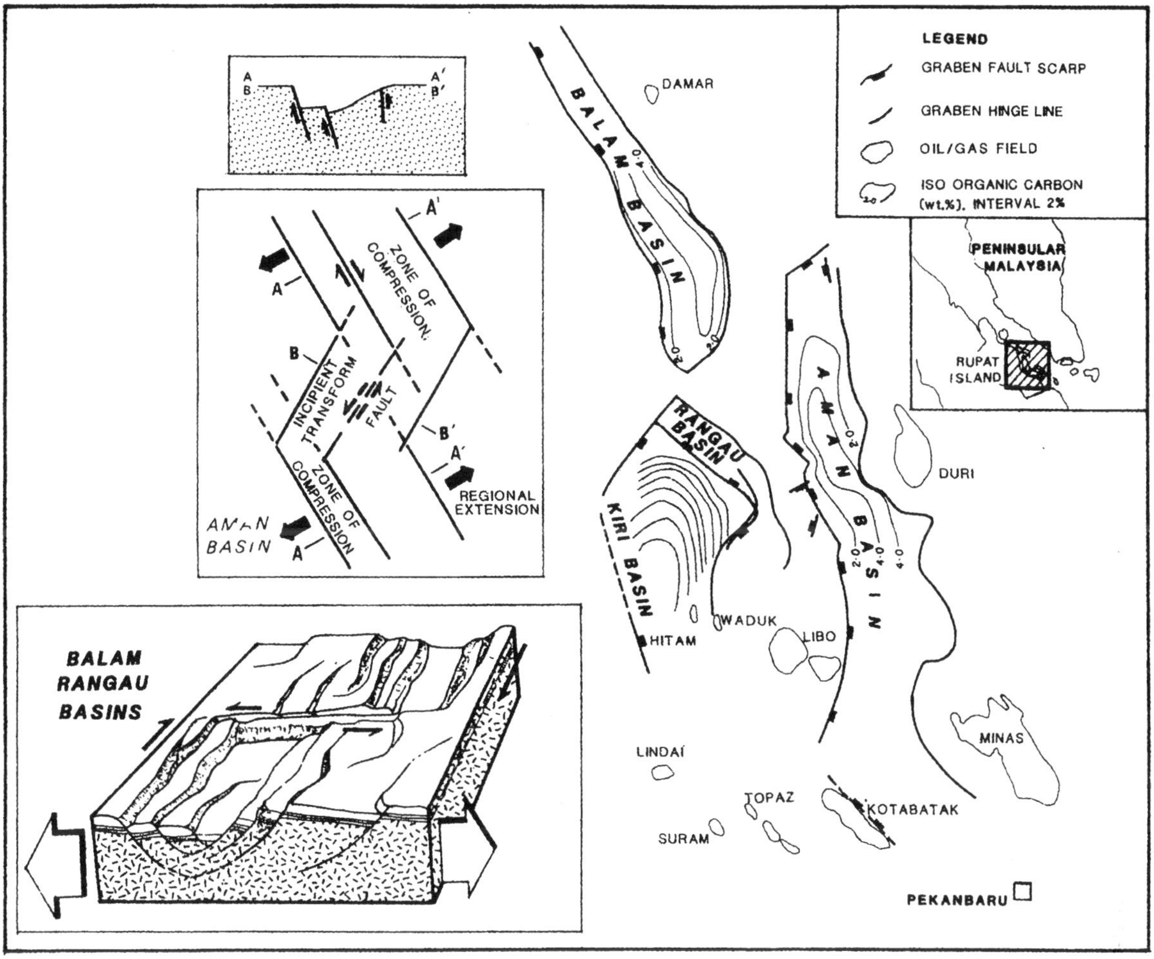

Fig. 4.11. Pematang rift basin location map of Central Sumatra showing source rock characteristics. Inset: Schematic plan of the Aman Basin dog-leg graben and zones of compression (after Williams *et al.*, 1985).

This Lower Miocene marine transgression came from the SW and NW, resulting in two facies—a deep water marine facies on the west, interfingering with shallow shelf to the E and NE (Mertosono and Nayoan, 1974) (Fig. 4.10).

4.4.1.4 Petani Formation

The Formation was deposited during uplift and is a Middle Miocene to Pliocene regressive formation becoming less marine upwards. Foraminifera diminish upwards. The rocks are mainly of grey shales with beds of sandstone. Subsequent uplift and erosion has left a rather incomplete record of the Petani Formation, which has been eroded completely from some areas.

4.4.2 Structure

Two main structural trends are present: a discordant NNW–SSE to N–S trend parallel to the main fabric of Peninsular Malaysia, cross-cutting the main fabric of Sumatra (see Fig. 4.6). This trend has controlled the main graben depressions within which the Pematang Group was deposited. The other major trend is NW–SE, which represents Tertiary faults parallel to the Indosinian tectonic grain of the pre-Tertiary basement. The NW–SE faults appear to be mostly normal, associated with horsts and grabens, but right lateral wrench faulting is invoked because the faults are parallel to the active Semangko Fault (Fig. 2.4; 4.6 and 4.7).

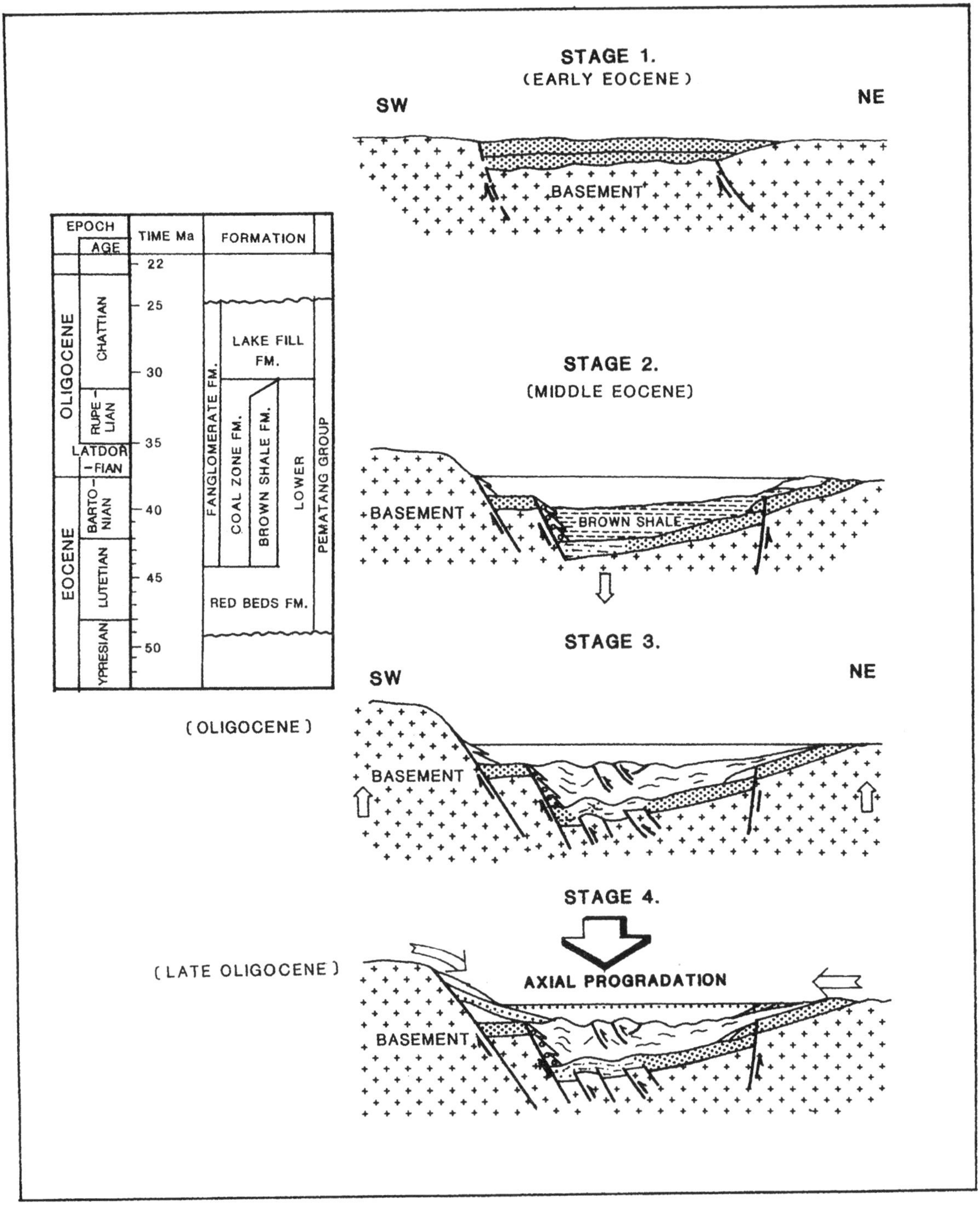

Fig. 4.12. Schematic cross sections showing the evolution of the Pematang basins of Central Sumatra. Inset: The Palaeogene stratigraphy of the basins (after Williams *et al.*, 1985).

The crests of the Tandun and Pungut fields (lying between Minas and Duri) are offset 8 km on opposite sides of a major right-lateral NNW–SSE to N–S fault. The vertical displacement of the wrench fault changes from reverse at Tandun to normal at Pungut. The Tertiary stratal thicknesses differ across the fault (Eubank and Makki, 1981). The fault was therefore rotational and active throughout the Neogene.

4.4.2.1 The grabens

These two trends are illustrated in Fig. 4.13, a map of the NNW–SSE to N–S trending Bengkalis Graben and the NW–SE trending horsts and grabens parallel to the basement grain of Sumatra. The Bengkalis graben is a NNW–SSE to N–S aligned graben 265 km long, aligned at approximately 102° 15′ E. It is the most spectacular of a series of approximately N–S grabens, which were developed upon the Sundaland continent and have been infilled by Pematang Group continental deposits (Moulds, 1989). The grabens are economically important as the sites of deposition of rich lacustrine source rocks. Smaller side grabens (trending NW–SE) or troughs pass outwards from the N–S grabens (Fig. 4.13). The Central Deep (Fig. 4.10) is the most prominent NNW–SSE to N–S graben, lying to the west of Pekanbaru. Like the Bengkalis Graben, it has local deeps of 2438 to 3657 m below land surface (Eubank and Makki, 1981). The Barumun sub-basin, beneath the coastline at 100° E, attains a depth in excess of 3657 m at its depocentre.

The Bengkalis graben halts abruptly at the NE faulted margin of the Tiga Puloh Mountains (Fig. 4.6). Although Moulds (1989) proposed that the graben must have continued through the Tiga Puloh Mountains and been eroded, this is unrealistic. The reasonable hypothesis is to conclude that the southerly extension of the Bengkalis Graben has been right-laterally offset by the faulted NE flank of the Tiga Puloh Mountains, in which case the NW–SE faults were created later than the NNW–SSE to N–S faults and the southerly continuation of the important Bengkalis Graben, which deepens southwards, has to be searched for in the NW in the neighbourhood of Sibolga.

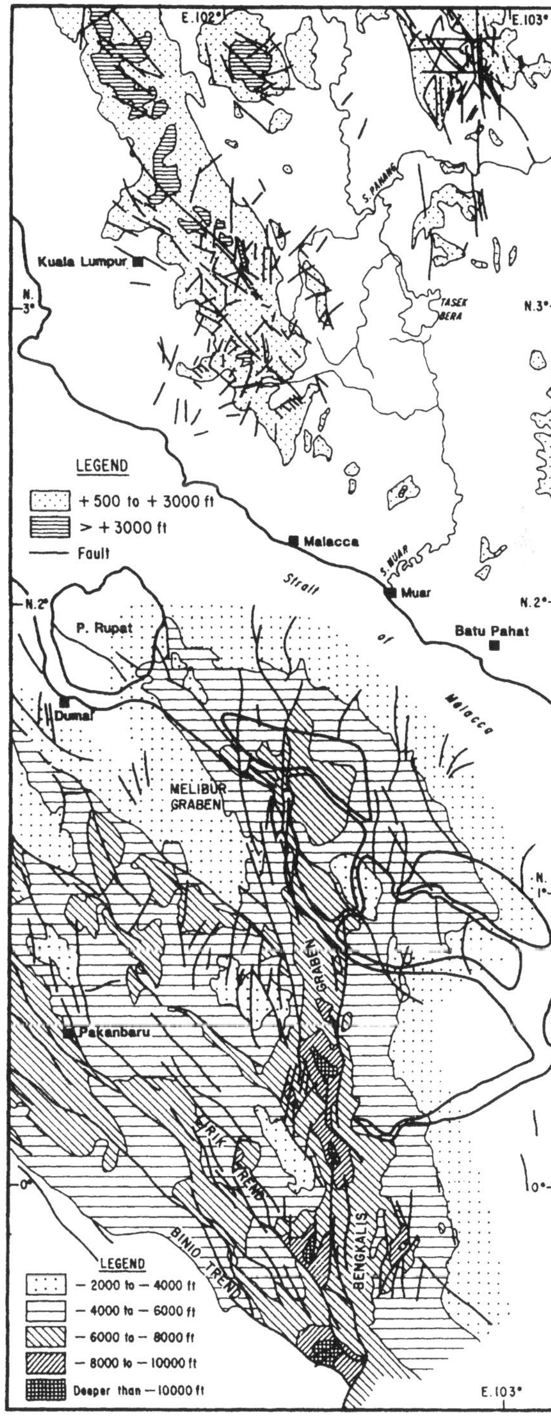

Fig. 4.13. Map of the Bengkalis Graben and adjacent area of Central Sumatra, showing depth to the pre-Tertiary basement in feet, after Moulds (1989). The northwards extension into Peninsular Malaysia shows elevation of the present day land-surface and the main rivers. The Pahang River once flowed southwards as the Muar River and found its way to the sea in south Sumatra along the Bengkalis Graben, before being captured and diverted eastwards towards the Penyu Basin.

To-date it has not been found. The NW–SE trending highs are usually devoid of or are overlain only by thin Pematang Group strata.

Moulds (1989) suggested a model of general E–W extension, utilizing strike-slip motion on the NW–SE lines of weakness of the basement (Fig. 2.4 inset). The basement extension resulted in NW–SE oriented pull-apart depressions or side grabens. Tjia (1985; 1989) presumed that the Palaeo-Tethys Bentong–Raub Suture continues south from Peninsular Malaysia and that the Bengkalis Graben is its location. This interpretation would make the suture discordant with the Indosinian orogenic fabric. This is not possible (Fig. 4.7) for the graben clearly cuts across the basement structure (Hutchison, 1993).

The region from north Peninsular Malaysia to Sumatra represents the Sundaland landsurface that was tilted southwards. Deeper tectonic levels are more deeply eroded and exposed in the north (Gobbett and Hutchison, 1973), and the basement progressively deepens southwards beneath the subsided Central Sumatra Basin (Fig. 4.13).

The map (Fig. 4.13) shows that the main river of Peninsular Malaysia, Sungai Pahang, must have once flowed southwards along what is now known as the Muar River. The present day Muar is a misfit. There is a deep valley beneath the flat alluvial coastal plain, but the river is now of minor importance. River capture, resulting in lakes such as the Tasek Bera, is obvious. The Muar makes a bend to the right, as if it knew where the Bengkalis Graben would lead it southwards to the sea. This suggests the Muar River dates back to the Eocene-Oligocene, and the bending is in keeping with a general right-lateral wrench motion on most of the NW–SE trending faults, one of which lies parallel to the coastline through Muar and Batu Pahat (Hutchison, 1989*a*, p. 59). Right-lateral motion would cause right-handed displacement of the river course. The Pahang–Muar River became dammed up from place to place along its course, causing lakes, which were formed in the deeps (Fig. 4.13). The lakes were caused by movement on the NW–SE oriented faults. The extant lake (Tasek Bera) is therefore worthy of full investigation, because it may date back to the Palaeogene and may overlie a Peninsular Malaysian Tertiary basin.

The major NW–SE faults through the Main Range of Peninsular Malaysia (Kuala Lumpur Fault Zone) belong to the same system as Sumatra (Fig. 4.13). However, whereas they parallel the basement tectonic grain of Sumatra, they are divergent to the Peninsular Malaysian grain (Gobbett and Hutchison, 1973). In Malaysia they were extensional and are commonly infilled by thick (up to 500 m) hydrothermal quartz reefs (dykes), with the spectacular example of Klang Gates near the Kuala Lumpur zoo (Hutchison, 1989*a*, p. 59). Hot springs are also commonly associated with these faults.

On the contrary, the tectonic grain of Peninsular Malaysia is NNW–SSE to N–S, and this direction extrapolates southwards through Sumatra as the Bengkalis Graben, and other cross-cutting faults (Figs 4.6, 4.10).

Movement on the various faults caused damming of the rivers from time to time and from place to place. Impounding was caused by major landslides off the fault scarps. After lakes were thus formed, the rivers eventually found an alternative route to the sea. Thus the Pahang River, which flowed southwards to the sea along the Bengkalis Graben in the late Eocene-early Oligocene, was forced into the alternative eastern route through Pekan to supply sediments to the Penyu Basin (Fig. 4.13).

4.4.2.2 Structure of the Pematang Group basins

The Central Sumatra Basin was formed in the Eocene–Oligocene as a series of half graben structures separated by horst blocks (Figs 4.10 and 4.13). They were filled with non-marine clastic and lacustrine sequences of the Pematang Group, which attains a thickness in excess of 1.8 km in some of the deep grabens. Widespread regional subsidence resulted in progressively more marine conditions throughout the Lower Miocene.

The early rift structures have a predominant half graben shape and N to NNW elongation, with a steep normal bounding fault on the west and SW margin and a gentle hinge margin on the eastern side, which in places may show some normal downfaulting (Fig. 4.12). The basins are characterized either by long linear grabens with dog-leg offsets at intersections with oblique faults, or by *en echelon* basin offset without a connecting dog-leg graben (Fig. 4.11). Major compressional structures occur at the inflection points of the dog-leg grabens. This compression is interpreted as a result of right-lateral motion on the main bounding faults associated with left-lateral movement on the oblique NE–SW faults.

The evolution of the basins may be summarized in four stages (Fig. 4.12):

4.4.2.2.1 Lower Eocene pre-graben stage Gentle crustal updoming followed by weak extension resulted in incipient block rotation along earlier developed planes of weakness. The Lower Red Bed Formation was preserved in these shallow rift basins.

4.4.2.2.2 Middle Eocene graben stage The Brown Shale and Coal Zone formations were deposited in the rapidly developing half grabens.

4.4.2.2.3 Oligocene Pematang structuring stage N-S right-lateral wrench movement resulted in left-lateral transform faulting along the NE–SW fault set, causing zones of compression.

4.4.2.2.4 Upper Oligocene–Lower Miocene lake fill stage Uplift and erosion of the highland areas resulted in rapid deposition of the Lake Fill Formation, culminating with a major unconformity at the end of the Pematang deposition.

4.4.2.3 Folding

A major tectonic event (Sabah Orogeny) occurred during the Middle Miocene, characterized by a hiatus in all but the deeper basins, and alkaline igneous events (see Section 4.3.4.4.).

The Tertiary strata of the Central Sumatra Basin were folded in Upper Pliocene time into a fairly simple sequence of open anticlines and synclines, whose fold axes are horizontal and oriented NW–SE (De Coster, 1974). These folds control the present day distribution of hydrocarbons, most of which are trapped in faulted anticlinal closures. A characteristic fold style has been called the Sunda Fold by Eubank and Makki (1981). Such folds occur in a wrench faulting environment, and may be classified as trans-pressional. The local structure is usually a graben below the fold, with a major bounding fault and smaller extension faults may be developed over the fold crest. There is a thickened wedge of sediments below the fold. The fold style is anticlinal at the crest, and fold amplitude decreases downward, with a half graben or syncline at the fold root. The thickness of the sedimentary unit decreases from the fold axis across the bounding fault or off flank of the fold. Several Sunda folds may exist along the same wrench fault system, with *en echelon* fold axes. The axes are frequently at an angle (transverse) to the wrench fault trace. The bounding fault reverses the sense of throw downwards, with shallow horizons of the fold upthrown and deeper horizons downthrown. Such folds are found in all Sumatran basins.

4.4.3 Source rocks

More than half of Indonesia's established oil reserves occur in the Central Sumatra basin. Most of the reserves are contained in Miocene reservoirs, and the waxy oil was sourced from underlying Palaeogene lacustrine shales. The lacustrine Brown Shale Formation is the primary source for oil, and the lacustrine Coal Zone Formation is gas and condensate prone, with minor oil potential. The average geothermal gradient of 61° C km^{-1} is the highest in South-East Asia and probably reflects hydrothermal activity along the basement fault complex. In contiguous Peninsular Malaysia, this has resulted in major quartz dyke infillings and hot springs associated with the NW–SE oriented faults. The high geothermal gradient would place the oil window at between 900 m and 2500 m.

The algal-amorphous facies of the Brown Shale Formation has been identified in the Balam, Aman and Rangau basins. It is of algal, bacterial and phyto-plankton origin (Williams *et al.*, 1985), characterized by high H/C ratios. The organic richness is indicated by total organic carbon (TOC) contents averaging 2 to 4 wt % in the Balam and Aman basins, rising to as high as 12%. This facies is mature for peak oil generation in the Balam and Aman basins. In the Rangau Basin the TOC averages 1 to 2%, and the Brown Shale Formation is overmature with H/C ratios less than 1.0. The algal-amorphous facies shows good to excellent correlation with the crude oils on the basis of organic geochemistry and stable carbon isotope studies (Williams *et al.*, 1985).

The carbonaceous facies occurs predominantly in the Kiri basin within the Coal Zone Formation. The TOC values are variable, from 1 to 43 wt %. The facies is hydrogen depleted, with low H/C ratios. This facies is generating and expelling gas and subordinate condensate and light oil in the Kiri Basin.

4.4.4 Reservoir and oil properties

The best reservoirs are in the delta front facies of the Lake Fill Formation. The measured porosities range as high as 17.5% with permeabilities ranging up to 122 mD. Log porosities as high as 20% have been noted in some fluvial sandstones.

The crude oils of Central Sumatra are paraffin based with an average composition of about 85% saturates, 10% aromatics and 5% NSO + asphaltenes. The gravities range from 16.5 to 47.0° API, and pour points from 4 to 49° C. Biodegradation and water washing have altered some of the oils. The wax content varies from 4 to 30 wt %. The viscosity varies from 0.5 to 2.0 cp and the sulphur contents are generally less than 0.2 wt %. From a comparison of the geochemistry and carbon isotope data, Williams *et al.* (1985) conclude that the bulk of the oils were sourced from the Eocene–Oligocene Brown Shale Formation lacustrine algal-amorphous facies with overprinting from the carbonaceous facies.

4.4.5 The Minas Field

The super-giant Minas Field, 35 km north of Pekanbaru (Fig. 4.10), was discovered in 1944. The field is a broad low amplitude anticline, 45 km long in a NW–SE direction and 24 km broad, cut by numerous N–S and NW–SE faults (Hasan *et al.*, 1977/78). The main N–S fault offsets the anticline left-laterally by 7 km. The presence of Pematang Formation on the west side of the Minas Fault, and its absence from most of the east side, indicates the fault was active during the Palaeogene. The fault continued to be active and even the Pleistocene Minas Formation gravels show fault offsets.

Minas is the largest known oilfield in South-East Asia. The main oil reservoirs are sandstones of the Lower Miocene Sihapas Group, which unconformably overlies the Pematang Formation and on-laps onto the pre-Tertiary basement horsts. The original oil column was 130 m. The Miocene marine formations (Telisa and Petani formations) are all relatively immature (Sundararaman and Teerman, 1991). Organic material in the Telisa Formation (depths about 610 m) have Ro reflectivities of only 0.44%. Thus, the oil of the Minas and neighbouring fields was not locally sourced, but has migrated up from lacustrine deposits in the contiguous deeper Central or Bengkalis grabens. The reservoir sands are massive and of porosity averaging 28%. Average permeabilities are good, up to 1500 mD. Production began in 1952 and reached a peak from around 200 wells of 400 000 barrels per day in 1969.

4.5 OMBILIN BASIN OF WEST SUMATRA

This is an intramontane basin of 1500 km^2 within the Padang Highlands. It has been well known for coal mining since 1891. Much of the basin area is covered by Recent volcanic flows. The Sinamar No. 1 exploratory well was drilled in 1983 to 3018 m into the Sinamar anticlinal closure.

Up to 4600 m of Tertiary sediments are present in the deeper parts of the basin (Koning, 1985). At least three major unconformities are recorded, and documented by major changes in the vitrinite reflectance values. Seismic and vitrinite studies indicate that in total about 4600 m of sedimentary section have been removed by erosion during the time of these unconformities.

The Tertiary sequence begins with breccias grading upwards into laminated lacustrine organic rich euxinic shales of Eocene age. These Sangkarewang Beds contain a rich fauna of freshwater fish. These lacustrine rocks are considered to be suitable source rocks. During drilling, the geothermal gradient was measured at only 29.6° C km^{-1}, considerably less than for the Central Sumatran Basin (average 60.3° C km^{-1}), despite being localized along the Sumatran volcanic arc. It appears that the heat flow from the volcanic arc is dissipated locally, and does not result in a high regional geothermal gradient. This would place the oil generating window rather deep, at between 3 and 4 km. Although Koning (1985) gave no scale for his vitrinite and spore colouration index, it must be assumed that the Sangkarewang lacustrine shales are well within the oil generation window in the basin deep. These rocks outcrop along the basin margins.

Concurrent with the lacustrine deposition in the basin centre, thick fan breccias, conglomerate and sandstones were deposited along the fault controlled basin margins (the Brani Formation). The lakes would have been impounded between active alluvial fans. A significant intra-Sangkarewang unconformity was encountered at 2700 m depth in the well, indicated by marked changes in vitrinite and spore properties.

The overlying Eocene Sawahlunto Formation is of shale, sandstone and coal, which is mined at Sawahlunto. This is overlain by Oligocene conglomerate, sandstone and shale of the Sawahtambang Formation, that was deposited by braided streams. Coals are rare and thin. A distinct marine influx is indicated near the base of this formation, and its top is also marine (Koning, 1985).

The overlying Lower Miocene Ombilin Formation is of marine calcareous shale, with local coral limestones. This formation would act as a caprock for reservoir sands of the underlying Sawahtambang Formation. Tertiary sediments younger than Lower Miocene are absent. Post-Lower Miocene erosion was therefore spectacular. The Ombilin Formation is overlain by Recent volcanic rocks.

Two oil-seeps are known, of 23° API biodegraded oil, with characteristics similar to the crude oils of the Central Sumatra Basin. A drill-stem test of the Upper Sawahtambang Formation recorded a minor flow of oil of 36° API, 35° C pour point, and gas. The oil is concluded to be sourced from either the Sangkarewang lacustrine shales or from organic rich sporadic intervals within the Sawahtambang and Sawahlunto Formations.

The Ombilin Basin is interpreted as a graben-like pull-apart structure resulting from right-lateral Lower Tertiary motion on the Semangko Fault, actually consisting of a set of at least six major faults which trend 140° and control the basin. Trans-tensional right-lateral faulting on a series of NW–SE trending faults is inferred to have occurred in the Palaeocene and to be responsible for the formation of the first asymmetrical graben (Situmorang *et al.*, 1991). This was followed by dextral motion on a set of two major N–S trending faults, resulting in a second and third graben, all migrating south-eastwards. Transpression related to on-going motion on these faults has produced complicated structures (Situmorang *et al.*, 1991).

4.5.1 Regional correlations

The lacustrine sediments of the Sangkarewang Formation are very similar to the Brown Shale Formation of the Pematang Group of the Central Sumatra Basin. The marine Ombilin shales are regarded as equivalent to the Telisa Formation of Central Sumatra and the Gumai Formation of South Sumatra.

4.6 STRAITS OF MALACCA

The pre-Tertiary basement of the Straits of Malacca slopes gently south-westwards. A number of northerly trending grabens have been mapped, connecting those of North and Central Sumatra as far as the coastline of Peninsular Malaysia (Liew, 1994). From south to north, they have been named: Kukup and Johor (both related to the Bengkalis Graben of Sumatra), Port Klang, Angsa and Sabak (all three are the northwards extensions of the Pematang–Balam Trough of Sumatra), Southern, Central, Eastern, East Penang, West Penang and North Penang (all located between the Asahan and Kepulauan arches), MSS-XA, Northern, Thai Border East and Thai Border West Graben (located north of the Tamiang and Yang Besar highs of Sumatra). The majority have bounding faults along their western margins. Their depth in Malaysian waters ranges from 830 m to 3960 m. Those related to the Central Graben are the deepest.

These grabens contain in their deepest levels alluvial and lacustrine strata equivalent to the Pematang Formation of Sumatra, which should have source rock potential. Seismic sections show the Malaysian grabens to be strongly faulted, which might explain their poor oil retaining potential.

4.7 SOUTH SUMATRA BASIN

There is no sharp boundary between the Central and South Sumatra basins; they are continuous and share many characters (De Coster, 1974). The arbitrary boundary is taken as the Bengkalis Graben extending north from the Tiga Puloh Mountains (Fig. 4.6). However there are some differences, which will now be described.

4.7.1 Structure

The NW–SE oriented basement horst and graben faults continue south-eastwards from Central Sumatra and also dominate South Sumatra (Fig. 4.14). However the depressions are even deeper and larger in the south. The prominent Palembang Depression is at a depth exceeding 4500 m (De Coster, 1974). The Benakat Gulley is flanked on the west by faults and is the only basement feature which maintains the N–S alignment so typical of the Central Sumatra Basin. It connects in the south with the Lematang Depression (Fig. 4.14).

There is a distinctive NE fault trend, which is not found elsewhere, and it takes the place of the NNW–SSE to N–S trend, which is less important in the South Sumatra Basin. The NE trend forms the margin of the prominent Jambi Depression and extends north-eastwards into what is known as the Corridor Area. The rugged horst and graben topography existed before and controlled the Eocene to Oligocene sedimentary pattern (Pulunggono, 1986).

Most of the strata were folded along NW–SE trending horizontal axes, just as in the Central Sumatra Basin. Compression began in the Middle Miocene and continued through to the Pliocene (Pulunggono, 1986). The surface outcrops show prominent anticlinoria trending NW–SE: that of North Palembang, trending SE through Palembang, the Pendopo–Limau anticlinorium trending SE through Prabumulih, and that of Muara Enim, which borders the Gumai and Garba Mountains (Pulunggono, 1986). However, a few small stable basement blocks avoided this deformation, notably the Palembang High, the region of the Kluang Limestone basement (Fig. 4.7), the Musi High and the NE Corridor area.

The Corridor Block lies on the coastal area of South Sumatra, immediately west of Bangka Island, and to the NW of Palembang. The most obvious structural elements are NW–SE trending folds, which resulted from the Plio–Pleistocene compression directed NE–SW (Zeliff *et al.*, 1985). The NW–SE fold pattern was superimposed on the pre-existing grain of the basement. The Block consists of four major NW–SE trending anticlinoria, cross cut by numerous NE–SW striking normal faults.

An old N–S structural grain exists beneath the shallow regressive clastic sequence (Gumai, Air Benakat, Muara Enim). The N–S structures in this basement have been reactivated, and right-lateral wrench movement of the basement faults can be elegantly held responsible for the NW–SE trending folds and the NE–SW striking normal faults.

4.7.2 Stratigraphy

4.7.2.1 Kikim Tuffs

The oldest rocks that overlie the pre-Tertiary basement have been encountered by drilling in the south and as outcrops in the Gumai Mountains. They are tuffaceous sandstones, conglomerates, breccias and clays, generally held to be Upper Cretaceous to Palaeocene, but they are not widespread (De Coster, 1974).

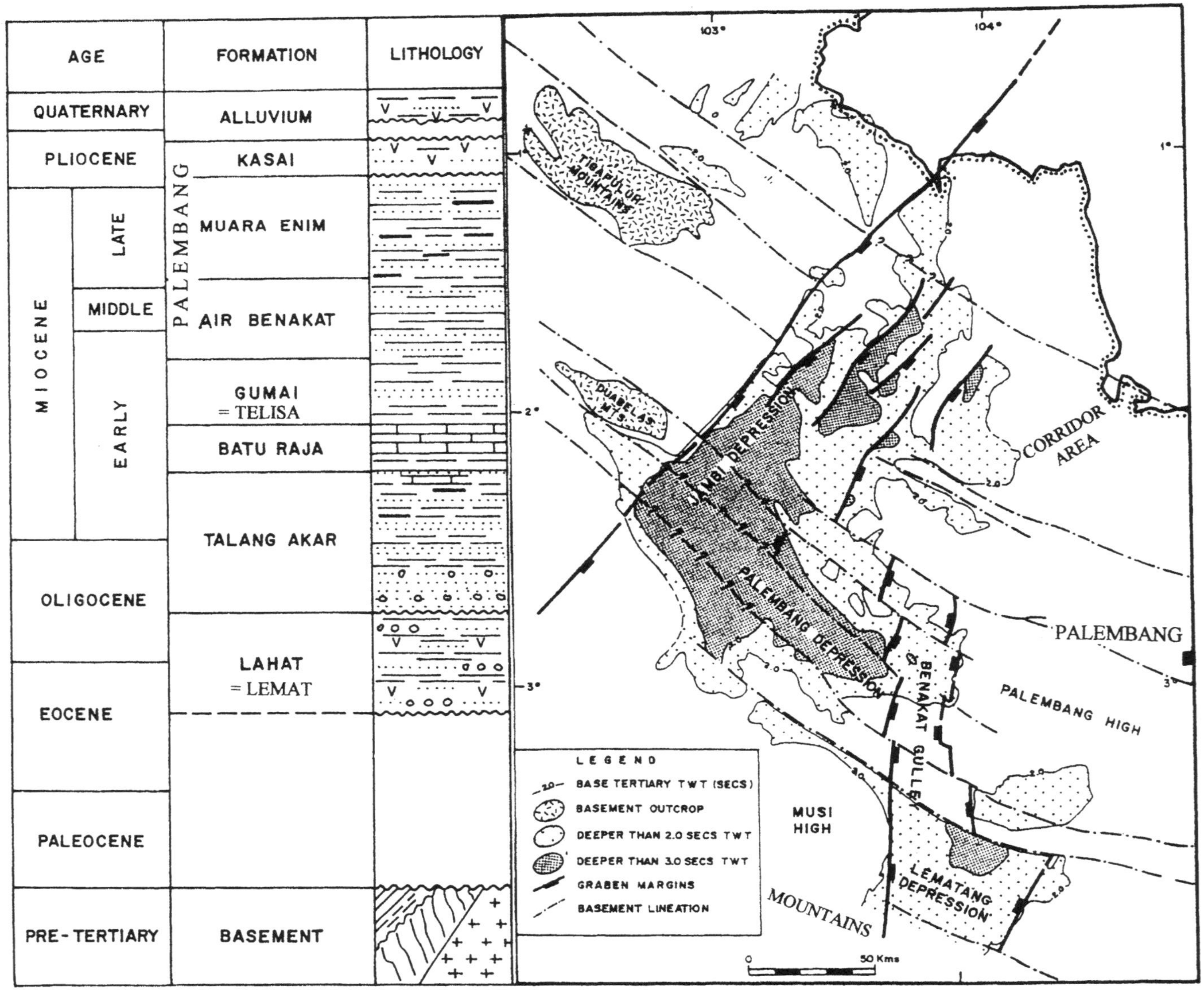

Fig. 4.14. Stratigraphy and basement morphology of the South Sumatra Basin, modified after De Coster (1974), Moulds (1989) and Zeliff *et al.* (1985).

.7.2.2 *Lahat (= Lemat) Formation*

n previous literature, the Kikim Tuffs was known as the Old Lemat', and what is now referred to as the Lemat or ,ahat Formation was previously known as the 'Young ,emat'.

The lowest levels are frequently described as 'granite wash'. The Lahat is a coarse clastic continental deposit of sandstones, clays, rock fragments, breccias, with occasional coal beds and tuffs. The Benakat Member of the Lahat Formation (which should not be confused with the Air Benakat Formation) occurs in the central portion of the basins, composed of grey-brown shales, with some beds of tuffaceous shale, siltstone and sandstone, and occasional thin coal beds, carbonate stringers and glauconite (De Coster, 1974). It was deposited in brackish water and lies conformably upon the coarse clastic lower part of the Lemat Formation.

The thickness of the formation is highly variable, generally more than 760 m but ranging from zero to 1070 m in one fault depression in the central part of the basin. The formation is probably mainly Upper Eocene to Oligocene. An unconformity marks its top.

4.7.2.3 *Talang Akar Formation*

This sandy formation unconformably overlies the Lahat Formation and onlaps onto the pre-Tertiary basement. It is conformable with the overlying Telisa Formation. The Talang Akar Formation is composed of delta plain sandstones, siltstones and shales that grade basinwards into marginal marine sandstones and shales. The thickness varies from zero around basement highs to 610 m in some deeper basin areas. Usually, where the Lemat is thick, the Talang Akar is also thick and *vice versa*. The age is Upper Oligocene to Lower Miocene (Fig. 4.14).

4.7.2.4 *Batu Raja Limestone*

The limestone was deposited in shelfal parts of the South Sumatra Basin, on and around platforms and highs. It is also known as the Basal Telisa Limestone. It is a platform carbonate topped by further build-ups in restricted localities. Thickness on the shelf areas averages 60 to 75 m, with an additional 60 to 120 m in the build-up areas. The Batu Raja Limestone outcrops in the Garba Mountains, where it is 520 m thick.

4.7.2.5 *Telisa (= Gumai) Formation*

This Lower Miocene formation is the most widespread, being deposited at the time of maximum marine transgression. It is typically a fossiliferous marine shale containing occasional thin beds of glauconitic limestone. On the basin rims and shelfal areas, it contains fine sandstones and siltstones. The overlying contact with the Lower Palembang (= Air Benakat) Formation is difficult to select consistently on seismic records. Thickness is variable, but may be as great as 2700 m in the trough areas. A minor hiatus occurs in the uppermost Telisa and the lower Palembang Formation on the basin edges, but is not discernable in the basin interiors.

4.7.2.6 *Lower Palembang (= Air Benakat) Formation*

This formation was deposited at the early stage of a regressive sequence. It is composed of shale, with glauconitic sandstone and occasional limestone, neritic to shallow marine upwards. The upper contact with the Middle Palembang Formation is marked by coal beds. Formation thickness is variable, usually ranging from 1000 to 1500 m.

4.7.2.7 *Middle Palembang (= Muara Enim) Formation*

This formation is shallow marine to brackish at the base, becoming non-marine upwards. It is composed of sandstones, mudstones and coal beds. The number of beds and thicknesses of the coal decrease from south to north. Maximum formation thickness may reach 750 m. The age is usually interpreted to be uppermost Miocene to Pliocene.

4.7.2.8 *Upper Palembang (= Kasai Tuff) Formation*

This Formation comprises tuffaceous sands, clays and gravels and occasional thin coal beds, with great thickness variation and composition (Fig. 4.14). The formation usually occurs in the synclines of folds formed during the Plio Pleistocene deformation.

4.7.3 Source rocks and reservoirs

The Talang Akar Formation contains the main source rocks, but some hydrocarbons have been sourced from the Gumai Formation (Hartanto *et al.*, 1991). The South Sumatra Basin has a considerably lower geothermal gradient than Central Sumatra, averaging 49° C km^{-1}, so that the oil window is at a greater depth.

Oil is found reservoired in several formations. Along the south flank of the South Palembang sub-basin, oil is produced from the highly faulted basement of granodiorite and quartzite, some 100 km SW of Palembang (Sardjito *et al.*, 1991). The principal source is the Talang Akar Formation, which directly overlies the basement.

Throughout South Sumatra, oil is reservoired in the Talang Akar, Gumai, Air Benakat, and the Lower Muara

Enim formations, while gas is reservoired mainly in the Batu Raja Limestone (Zeliff *et al.*, 1985). The Batu Raja Limestone commonly has a porosity of 15% (Hartanto *et al.*, 1991). In the Corridor area, 78 km NW of Palembang, oil is reservoired in and produced from a simple anticline in the Talang Akar Formation sandstones (Hasan and Soebandrio, 1988), which are still largely above the oil window.

4.8 SIBOLGA FORE-ARC BASIN

This basin, usually called the West Sumatra Basin, was explored from 1968 to 1978 (Rose, 1983). Nineteen wells were drilled, and six tested gas. However the basin was abandoned. There was renewed interest in 1994 and additional wells have been drilled to the east of Nias island, witnessing the greater regional interest in gas. Caltex drilled a successful gas well in 1994. The low geothermal gradient means that potential source rocks are rather deep. It is interesting that the tendency for South-East Asian limestones to be gas prone is unrelated to geothermal gradient (c.f. North Sumatra Basin, South Sumatra Basin and Luconia Province).

4.8.1 Structure of the basin

West of Sumatra, the outer arc ridge forms a chain of islands. Nias island, for example, is composed both of tectonically mélanged accreted material (Oyo Complex), and slope strata of Middle Tertiary age (Moore and Karig, 1980). Coherent slope strata occur between belts of chaotic mélange and are strongly folded and faulted. A steep monoclinal flexure along the eastern side of Nias marks the seaward flank of the Sibolga fore-arc basin (Fig. 4.15). Sediments of the fore-arc basin are structurally upwarped against the outer arc ridge. This seems to suggest that the fore-arc basin became downwarped as the outer arc ridge rose. Moore *et al.* (1980) do not support the idea of Hamilton (1979) that arc-directed gravitational flow from the outer arc ridge has deformed the fore-arc basin sediments. The fore-arc basin strata overlie a seaward-dipping basement of seismic refraction velocity 6.0 km s^{-1}. A 30 mgal free-air gravity low over the fore-arc basin may be modelled best if pre-Miocene mélange or continental crust underlies the basin, and it seems unlikely that the basin can be underlain by a large thickness of oceanic crust (Kieckhefer *et al.*, 1981). Sedimentation in the fore-arc basin was probably predominantly by longitudinal flow along the basin axis.

The Sibolga Basin is a strongly subsiding fore-arc trough bounded by the growing and rising accretionary prism (outer arc ridge) on the seaward side, and the elevating continental Sundaland margin (Karig *et al.*, 1980). Their interpretation of the data shows the progressive growth of the subduction accretionary wedge, largely constructed of scraped-up and deformed Nicobar Fan sediments, and the uplifting continental margin, while the intervening fore-arc basin continued to deepen (Fig. 4.15). Up to 4 km of Miocene to Recent strata on the eastern flank of the basin have been transgressed over an unconformity cut across a Palaeogene continental margin that was uplifted during the Oligocene. The disruption of the continental margin can be attributed to right-lateral wrench faults that splay across the fore-arc basin from the Semangko transform fault. Offsets along the fault range up to 100 km or more, producing small turbidite filled sub-basins trapped behind the growing accretionary prism. The larger of these re-entrants may be floored by oceanic crust, but most of the fore-arc basin is thought to be floored by continental crust. Folding of the basin strata occurs only adjacent to large displacement faults (Rose, 1983).

4.8.2 Stratigraphy

The sedimentary fill of the basin is greater than 3 km. There is a prominent angular unconformity between the Palaeogene and Neogene, readily identified on seismic sections. It has been drilled in several places. The unconformity is more noticeable in the north. The basin fill began with a Middle Miocene mudstone and sandstone non-marine to transitional marine 'basal clastic unit', which unconformably overlies pre-Neogene basement of bathyal mudstone with minor siltstone and sandstone (Rose, 1983). Chalky microcrystalline limestone is important in the Middle Miocene, forming a widespread shelf with scattered reef build-ups. It is overlain by interbedded micrite and bioclastic packstone. A Middle Miocene 'Transgressive unit I', of deeper water mudstone with thin sandstones, overlies the carbonate unit. Carbonate again becomes important in the Upper Miocene, of interbedded micrite and bioclastic packstone with reefal build-ups at the top. Reefs are fewer in the south. There is a lateral facies change to quartz sandstone and thin packstone beds (Rose, 1983). The carbonates form an extensive shelf with isolated reefs. The overlying Pliocene 'regressive clastic unit' is of widely deposited sand, clay, silt with some limestone and coal beds and delta deposits.

4.9 MENTAWAI FORE-ARC BASIN

This basin was explored between 1969 and 1974, and only superficial descriptions have been published (Hariadi and Soeparjadi, 1975) and 2 exploratory wells were

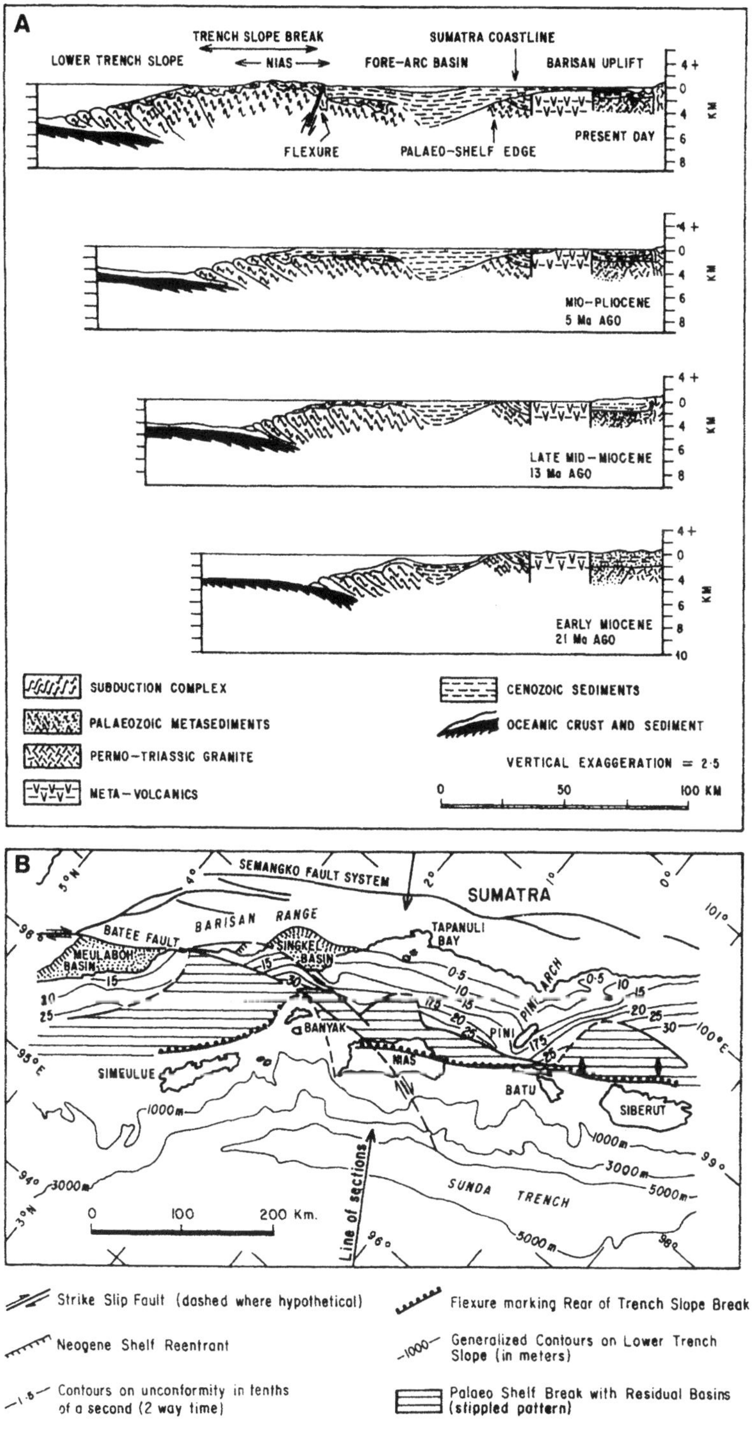

Fig. 4.15. **A**: An interpretative geological cross section across the Sumatran continental margin showing growth of the accretionary prism and Sibolga fore-arc basin in the vicinity of Nias (after Karig *et al*., 1980). **B**: Tectonic elements of the Sibolga fore-arc basin and the continental margin of the Nias area of Sumatra.

drilled. In 1990–1993 a consortium of companies held an exploration contract for an area offshore Bengkulu. One oil seep has been mentioned onland near Bengkulu town.

Four different horizons have been differentiated in the seismic records. The upper horizon is referred to as the Young Transgressive Series, of marl, sand, clay, silt and reefal limestone, thought to be of Upper Miocene and Pliocene age. In the coastal belt it is known as the Eburna Marl, of sandy clay, conglomerate, and tuff with coal seams. The underlying Limestone–Marl Series, of average 600 m thickness, is considered to be of Middle Miocene age. It is of lime clay, shale, marl, and tuffaceous sandstone. It is overlain in the coastal belt by the Serangai Series, which contains conglomerate and coal beds. The underlying Tuff–Marl Series, about 1000 m thick, is of tuffaceous clay with minor sandstone and marl, considered to be Lower Miocene. There is an equivalent Brown Series in the Bengkulu coastal belt. The Oligocene Tuff Series of 1000 to 1400 m is of tuffaceous sandstone, clay, breccia, and conglomerate. Underlying this is the undifferentiated basement (Hariadi and Soeparjadi, 1975).

The basal section contains carbonaceous shales, coals and sandstones and is of non-marine to delta facies. Very little information has been published on this basin. Possible source rock shale of Lower Miocene age was drilled at a depth of 1812 m in the Bengkulu X2 well.

4.10 SUNDALAND SOUTH OF THE RAJANG BASIN

The Palaeocene landmass of Sundaland extended eastwards from the Malay Peninsula and Sumatra at least as far as Kangean, eastern Borneo and West Sulawesi (Van de Weerd and Armin, 1992) (Fig. 2.2). The eastern margins of this landmass are difficult to define, for they have been disrupted by many marginal basins, such as the Eocene Celebes Sea, and younger Makassar Straits and Sulu Seas.

In western Sulawesi, Cretaceous ophiolite and flysch strata are unconformably overlain by the terrestrial Malawa Formation (Hasan, 1991; Coffield *et al.*, 1993). It contains abundant quartz sandstones, siltstones and coal beds. As in Borneo, the Lower Palaeogene clastic strata, locally of redbed facies, are widely transgressed by Oligocene to Miocene Tonasa Formation platform carbonate (Hasan, 1991).

The now uplifted Rajang Group of Sarawak and Sabah, in Palaeocene time was either a major marginal sea or a gulf of the Pacific Ocean. Its oceanic or marginal sea ophiolitic basement is overlain by Lower Cretaceous (Barremian) chert, which overlies ophiolite, is now exposed along the Lupar Line and in N.E. Sabah (Hutchison, 1991*a*; Rangin *et al.*, 1990). The sedimentary fill of this marginal sea or gulf is represented by the predominantly turbiditic Belaga Formation of Sarawak, the Trusmadi Formation and its correlatives of Sabah. This basin persisted until the late Eocene in Sarawak and contiguous Kalimantan, but the Crocker Basin persisted till the early Miocene in Sabah, when it was compressed and uplifted to form the provenance for the Baram and other deltas which prograded northwards. A possible scenario for the early development of the Rajang Group is given by Ru and Pigott (1986), though they did not recognize the older age of the chert associated with the ophiolite, or by Taylor and Hayes (1983), who more generally referred to the basin as being floored by 'Mesozoic oceanic crust', representing a gulf of the Pacific Ocean into Sundaland.

The pre-Tertiary basement of Sumatra and Peninsular Malaysia is of Palaeozoic and Mesozoic rocks. This part of the Palaeocene Sundaland landmass may accordingly be considered to have been cratonic. However, the Sundaland peninsula extending eastwards from the southeast coast of Sumatra did not appear to have been cratonic, for the pre-Tertiary formations are dominantly Cretaceous, many of uplifted ophiolite, mélange and turbidite (Pieters and Supriatna, 1990). It may represent terrains accreted to Sundaland during the late Mesozoic to have become part of the extensive Palaeocene landmass.

4.10.1 The nature of the pre-Eocene landsurface

The dismembered Meratus–Bobaris ophiolite outcrops predominantly as variably serpentinized peridotite and banded gabbro, with plagiogranite; basalt is of minor occurrence. A meta-dolerite dyke was K:Ar dated 116 Ma (Sikumbang, 1986). Associated metagabbro and amphibolite yielded a K:Ar date of 108 Ma.

The ophiolite is in tectonic contact with the Pudak Formation of the Alino Group, and mélange contains clasts of ophiolite and chert. The chert has been dated by radiolaria as Lower Cretaceous, and it is associated with turbidite sandstone. Although field relationships are highly faulted, there can be little doubt that the ophiolite and associated deep water sediments are of Lower Cretaceous age, a similar situation to the widespread ophiolite and chert-spilite association throughout Sabah, Kalimantan and W Sulawesi.

As shown by Sikumbang (1986), Supriatna *et al.* (1980), Umar *et al.* (1982) and Rustandi *et al.* (1981), the ophiolite is extensively overlain by the Upper Cretaceous (Turonian) Pamali Formation of the Manunggul Group. The basal conglomerate directly overlies the ophiolite and has long been wrongly known as the Pamali Breccia. Thirteen small diamonds were recovered from a 10 ton bulk weathered sample (grade 35 carats per 100 t). Subsequent work has clearly shown that this rock is a sedimentary conglomerate and is not the primary source of diamonds (Bergman *et al.*, 1987). In thin-section, the Pamali conglomerate contains predominantly serpentinite and pyroxenite clasts, with lesser norite, chert and gabbro. The matrix is largely formed of polycrystalline calcite, with some finely divided ultramafic material. Accordingly, it must be concluded that the Meratus–Bobaris ophiolite and chert of the Alino Formation were uplifted and eroding by late Cretaceous time. The serpentinite conglomerate and associated serpentinite sandstones of the Pamali Formation have an analogy in the Bidu-Bidu Hills of Sabah (Hutchison and Tungah Surat, 1991).

The Pamali Formation is interpreted as deposited in a shallow marine environment, and the strata contain shallow water bivalves and gastropods (Sikumbang, 1986).

The Meratus complex also includes Upper Cretaceous volcanic rocks, which together with the ophiolite, indicate a Cretaceous convergent plate margin environment. Sikumbang (1986) proposed collision, followed by strike-slip slicing up of the complex. The tectonic complexity makes it difficult to decipher, but there is no doubt that the ophiolite and its sedimentary and volcanic envelope were uplifted and eroding by late Cretaceous time, and cannibalized into the late Cretaceous Pamali and Lower Eocene Tanjung conglomerates and coarse sandstones. All rocks of the Meratus complex were equally inundated by the Tertiary sediments, so that the Meratus Mountains foundered through the Tertiary and re-emerged as a mountain range only in the Upper Miocene, so that only the Pliocene–Pleistocene coarse clastic Dahor Formation does not appear to have been deposited on the uplifted Meratus.

4.10.2 South Kalimantan tectonics

In Palaeocene time all of Borneo, south and south-east of the Rajang Group turbidite basin (Danau Sea), was an extensive landmass, contiguous with West Sulawesi and the area now below the Java Sea, and dominated by the Schwaner Mountains High (Fig. 2.2).

The outcropping rocks of the Palaeocene landsurface were predominantly of Cretaceous age. The highlands of the Schwaner Mountains were dominated, as today, by Lower Cretaceous I-type granitoids and Upper Cretaceous volcanic rocks (Williams *et al.*, 1988). Outside of the mountain zone, the outcrops were also predominantly of Cretaceous rocks, composed of abundant ophiolite suite, deep water turbidite strata, volcanic rocks and extensive mélange (Pieters and Supriatna, 1990; Wain and Berod, 1989; Sikumbang, 1986). There was no sign of a Phanerozoic cratonic stratigraphic continuity and the general impression gained from details from the Semitau High (Williams *et al.*, 1988), Meratus Mountains (Sikumbang, 1986) and the Upper Kutei area (Wain and Berod, 1989) suggest that this part of Sundaland was built by Cretaceous accretion onto Sundaland of marginal basin lithosphere and turbidite sedimentary fill, island arcs and numerous mélange terrains.

I am also proposing that Sabah east of the Crocker Range (Rajang Basin) was also an integral part of this non-cratonic Sundaland landmass. There is a large hiatus in the sedimentary record of Sabah. The basement is of marginal sea or oceanic lithosphere overlain by Barremian (Lower Cretaceous) chert (Hutchison, 1991*a*) and undated turbidite sandstones of the Chert Spilite Formation. Uplift and erosion of this basement is indicated by serpentinite and chert-bearing sandstones and conglomerate in the Bidu-Bidu Hills (Newton-Smith, 1967; Hutchison and Tungah Surat, 1991). The serpentinite sandstones and conglomerates of Sabah occupy a similar tectonic position to the ophiolite-derived coarse clastics of the Pamali Breccia of the Meratus Mountains (Sikumbang, 1986; Bergman *et al.*, 1987). The latter is an Upper Cretaceous conglomerate, which contains detrital diamonds, scavenged by the rivers from cratonic Sundaland, and carried to the outer non-cratonic accreted part of the Sundaland land surface. This complex landmass, composed predominantly of Cretaceous complexes, was widely unconformably overlain by continental formations of Eocene age (Fig. 4.16).

4.10.2.1 Eocene extension phase

The early graben phase is best described in the Meratus area (Figs 4.17, 4.18), in which the Tanjung Formation was deposited, consisting of basal conglomerate, sandstones, shales and coals, with local basalt intercalations, regarded as Eocene by Van de Weerd and Armin (1992) but as old as Palaeocene by Kusuma and Darin (1989). The lacustrine coal-rich deposits provide important source rocks for the Barito Basin.

Similar alluvial and marginal marine strata occur in the Ketungau, Melawi (Fig. 4.16) and Mandai basins (Pieters *et al.*, 1987). In the Upper Kutei Basinal area (Fig. 4.16), up to 1500m of Kiham Haloq Formation outcrops in the Mahakam rapids. It is of cross-bedded and massive conglomeratic sandstone overlying a basal red-chert conglomerate (Wain and Berod, 1989), interpreted as an alluvial fan deposit. The abundance of chert in the basal conglomerate has been provenanced from Lower Cretaceous ophiolite and its deep water sedimentary cover, which formed an important part of the pre-Tertiary landsurface. Whereas Wain and Berod (1989) consider

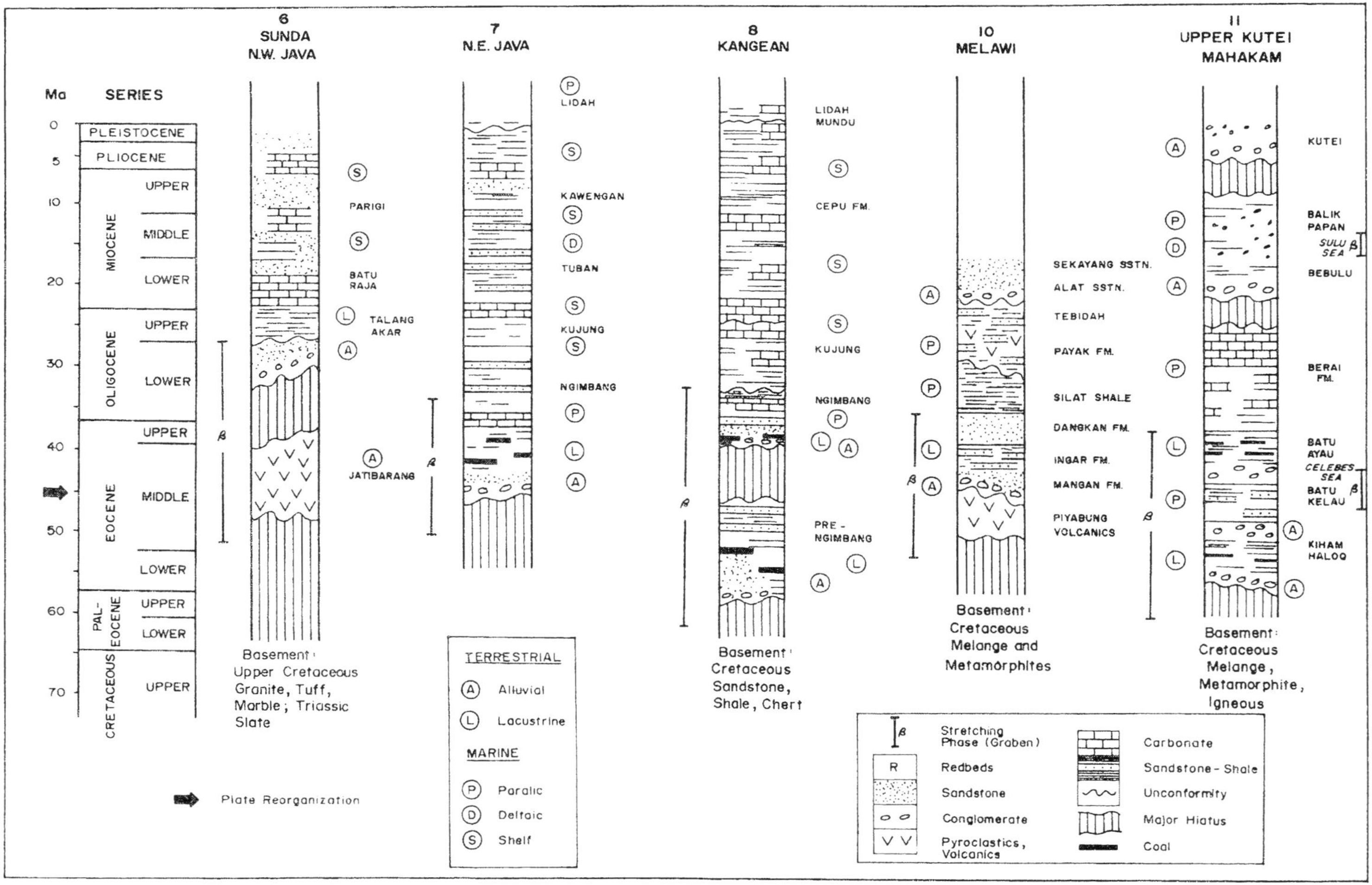

Fig. 4.16. Some typical stratigraphic sections of Tertiary basins of Sundaland south of the Rajang Group of Borneo. **6**: after Gordon (1985) and Reminton and Pranyoto (1985). **7**: after Soeparyono and Lennox (1989). **8**: after Kohar (1985) and Phillips *et al.* (1991). **9**: (given in Table 3.2), after Kusuma and Darin (1989). **10**: after Pieters *et al.* (1987). **11**: after Wain and Berod (1989) and Van de Weerd and Armin (1992). See Fig. 2.2 for localities.

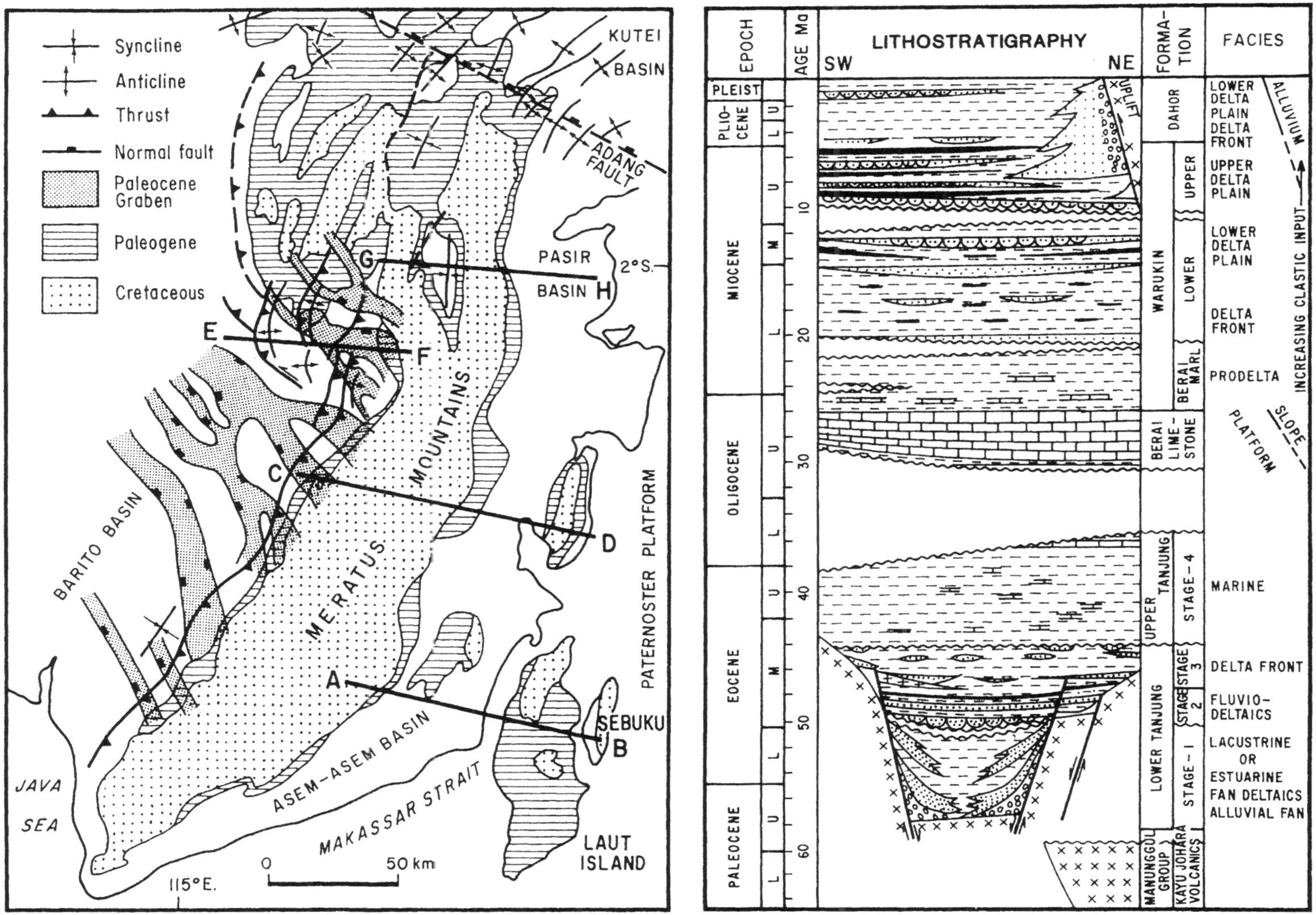

Fig. 4.17. The Palaeogene grabens and lithostratigraphy of the Barito Basin (**9** on Fig. 2.2), S.E. Kalimantan, based on Kusuma and Darin (1989) and Van de Weerd and Armin (1992). The located cross sections are given in Fig. 4.18.

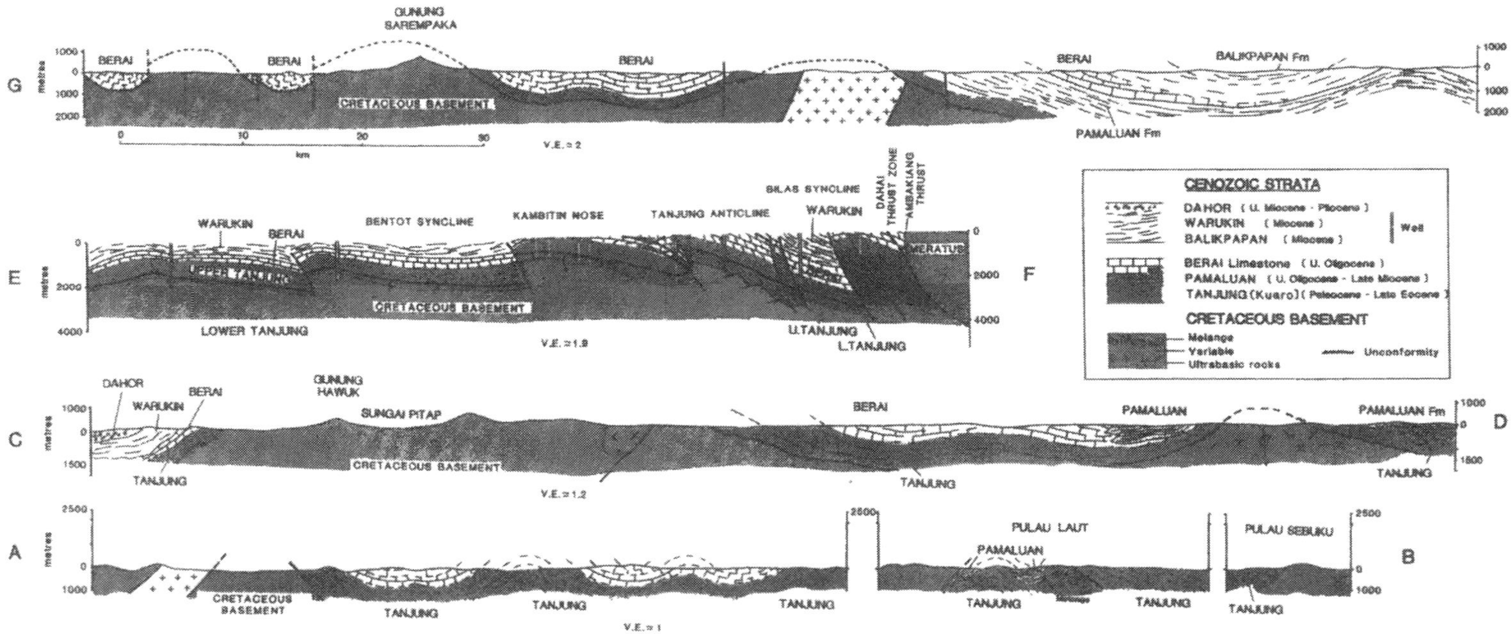

Fig. 4.18. Typical cross sections across the eastern Barito Basin (**9** on Fig. 2.2), the Meratus Mountains uplift, and the Pasir Basin of S.E. Kalimantan. **A-B** after Rustandi *et al.* (1981); **C-D** after Supriatna *et al.* (1980); **E-F** after Kusuma and Darin (1989) and **G-H** after Umar *et al.* (1982). Location of cross sections given in Fig. 4.17.

the Kiham Haloq Formation to be as old as Palaeocene, Van de Weerd and Armin (1992) regard it as Eocene.

In Sabah and contiguous north Kalimantan, the graben development stage was delayed to the late Lower and Middle Miocene, well displayed in the Tarakan Basin (Van de Weerd and Armin, 1992) and the connecting rift system of Sabah, which received the Tanjong and Sandakan formations (Hutchison, 1989*a*). This rift system of North Kalimantan and Sabah is the onland extension of the Sulu Sea marginal basin rifting, recently documented by Hinz *et al.* (1991).

4.10.2.2 Marine transgression phase

By late Oligocene time, Sundaland south and east of the Rajang Basin was extensively inundated by a shallow marine carbonate platform (Figs 2.5, 4.16). It is known as the Berai Limestone in the Barito, Asem-Asem and Pasir basins (Figs 4.16, 4.18) and in the Upper Kutei–Mahakam area (Wain and Berod, 1989).

The carbonate platform is absent from the more pronounced rifts which became the focus of major river and delta outgrowths. The siliciclastic-dominated areas are mainly the Eocene Kutei Basin, extrapolating WNW to the Melawi and Ketungau basins (Van de Weerd and Armin, 1992) and the Miocene rift system of the Tarakan Basin and contiguous Tanjong Formation basins of Sabah.

4.10.2.3 Basement uplift phase

The Meratus Mountains did not represent a topographic high from Palaeocene through Middle Miocene. This is well documented by the cross sections constructed across the Meratus Mountains (Fig. 4.17). The Upper Oligocene Berai Limestone occurs on both sides of the Meratus (Barito and Pasir basins) as well as on top of the Meratus (Fig. 4.18). There are no E–W facies and thickness changes and therefore the Meratus was not a palaeo-topographic feature. Throughout the period from Oligocene through Middle Miocene, the Barito was continuous with and was one and the same with the Asem-Asem and Pasir basins (Fig. 4.18). Because these basins were superimposed on the uplifted and peneplained Meratus, they should not be classified in relation to any Cretaceous or earlier tectonic interpretation of the underlying Meratus assemblage.

Uplift of the Meratus ophiolitic mélange took place in the Upper Miocene–Pliocene, as recorded in the thick non-marine Dahor Formation, which is found only in the Barito Basin, but not on top of the rising Meratus Mountains (Figs 4.17, 4.18). Its clast petrography indicates derivation from erosion of the uplifted Meratus.

The rapid Upper Miocene to Pliocene uplift of the Meratus could be held responsible for the steeply dipping westwards overthrusts in the Tanjung Field (Fig. 4.18, section E-F). However, the cause of uplift remains obscure. The mélange formation of the Meratus may well have been developed during this Neogene uplift phase. By analogy, it may be suggested that the Semitau Ridge, predominantly composed of mélange, may likewise have risen up to divide the Melawi from the Ketungau and Mandai basins, which in the Palaeogene may have been one and the same basin. To classify such basins as fore-arc is misleading. The Boyan Mélange even contains clasts of granite, unlikely to be found in an accretionary wedge, as interpreted by Williams *et al.* (1988). Like the Barito, Asem-Asem and Pasir, these basins were formed by rifting of a landsurface composed of uplifted Cretaceous ophiolite, deep water sediments and mélange.

4.10.3 Barito Basin

The Barito Basin, located in SE Borneo, is now separated from the Asem-Asem and Pasir areas by the Meratus Mountains. The stratigraphy of these basins (Figs 4.17, 4.18) indicates that the three were from Eocene through Miocene formerly parts of the same single basin. The prominent Meratus Mountains were not a topographic feature in early Tertiary time. The whole region was an integral part of a very extensive peneplained Upper Cretaceous landmass, representing a large peninsular extension of Sundaland into the South-East Asian seas. It can be shown to have extended across the whole of central Borneo as far NE as Sabah and as far SE as eastern Java, the Java Sea and W Sulawesi, where Eocene continental sandstone and conglomerate unconformably overlie a Cretaceous basement (Djuri and Sudjatmiko, 1974).

4.10.3.1 Stratigraphy

The first sediments to accumulate on this landsurface are difficult to date. They are predominantly Eocene, but may extend back to the Palaeocene. They are of alluvial facies (the Tanjung Formation of the Meratus), followed by important lacustrine deposits (Fig. 4.17). The Lower Tertiary landsurface therefore had a well developed fluvial system, which flowed from the interior highlands of Sundaland lying to the west and NW (Peninsular Malaysia, Sumatra and Thailand). The Lower Tanjung Formation accumulated preferentially in grabens and half grabens. This period of faulting caused the initiation of sediment accumulation in depressions, and rivers were dammed by active faulting to create important lakes, which accumulated source rocks for the oilfields. The

Lower Tanjung Formation is of basal conglomerates, sandstones, coals, shales and local basalt. The graben phase was localized and was rapidly followed by regional sagging, eventually leading to widespread marine incursions over the former landmass.

The overlying marine sequence consists of Upper Eocene shales and sandstones and uppermost Eocene, Oligocene and Lower Miocene carbonates of the Berai Formation. Both the Tanjung and the Berai formations are about 1100 m thick near Tanjung.

Regressive deltaic deposits of the Warukin Formation overlie the Berai Formation. They are probably of Lower to Middle Miocene age. They were probably provenanced from the north and NW and are up to several thousands of metres thick near the Meratus Mountains depocentre. Seismic profiles show the presence of angular unconformities near the top of the Warukin Formation. These unconformities, and thick non-marine coarse clastics of the unconformably overlying Upper Miocene to Pliocene Dahor Formation, record the emergence of the Meratus Mountains to form a ridge subdividing the basins.

4.10.3.2 Source rocks and reservoirs

Non-marine Eocene sandstones near the base of the sequence produce oil in the Tanjung field (Siregar and Sunaryo, 1980). Geochemical and bio-marker studies of the oils suggest they are derived from a non-marine source and can be correlated to the high-quality source rocks in the non-marine Eocene sequence, present in many places in the Barito, Pasir and Asem-Asem basins (Fig. 4.17).

Thick channel sands near the base of the Miocene Warukin Formation produce oil in the Warukin and Tapian Timur fields in the northern Barito Basin (Van de Weerd and Armin, 1992). Analyses suggest that these oils are also derived from higher plants of a non-marine source. In this case they may be from nearby mature Miocene coals and coaly shales. About 180 million barrels of recoverable oil have been found in this area, most of which has been produced, but there still exists a potential for undiscovered small fields.

4.10.4 Pasir and Asem-Asem basins

Upper Eocene and Oligocene deep marine shales with rare basalt flows overlie Middle Eocene non-marine to shallow marine clastics in the Pasir Basin (Baumann, 1972). The Eocene–Oligocene shallow water Tanah Grogot Limestone platform occurs just east of this deep water facies. The sequence is overlain by up to 1000 m of shallow to marginal marine Middle Miocene clastics. The Lower Miocene carbonate shelf edge of Labangka marks the rapid transition of the shallow Pasir Basin to the deep Kutei Basin, and it continues SE as the Adang Fault (Fig. 4.17).

The Asem-Asem Basin is now separated from the Pasir Basin by the basement uplifts of Kukusan and Laut Island (Fig. 4.18). Its sedimentary sequence is similar to that of Pasir and Barito. However, the Eocene–Oligocene deep marine facies of Pasir did not extend into Asem-Asem. There is an absence of Oligocene tectonic events in the Barito, Pasir and Asem-Asem basins.

4.10.4.1 Oil and source rocks

The Kendilo 1 well in Pasir tested oil from Miocene sands. Analyses suggest that the oil had migrated from early mature non-marine Eocene source rocks. There are commercial Eocene coal beds in both the Pasir and Asem-Asem basins.

4.10.5 Paternoster Platform

The offshore Paternoster Platform is contiguous with the Pasir Basin. The northern part of the Platform has a relatively thin Tertiary cover of up to 1800 m. Drilling indicates the presence of thin basal non-marine Eocene clastics overlain by Oligocene–Miocene sediments of mixed clastic-carbonate facies. There are good Miocene reefs, and at least two have been drilled (Van de Weerd and Armin, 1992).

4.10.6 Kutei Basin

The greater Kutei Basin is informally subdivided into the Lower and Upper Kutei basins. All oilfields are in the Lower (Fig. 4.19), embracing coastal eastern Kalimantan and the shallow marine platform facing the deep Makassar Straits. The upper basin has been uplifted during Miocene tectonic inversion and is deeply eroded. Only Miocene sections have been penetrated in the Lower Kutei Basin, and its Palaeogene history can be inferred only from comparison with the uplifted and eroded Upper Kutei regions.

4.10.6.1 Teweh–Kerendan area

Eocene strata rest unconformably on a basement of Mesozoic intrusives, Jurassic strata and Permo-Triassic metasediments (Van de Weerd *et al.*, 1987). They are succeeded by the Middle to Upper Oligocene Berai Limestone. The uppermost carbonate sequence contains gas and condensate in the Kerendan gas field. The carbonates give way laterally to deeper marine shales of the

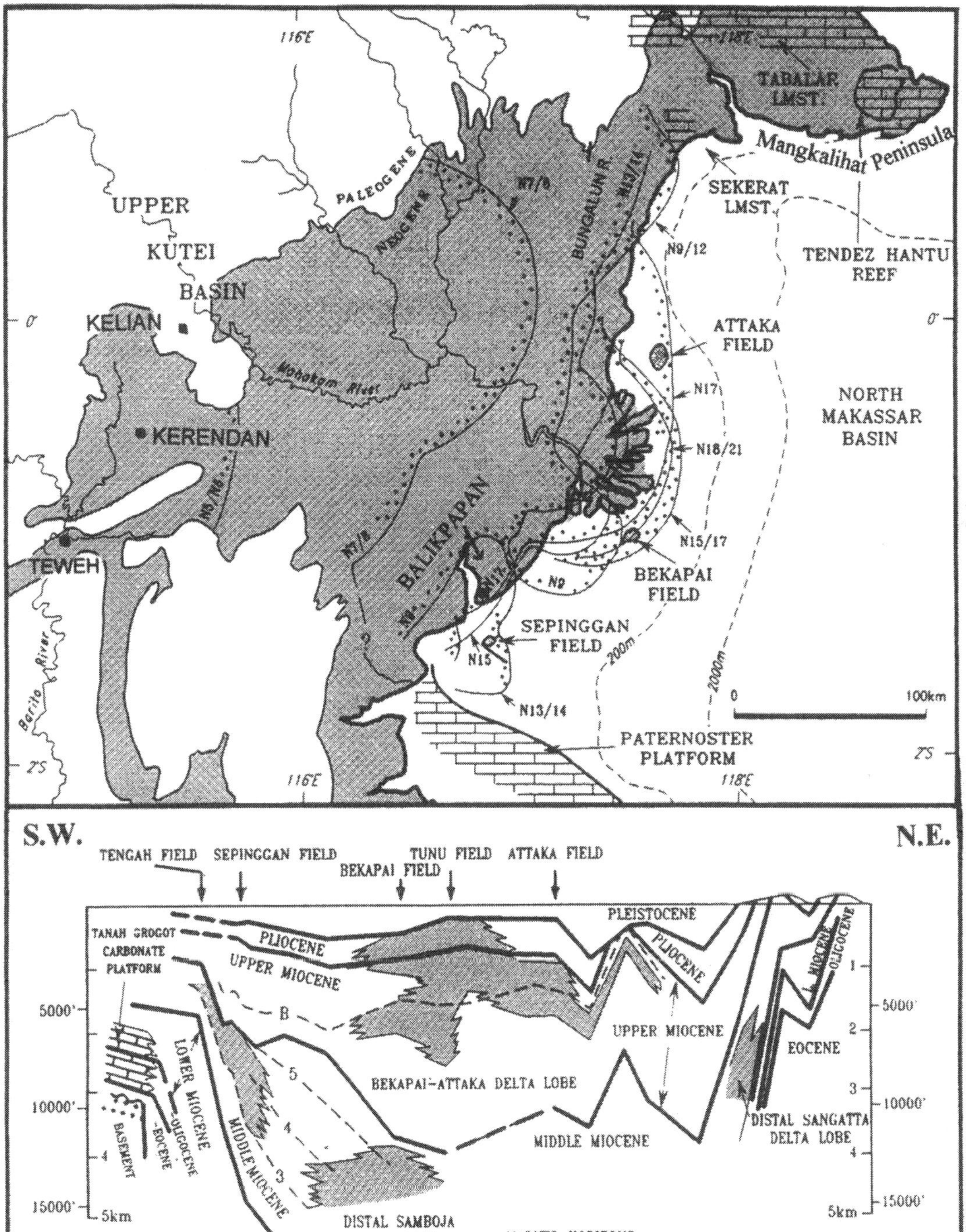

Fig. 4.19. Upper: Approximate positions of delta fronts in the Kutei Basin of Kalimantan for successive foraminiferal zones. Up to N8 = Lower Miocene, N9 to N14 = Middle Miocene, N15 to N17 = Upper Miocene, N18 to N21 = Pliocene. **Lower**: Schematic SW–NE cross section across the coastal zone of the Kutei Basin. Shaded zones indicate sand-rich delta lobes (after Van de Weerd and Armin, 1992).

Bongan Formation, which contains volcanic breccias and tuffs. Up to 4000 m of deltaic clastics prograded over the Oligocene carbonates. They grade upwards into marginal marine shales with thin channel sands and overlying non-marine coal bearing fluvial deposits. Volcanic rocks have been K:Ar dated 24 to 14 Ma (Upper Oligocene to Lower Miocene).

Miocene uplift and erosion were important, and Plio-Pleistocene basalt flows and cones (Metulang Volcanics) are common, extending into interior Kalimantan.

4.10.6.2 Upper Mahakam River

Eocene strata rest unconformably on Cretaceous turbidites, granites and tectonic mélanges (Wain and Berod, 1989). Up to 1500 m of conglomerates and coarse pebbly sands at Kiham Haloq form the base of the Eocene sequence, interpreted as an alluvial fan deposit. It is overlain by deltaics and shallow marine Batu Kelau and Batu Ayau formations, up to 1500 m thick (Fig. 4.16). Eocene pyroclastics of minimum thickness 300 m occur in the Kelian area (Van Leeuwen *et al.*, 1990). The Batu Ayau is overlain by up to 700 m of Lower Oligocene platform carbonates of the Batu Belaq Formation. In more deeply subsiding parts of the Kutei Basin, the Batu Ayau grades upwards and laterally into deep marine shales and turbidites of the latest Middle Eocene to Oligocene Ujoh Bilang Formation, which includes volcaniclastic turbidites of the Len Muring Sandstone Member. Miocene intrusives have been reported from the Kelian River area (Van Leeuwen *et al.*, 1990).

4.10.6.3 Central parts of the Upper Kutei Basin

This area is poorly known only from reconnaissance reports. Several thousands of metres of Middle Eocene to Lower Oligocene deep marine shales, with turbiditic, siliciclastic and volcaniclastic sands and calci-turbidites, are in fault contact with the pre-Tertiary basement. A strong angular unconformity exists between Oligocene strata and overlying marine Lower Miocene strata (Wain and Berod, 1989). Basaltic flows have been K:Ar dated 23.2 to 12.4 Ma.

Several thousands of metres of non-marine to shallow marine Eocene shales, sandstones, thin coals, and tuffaceous sandstones, are present near Bungalun River (Fig. 4.19). Basaltic rocks and Eocene limestones up to 800 m thick occur locally. South of Bungalun River, up to 4000 m of Upper Oligocene to Lower Miocene clastics are present. Lower Oligocene and Miocene carbonates have been preserved north of the river. Oligocene pyroclastic flows and basalts have also been reported (Van de Weerd and Armin, 1992).

4.10.6.4 Lower Kutei Basin

A formal Miocene stratigraphy has been worked out by Marks *et al.* (1982), Loiret and Mugniot (1982) and Vallet (1983). The Neogene section is up to 6 km thick offshore and in the coastal area, and it thins both SW and NE along the coast, and inland towards the NW (Fig. 4.19).

4.10.6.4.1 Stratigraphic nomenclature There are five major sandstone-shale cycles. The nomenclature of Marks *et al.* (1982) is: (1) an Upper Oligocene to Lower Miocene Pamaluan Formation, which is generally outer neritic to bathyal; (2) a Lower to Middle Miocene Bebulu Group, composed of a shallow marine carbonate shelf (Maruat Formation) and inner neritic and deltaic Pulau Balang Formation; (3) a Middle Miocene Balikpapan Group, composed of the paralic to deltaic Mentawir Formation, the middle neritic Gelingseh Formation, and its Klandasan Tongue Member, of inner shelf and carbonate reefs; (4) a Middle Miocene to Upper Pliocene Kampong Baru Group, composed of the Batu Formation (paralic-deltaic with fluvial channels) and the Sepinggan Formation (inner neritic with reefs and shoals); (5) a Pleistocene to Recent Mahakam Group, composed of the Attaka Formation (neritic to shallow marine) and the Handil Dua Formation (mixed fluvial to tidal deltaic).

4.10.6.4.2 Sedimentation pattern Deltaic sedimentation, that started in the latest Oligocene in the Teweh area, prograded eastwards and by late Lower Miocene, regressive delta-plain sediments were built out into the Samarinda and Balikpapan areas (Nuay *et al.*, 1985). Younger deltas subsequently prograded into what is now the offshore area (Fig. 4.19). The oldest delta units built out over Lower Miocene clastic shelf deposits with locally thick carbonate interbeds, known as the Bebulu Limestone. Some of the lowermost deltaic sediments contain abundant volcaniclastics (Van de Weerd and Armin, 1992).

Benthonic foraminifera and stratigraphy indicate an eastwards prograding deltaic sequence during early Middle Miocene in the southern Kutei Basin near Balikpapan (Nuay *et al.*, 1985). A thick sequence in excess of 1300 m of barrier bar, nearshore and beach sediments was deposited in Middle Miocene time in the Klandasan area. This complex was probably deposited by nearshore currents along the coast line in front of the delta. The sandstones of this sequence indicate multicycle sedimentation. Their provenance was probably from an earlier progradational cycle, ultimately derived from Sundaland.

The direction of the delta progradation may have been controlled by the shelf edge position in the Samboja area.

Its position and orientation were closely related to the NW trending growth faults which occur in the southern part of the Kutei Basin.

Several Middle to Upper Miocene delta lobes have been recognized (Fig. 4.19) and several shelf margins have been identified on seismic profiles in the offshore area, parallel to the present day coastline. Carbonates (for example the Dian Limestone) developed locally on the shallow marine shelf when deltaic influence waned.

Middle and Lower Miocene strata are exposed onshore towards the west. Vitrinite reflectance indicates that several thousands of meters of sediments were uplifted and eroded (Van de Weerd and Armin, 1992). Deposition of Middle to Upper Miocene sediments in the Lower Kutei area was therefore concomitant with basin inversion and cannibalization of the Upper Kutei area.

4.10.6.4.3 Structural and tectonic events Lower Miocene extensional faulting has been observed on seismic profiles near Balikpapan and offshore south of Mangkalihat Peninsula. The NW–SE trending Balikpapan Fault zone (through the Sepinggan Field of Fig. 4.19), near the south-western boundary of the Kutei Basin, was intermittently active since early Miocene, resulting in local unconformities.

The main anticlinal structures of the Lower Kutei Basin trend approximately parallel to the present day coastline. Fold axes trend NE–SW in the Balikpapan area, and swing N–S towards the Bungalun River and the Mangkalihat Peninsula. Middle and Upper Miocene strata are stratigraphically thinner over anticlinal structures, indicating that folding was syn-sedimentary (Van de Weerd and Armin, 1992).

4.10.6.4.4 Hydrocarbons Oil and gas fields of the Lower Kutei Basin produce from Middle and Upper Miocene deltaic sandstone reservoirs The total recoverable reserves of oil and oil-equivalent gas may be 5 billion barrels.

Throughout the explored depth of the Mahakam Delta, only humic coaly organic material of Type III is encountered, which is usually considered to be capable only of producing gas, and generally held not to be a viable source for crude oil (Kelley *et al.*, 1985). Yet the delta is one of the major oil producers of Indonesia (see Table 1.2). The humic organic matter uniformly increases in carbonization with depth and reaches the principal stage of oil formation between 3 and 4 km depth. The principal zone of gas formation can be estimated to extend below 5 to 6 km. Extracts from mature samples both compositionally and isotopically resemble the oils produced at shallower depths and gas collected at depths of around 2 km have the signatures of overmature gas (Schoell *et al.*, 1983).

There is an uneven distribution of oil and gas in the Mahakam Delta. Whereas Handil is an oilfield, Nilam and Badak are predominantly gas producers. This uneven distribution is controlled by the position of the oil window relative to an underlying over-pressured zone, which normally occurs at a depth between 4 and 5 km. If the oil window is immediately above the over-pressured zone, oil can be expelled and migrate as is the case for the Handil field. If the oil window is within the over-pressured zone, liquid hydrocarbons cannot be expelled and only gas can migrate (Schoell *et al.*, 1985). The over-pressured zone may also be held responsible for mud diapirism in the delta. One circular synclinal sub-basin is known in the Lower Kutei Basin, but it has not been described in publications (see Section 4.10.7.2).

Generally only gas and formation water are encountered in the over-pressured zone. It is conceivable that the active formation of gas in the principal zone of gas formation causes over-pressure to build up below the principal zone of oil formation. Migrating methane, through solution and dissolution, therefore probably is the main cause of oil migration in Cainozoic delta systems. The finding of mature to overmature gas in relatively shallow reservoirs is strong support for this migration model.

4.10.7 Tarakan and Berau–Muara Basins

The Tarakan Basin occupies the coastal zone immediately south of the Kalimantan border with Sabah. The Berau Basin lies south of it, and the Muara Basin lies adjacent to the Mangkalihat Peninsula in eastern Kalimantan. The older strata are found to the south, and the Tarakan Basin (*sensu stricto*) has an explored succession extending back to the Lower Miocene.

The Mangkalihat Peninsula is of older ophiolite and Mesozoic volcanic rocks, overlain by folded Lower Eocene Sembakung Formation, of red sandstone, conglomerate and mudstone (Achmad and Samuel, 1984).

The Berau/Muara basinal succession began with the Upper Eocene Sajau Formation, which is unconformable on the older formations. It is of folded conglomerate, sandstone and volcanics. Coal seams and limestone intercalations are common. The conformably overlying Lower Oligocene Seilor Formation is of micritic platform limestone of thickness variable up to 500 m. It is gradually succeeded upwards and basinwards (eastwards) by thick massive Oligocene marls of the Mangkabau Formation. The succeeding Upper Oligocene Tempilan Formation is of thin sandstone, tuff and shale sequences with coal seams. The Lower Miocene Tabalar Formation is of extensive limestone platform with local reefs. The limestone grades upwards and laterally into shale and marl with some limestone beds, known as the Birang

Formation. The latter is known as the Naintupo Formation, the oldest penetrated rocks of the Tarakan Basin to the north (Achmad and Samuel, 1984). These Lower Miocene formations contain volcanic rocks dated 16 to 17 Ma.

The unconformably overlying Middle Miocene formations are of cross bedded sandstones and shales of the Latih Formation, known as the Meliat Formation in the Tarakan Basin. Some of this formation has been eroded during Upper Miocene uplift. The Middle Miocene Menumbar Formation is of calcareous mudstone, marl and limestone occurring in the west. In the Tarakan Basin, its equivalent is the Tabul Formation, of sandstone, silt and shale interbeds formed in an eastwards prograding delta complex, which conformably overlies the Meliat Formation. The Santul Formation is of sandstone, shale and coal of Upper Miocene delta front to delta plain deposits. The overlying Sajau or Tarakan Formation is of sands and shales deposited in a Pliocene delta system. The Pleistocene Bunyu Formation is of delta plain to fluvial sands, shale and lignite.

4.10.7.1 Hydrocarbon occurrences

There are four major oilfields in the Tarakan Basin that have cumulatively produced more than 320 million barrels of oil (Wight *et al.*, 1993). There is also one large gas field and five minor oil accumulations. The reservoirs are mainly in stacked, predominantly fluviatile, thick sandstones of Pliocene and Pleistocene age. The traps are mainly downthrown, independently closed, roughly N-S trending anticlines, ranging in size from 960 to 2600 acres. Most major accumulations are on Tarakan and Bunyu islands.

The main fields of Pamusian and Bunyu were discovered in 1901 and 1923 respectively. Both structures were easily identified by surface expressions and oil seeps. The reservoiring in Pleistocene sandstones indicates late migration or re-activation from Pliocene or older reservoirs. In addition to hydrocarbons, the Tarakan Formation contains important outcropping coal deposits (Koesoemadinata, 1978).

4.10.7.2 Tarakan Basin extension into Sabah

The Tarakan Basin continues onland into Sabah as far northwards as the Maliau Basin (Collenette, 1965; Tjia *et al.*, 1990). Here the strata are of Lower to Middle Miocene Tanjong Formation, deformed into a unique circular or saucer-shaped basin, and strongly uplifted. From the Maliau Basin, another line of Tanjong Formation outcrops extends north-eastwards towards Sandakan, on the Sulu Sea coast, subsequently disrupted into separate smaller circular basins, a little understood characteristic feature of Sabah (see Hutchison, 1989*a*, page 190). The Tanjong Formation was deltaic and the currents were predominantly from the north-east (Tjia *et al.*, 1990). The lower part of the sequence is predominantly of mudstone with channel sands. Upwards the formation becomes more sandy, and the sandstones form prominent scarp ridges. Good quality coal seams have been mapped. The Maliau Basin was explored for hydrocarbons, but the circular-shaped morphology, synclinal in every cross section, and the general absence of unfaulted and non diapiric truly anticlinal structures, obviated good hydrocarbon trapping conditions. The uplift of the Maliau area resulted in the inward-dipping sandstone scarp ridges in places reaching up to 1606 m above sea level.

4.10.8 Basins of Java and Java Sea, Indonesia

The N–S elongated grabens, which developed on the pre-Tertiary Sundaland landmass of northwest Java, first received the Palaeogene Jatibarang Volcanic Formation (Adnan *et al.*, 1991), which acts locally as an oil reservoir (Fig. 4.20). The main reservoir lithologies are fractured massive tuff and conglomerate (Reminton and Pranyoto, 1985). The main source rocks are formed by the Oligocene lacustrine Talang Akar Formation (Gordon, 1985), which was deposited in grabens (Fig. 4.21). The formation represents a progression from continental to deltaic and shallow marine. The basal grits are preserved in graben depocentres, followed by coals and marine shales with thin limestones.

In north-east Java, the Middle Eocene to Lower Oligocene Ngimbang Formation rests with a pronounced unconformity on the pre-Tertiary basement. Its base is of coarse grained sandstone and shale with minor coal seams, grading upwards into fine grained limestones, minor marls and sandstone, representing a transition from terrestrial to shallow marine (Soeparyono and Lennox, 1989).

Similar geology extends at least to the Kangean region, NE of Bali (Figs 2.2, 4.16). The basement is of steeply dipping Cretaceous weakly metamorphosed sandstone, siltstone, shale and chert (Kohar, 1985; Phillips *et al.*, 1991). This basement is unconformably overlain in palaeo-lows by the pre-Ngimbang Formation. Sparse fossils indicate a Palaeocene to Middle Eocene age. It is composed of sandstone, siltstone and shale with some coal, deposited in a continental to shallow marine environment. The Upper Eocene Ngimbang Formation rests unconformably on the pre-Ngimbang. The basal clastic member is of sandstone and conglomerate, grading upwards into a sequence of sandstones, shales and coals, interpreted as terrestrial to marginal marine. Marine transgression is indicated by the Upper Eocene Ngimbang carbonate (Phillips *et al.*, 1991).

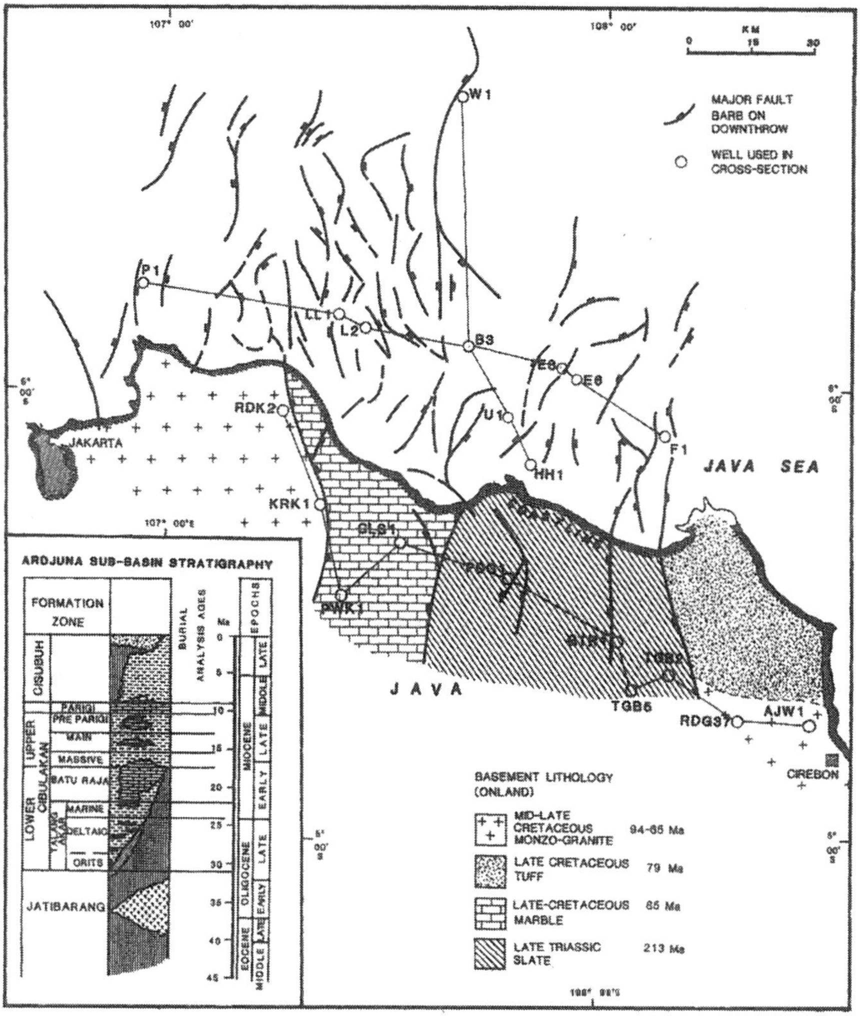

Fig. 4.20. Major faults of the Ardjuna sub-basin of N.W. Java, showing the position of the cross sections of Fig. 4.21 and the onshore basement lithologies as determined by drilling. **Inset:** basin stratigraphy (after Gordon, 1985; Reminton and Pranyoto, 1985).

4.10.8.1 Ardjuna Basin

The Ardjuna sub-basin extends northwards from onland NW Java (Fig. 4.20). The sub-basin originally contained nearly equal reservoired gas and oil, but production has been mainly of oil, with 1985 daily production figures of more than 100 000 barrels oil and more than 3.4×10^6 m^3 of gas.

The basin has a sedimentary fill averaging 2440 m, with over 3050 m in the depocentres. After the initial Upper Oligocene fluvio-deltaic Talang Akar formation, the basin has continued marine to the present day. Much of the clastic deposition was of outer marine shelf (Gordon, 1985). The clastic supply was predominantly from Sundaland.

The Talang Akar Formation is of three members, representing a progression from continental, through deltaic to shallow marine. The basal grit is of coarse alluvium, and is preserved only in the depocentres. The deltaic coal sequence thins and directly overlies the basement highs. The upper part of the formation is of marine shales with carbonate stringers.

4.10.8.1.1 Hydrocarbons Pyrolysis analyses of Talang Akar deltaic samples indicate that only the coals have an oil source potential. Oil is intimately associated with coal in a few wells. Bitumen was microscopically observed in the microfractures of coal cuttings. The evidence suggests that the waxy oil-bitumen associated with the coals was generated *in situ*. Geochemical analyses allow the interpretation that the coals are the source of the Ardjuna oil.

Most wells in the sub-basin depocentres have overpressured Talang Akar shales, and this is likely to be the result of active hydrocarbon generation.

4.10.8.1.2 Maturation The onset of maturity has been shown to be at 100° C. All the areas underlying and/or adjacent to all the discovered oil accumulations are indicated to have source material that is mature (Gordon, 1985). Where the coal sequence directly overlies basement, it has access to maximum heat flow from the basement, whereas if it is underlain by the grit sequence, as in the deeper parts of the sub-basin, it has been partly insulated from the heat flux. Accordingly the iso-maturation levels generally conform to the structures (Fig. 4.21). Numerous small stocks intrude the basins and appear to have given only very restricted thermal effects, but they may have contributed to the overall high geothermal gradients within the range 45 to 55°C km^{-1}, giving a maturity onset at a shallow depth of 1210 m to 1830 m. The maturity onset is reached with the first occurrence of significant ditch gases (especially wet ethane-pentane which indicate the formation is oil prone) associated with the coals.

The time–temperature index (TTI), as a measure of maturity of the source rocks, has been measured using the method of Waples (1980). Gordon (1985), however, has set the beginning of maturity at a TTI of only 3, whereas Waples defined it at 15, equivalent to a vitrinite reflectivity (R_o) of 0.65 and thermal alteration index (TAI) of 2.65. Gordon (1985) states that the TTI of 3 correlates with the first occurrence of wet gas from coal in various wells and also correlates with a temperature of 100°C (usually at about 1.52 km). The burial history analysis of well MA–1 indicates an overall accumulation rate for the entire sedimentary sequence of approximately 82 m Ma^{-1} with a heating rate of the entire sequence of about 3.25°C Ma^{-1} (Fig. 4.21). The coal sequence has been exposed to enough heating to induce moderate maturity (TTI = 33), whereas the peak oil generation is considered to be at 75 (Waples, 1980). The onset of maturity may be calculated to have begun only 11 Ma ago.

The timing of onset of maturity over the whole sub-basin appears to have been at 13 Ma ago, and peak oil generation has been reached only in very limited areas. The end of the oil generation stage (TTI = 160) has nowhere been reached within the Ardjuna sub-basin (Gordon, 1985).

The relatively recent onset of maturity and its ongoing nature indicate that most of the traps existed before maturation and that most of the oil generated has been trapped, with little escape to the surface. The mildly compacted Upper Cibulakan claystones overlying the Main Formation have made a very effective barrier to upwards migration, and even the shallow faults have been sealed. Only rare occurrences of oil have been found in the Parigi Limestone. The uncompacted Cisubuh Formation clays have provided a good barrier to gas migration, but some gas has escaped to higher levels. Most of the known hydrocarbon accumulations have associated faults connecting them to the generating source rocks, and are either on basement highs, within the sub-basin, and/or in close proximity to the source on the adjacent platform areas. The faults have therefore served as major migration conduits. Biodegradation of the reservoired oils has taken place only on the cool platform areas, whereas temperatures in excess of 72°C have been present in the reservoir since the onset of oil generation and have prevented biodegradation.

4.10.8.2 North-west Java

The pre-Tertiary basement of the onshore NW Java Basin has been documented from drilling results by Reminton and Pranyoto (1985). The Tertiary succession begins in some of the deeper grabens with the Palaeogene Jatibarang Volcanic Formation, which reservoirs oil and is a producer. The two different reservoir lithologies are: massive tuff, in which the porosity and permeability are a result of fractures, and the second is a conglomerate.

The Lower to Middle Miocene Cibulakan Formation unconformably overlies the Jatibarang Volcanic Formation and onlaps onto the basement. The Lower Cibulakan consists of a lower part, or Talang Akar Member (carbonaceous shales and sandstones) overlain by the Baturaja Member (bedded limestone and reefs). The Talang Akar Member is the main source of hydrocarbons and the Baturaja contains good producing reservoirs. Some Talang Akar sandstones also produce oil.

The Middle to Upper Miocene Parigi Formation, ranging in thickness from 50 to 300 m, is of limestone with local reefs. This formation produces hydrocarbons from the Tugu Barat, Gantar and Pasirjadi structures.

The Upper Miocene to Pliocene Cisubuh Formation is of interbedded clay, shale and sandstone. None of its structures produces hydrocarbons.

4.10.8.2.1 Source rocks The Talang Akar Member is the prime source rock in NW Java. Most of the Lower Cibulak Formation is now within the oil generation window (Fig. 4.22). The Talang Akar shales are good source rocks, with

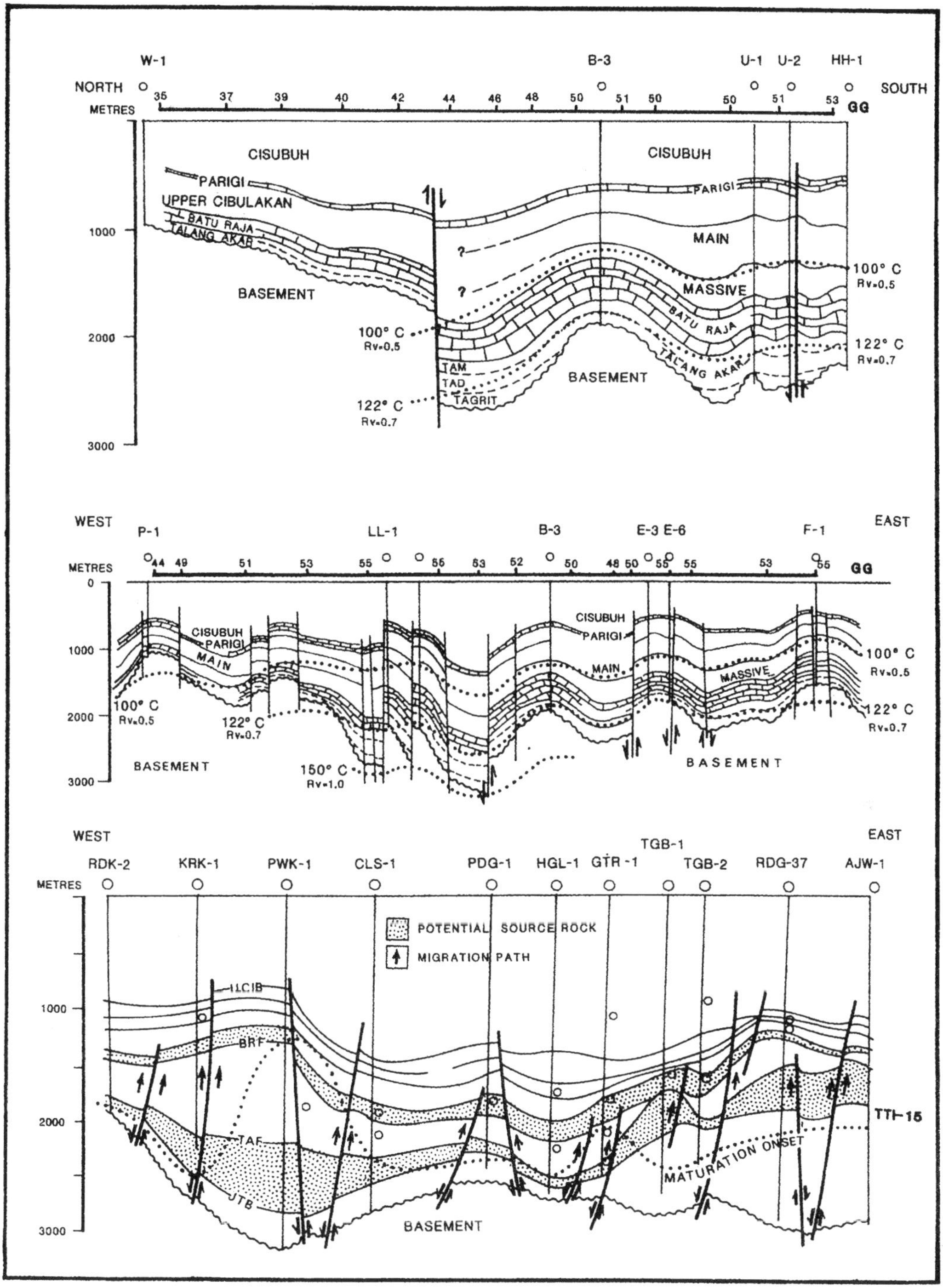

Fig. 4.21. Cross sections across the Ardjuna sub-basin of N.W. Java, showing iso-maturity horizons. **Top**: N–S section, **Middle** and **Bottom**: E–W sections (after Gordon, 1985; Reminton and Pranyoto, 1985). **GG** = geothermal gradient in °C km^{-1} measured at 1524 m depth. $\mathbf{R_v}$ = vitrinite reflectivity (%), **TTI** = time–temperature index (Waples, 1980). See Fig. 4.20 for locations.

a total organic carbon content of 0.5 to 20% (Reminton and Pranyato, 1985; Wahab and Martono, 1985) . The Baturaja shales and liptinic coals are good source rocks. The Upper Cibulakan Member shales are potentially good source-rocks (Reminton and Pranyoto, 1985).

4.10.8.2.2 Geothermal gradient The Rengasdengklok and Arjawinangun areas have high gradients of 44 to 49° km^{-1}. The gradient is lower in the Cilamaya–Kandanghaur areas, and in general decreases to the south (Reminton and Pranyato, 1985).

4.10.8.2.3 Maturity of the source rocks The depth of TTI = 15 lies between 1800 and 2100 m in the Rengasdengklok and Arjawinangun areas, and lies deeper in the Cilamaya–Kandanghaur areas and towards the south. The oil window of the Talang Akar and Baturaja members commenced in the late Miocene (Fig. 4.22).

The TTI iso-maturity contours indicate that the base of the Jatibarang Volcanic Formation (top of the basement) has reached the onset of maturity (TTI = 15) over the whole NW Java Basin, and the end of oil generation (TTI = 160) has been reached in Purwakarta-1 well in the Jatibarang Volcanic Formation.

The lower parts of the Talang Akar Member have generally been buried deep enough to now be within the oil window, but in some areas it is still immature (Gordon, 1985; Wahab and Martono, 1985).

It may be concluded that the CO_2 produced in the NW Java Basin is not a result of overmature conditions within the Tertiary sequences, and may have its source in the limestones.

4.10.8.3 Sunda Basin

The Sunda Basin is elongated N–S. It is bounded on the west by the Lampung High, on the east by the Seribu Platform and Saleha Ridge, and on the south by the Seribu Platform. The northern boundary is poorly defined, and the Tertiary strata gradually thin onto the Sunda Shield.

There are six major half graben depocentres. The overall fault pattern is dominantly N–S and NE–SW, with a secondary E–W set.

The stratigraphy has been described by Wight and Hardian (1982).The Lower to Middle Miocene reef is a carbonate build-up within the Baturaja Formation. The best reservoirs are in the skeletal packstone and wackstone facies (Djuanda, 1985). They exhibit extensive mouldic and vuggy porosity, together with some fracturing, and contain producible hydrocarbons. Porosities generally exceed 30%, but core permeabilities generally range only from 10 to 150 mD. The porosity was probably produced by freshwater leaching. The back-reef lime mudstones and marly limestone contain nearly half the total oil, which cannot be produced because of poor reservoir characteristics. Although the logs indicate porosities between 30 and 40%, the mudstones and marly limestones have typically low permeabilities.

4.10.8.3.1 Source rock potential The source rock potential of the Batu Raja Formation is only fair. TOC values range from 1.5 to 3.0 % in the northern and central basin. The organic matter is woody and coaly.

The Transition Member of the Talang Akar Formation contains large concentrations of organic matter associated with coals and shales, with TOC values in excess of 30% in the central and southern parts of the basin (Molina, 1985). In the north, the maximum is 15%. The organic matter is mainly woody and coaly material, and the Talang Akar coals are of type II kerogen (exinite). These coals are significant source rocks.

The Grit Sand Member of the Talang Akar Formation contains TOC values exceeding 8% in the south, and up to 4% in the north. Woody and coaly materials predominate. The Lower Talang Akar Formation contains TOC maximum values in excess of 6% in the basin centres.

4.10.8.3.2 Thermal maturity The onset of the oil window is generally taken at R_o 0.5%, but oil and condensate have been generated from terrestrially derived organic matter at reflectivity values as low as R_o 0.4%. The Batu Raja rocks are mature ($R_o > 0.5\%$) only over a narrow zone along the centre of the Sunda Basin. Talang Akar Transition Member sediments are mature over a greater area, but immature over the basin flanks and within the Seribu depocentre. Lower Talang Akar sediments are almost everywhere mature where they have been preserved. The results from vitrinite reflectivity have been supported by pyrolysis analysis.

Analyses of oils from the Sunda Basin show that they were generated from two different major organic matter types (Molina, 1985): terrestrial and aquatic (lacustrine, marine or brackish), and most oils were generated from mixed terrestrial and aquatic kerogens. As a result of biodegradation, the oils become progressively heavier towards the W, SW and S shallower margins of the basin.

4.10.8.3.3 Generation timing The onset of maturation has been determined by Lopatin's method for a number of wells. A TTI of 3 has been taken as the onset. This onset of oil production began in the Middle to Upper Miocene, and the source rocks remain today in the oil window. The peak of oil generation (TTI = 75; R_o = 1.0) has not yet been reached in most of the well areas. Generation is incomplete and ongoing, and hence late migration has occurred in the Sunda Basin. Most of the oil generation has been from the basin depocentres.

4.10.8.4 Basins connecting Java to Kalimantan

The NE-trending horst and graben structures, which connect Java to Borneo beneath the Java Sea, have been

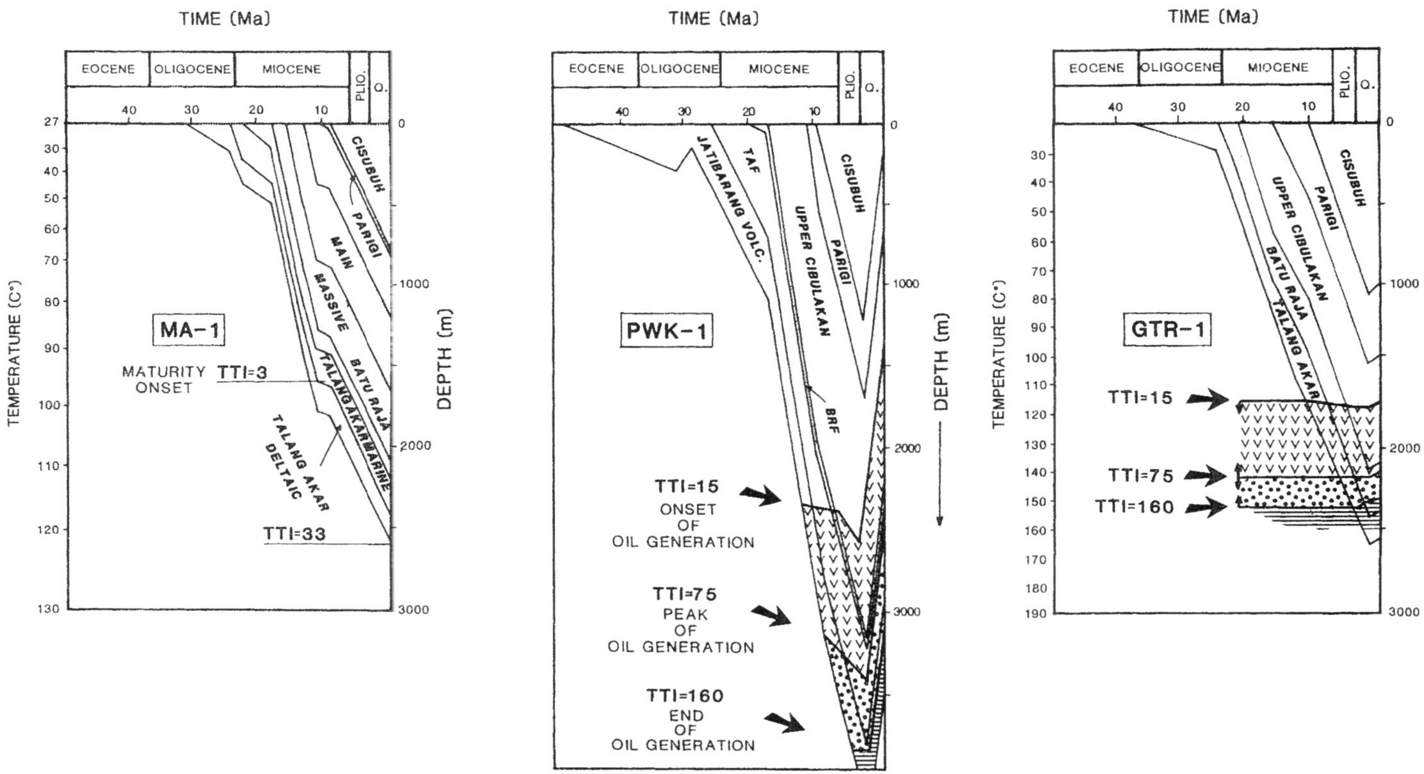

Fig. 4.22. Burial history and time–temperature index for several wells of the NW Java basin, using the method of Waples (1980). **MA-1** (after Gordon, 1985), **PWK-1** (after Reminton and Pranyoto, 1985) and **GTR-1** (after Wahab and Martono, 1985).

described by Bishop (1980). The pre-Tertiary basement has been widely drilled and shown to be of a variety of volcanic, plutonic, metamorphic and ophiolitic rocks, many of which are Cretaceous and represent a SW continuation of the Meratus Mountains geology, which remained depressed below sea level.

Palaeocene to Eocene strata were deposited in grabens developed on an erosional landsurface. These terrestrial (lacustrine and fluvial) sediments include pebbly sandstones and tuffaceous mudstone. As in Borneo and western Sulawesi, they are widely transgressed by massive 'CD Limestone' of Middle Oligocene to Lower Miocene age, comparable to the Berai Limestone (Bishop, 1980).

4.10.8.5 Kangean Block

The Kangean Block of pre-Tertiary block-faulted basement (predominant E–W trend), of low grade metamorphosed slate or meta-siltstone, lies 100 km north-east of Bali. It is unconformably overlain in places by a thin succession of Eocene siliciclastics, composed of interbedded conglomerate, sandstone, shale and coal, deposited in a continental to shallow marine environment (Kohar, 1985). The pebbles are of quartzite and grey siliceous rock. The shale contains nodules of pyrite. The origin of these quartz-rich clastics is unknown, but must have a provenance from Sundaland to the west. An Upper Eocene carbonate complex overlies the clastic strata (Fig. 4.16). It was deposited in inner to outer sub-littoral environments. The lower part is of shelf or bank limestone, while reefs and build-ups characterize the upper part. They developed along E–W trends on basement highs along a presumed palaeo-shelf edge. The section is in places more than 350 m thick and is widespread. Secondary solution porosity is excellent, and these carbonates form good reservoirs.

The overlying Upper Eocene to Oligocene is of deep marine calcareous claystones containing a few limestone streaks. The thickness of this formation ranges from 600 to 920 m. This sequence is overlain by 75 to 450 m of shallow marine calcareous claystone, sandstone, and a few thin limestones. The overlying Pliocene sediments are of claystone, sandstone and thin limestones of shallow marine origin. This unit is very thin or absent over most of the area.

Tertiary anticlinoria trend dominantly E–W. The major southern E–W striking fault is the Sepanjang normal fault, which strongly downfaults the Lower Eocene carbonates to the south.

The Kangean source rocks are the Eocene Ngimbang and pre-Ngimbang Clastics. The underlying Cretaceous clastics have poor source rock potential (Phillips *et al.*, 1991).

4.11 SUMMARY

The region described in this chapter was an integral part of the South-East Asian Sundaland landmass at the beginning of the Tertiary, but unlike the Sundaland core described in Chapter 2, its basement is not cratonic. The southern and eastern coastline extent of the Sundaland peninsular landmass lay along the southern coast of Java and along central Sulawesi. Although mostly land, there were marine embayments into Sundaland created by marginal seas whose spreading histories ranged from Eocene to Miocene, and the Kutei Basin of eastern Borneo represents a major Tertiary delta growing eastwards into one of these seas.

Over the Sundaland landmass, sedimentation began in the Upper Eocene or Lower Oligocene within basement-controlled grabens, firstly as alluvial fans and then as lacustrine deposits, which provided the major source-rocks for the basins, they then became marine in the late Oligocene through to the Miocene.

The provenance of quartz sandstones remains a problem, for the basement is generally devoid of granitoid rocks, which are dominant only in the Schwaner Mountains of western Borneo and in northern cratonic Sundaland.

Sumatra Island is of intermediate character between northern cratonic Sundaland and the outer non-cratonic peninsula. Its oldest known basement is of Carbo-Permian rocks, and there is also a younger accreted terrain in the NW. The Tertiary basins of Sumatra were established on a basement which has been dominated by right-lateral wrenching, resulting in an array of NW–SE trending horsts and grabens. Upper Eocene source rocks are rich in Sumatra; indeed they are the best in SE Asia. Sumatra has benefited from close proximity to outcropping granites in Peninsular Malaysia, which have been a provenance for the excellent Miocene reservoirs, fortunately overlain by good sealing strata.

Outer peninsular Sundaland is characterized by a basement predominantly of accreted terrains of island arc volcanics, ophiolite, mélange, and turbidite fill of marginal basins. This basement has been more tectonically unstable than that of interior cratonic Sundaland. The accreted terrains formed the extensive peninsular landmass at the beginning of the Tertiary, and rifting resulted in initial alluvial fill followed by lacustrine deposits containing source-rocks. Marine platform carbonate became widespread during the late Oligocene and early Miocene except over the Kutei Basin which continued its clastic delta progradation south-eastwards, but sandstones may have provided reservoir rocks in other selected places and they reached the Sundaland coastline in places. Upper Miocene uplift of the predominantly Cretaceous basement was spectacular, with the best example in the Meratus Mountains.

5

Basins of Eastern Indonesia

There are several basins lying between the eastern margin of Sundaland and the northern margin of Australia. The Savu Basin lies in a fore-arc setting and the Bone Basin is a Tertiary rift basin. The basins of Irian Jaya are unlike the basins described in the foregoing chapters and are integral parts of the continental shelf of Australia.

5.1 BONE BASIN

The Gulf of Bone separates the eastern and western arcs of Sulawesi. It is thought to be the result of extension (Hamilton, 1979; Bowin *et al.*, 1980). The sea is deeper than 2 km, and the sedimentary fill exceeds 5 km, and probably rests on oceanic crust in the deeper parts. This is supported by gravity data. A 3 km deep well at the head of the Gulf of Bone bottomed in Middle Miocene clastic strata. The measured geothermal gradient is extremely low, only 8° C km^{-1}, and the exact significance of this is uncertain. The opening of the Gulf to form the basin may be related to transcurrent movement on faults that trend WNW across the SE arm of Sulawesi.

During the early stages, shallow marine clastics were deposited, intermixed with volcanic derived material of the Eocene Langi Formation (Fig. 5.1). There appears to be a stratigraphic gap between Oligocene to Lower Miocene (CCOP, 1991). Shallow marine shales and sands of Middle to Upper Miocene age followed the unconformity. Carbonates, up to a thickness of 700 m were deposited in the Late Miocene. In Pliocene time shallow water sands and shales were succeeded by deeper water sands and shales as the basin deepened.

The prominent fault direction trend is N–S within the basin (CCOP, 1991). Most of the faulting occurred prior to deposition of the Middle Miocene Camba Formation. N–S trending wrench faults, which have been active during the Pliocene, occur along the western margin of the basin.

A number of exploration wells have been drilled, but no results have been published.

5.2 BULA BASIN

The first exploration well was drilled in 1897 near a prolific oil seepage on the coastal region of Seram Island (previously Ceram). This was the seventh oilfield to be discovered in Indonesia by the Dutch. It initially flowed 200 barrels oil per day from Pleistocene sandstone reservoirs of the Fufa Formation (Fig. 5.2). Most of the oil has been produced from the Fufa Formation, but in 1983 production was begun from Triassic sandstones.

5.2.1 Tectonic setting

The general setting of the oilfield is in the fore-arc, between the recently extinct Seram trough and the volcano-plutonic arc through Ambon, where cordierite-bearing Upper Pliocene granites (ambonites) indicate the presence of underlying continental crust (Hutchison, 1989*a*). Subduction at the Seram Trough has now been replaced by important left-lateral transcurrent faulting sub-parallel to the major E–W Sorong Fault.

5.2.2 Stratigraphy

The oldest sedimentary rocks of the oilfield region are the deltaic sandstones and shales of the Middle Triassic to Lower Jurassic Kanikeh Formation, formerly called the Wakuku Beds. The sandstones are quartzose to lithic, fine to medium grained, carbonaceous, micaceous and glauconitic. The shales are coaly and contain plant fossils closely related to the flora of northern Australia. The Upper Triassic to Lower Jurassic Manusela Formation consists predominantly of reefal, bioclastic and oolitic limestone, which is bituminous in places and is interbedded with dark grey marls. The depositional environment ranged from nearshore with restricted bays, to outer shelf. The Middle Triassic to Lower Jurassic Saman Saman Limestone contains abundant planktonic fauna and its environment is interpreted as bathyal (Fig. 5.2).

The Nief Beds extend from Upper Jurassic to Middle or Upper Miocene. They consist of calcareous claystone at the base, sandy calcareous siltstone and marl in the centre, grading upwards into upper reefal limestone. The environment therefore appears to range from bathyal at the base gradually shallowing to an inner shelf environment at the top.

From late Miocene onwards, the geological evolution has been dominated by subduction at the Seram Trough,

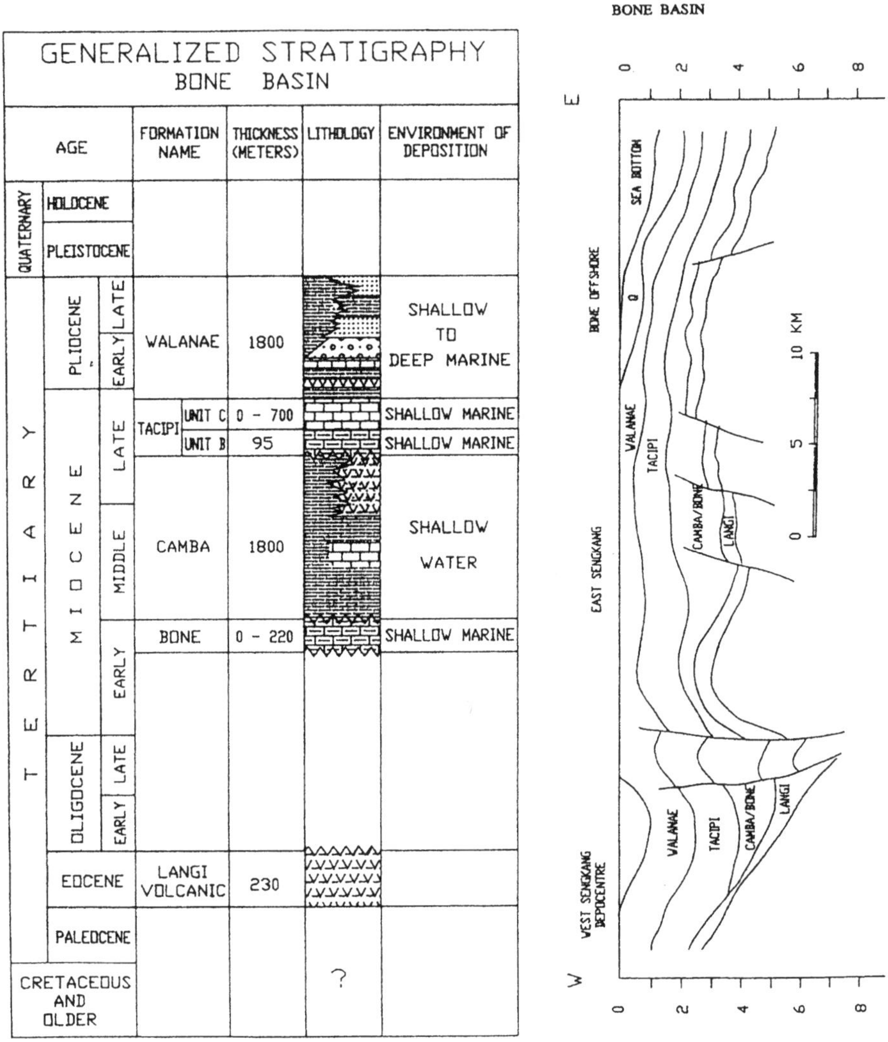

Fig. 5.1. Generalized stratigraphy and E–W cross section of the western margin of Bone Basin, Sulawesi after CCOP (1991). The vertical scale is in km.

choking of subduction by collision of Australian continental crust, followed by isostatic uplift and important left-lateral transcurrent faulting.

Upper Miocene to Lower Pliocene deep water clay with exotic blocks, the youngest of which are Upper Miocene, sporadically outcrops in eastern Seram. It is known as the Salas Block Clay, interpreted as an olistostrome, which was subsequently incorporated from early Pliocene to Recent into successively younger imbricate wedges towards the Seram Trough.

The Upper Pliocene Wahai Beds and the Pleistocene Fufa Formation were deposited in the Wahai and Bula basins formed in depressions within the accretionary prism. From the Pliocene to the present day, Seram island has been rapidly uplifting, so that the Wahai Beds are of bathyal mudstone and siltstone, whereas they are overlain by the Pleistocene shallow water neritic Fufa Formation beach and bar conglomerate, reefal, and other shallow water carbonates, in a sequence that contains muds and lignite beds (Zillman and Paten, 1975*b*). The Fufa Formation is the principal oil reservoir, divided into five units, from oldest to youngest: E to A. Units B and D are predominantly claystones, and A, C and E are of conglomerate, sandstone and carbonates.

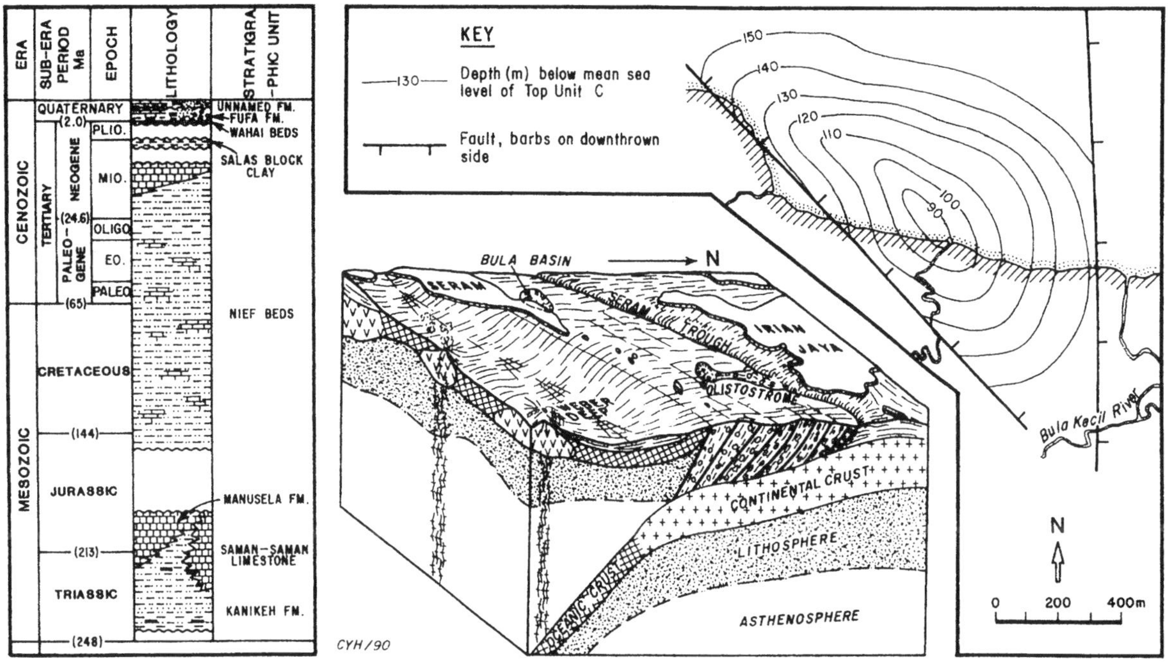

Fig. 5.2. The Bula Basin of NE Seram, Indonesia (after O'Sullivan *et al.*, 1985). **Left**: Stratigraphic sequence of central and NE Seram. **Centre**: Interpretative plate tectonic setting of the region. Subduction at the Seram trough is now extinct, choked by continental crust underthrust from Irian Jaya. **Right**: Simple shallow structure of the Pleistocene Fufa Formation, the main producing reservoir.

5.2.3 Hydrocarbon occurrences

Bituminous Upper Triassic to Lower Jurassic limestones of the Manusela Formation occur in central and eastern Seram. The source rocks are presumed to be marls of the same formation. O'Sullivan *et al.* (1985) have proposed that these rocks are the source of all the oil in Seram. Geochemical studies indicate that the Seram oil has a carbonate source. Numerous oil and gas seeps occur in eastern Seram, located in at least three formations (Fufa, Manusela and Kanikeh).

Most of the Bula oil has been produced from sandstones and carbonates of Unit C of the Pleistocene Fufa Formation. The traps are of rollover anticlines associated with normal faults (Fig. 5.2), as well as stratigraphic. The shallow Unit C reservoirs are stacked barrier bar and beach sand bodies, while reefal limestones predominate to the NW. The patch reef reservoir facies has high permeability and high production rates. Total original recoverable oil reserves for the Unit C reservoirs of Bula have been estimated at 15 million barrels.

A small production was made from wells drilled into the Manusela and Kanikeh formations during the early years of this century. A new field has been discovered at Bula Tenggara, in which the production is from sandstones of the Triassic Kanikeh Formation. Anticlinal and fault block trapping closures are present. This field has half a million barrels of oil reserves, but the figure may increase with further field development.

The oil from the Fufa Formation is napthenic, with 3% sulphur, and of low API gravity (20°), while that from the Kanikeh Formation is paraffinic, only 0.6 % sulphur content and has a high API (40°). Geochemical analysis has shown that the Fufa and Kanikeh oils have closely related origins. The differences may be related to biodegradation within the cold high level Fufa reservoirs. It may therefore be concluded that the oil was originally reservoired in the Triassic formation, and recently migrated to the high level Pleistocene reservoirs, in which the biodegradation took place (O'Sullivan *et al.*, 1985). Migration is probably still taking place. Zillman and Paten (1975*a*) maintained that the oil is indigenous to the Plio–Pleistocene succession. This is totally unrealistic because of the low geothermal gradient of only 17° C km^{-1} and the very shallow depth, less than 200 m, of the reservoirs (Fig. 5.2).

5.3 SALAWATI BASIN

The basin is the westwards dipping Atlantic-type margin of continental Irian Jaya, and is floored by older continental Australian basement (Fig. 5.3). The basin is abruptly bounded on the north by the Sorong Fault zone, which contains a 7 km wide brecciated zone and delimits the northern margin of the Australian cratonic basement.

The Salawati Basin has a sedimentary sequence in excess of 6 km of marine Tertiary sediments (Fig. 5.3). The Palaeocene through Eocene sediments are of shelf carbonates with minor shallow marine shales and sandstones (Vincelette, 1973). Miocene subsidence created the Salawati Basin, in which dark grey basinal limestones (the Klamogun Formation) were deposited together with local calcareous shales. Eastwards the basinal limestones change to the Kais Formation, of light grey calcarenite with chalk or limey mudstone beds. A platform of Kais shelf facies built out onto the shallowing basin. Pinnacle reefs developed on this limestone platform. They continued to grow as the sea level rose. The pinnacle reef build-ups attained a height of 300–750 m above the platform. The intervening areas were filled with shale and calcareous mudstone of the Klasafet Formation. The calcareous shale of the Klasafet Formation eventually buried the reefs and caused their extinction. A Pliocene fall in sea level resulted in the regressive Klasaman Formation of glauconitic shales grading up into alluvial-deltaic sands, shales and coal beds (Vincelette, 1973). The basin has an average geothermal gradient of 36° C km^{-1}.

5.4 BINTUNI BASIN

This basin is similar to the Salawati Basin, and separated from it by the Ayamaru Platform. The two basins have shared similar stratigraphic histories from Eocene onwards; the Miocene New Guinea Limestone (Kais Formation) continues across the platform into both basins. Both are deep basins located on the Australian miogeoclinal

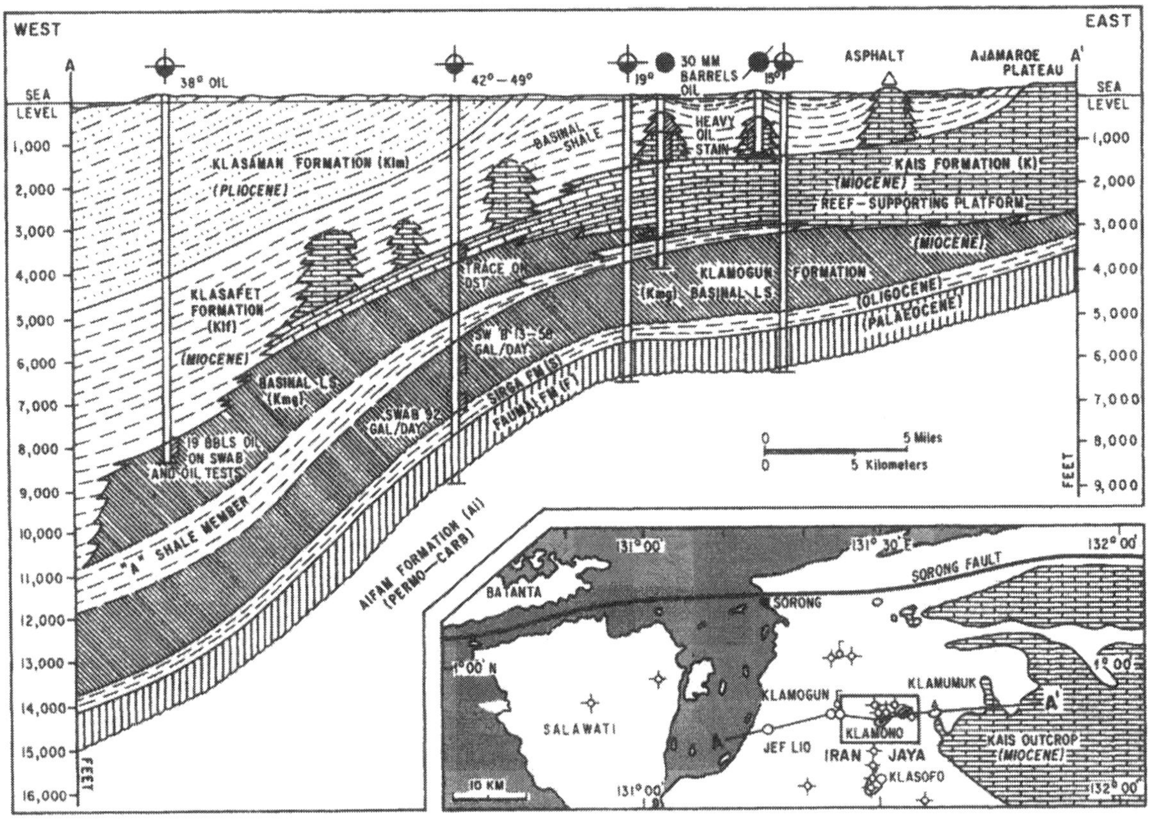

Fig. 5.3. Cross section across the Salawati Basin on the Irian Jaya sector of the Australian miogeoclinal shelf (after Vincellette, 1973).

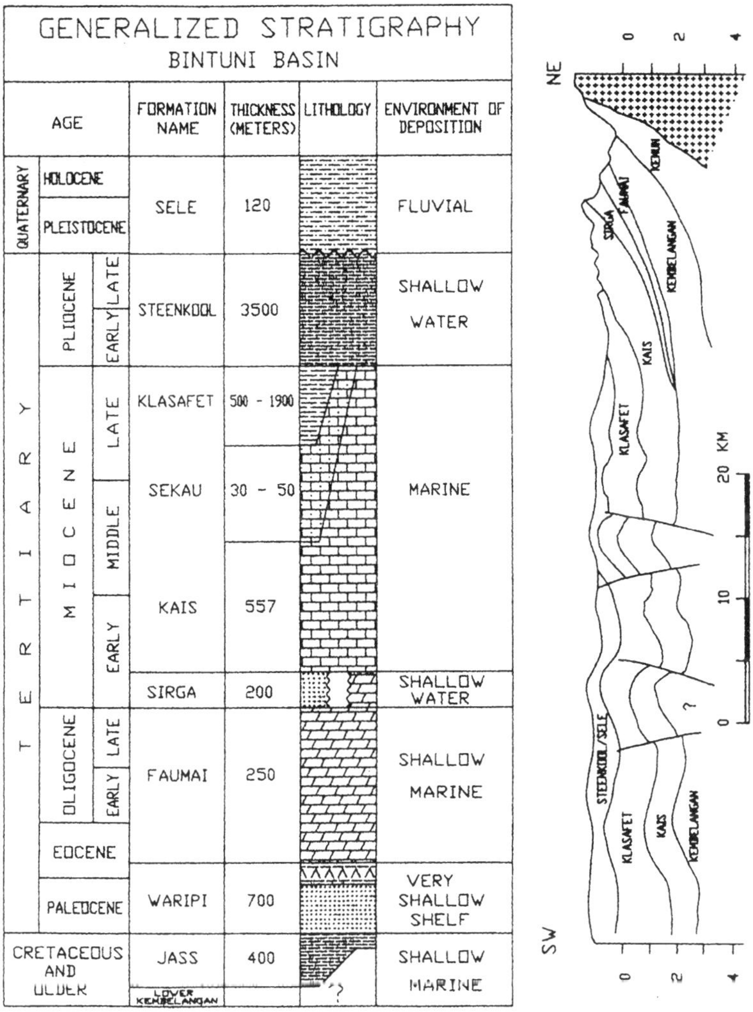

Fig. 5.4. Generalized stratigraphy and SW–NE cross section across the Bintuni Basin (after CCOP, 1991). The vertical scale is in km.

platform. The Bintuni Basin is deeper than the Salawati, and its sedimentary fill began earlier. A major part of its 4 km thickness of strata is made up of carbonates (Fig. 5.4). The basin has been gradually sinking during most of the Tertiary. The Permo-Triassic Tipuma Formation occurs on the west and is of 500 m of green-red claystone and dolomite, with a facies change eastwards into the Brug Formation dolomite (Hermes, 1974). The Jurassic to Cretaceous Kembelangan Formation is of 1.5 km of sandstone, siltstone and shale. The latter is overlain by the Eocene Faumai Formation, which is common to both basins (Koesoemadinata, 1978). The Bintuni Basin has an average geothermal gradient of 30° C km^{-1}. Pigram *et al.* (1982) have proposed that the Bintuni Basin and the associated Lengguru foldbelt are young features formed in late Cainozoic time by collision between the West Irian microcontinent and the north-west Australian continental shelf.

Potential source rocks exist as shaly layers in the Mesozoic sequence. The main established reservoir rocks are Eocene and Middle Miocene carbonates, which are assumed to form stratigraphic traps where they are sealed by shales or compact carbonaceous shales.

5.5 SAVU BASIN

The Savu Basin is a fault controlled fore-arc basin lying between Timor, Flores and Sumba (Fig. 1.1). It occurs in a situation complicated by continental fragments and collision. The stratigraphy of the basin, as indicated by outcrops and one offshore well, is composed of Upper Miocene to Pleistocene interbedded deep water chalky limestones and mudstones with only minor amounts of clastic rocks, which are mainly volcanic tuffs (Crostella, 1977). About 500 m of chalky limestone is exposed on Savu island, and up to 600 m of interbedded limestone and marlstone on Sumba island, 900 m of which were penetrated in an offshore well, which bottomed in Upper Cretaceous sediments, and gave a geothermal gradient of 19.2° C km^{-1}.

The seismic data suggest a maximum thickness of 1.5 km of Tertiary strata. Little terrigenous detritus entered the basin.. Although the strata are generally flat lying, gentle folds are present. It appears that the deeper parts may have been filled by turbidity currents (Hamilton, 1979).

6

Basins of the Philippine Archipelago

Numerous volcanic arc-related basins occur in the Philippines (Fig. 2.1). Their details have been summarized by the Bureau of Mines (1976, 1981). They have formed by rifting of a pre-Tertiary basement of ophiolitic and volcanic rocks and the proximity of an active volcanic arc has resulted in the sedimentary fill commonly being tuffaceous. Plutons and volcanic flows may also be found within the basinal sequence. These basins have been explored but are not productive.

6.1 CAGAYAN VALLEY BASIN

This north–south elongated basin is bounded by the Sierra Madre Range on the east and the Central Cordillera on the west. The Cagayan River flows along its axis. The bounding ranges are of block faulted and tilted Tertiary volcanic rocks. The basin assumed its present shape in the late Miocene. The flanks of the basin are characterized by large anticlinal features. The basement is of pre-Tertiary metamorphic and plutonic rocks (Fig. 6.1).

A thick marine clastic and carbonate section, ranging in age from Oligocene to Recent, occupies the main part of the basin. Thickness varies from 5 to 10 km along the flanks, but attains a maximum of over 12 km at the centre (Fig. 6.1). The Oligocene is of basic lava flows, agglomerate, tuff breccia, and tuffaceous sandstones, localized in the south-west where they unconformably overlie basement and attain a maximum thickness of 600 m.

The lowermost Miocene is of massive biohermal limestone of 150–600 m thickness. Conformably overlying the Ibulao Limestone are Lower to Middle Miocene sands, shales and silts of the Lubuangan Formation, variable from 1.5 in the east to 6 km thickness in the west. In the north-east the Middle Miocene Callao Formation is of reef complex grading basinward into calcarenite. The Upper Miocene to Pliocene Cabagan Formation consists of shale with limey sandstone and conglomerate intercalations. It is a widespread formation with maximum thickness of 3 km in the west. The most widespread formation is the Pliocene to Pleistocene Ilagan Formation, consisting of lower marine shale and sandstone and an upper marine sandstone, becoming continental with siltstone and conglomerate. Upper Pleistocene to Recent Awiden Formation occurs in the central part of the basin and is entirely non-marine. The geothermal gradient is low on average, with a value of only 22° C km^{-1}.

6.2 SIERRA MADRE BASIN

This offshore basin lies east of the Sierra Madre Range and is separated from the East Luzon Trough by the Isabella Ridge (Lewis and Hayes, 1983). The basin contains up to 4.5 km of sediments. North-north-east striking basement ridges may be traced into the deep ocean environment of the adjacent West Philippine Basin. The East Luzon Trough, the Sierra Madre Basin and the Isabella Ridge may, respectively, represent the trench, fore-arc basin and accretionary complex of a relict subduction system which was active during the Oligocene along eastern Luzon. There are no oil company explorations to elucidate the stratigraphy of this basin.

6.3 LUZON CENTRAL VALLEY

The Luzon Central Valley represents a fore-arc basin between the uplifted and tilted Zambales ophiolite on the west and the volcanic arc of the Cordillera Central. The Cainozoic history of the 14 km thick sequence illustrates the development of a fore-arc basin (Bachman *et al.*, 1983). The basin is floored by oceanic crust on the western side and by older accreted terrains on the eastern arc side. Initial sedimentation on the Zambales oceanic crust occurred during Lower Tertiary northwards translation and emplacement of the crust as an ophiolite along a strike-slip or oblique-slip zone.

Pelagic and hemipelagic sediments were deposited directly on the basaltic part of the Zambales ophiolite and contain a nearly continuous depositional record of the original setting and emplacement history of this large ophiolite from Upper Eocene through Miocene. Pelagic limestone with ash layers (the Aksitero Formation) caps the volcanic complex of the ophiolite (Schweller *et al.*, 1984). There are no cherts and limestone was directly deposited on the oceanic crust. Calcareous nannofossil stratigraphy allows the sedimentation rates to be calculated at 3–5 m Ma^{-1} from Upper Eocene through Lower Oligocene. The rate increased to 10 m Ma^{-1} or more in

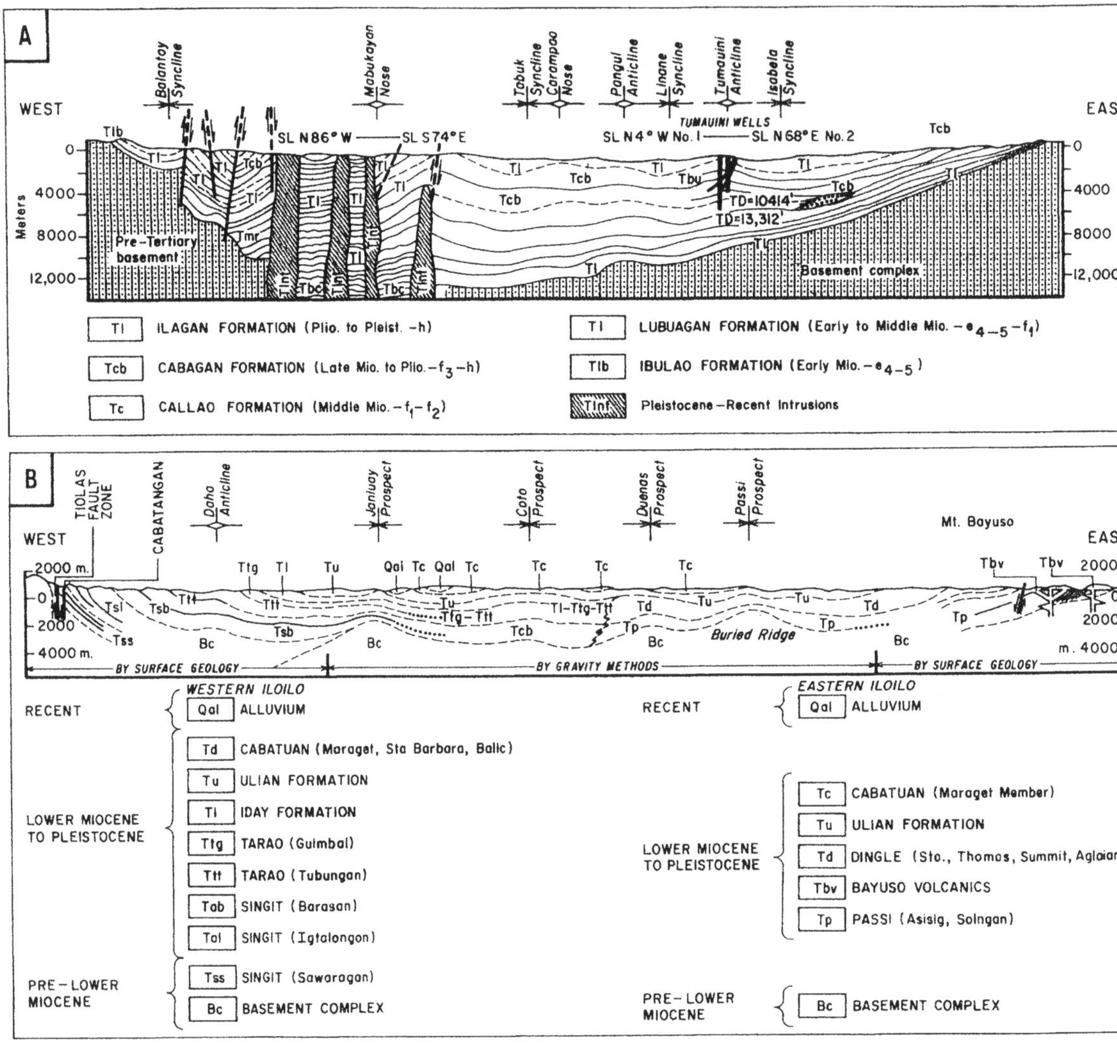

Fig. 6.1. Cross section across (A) the Cagayan Valley Basin, Luzon; (B) the Iloilo Basin, Philippines (after Bureau of Mines, 1981).

the Upper Aksitero Formation, where sandy turbidites appear in the Middle and Upper Oligocene (Fig. 6.2). Lower Miocene mudstone, sandstone, and conglomerate of the Moriones Formation were deposited at much higher rates; this formation includes channels and debris-flow deposits characteristic of deep-sea fans. Oligocene sandstones are predominantly volcaniclastic, whereas sandstones in the Lower Miocene section contain serpentinite and other components derived from the uplift and progressive dissection of the ophiolite. Sedimentary facies and sandstone composition show that the Zambales was deeply eroded by early Miocene and probably first emerged above sea level in Middle to late Oligocene, only 10–15 Ma after it was formed as new oceanic lithosphere (Schweller *et al.*, 1984). A comparison of the Aksitero Formation with ODP sites in the Western Pacific suggest that the Zambales ophiolite was originally part of a marginal basin and not an island arc.

A sequence of arc-derived sediments at least 8 km thick was shed into the eastern arc side of the basin during the Upper Palaeogene to Quaternary convergence along the western margin of Luzon.

By Middle Miocene, the Central Valley became a continuous elongate basin fringed by extensive shelf deposits

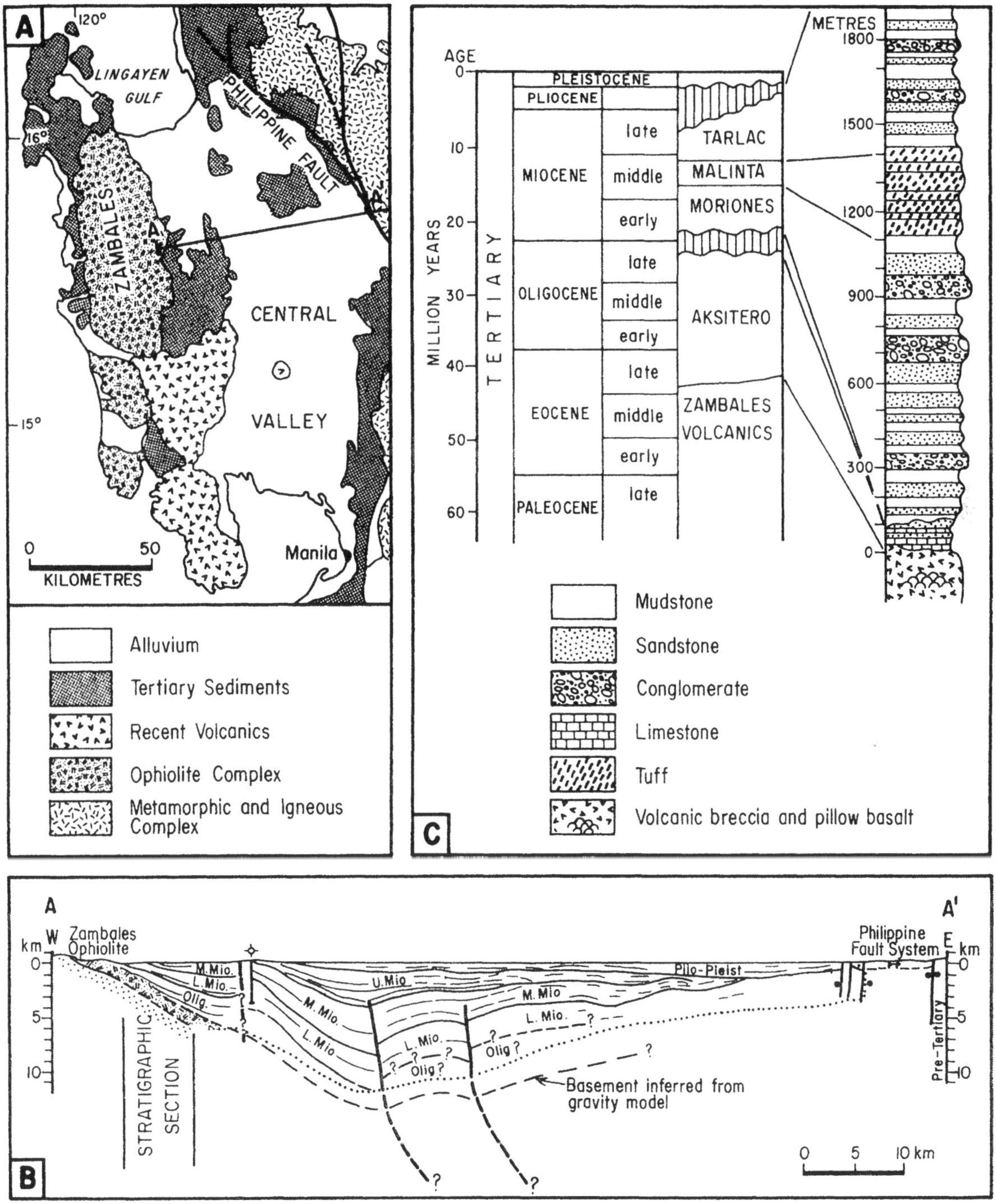

Fig. 6.2. The Luzon Central Valley fore-arc basin. (A) Generalized geological map. (B) Cross section based on outcrop, well data, seismic, and gravity modelling. (C) Stratigraphic column for the sequence which overlies the eastern flank of the Zambales ophiolitic complex (after Bachman *et al*., 1983; Schweller *et al*., 1984).

along both the uplifted Zambales and eastern arc sides of the basin. Detritus shed from both flanks filled the subsiding basin and resulted in progressively shallower water depths (Bachman *et al.*, 1983). Non-marine deposition began in central portions of the basin in the Pliocene and migrated with time both north and south along the basin axis.

Upper Miocene to Holocene movement along the Philippine Fault zone caused uplift and folding of adjacent parts of the basin. Isolated volcanic plugs occur generally near the basin margins, but the recently extinct Mount Arayat lies near its centre.

Most source rocks in the basin are thermally immature because of the low geothermal gradients of 16–24° C km^{-1}.

6.4 RAGAY–SAMAR (SW LUZON) BASIN

The Bondoc Peninsula is transected by branches of the Philippine Fault system and sediments are folded and mildly metamorphosed (Bureau of Mines, 1976, 1981). Tight north-west trending folds and major faults trend parallel to the peninsula. There is a section of Miocene to Pliocene exceeding 5 km which includes a 2.4 km thick basal Lower to Middle Miocene bathyal to neritic sandstone-shale sequence with carbonaceous layers. This Vigo Formation gives oil and gas seepages in the south. The Upper Miocene is of unconformably overlying clayey sandstone and shale with minor limey sandstone interbeds. An Upper Miocene conglomerate occurs in the north, overlain by Pliocene siltstone. Widespread Pleistocene coral limestone disconformably overlies the older formations.

Bicol Peninsula of southern Luzon has a section of more than 5 km, including an Eocene thin fractured basal limestone, unconformably overlain by a thick Miocene bathyal to neritic clastic and carbonate sequence. Unconformably overlying this is an Upper Miocene carbonate-siltstone-sandstone series of 2.5 km thickness. The Pliocene–Pleistocene is of limestone, sandstone, marl and pyroclastic rocks, overlain by the San Roque Tuff.

The Ragay–Samar basinal region has a relatively high geothermal gradient of 41° C km^{-1} as also has the nearby Leyte Basin, about which little is known.

6.5 VISAYAN BASIN

This major basin probably has an offshore sediment fill of 10 km, deposited over a basement complex of ophiolitic rocks and andesites associated with limestones and metamorphic rocks. It is overlain by Palaeocene to Lower Oligocene limestones, tuffs and basalts. This is unconformably overlain by 3 km of Oligocene to Lower Miocene clastic and limestone formations. The Middle to Upper Miocene is of bathyal clastic rocks. The Upper Miocene to Lower Pliocene is of limestone and marl, followed by Pliocene to Pleistocene limestone and a sandstone-shale sequence, and an upper tuff section. Facies changes within the basin are complex and probably are related to movements on the Philippine Fault. Considerable drilling has been carried out, showing an average geothermal gradient of 31° C km^{-1}.

6.6 ILOILO BASIN

The basement is of serpentinite, basalt, slate and metavolcanic rocks. The sedimentary fill is of Upper Oligocene to Miocene sandstone, shale and reefal limestone (Fig. 6.1). Facies changes are complex and poorly understood. The facies include Middle Miocene limestone, a Middle to Upper Miocene sandstone-siltstone unit, and an Upper Miocene sandstone-conglomerate formation. The Lower Pliocene is of calcareous mudstone. The Pliocene-Pleistocene is represented by limestone, becoming shaly upwards and containing some coarse grained sandstones. Correlation of Miocene and Pliocene formations is not yet possible across the basin because of radically different lithologies and dissimilar faunal assemblages (Bureau of Mines, 1976). The wells drilled indicate a very low geothermal gradient of 21° C km^{-1}.

6.7 AGUSAN–DAVAO BASIN

The basin appears to be a half graben, the sedimentary section is regionally tilted, dipping 15–20° (Bureau of Mines, 1976). It is believed that the Philippine Fault is responsible for the major morphology of the basin. The basement is of ophiolite and island arc volcanics. The strata range from Eocene to Pleistocene, with a maximum thickness of 4 km. The Eocene is of massive reef limestone with bedded carbonates, overlain unconformably by a Lower to Middle Miocene marine sandstone-shale-conglomerate and limestone sequence, associated with coal beds. The Upper Miocene is predominantly of limestone overlain by a sandstone-shale sequence. The uppermost Miocene through Pleistocene is very thick (2.5 km), of shale, sandstone, limestone, mudstone and conglomerate.

6.8 COTABATO BASIN

The sedimentary fill is of Upper Oligocene to Pleistocene of more than 8 km maximum thickness. The ısement is probably of arc andesite and basalt, with ›me metasediments. The Upper Oligocene Maganoy ormation is of marine limestone and conglomerate conıining ophiolite clasts. It is succeeded by Lower ⁄liocene greywacke, sandstone, and conglomerate, conaining limestone members. Middle Miocene limestone :ontains interbeds of mudstone and siltstone. The Lower ınd Middle Miocene strata are intruded by basic pluonic rocks. The Upper Miocene is of a thick marine ;ection that becomes non-marine upwards, containing :onglomerates and tuffaceous sandstones. Limestones are also locally present as thick reefs. The Pliocene–Pleistocene is of fluvial to shallow marine sands, silts, and conglomerate. Details of the offshore basin are not available, and drilling has not been successful in oil production. The measured geothermal gradient is extremely low, only 18° C km^{-1}.

6.9 MINDORO BASIN

The Mindoro Basin is part of the Calamian microplate. Two major tectonic styles are recognized in the basin. Extension during the Palaeogene created rift depressions. Middle to Upper Miocene compression resulted in NW–SE trending thrust fold belts that partly overstepped the older fault blocks (CCOP, 1991).

Middle Eocene marine shales rest unconformably upon Cretaceous rocks. The Upper Eocene consists of fluvial to shallow marine shales and sandstones, while pisolitic

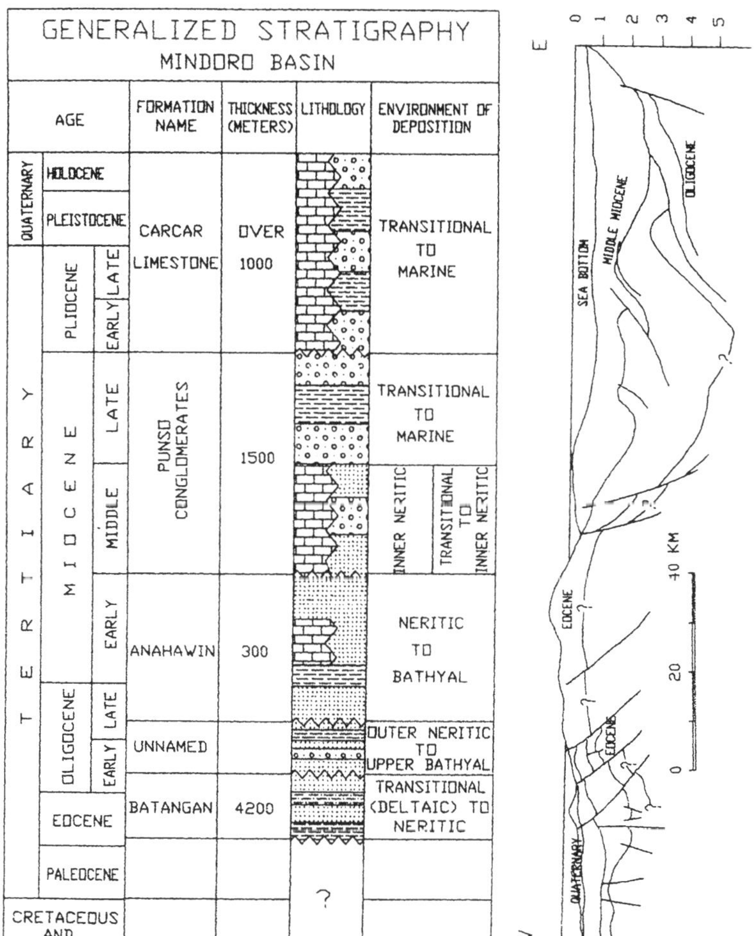

Fig. 6.3. Generalized stratigraphy of the Mindoro Basin, SW Philippines (from CCOP, 1991). The SW–NE cross section has a vertical scale in km.

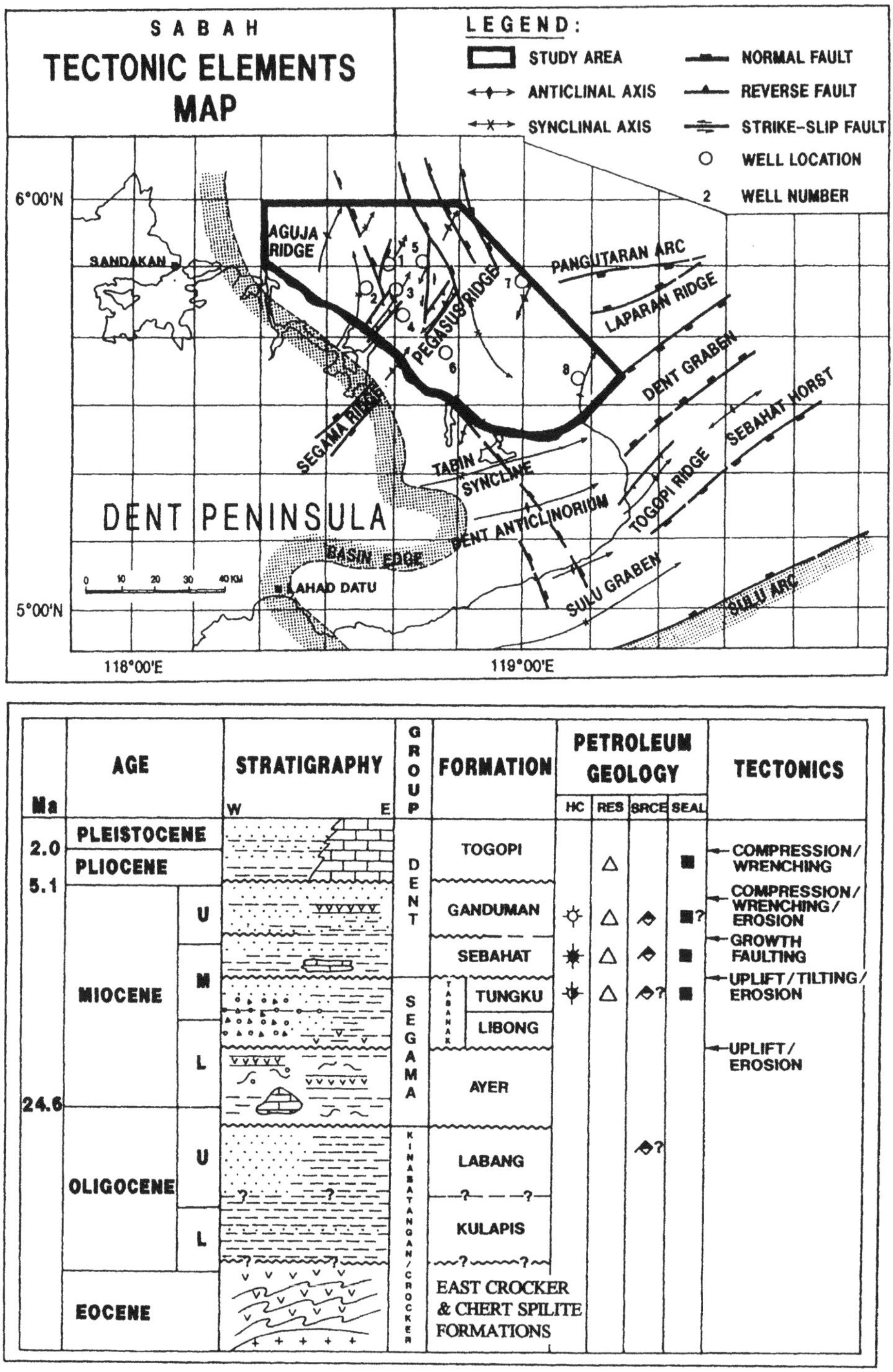

Fig. 6.4. Sandakan Basin of the SW Sulu Sea. **Top**: the major tectonic elements and drill sites of the basin. **Bottom:** stratigraphy and hydrocarbon occurrences in the basin (after Wong, 1993).

limestones occupy the uppermost part (Fig. 6.3). During the Oligocene, marine conditions prevailed with deposition of shales, sandstones and conglomeratic turbidites. Deep marine conditions continued through the early Miocene. In the Middle Miocene, shelfal limestones and reefs developed in a shallow marine environment. The regression continued into the Upper Miocene with deposition of terrestrial facies in some places. During the Pleistocene, coral reefs formed in a very shallow neritic environment with little influx of clastic material.

The first exploratory well was drilled on Mindoro island in 1959, but since then only 3 additional wells have been drilled. Numerous oil and gas seeps occur on SW Mindoro, but no commercial discoveries have been made. The stratigraphy contains several possible source rocks and suitable reservoir rocks.

6.10 SULU SEA BASINS

The western part of the Sulu Sea has several basins which have been explored. Their stratigraphic nomenclature has been extrapolated from onland Sabah (Fig. 6.4). The major tectonic event in the region was the Middle Miocene rifting followed by sea-floor spreading to form the oceanic SE Sulu Sea Basin (Fig. 1.5). The pre-rift formations have all been faulted and broken and incorporated into extensive Middle Miocene marine mudstone-matrix mélanges, such as the Garinono and Ayer Formations (Clennell, 1991, 1992). The marginal basin was created by rifting of an active volcanic arc; the record is to be found in the Ayer Formation and Tungku Formation (Hutchison, 1992*a*).

There are three basins which have been investigated in the SW Sulu Sea. They are the Balabac, Bancauan and Sandakan basins (Fig. 1.5). The Balabac Basin is bounded by the NE-trending Palawan island and the Banggi Ridge. The Bancauan Basin is bounded on the SE by the Keenapusan Ridge. The Sandakan Basin is bounded on its SE by the Laparan Ridge, and finally the Dent Graben is bounded by the Sebahat Horst (Fig. 6.4). These horsts and grabens are a seaward continuation of the outcropping geology of the Dent Peninsula of Sabah (Bell and Jessop, 1974).

There has been no oil or gas exploration work done in the 4000 km^2 Balabac Basin since the mid 1970s. Only one test well (Coral 1) was drilled (Beddoes, 1976). There are other anticlines which could have been investigated (Durkee, 1993). Seismic interpretation in the Bancauan Basin demonstrates that the only two wells drilled were in reality no more than stratigraphic tests (Durkee, 1993).

The Sandakan Basin has been more thoroughly investigated (Wong, 1993; Walker *et al.*, 1992). The stratigraphy of the Dent Peninsula of Sabah (Haile *et al.*, 1965) continues offshore with a general eastwards dip. The Middle Miocene Sebahat Formation is shaly and has source rock potential. The sandy Upper Miocene Ganduman Formation is deltaic and the offshore area shows growth faulting which could offer trapping potential. The Plio-Pleistocene Togopi Formation onshore is of rubbly limestone. It thickens and occurs extensively offshore, where in places it has precluded the obtaining of good seismic records. The stratigraphic section thickens seawards. Deltaic sedimentation with outer-shelf reef growth dominated the Upper Neogene, and there are good reservoirs and seals. Source rocks are dominantly terrestrial.

The Sandakan Basin can be divided into northern and southern structural provinces, separated by the NE–SW Segama–Pegasus High (Fig. 6.4). The sediments in the northern province were deposited in a more landward environment and have undergone several phases of deformation from Middle Miocene to Pleistocene. The western part of this province is characterized by NE–SW trending wrench-related flower structures, whereas the eastern part is dominated by growth faults (Wong, 1993).

In contrast, the basinward southern province encompasses a zone of relatively undisturbed sediments. Subsidence and sedimentation, with little or no structuring, characterizes this southern province.

7
Coal and lignite deposits

Throughout the region there are numerous Tertiary basins which contain lignite and sub-bituminous coal. However, within the older continental core of Sundaland, Palaeozoic and Mesozoic formations contain coal and anthracite. The distribution of the main deposits is shown in Fig. 7.1.

7.1 MYANMAR

The coal deposits of Myanmar (formerly Burma) are not of great importance. Occurrences are known in the Shan states, in the Central Tertiary basins and along the Arakan coastal region (Bender, 1983).

7.1.1 Shan States

The Lower Silurian Wabya Formation contains a mylonitized 80 cm thick anthracite seam 10 km north of Heho. Jurassic coal occurs 5 km east of Kalaw in the Loi-An Series of the southern Shan states, but the beds are thin and discontinuous and are considerably faulted and folded.

Miocene and Plio-Pleistocene lignite occurs in intermontane basins and detailed exploration would be justified in view of the success in adjacent Thailand. The known occurrences are in the Lashio district and towards the south in the Shan states. Farther south, occurrences have been recorded in the Theindaw–Kawmapyin basin of the Mergui district.

7.1.2 Arakan coastal district and Indo-Burman Ranges

Numerous thin lignite seams and streaks occur in the Eocene sequence in the coastal area and offshore islands, but they have no economic value (Bender, 1983). They occur on the islands of Ramree and Cheduba. Coal of Eocene age is exposed in seams usually less than 0.6 m thick. Near Kalewa (Fig. 7.1) some seams of maximum thickness 2.7 m are mined. The production is of around 10 000 t per year and the reserves are around 123 Mt. Highly deformed and fractured coal occurs at Kywezin, but attempts to mine it have been abandoned.

7.2. THAILAND

The deposits in Thailand are mainly of low quality, ranking from lignite to high-volatile bituminous coal (Ratanasthien, 1986). They occur in more than 30 of the Cainozoic basins which are scattered from north to south Thailand, mostly in the western half of the country (Fig. 7.1).

7.2.1 Northern Thailand

Only six mines are in active operation. The seams occur in lacustrine or intermontane forest swamp deposits. The Cainozoic strata of the basins rest unconformably on Mesozoic and older basement. Generally the successions begin with conglomerate, grading upwards to sandstone, mudstone, coal and oil-shale. The upper sections contain conglomerate and redbeds with some gypsum and laterite cappings.

The mines all lie south and west of Lampang. The Mae Moh is the main deposit, and has two lignite seams up to 30 m thick, of synclinal geometry, faulted on one side of the basin. The grade is lignite to sub-bituminous. Total reserves are around 650 million t, with 300 to 400 t per day production (Asnachina and Pitragool, 1978). The Li deposit contains lignite seams 7 m thick with 15 Mt of reserves.

7.2.2 South Thailand

The best lignite deposit is at Krabi (Fig. 7.1). It lies on the coastal plain facing the Andaman Sea. The lignite seams total 28 m thickness, and are associated with Upper Tertiary sand, clay and marl, some of which are marine. The reserves are around 100 Mt.

7.3 VIETNAM

Anthracite is by far the most important commodity mined in Vietnam. Its northern coal fields occupy the E–W trending Upper Triassic belt of the Quang Ninh anticlinorium, forming the coastal zone of the Ha Long Bay east of Hai Phong, the port of Hanoi (Fig. 7.2). Production is around 5 Mt per year, mostly of anthracite, with less than 10% bituminous coal.

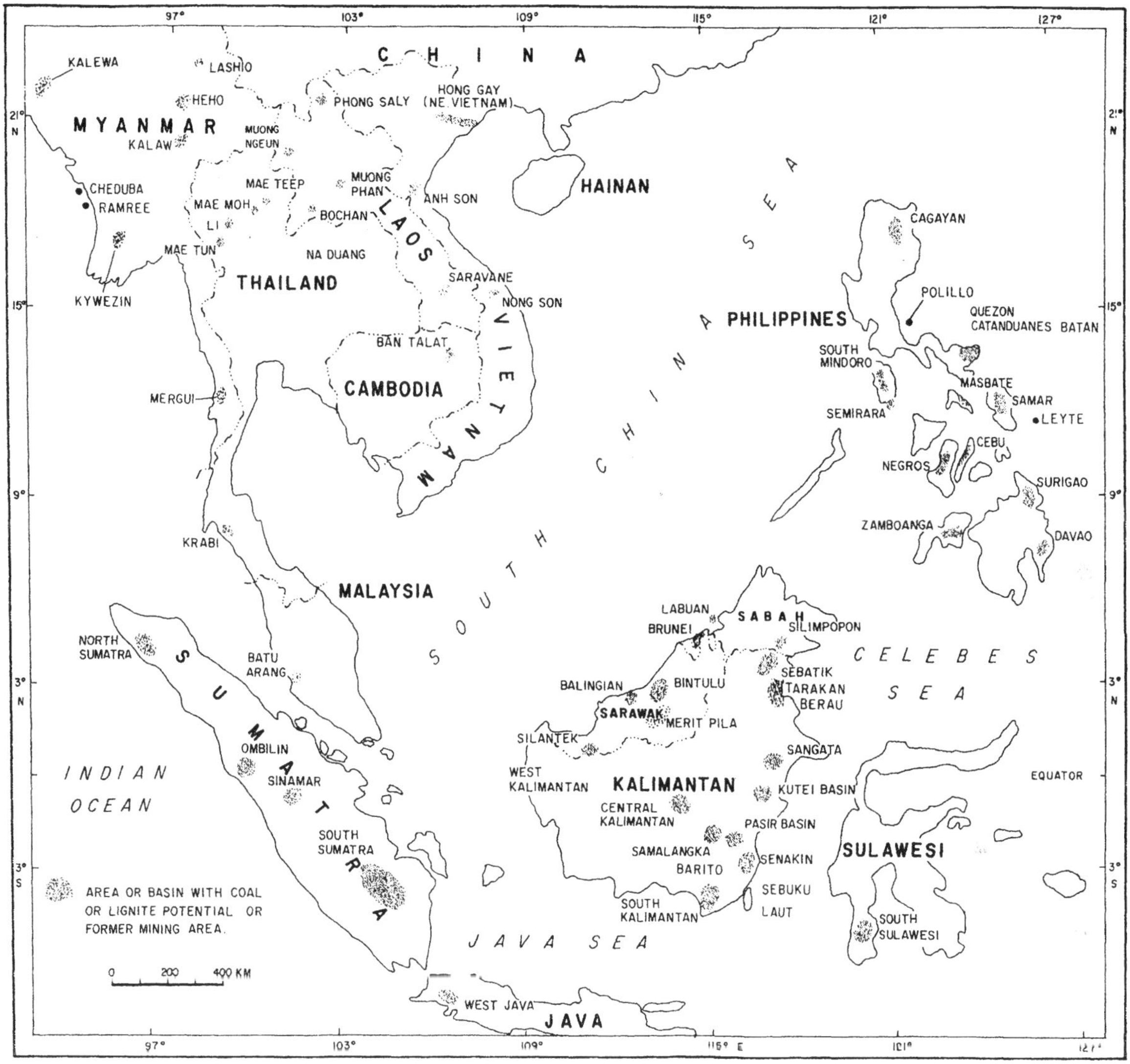

Fig. 7.1. Distribution of the main coal and lignite deposits of South-East Asia. Based on Kendarsi Roeslan (1984), Fontaine and Workman (1978), and Bender (1983).

The Hon Gay district, east of Hai Phong (Fig. 7.2) is the foremost anthracite mining district. It has reserves exceeding 1000 Mt and about 20 individual mines are in operation. The coal is mined from Upper Triassic strata, mainly Norian and possibly ranging from Carnian to Rhaetic. The coal beds have yielded a well described Tonkin flora from the Hon Gay coal field (Zeiller, 1903). Possibly coeval flora has been described from Quang Nam in the Trungson area by Counillon (1914). The Krusin flora of the Triassic Sadong Formation of west Sarawak has close similarities with the Hong Gay flora of Tonkin, thus proving a close proximity of the two terrains in late Triassic time.

There are four principal coal basins north and east of Hanoi, extending over a total E–W distance of 230 km (Fromaget, 1941). Most of the mines occupy the southern margins of the basins. The most important field, Hong Gay, is 35 km long and averages 10 km width. It contains up to 15 seams or composite seams of anthracite. Individual seams range from less than a metre to 10 m

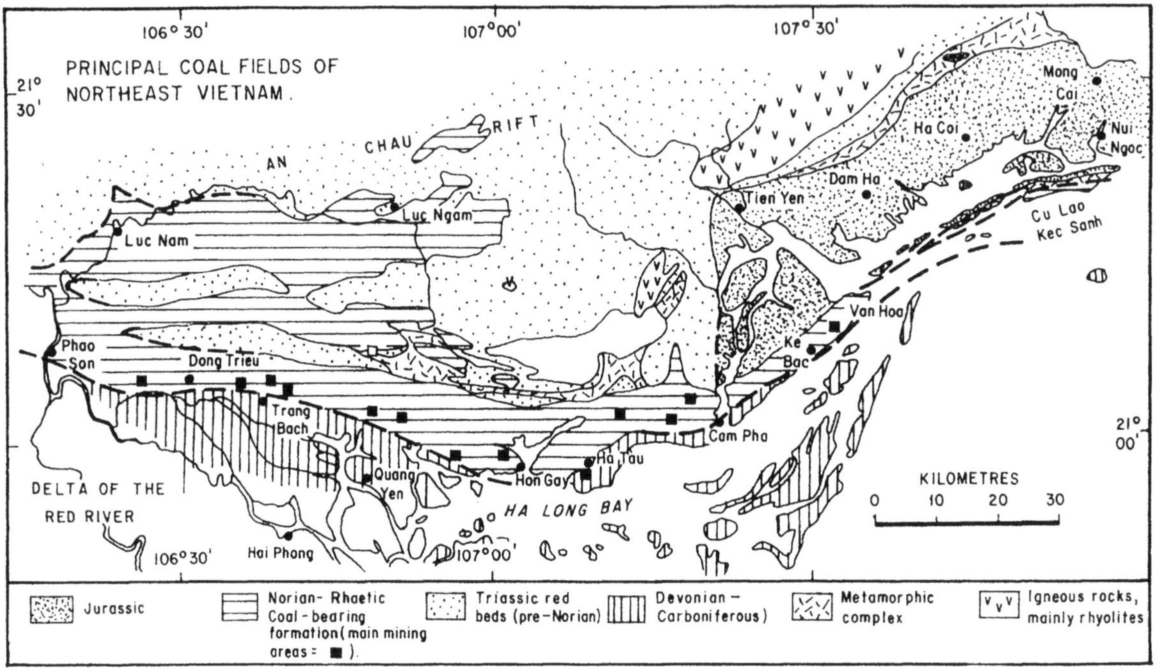

Fig. 7.2. Simplified geological map of the Quang Ninh anticlinorium of northern Vietnam, showing the position of the Hong Gay and other major Triassic anthracite mines (after Fontaine and Workman, 1978).

thickness. The Hong Gay field itself has reserves of the order of 200 Mt.

The principal coal mine elsewhere in Vietnam is at Nong Son, 45 km SW of Da Nang. The coal occurs in an E–W elongated basin near the base of the Mesozoic strata and is of Rhaetic age. The main coal seam has a vertical thickness of up to 42 m in the central part of the basin. Production reached 104 000 t in 1963 and the reserves are quoted at 8 Mt (Fontaine and Workman, 1978).

Around the city of Vinh there are a number of coal and anthracite occurrences. Neogene lignite deposits are reported along the Red River valley.

7.4 LAOS

Coal occurs in three regions, from Saravane in the south to Phong Saly in the north. The best prospects are NE of Saravane, in an inlier of folded Carbo-Permian strata, overlain and surrounded by gently dipping Mesozoic redbeds. The coal occurs in Upper Carboniferous beds and the seams are strongly folded. Drilling has revealed at least three anthracite beds (Banthay *et al.*, 1986).

Coal occurs NW of Vientiane in an Upper Carboniferous series. The beds are strongly folded and overlain unconformably by gently dipping Triassic redbeds in the Nam Thong Valley. Drilling has revealed a 7 m thick anthracite seam. Coal occurs in northern Laos in Upper Carboniferous to Upper Triassic formations, mainly around Phong Saly. Several seams about one metre thick occur in sub-horizontal Triassic strata (Fromaget, 1941).

The border region between Laos and Vietnam contain several Cainozoic lacustrine basins, which contain lignite (Banthay *et al.*, 1986).

7.5 CAMBODIA

No definitely exploitable seams have been reported, but thin coal beds occur in Permian, Triassic and Jurassic strata. The main occurrence is in north-east Cambodia where reserves of 7 Mt have been reported (Fontaine and Workman, 1978).

7.6 MALAYSIA

7.6.1 Peninsular Malaysia

With the exception of a few very thin coal layers in Viséan strata along the coastal regions north of Kuantan, which contain plant fossils (Jennings and Lee, 1985), and which are totally uneconomic, all Peninsular Malaysia's coal potential is in the form of lignite seams within small lacustrine Palaeogene basins (Stauffer, 1973*b*).

At Batu Arang, NW of Kuala Lumpur, mining started in 1915 and continued until 1960. The beds are of high grade lignite and sub-bituminous coal as streaks in dark shales and as seams up to 2.4 m thick. Two prominent seams occurred along the eastern side of the basin and were mined down dip westwards. The basinal strata are overlain by Pleistocene boulder beds.

7.6.2 Sarawak

Several lignitic coal-bearing basins have been identified (Fig. 7.1), and some have supported past mining activity.

The Silantek coal field lies 4 km north of the Sarawak–Kalimantan border, about 100 km SE of Kuching (Haile, 1954). The basin extends into Kalimantan (Zeijlmans van Emmichoven, 1939). The sediments are of Eocene neritic to continental strata, overlain by the undated continental Plateau Sandstone Formation. They consist of siltstone, shale and sandstone. One coal seam has a thickness of 1.9 m. The formation is locally intruded by Miocene stocks and dykes, which in places have metamorphosed the lignite to anthracite. Fifty Mt of coal are estimated on the Sarawak side of the border. The Sadong colliery produced 876 000 t during its operations from 1874 until 1931.

The Merit–Pila deposit is located 75 km upstream from Kapit in the Rajang River basin (Kirk, 1957; Johari and Mohamad, 1994). It is the biggest coal-bearing basin in the country, and covers 160 km^2, elongated in an E–W direction. The coal is contained within this the largest outlier of the Upper Oligocene to Miocene Nyalau Formation, which unconformably overlies the Rajang Group orogenic belt of the Sibu Zone. The Miocene rocks include conglomerate, sandstone, mudstone and coal seams. Two or three seams range in thickness from 1.5 to 3.8 m and have commercial potential. An estimated 300 Mt have been reported from drilling on the western side of the basin and 50 to 60 Mt may be mined by open pit, now in operation. The rank is high grade lignite.

The Balingian coal area occupies the coastal plain between Mukah and the Balingian River (Wolfenden, 1960). The coal-bearing formations are the Miocene Balingian and Pliocene Liang formations. Lignite seams occur within the Miocene strata, and three seams of 1.5 to 5 m thickness occur in the Liang Formation, both investigated by drilling. Reserves are estimated to be of the order of 100 Mt of lignite grade.

The Bintulu coal deposit occurs in Miocene paralic strata (Kho, 1966). About 2 Mt of bituminous coal are regarded as mineable, but less than half is accessible by open pit.

7.6.3 Sabah

The Labuan coal mine occupied the north-eastern tip of Labuan island, extending just 2 km west from Kubang Bluff. Coal was mined from 1848 until 1912 (Wilson, 1964). The coal measures are of Miocene age, occurring in the Belait Formation, in a similar setting to the deposits in Brunei. Four seams were mined, with thicknesses of 1.2 to 3.3 m. The grade was variable from black sub-bituminous to brown lignite. Unmined reserves are estimated to exceed 9 Mt.

The Silimpopon coal field occurs 40 km west of Tawau. The strata belong to the Miocene Kapilit Formation. Three or four coal seams have thicknesses up to 1.8 m of sub-bituminous coal. The deposits were mined from 1904 until 1932 and remaining reserves are about 7 Mt (Collenette, 1965). Coal is exposed elsewhere in the Tanjong Formation, usually of good quality lignite, currently being evaluated by BHP from their Kota Kinabalu office.

7.7 BRUNEI

The eastern limb of the NE-plunging Berakis Syncline, which steeply dips NW, locally vertical, contains a line of coal deposits extending from Brunei Town (Bandar Sri Begawan) to Muara. The coal beds occur within the Upper Miocene to Pliocene Belait Formation (Wilford, 1961). The Brooketon Colliery, 2 km WNW of Muara, operated and was regionally important from before 1883 and closed in 1924. The coal beds dip vertically to 65° W or WNW, interbedded with soft massive sandstone and mudstone. Five coal seams range up to 8 metres thick. The grade is high grade lignite, described as 'bright-looking lignite with a shaly fracture'. During mining it was very susceptible to spontaneous combustion, a major reason for eventual closure of the colliery.

7.8 INDONESIA

Although pre-Tertiary thin coal beds are known in Indonesia, they have not warranted investigation, and the known coal potential of the country is in Tertiary basins in which the coal beds range from Eocene to Miocene age. The deposits and reserves have been reviewed by Koesoemadinata (1978), Bainton (1978) and Kendarsi Roeslan (1984).

The rank usually is high grade lignite to sub-bituminous, but may have been ameliorated through contact metamorphism around igneous intrusions. The distribution is shown in Fig. 7.1.

7.8.1 Sumatra

North Sumatra is currently being investigated in areas where lignite beds outcrop.

The Ombilin region began producing coal in 1892, but production has now declined from 665 000 to 85 000 tonnes per year. The basin sequence is of alluvial fan on the margins, with lacustrine deposits in the centre (see Section 4.5). Three coal seams have thicknesses variable up to 8 m. The quality of the Eocene coal is sub-bituminous. The mine is still in operation and reserves are around 185 Mt.

The Sinamar coal field contains three main coal seams, of thickness variable up to 8 m. They occur in a sequence of conglomerate, sandstone and occasional thin limestone.

In South Sumatra, the coal measures occur within Miocene–Pliocene formations. Coal mining has been in progress at Bukit Asam since 1919, and the Air Laya opencast mine since 1942. Coal production reached 650 000 t but declined to 85 000 t in 1974. The coal is hard lignite, which has been ameliorated to various grades through bituminous to anthracite by post-Pliocene andesite intrusions. There are three main composite seams, separated by clay-sandstone sequences. Several individual areas in South Sumatra give a cumulative total reserve exceeding 1000 Mt (Kendarsi Roeslan, 1984).

7.8.2 Kalimantan

The prominent coal areas in eastern Kalimantan are Berau, Tarakan, Kutei Basin, Pasir Basin, Senakin, Samalanga and Sangata. There is some mining, and investigations may lead to greater future activity.

In the Kutei Basin, coal bearing strata, known from outcrops and extensive oilfield drilling, occur in Miocene and Pliocene beds, deposited in a deltaic environment. They are thin and usually are less than 1 metre. The Upper Balikpapan Beds, however, contain four lignite seams of between 1.6 and 3.1 m thickness. Thick Middle Miocene delta plain coals outcrop and are mined from large open pits inland west of the Kutei Basin oilfields (Land and Jones, 1987).

Three seams ranging from 4.4 to 6.9 m thickness have been identified in Eocene strata of the Pasir Basin. Eocene coal is also reported in the Senakin Peninsula. Pulau Laut and Pulau Sebukku also have Eocene coal. The Sebatik coal measures are thought to have a genetic relationship with the Silimpopon area of Sabah.

In southern Kalimantan, there is coal potential in the Barito Basin, and in western Kalimantan, from continuations of the Silantek Formation.

7.8.3 Java

Of all the coal occurrences found, the prominent areas are at Bayah and Cimandiri in west Java, where the coal is of Eocene age. However, it is intensely folded and faulted, and seams rarely exceed 1 metre. The quality is good and the rank bituminous.

7.8.4 South Sulawesi

Coal is found in a few areas, but seams are thin, and commonly covered by thick overburden, so that the commercial potential is limited. Five Lower Eocene seams have been identified at Malawa.

7.9 PHILIPPINES

Philippine coal occurs in several Tertiary basins (Fig. 7.1), ranging from Eocene to Pliocene. 42% are lignite to sub-bituminous and 55% sub-bituminous to bituminous with 3% local higher grades (Bureau of Mines, 1986).

The Cagayan Basin contains Pliocene coal within the Ilagan Formation, the youngest in the Philippines. The basinal strata are flat lying to gently dipping, and the coal grade ranges up to sub-bituminous. Reserves have been estimated for the basin of 336×10^6 tonnes.

Polillo island provides Miocene coal from the Burdeos Formation. The basinal strata dip 30 to 50°. The coal is sub-bituminous to bituminous, but reserves are less than half a million tonnes.

Cataduanes contains Eocene coal in the Payo Formation. The basin is structurally complex, and the reserves are small, but the grade is bituminous.

Batan contains Miocene to Pliocene coal within the Bilbao and Liguan formations. The strata are flat lying to low dipping. The coal is sub-bituminous but reserves are less than 10×10^6 tonnes. Coal-bearing strata of Miocene age occur on Samar and Leyte islands. The grade is sub-bituminous.

Miocene coal beds occur in the Surigao Basin within the Bacuag Formation. In the southern part of the basin, coal is also found in the Mekoupe and Pampayan formations, of bituminous to sub-bituminous grade.

Miocene coal beds occur within the Lumbog Formation of the Zamboanga Basin. At Malanges, the coal is upgraded from sub-bituminous to semi-anthracite because of contact metamorphism from nearby igneous intrusions.

Coal measures on Cebu range from Oligocene to Miocene in age. The basin is strongly faulted, and dips may attain 70°. The grade is sub-bituminous to bituminous, but the reserves are less than 3×10^6 tonnes.

Sub-bituminous coal of Miocene age occurs on Negros. Miocene coal occurs in NE Masbate island. The beds are highly disturbed by the Philippine Fault Zone. The coal is lignite to sub-bituminous.

The most important coal deposits of the Philippine Archipelago lie within the Semirara Basin, and the potential may reach 500 Mt. The coal beds occur within the Miocene Semirara Formation. The structure of the coal measures is anticlinal, with dips between 5 and 30°. The grade is lignite to sub-bituminous.

Southern Mindoro has Miocene coal seams of sub-bituminous rank, occurring within strata which dip from 0 to 30°.

8

Mineralization associated with ophiolite

8.1 DEFINITIONS

The exact definition of ophiolite was formalized by Anonymous (1972) and has generally been followed since. Ophiolite is a distinctive assemblage of mafic to ultramafic rocks. In a completely developed ophiolite the rock types occur in the following sequence:

4. Mafic volcanic complex, commonly pillowed (at the top of the sequence)
3. Mafic sheeted dyke complex
2. Gabbroic complex, ordinarily with cumulus textures commonly containing cumulus peridotite and pyroxenite and usually less deformed than the underlying ultramafic complex.
1. Ultramafic complex, consisting of variable proportions of harzburgite, lherzolite and dunite, usually with a metamorphic tectonite fabric and more or less serpentinized (bottom).

Associated rock types include: (i) an overlying sedimentary section typically including ribbon cherts, thin shale inter-beds, and minor limestones, (ii) podiform bodies of chromite, generally associated with the dunite, and (iii) sodic felsic intrusive and extrusive rocks.

Faulted contacts between mappable units are common. Whole sections may be missing. An ophiolite may be incomplete, dismembered, or metamorphosed (Hutchison, 1975). Although ophiolite is generally interpreted to be oceanic crust and upper mantle, the use of the term should be independent of its supposed origin.

The most common occurrence of ophiolite is along the margins of marginal basins, where the obducted ophiolite has not yet been compressed into an orogenic suture zone. Notable examples occur in the Philippines, eastern Indonesia and Sabah. Within the continental sector of SE Asia known as Sundaland, ophiolites are more restricted, more dismembered and incomplete (Hutchison, 1975) and confined to narrow suture zones between different continental lithospheric terrains.

Podiform chromite concentrations within upper mantle dunite or harzburgite are referred to as chromitites. Cyprus-type massive copper sulphide deposits are interpreted to have formed by submarine hydrothermal exhalations directly onto the ocean or marginal basin floor. They are associated with pillow basalts and the overlying sedimentary rocks of the ophiolite association. Manganese deposits associated with ophiolite may also have resulted from submarine exhalations and may be likened in places to manganese nodules which are common on the deep ocean floor. Once uplifted, the ultramafic part of an ophiolite is susceptible to deep subaerial weathering, resulting in blankets of residual nickel and iron rich laterite. The weathering and serpentinization may also liberate precious metals.

8.2 GENERAL DISTRIBUTION IN SE ASIA

Chromitites are found in economic concentrations in ophiolites derived from marginal basin lithosphere of the now inactive Celebes, Sulu and South China seas (Western Philippines, Sabah, Meratus Mountains and Taiwan in part). Ophiolites derived from Indian Ocean lithosphere (Timor, Andaman–Nicobar, and possibly Myanmar in part) have been non productive to date. The distribution of the most important ophiolite-related ore deposits is shown in Fig. 8.1, after Hutchison and Taylor (1978) and Taylor and Hutchison (1979). Nickel sulphides and platinum occur locally within dunite and serpentinite in the Acoje mines of the Philippines. Nickel has also been remobilized and deposited as sulphides in quartz-marcasite epithermal veins within andesitic rocks in northern Leyte.

Deep tropical weathering of the ultramafic parts of ophiolite bodies can result in valuable nickel ore in the form of blanket residual deposits of nickel-bearing laterite, which has become the increasingly important ore of nickel. The laterite is strip mined if it is developed on a flat surface such as a peneplain or plateau. The typical nickel oxide content of fresh peridotite is 0.44% or less. Tropical weathering may result in a seven-fold enrichment to give typical mining values of 2.7%.

The major units of a laterite profile are, from top to bottom: the limonite zone, dominated by goethite; the smectite-quartz zone consisting of soft nontronite clay and quartz; and the saprolite zone consisting of altered bedrock in which the bedrock texture is preserved

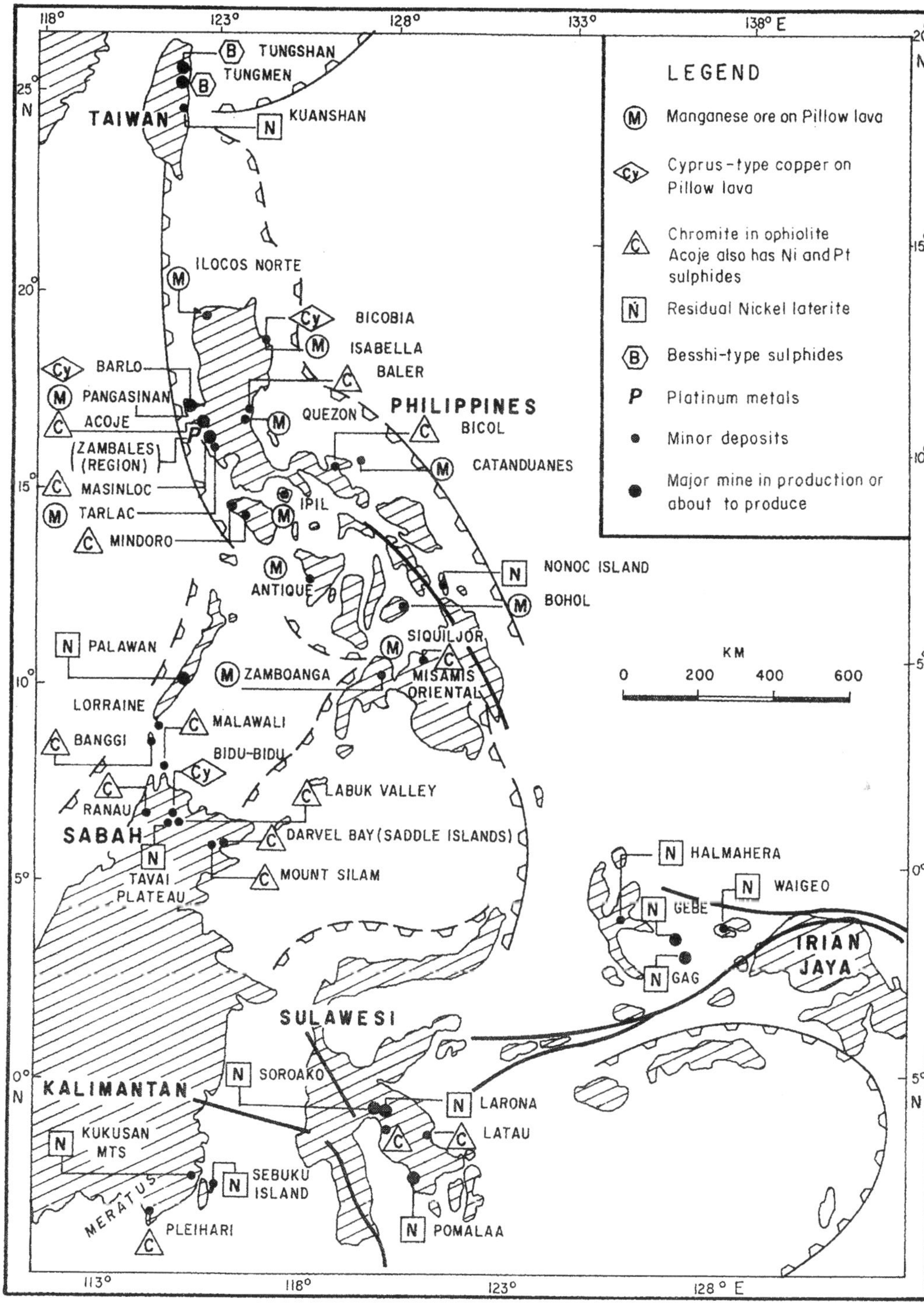

Fig. 8.1. Locality map of the most important ophiolite-related ore deposits of SE Asia (after Hutchison and Taylor, 1978).

(Golightly, 1979*a*). Serpentinized peridotite alters to montmorillonite, aluminous goethite and quartz. Stable chromitiferous chlorite becomes concentrated in the weathering profile. There is no kaolinite and the alumina occurs within the goethite. In the upper layers, serpentinite and montmorillonite disappear completely and the aluminium content of the goethite diminishes. With increasing depth, iron oxide (Fe_2O_3) and aluminium oxide (Al_2O_3) contents fall, while silica (SiO_2) and magnesium oxide contents increase. The trace elements nickel, cobalt, manganese, copper and chromium are all concentrated during weathering. There is strong evidence that much of the nickel is contained in the goethite.

There are two types of laterite horizons. One is high in silica and magnesia and has low iron oxide contents. It occurs low in the weathering profile, is green and distinct nickel-bearing minerals are visible. The second group is low in magnesium oxide and silica and high in iron oxide (Fe_2O_3). It occurs high in the weathering profile, is red or reddish-brown and no nickel-bearing minerals are visible. Both types occur within any deposit. The three nickel-bearing minerals of common occurrence are pimelite (a montmorillonite-like mineral), schuchardite (a nickel-containing chlorite) and nickel-containing serpentine. The name garnierite is used loosely for these hydrous nickel minerals. The main bulk of the nickel is associated with goethite, and high nickel values are often associated with the Mn-Ni-Co nodules that occur in some laterite profiles. The concentration of nickel can be seen as a result of percolating ground water, causing release of the nickel that was fixed in the newly formed minerals.

In many nickel-containing laterites, the nickel is richest just above the fresh bedrock. For example, Reynolds *et al.* (1973) presented analyses for fresh peridotite core boulders in a laterite. They contain 0.48% nickel, while the weathered rim contains 1.76% Ni. Likewise, cobalt increases from 0.013% to 0.02%. Core boulders are tumbled to remove the nickel-rich rims before being discarded. If the nickel is fixed in the aluminium-containing goethite high in the weathering profile, then downward migration of nickel to concentrate just above the fresh bedrock will be severely restricted. How this concentration takes place in rich commercial deposits is not yet understood. Nickel-containing laterites are now being mined in the south-west Pacific at Soroako and Pomala in south-east Sulawesi, on Nonoc island in the Philippines, and in New Caledonia (Hutchison, 1983*a*).

The richest lower zones in the Philippines typically have values of Ni 2.0, Co 2.0 and Cr 3.0 wt %. All current nickel-laterite mining is derived from ophiolites formed from obducted Pacific Plate and related marginal basin lithosphere (Sulawesi, eastern Philippines, Halmahera, Irian Jaya and eastern Taiwan). In addition to being concentrated in residual laterite, chromite forms an important constituent of placer and beach sands (Bureau of Mines, 1986).

8.3 PHILIPPINES

8.3.1 Zambales chromitite

The Zambales complex is divided into the Masinloc, Cabangan and San Antonio massifs (Bacuta, 1979), each of which is bounded by faults. Fernandez (1960) proposed a further subdivision of the Masinloc massif into the Acoje ultramafic belt, which is associated with the middle gabbro belt, and the Coto ultramafic belt, associated with the eastern gabbro belt (Fig. 8.2). Hawkins and Evans (1983) refer to the two peridotite-gabbro units, which are separated by the Lawis Fault, as the Acoje and Coto blocks. The Coto Block reaches a thickness of approximately 5 km and is interpreted as originating in a back-arc basin, whereas the Acoje Block is approximately 9 km thick and may have originated in a young island arc environment (Hawkins and Evans, 1983). The Zambales complex has been classified as a supra-subduction zone ophiolite, related to back-arc marginal basin formation (Yumul and Balce, 1994).

With 62 chromite deposits, the Zambales Range is the major source of chromium for the Philippines, which once accounted for about 6% of the world's production. From 1974 onwards, the total Philippine chromite production annually averaged 500 000 t. In 1982 the estimated reserves were: metallurgical grade (average 34% Cr_2O_3) 18 832 000 t and refractory grade (average 31% Cr_2O_3) 11 367 000 t (Bureau of Mines, 1986). 80% of these reserves occur in the Zambales Range. The geology of the deposits, regional distribution of refractory and metallurgical ores, and their mineralogy and geochemistry have been summarized by Fernandez (1960), Stoll (1962), Bryner (1969), Gervasio (1973), Bacuta (1979) and Hock and Friedrich (1985). The serpentinized ultrabasics of the Acoje Mine are overlain by 1 million tonnes of laterite containing 0.47% Ni. The mineral figures show a production of platinum only for the years 1970–1974, and since then it has declined to zero. The peak was in 1972, when 213 200 kg of platinum and palladium metals were produced from the Zambales (Bureau of Mines, 1986). The platinum group element abundance and mineralization in the ultramafic rocks of the Acoje Block have been described by Bacuta *et al.* (1987) and Orberger *et al.* (1987). The Acoje was also the only mine to have produced nickel sulphide from the chrome ore. The nickel concentrations are related to Ni-Cu sulphides within 1 to 25 m thick lenses

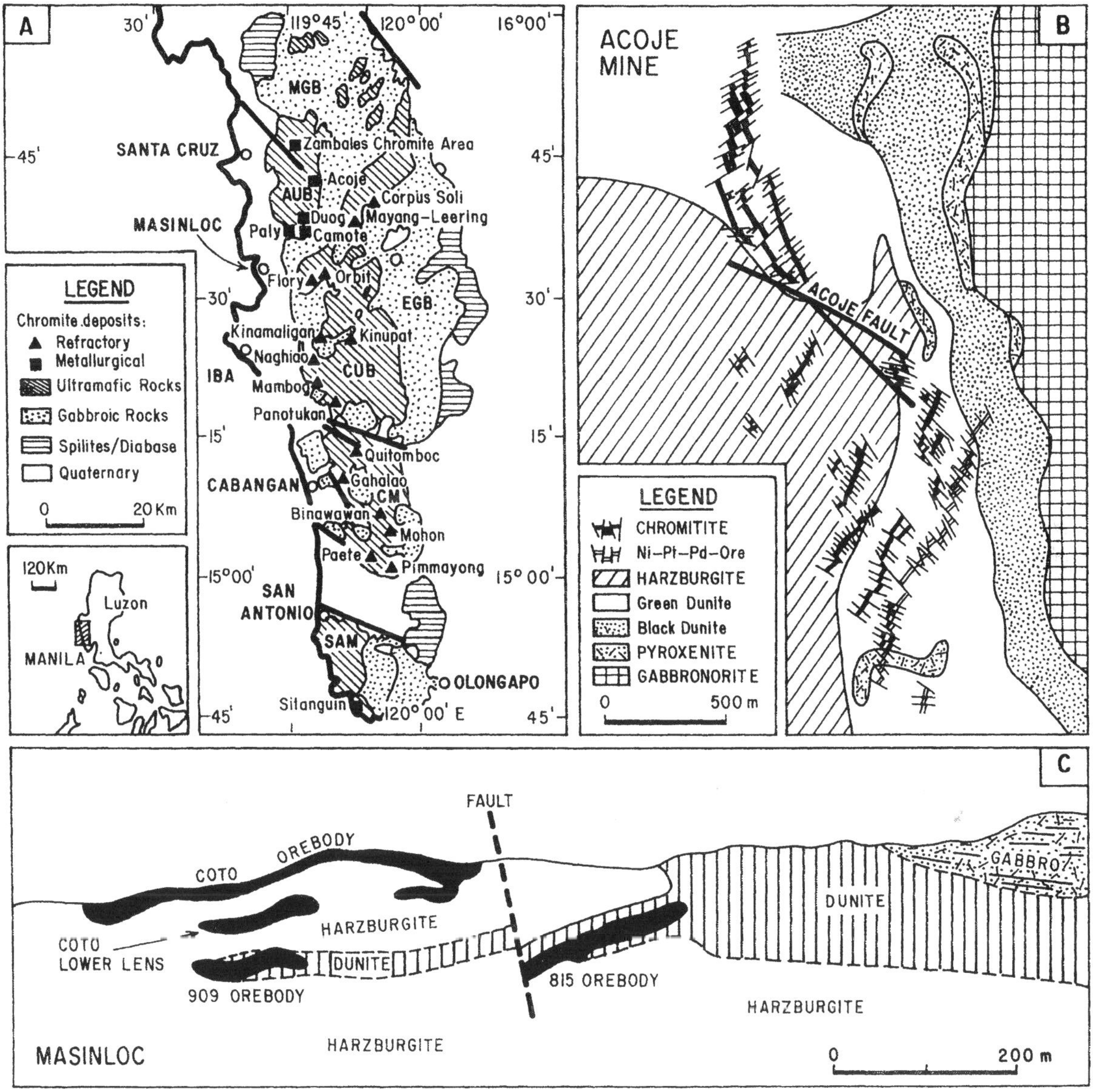

Fig. 8.2. Chromitite deposits in the Zambales ophiolite complex. **A**: geological map of the structural units and some mine localities. AUB = Acoje ultramafic belt, CUB = Coto ultramafic belt, MGB = Middle gabbro belt, EGB = Eastern gabbro belt, CM = Cabangan massif, SAM = San Antonio massif; **B**: geological map of the Acoje area with projection of sub-surface mined chromite bodies onto the surface geology (after Hock and Friedrich, 1985); **C**: cross section through the Coto orebody (after Bryner, 1969).

of black serpentinite occurring within the more common green serpentinite. The black colour is due to initial weathering of the dunite rather than to serpentinization (Orberger and Alleweldt, 1994). Palladium and platinum have been exploited as a by-product of Ni mining of the black serpentinite lenses.

Cr-rich metallurgical ore occurs in the Acoje and San Antonio massifs, where it is associated with a sequence of peridotite, pyroxenite and gabbro-norite. Hawkins and Evans (1983) deduce from bulk rock compositions that the Acoje unit was derived from a part of the mantle which has been through one or more previous episodes of fractional melting. Chromite seams are numerous. Folded chromite layers are common at Acoje, conformable with the foliation of the enclosing peridotite. Massive chromitite pods are concentrated in the fold hinge zones (Bacuta, 1979). Post-mineralization faulting is common, resulting in brecciated ore. Satellitic chromite strings occur approximately 700 to 1000 m below the peridotite-norite transition (petrological Moho). The chromite-bearing rock is predominantly dunite with interlayered harzburgite and pyroxenite. Detailed studies show that the Cr-rich pods are located near the lherzolite-harzburgite transition (Leblanc and Violette, 1983).

Al-rich refractory ore occurs in the Coto belt and in the Cabangan massif (Fig. 8.2), where it is restricted to a sequence of peridotite, troctolite and olivine gabbro. The chromitite-bearing sequence occurs approximately 250 to 350 m below the peridotite-gabbro transition and consists of either harzburgite with intercalated layers of feldspathic peridotite (picrite) or lherzolite. The Al-rich pods are distributed along the harzburgite-dunite transition zone (Leblanc and Violette, 1983).

Masinloc is the world's greatest producer of refractory ore. There is a big lens and 12 smaller ones, all strung out in a belt 2.5 km long (Fig. 8.2). A rough parallelism exists between ultramafic layering and the string of chromitite lenses, which all have a layered structure. Of the orebodies, the Coto is remarkable for its size. According to Stoll (1962), it is an oval flat-lying lens about 600 m by 300 m and is estimated to have been 80 m thick before erosion. The belt of chromitite bodies coincides with a zone of intense serpentinization and fracturing. Many of the fractures are occupied by diabase dykes which cut the chromitite bodies.

The chromite deposits may be divided into three types (Hock and Friedrich, 1985): concordant tabular, strings of pods, and pocket-like deposits. This classification corresponds to increasing deformation of the chromitites.

8.3.1.1 *Concordant tabular deposits*

Concordant chromitites in the Acoje Mine are parallel to the country rock layering, and the ore zone can be traced along strike for at least 8 km. The transition from chromitite to host dunite follows the following scheme: massive ore at the basal orebody contact, schlieren ore, disseminated ore, and finally dunite at the top wall. The maximum thickness reaches 10 m. Tectonized chromitites close to the Acoje Fault may however be as thick as 40 m. The chromitites are cut and dislocated by faults. The trend of the mineral lineation is parallel to the ridge axis of the Acoje block.

Concordant tabular chromitite deposits are the most easily mapped, and hence have the largest reserves because they are more continuous and more easily exploited. The Acoje Mine has reserves of 1.4 Mt.

8.3.1.2 *Strings of pods*

Like the concordant deposits, the strings of pods occupy distinct stratigraphic positions close to the 'petrological Moho'. Individual pods range from a few tonnes to several 10 000 t of chromite. Like concordant chromitites, some pods show grading from ore to host, but others are bordered by thin shear zones of flaky serpentinite (Hock and Friedrich, 1985). Generally the pods are boudinaged, possibly as a result of flowage in the Mantle, or they are cut by faults, and some pods are folded. In the Kinamalingan deposit the chromite pods are oriented parallel to the layering of the host rocks, indicated by cm thick layers of chromite in dunite or clinopyroxenite seams in harzburgite, and parallel to layering in olivine gabbro.

Strings of pods are less easy to exploit than the tabular bodies, and have the next best reserves, in the range 160 000 to 400 000 t.

8.3.1.3 *Pocket-like deposits*

These have faulted contacts. Some are rotated relative to the regional strike of the host-rocks. Chromitites and host rocks are brecciated or mylonitized. These ores may contain secondary chrome-bearing minerals such as uvarovite, smaragdite or kaemmererite. The ore in the harzburgite-hosted Gahalao Mine is brecciated, filled with smaragdite, and cross cut by pegmatoid gabbro dykes defining a net structure. The dykes are assumed to represent the mobilized silicate rock matrix of the primary chromite cumulates (Hoch and Friedrich, 1985).

Pocket-like deposits are difficult to exploit and have the lowest reserves, of the order of only 40 000 to 75 000 t.

8.3.1.4 *Chromite chemistry*

Acoje chromite has Cr/(Cr + Al) in the range 0.60 to 0.81 (average 0.73), low Fe^3/Fe^2, and Mg/(Mg + Fe^2) of 0.5 to 0.6. Molecular ratios average Cr:Fe 2.7:1, MgO:FeO 1.8:1, and Cr:Al 2.6:1. Disseminated intergrowths of nickel and platinum sulphides (pyrrhotite, troilite, pentlandite and violarite) have been mined from the lower levels, occurring as irregular blebs in dunite which contains an average 1.5% nickel.

By contrast the Coto refractory ore has Cr/(Cr + Al) in the range 0.39 to 0.62 (average 0.47). Average molecular ratios are MgO:FeO 2.8:1, Cr:Fe 2.2:1, and Cr:Al 0.8:1.

8.3.2 Nickel laterite

Nickel laterite, overlying ophiolite obducted from the West Philippine Sea Plate, is mined on Nonoc island, Surigao, where Marinduque Mining operates the Surigao nickel refinery (Fig. 8.1). At the end of 1981, reserves of 60 492 605 t of laterite ore of average grade 1.22% Ni, 0.101% Co and 38.8% Fe were assessed (Bureau of Mines, 1986). The production of 22 731 t nickel was attained in 1980, and since then production has steadily declined. The Rio Tuba and Toronto deposits have an average thickness of laterite and ore zone about 7 m, developed on serpentinized dunite parent rock (Ogura *et al.*, 1981). The saprolitic nickel ore zone lies immediately above the serpentinized bedrock. It has a nickel content of 2.2 to 2.6% in contrast to 0.3% in the underlying serpentinite. The cobalt content is less than 0.05%. Higher up in the iron rich residual laterite, the Ni content falls to 0.73%.

Residual laterite has been mined since 1981 on East Samar. The laterite is mainly residual over ultramafic rocks, but has also been redeposited in karstic depressions. The reserves are estimated at 2.8 Mt of non-magnetic heavy mineral concentrates in 56.7 Mm^3 of laterite. The concentrates average 45% Cr_2O_3.

8.3.3 Cyprus-type

The Barlo mine at Pangasinan in west Luzon, at the north end of the Zambales Ophiolite, is of Cyprus-type low-grade fine-grained pyrite surrounded by small random zones of chalcopyrite and sphalerite ore containing up to 8% Cu and/or 6% Zn. The gangue is of quartz, gypsum and barite and the ore bodies were overlain by limonitic gossan. The ore is closely associated with pillow basalt and quartz keratophyre but has subsequently been localized as lenses within a breccia zone (Bryner, 1969). The breccia zone contains blocks of silicified country rock and chert and is intensely argillitized. The mine opened in 1962 and closed in 1981, by which time the toal production had been 379 t of Cu, 186 kg Au and 7961 kg Ag. Remaining reserves were estimated at 320 000 t of ore of average 1.35% Cu (Bureau of Mines, 1986).

8.3.4 Manganese

Manganese ore lenses and fracture fillings in pillow joints and interstices and in reddish volcanic shale, sometimes limy, are found in limited quantity within the chert-spilite association and more commonly the spilitic pillow lavas (Gervasio, 1973). The annual production of manganese ore has waned to less than 4000 t from the 1967 peak of 80 000 t.

Their thickness is usually less than 5 m. Most of the primary deposits have been enriched by tropical lateritization, resulting in grades of 48 to 55% Mn. The ore is commonly of braunite, cryptomelane and psilomelane, with secondary supergene pyrolusite. The grade averages 30 to 40% manganese. The internal deformation of the uplifted spilite sequence and lenticularity of the ore bodies hampers exploration. Occurrences are known from Vintar, Iliocos Norte, Luar and Malinao, Quezon Sierra Madre in Isabela, and Catanduanes. Low-grade ore (20 to 35% manganese) has been mined from bedded deposits interstratified with Miocene tuffs on Siquijor Island. On Busuanga Island, NE of Palawan, bedded high grade (48 to 55% manganese) deposits are conformable with bedded chert. In the Anda Peninsula of Bohol Island, low-grade ore (25 to 30% manganese) is associated with basalt and Miocene limestone (Sorem, 1956). The Philippine manganese ore may have been exuded onto the basaltic sea floor before uplift of the ophiolite sequences.

8.4 SABAH–SULU SEA

Ophiolite sequences occupy the southern part of Palawan Island, some of the Sulu islands and considerable parts of eastern Sabah, where the chromite deposits are too small for commercial mining. The reason for the absence of large deposits in Sabah is that only the upper layers of the ophiolite are exposed, and the chromitite layers are thin and apparently represent the lower ultrabasic cumulate part of the gabbroic layer. Hutchison (1972) made a systematic study of the Sabah chromite. It shows a range of Al/Cr substitution and the unit cell (a, Å) edge varies systematically thus: Al cations per unit cell = 407.3 – 48.6a. Reflectivity (%) at 590 nm wavelength varies linearly with composition thus: Al cations per unit cell = 27.93 – 1.82 × reflectivity. Hardness varies from 1200 to 1500 VH.

Nickel laterite occurs on Palawan, in the Tavai Plateau of Sabah and the Bidu-Bidu Hills of Sabah and on Palawan. Coastal alluvial and beach placers have been mined in Central Palawan since 1980. The reserves are estimated at 2.6 Mt containing an average 5.4% chromite (Bureau of Mines, 1986).

8.4.1 Lorraine deposit

The Lorraine ore body of Balabac Island (Fig. 8.3) is a massive cupriferous pyrite deposit averaging over 4% Cu (John, 1963; Hutchison, 1983*a*). The massive sulphide occurs within mudstone mélange lying immediately above spilitic pillow lava closely associated with chert (Fig. 8.3). The mélange contains blocks of serpentinite,

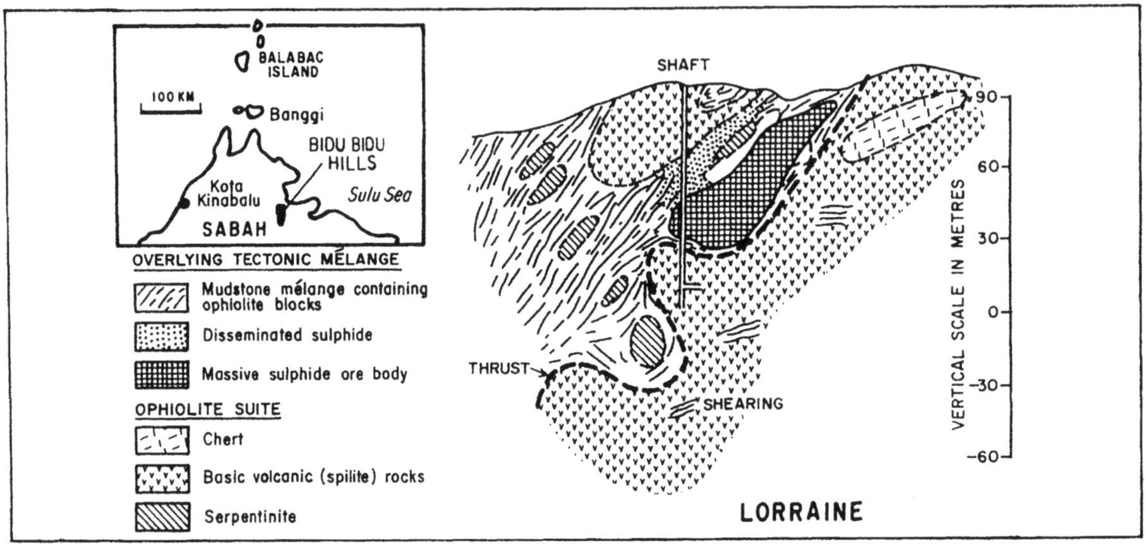

Fig. 8.3. Cyprus-type Lorraine ore-body of Balabac Island, in the Sulu Sea north of Borneo (after John, 1963). Similar type deposits have been investigated and one will be mined on the eastern flank of the Bidu Bidu Hills of nearby Sabah (Labuk Valley on Fig. 8.1).

basalt and chert. The contact with the country rock is sharp and there is no hydrothermal wall-rock alteration around the ore body, showing that it was not emplaced directly into its present setting. It is likely that the tectonic mudstone mélange, which contains huge ophiolite bodies, has also incorporated the ore body with its associated spilite. The ore body is 122 m long with a maximum down dip dimension of 95 m and an average width of 15 m. At its widest, it is 25 m thick. Pyrite is predominant, followed by chalcopyrite and subordinate sphalerite. There is local haematite concentration and the body is capped by gossan containing azurite and malachite.

8.4.2 Others

The search has been made for similar Cyprus-type sulphide deposits in Sabah (Lee and Weber, 1986; Weber and Lee, 1990). Geochemical exploration was followed by drilling in the eastern flanks of the Bidu-Bidu Hills. Massive sulphides were encountered. The ore occurs as pockets associated with shale and spilite of the upper parts of the Lower Cretaceous ophiolite complex (Chert-Spilite Formation). At Sualog, the massive sulphide bodies are closely associated with intensely chloritized and zeolitized pillow lavas of the ophiolite sequence. They are associated with quartz veining and silicification leading to an interpretation as 'stockwork' type, frequently found underlying massive sulphide deposits of submarine exhalative origin (Weber and Lee, 1990). They are considered to be syn-volcanic and genetically related to the basalt of the ophiolite suite. The volcanic rocks are tholeiitic and Ti, Cr and Ni plots indicate an ocean floor origin of MORB characteristics (Muff, 1990). The massive sulphide, which readily weathers to sulphate upon exposure of only a few weeks, occurs at the top of the basalt near or at its contact with overlying shale.

The ore minerals are predominantly pyrite, accompanied by marcasite, pyrrhotite, chalcopyrite, sphalerite, bornite, covellite, chalcocite, chromite and magnetite (Muff, 1990).

The small Leadstar deposit has been evaluated on the eastern flank of the Bidu-Bidu Hills, near the serpentinite conglomerate and sandstone locality described by Hutchison and Tungah Surat (1991), but production has been delayed until environmental approval is obtained. A similar massive sulphide body occurs at the contact of overlying shale with basalt at Kiabau (Mylius, 1990).

8.5 TAIWAN

The Tungshan stratabound massive sulphide deposits occur within low-pressure high-temperature schists in the Central Range. All are lenticular and of pyrite, with subordinate pyrrhotite, chalcopyrite, sphalerite and galena. They have been correlated with the Besshi-type deposits of Japan. However, only the Hungyehchi occurrence of Taiwan is contiguous with the blueschist belt.

8.6 INDONESIA

Lateritic nickel deposits are mined at Pomalaa and Soroako in eastern Sulawesi. The eastern part of the island is of ophiolite. On the eastern arm of the island, the ultramafic rocks are mostly massive peridotite, largely harzburgite, with little associated gabbro and basalt (Van Bemmelen, 1970). Local zones of serpentinization occur. The ultramafic rocks represent the lower part of an ophiolite that is in fault contact with Cretaceous and Tertiary limestones. Mélange south of Lake Matano has a serpentinite matrix (Silver *et al.*, 1978).

Geophysical studies indicate a westwards thickening of the ultramafics (Hamilton, 1979). Their eastwards thinning, their apparent dip beneath the schist belt and their close association with the mélange belt all fit the model that the ophiolite is composed of fault slices concordant with the direction of subduction (Silver *et al.*, 1978). Parkinson (1991) has demonstrated that the ophiolite has been thrust westwards over Mesozoic limestones, schists and mélange and that it is underlain by a metamorphic sole. This indicates that subduction was eastwards prior to collision of the western Sulawesi outer part of Sundaland with the Banggai–Sula micro-continental islands. Isostatic readjustment after collision also assisted uplift of the ophiolite.

Virtually all the ultramafic rocks are overlain by nickeliferous laterite (Golightly, 1979*b*). The weathered profiles commonly range from 5 to 30 m thickness, with the lower saprolitic layers rich in nickel. The saprolite contains core boulders of unserpentinized harzburgite surrounded by yellow to orange rims and fracture fillings of garnierite and manganese oxides. The garnierite has been determined to be $(Ni,Mg)_3Si_4O_{10}(OH)_2.H_2O$, with nickel in the range 10 to 25% by weight. The nickel occurs within garnierite and chrysoprase in the residual blanket which contains 3% Ni. It is contained predominantly in silicates in the saprolite zone and in goethite in the overlying limonite zone. A mass balance calculated from selected weathering profiles indicates that *in situ* leaching of a column of peridotite 20 to 50 m thick has taken place with only minor lateral migration of nickel in solution (Golightly, 1979*b*).

Nickeliferous laterite is also present on Halmahera island, Gag, Waigeo and Gebe and in New Guinea at Cyclops.

Platinum group minerals occur along with gold and diamonds in alluvium adjacent to the ophiolitic Meratus Mountains of south-east Borneo (Burgath, 1987).

8.7 THAILAND–PENINSULAR MALAYSIA

There was a small production of chromite from serpentinized peridotite at Amphoe Tha Pla, in the Uttaradit Province. Chromite was also mined at Narathiwat Province in Peninsular Thailand along the Bentong–Raub suture in a gabbro-serpentinized peridotite association. However, sporadic occurrences along the Bentong–Raub suture in Malaysia are too small for production (Hutchison and Taylor, 1978).

8.8 INDOCHINA AND CHINA

The only chromite mining is at Co Dinh, 20 km SW of Thanh Hoa in north Vietnam. The chromite is in placers derived from weathering of serpentinized peridotite in the Nuinua Complex, which zone extends 400 km to the NW towards Dien Bien Phu. Chromite also occurs in disseminated form in serpentinized dunite and harzburgite along the Song Ma suture zone. The north-west extension is the main Chinese source of nickel and chromium (Ikonnikov, 1975).

8.9 MYANMAR

Near Mwetaung in the Chin Hills, eluvial chromite occurs on the slopes of serpentinite hills, and in the jadeite workings, near Tawmaw in Myitkyina district.

Garnierite-bearing laterite caps some ultramafic bodies in the Arakan–Chin belt, especially at Mwetaung near Kalemyo (Goossens, 1978).

9

Mineralization associated with volcanic arcs

9.1 DEFINITIONS

Volcanic arcs can be characterized by a spectrum of mineralization types, from epithermal vein deposits, genetically related to near-surface volcanogenic fracture systems, to higher temperature vein and dissemination deposits associated with epizonal plutons, which have intruded older volcanic and sedimentary formations (Hutchison, 1983*a*).

Porphyry copper deposits are large low-grade (average 0.4% Cu) deposits of disseminated copper and iron sulphides often associated with gold, molybdenum and silver, which occur in and around the roof zones of cupolas within the deep dioritic to granodioritic roots of calc-alkaline volcanoes, both in island arc and cordilleran plate margins. The mineralization is grossly disseminated over large volumes of rock. Mining is from opencast pits of large size, and the whole mineralized zone is removed as ore by large-scale bulk mining. Depending upon the level of emplacement of the pluton responsible for the mineralization, porphyry copper deposits may be classified as volcanic, hypabyssal or plutonic. Porphyry molybdenum deposits are similar to porphyry copper deposits, but are thought to require continental rather than island arc lithosphere as the underlying magma source. Skarn deposits are similar to porphyry copper deposits, occurring where the intrusions are in contact with limestone country rocks.

Kuroko deposits are zoned massive base-metal deposits that are largely stratiform within volcano-sedimentary sequences dominated by felsic tuffs. They are of submarine origin in contrast to porphyry copper deposits, which are subaerial. The concept of a genetic relationship between porphyry copper and Kuroko deposits was expounded by Colley (1976). He proposed a classification based on distance of deposition from a felsic volcanic centre, proximal to distal. It is logical to include porphyry copper as Type 1 (subaerial). The five types of Kuroko deposits were formed more and more distally from the volcano. Type 2 forms within a submarine volcanic conduit. Type 3 is formed where the volcanic conduit reaches the sea bottom. Types 4 and 5 are near horizontal and stratiform. Type 6 is formed distant from the volcano, characterized by fine ore precipitation within sedimentary rocks showing little volcanic influence.

Besshi-type deposits are strata-bound base metal deposits which are low in zinc compared with the Kuroko. They occur mainly in the Sanbagawa metamorphic belt in Japan, but otherwise have similarities to the Kuroko-type deposits in being stratiform.

Epithermal vein deposits were hydrothermally emplaced in near-surface fracture systems, normally within or associated with volcanic or epizonal porphyritic stocks. This style of mineralization is the world's major source of gold and silver. Epithermal vein deposits were emplaced in a near-surface environment, at a higher structural level than porphyry copper deposits. Degree of erosion is therefore important in determining if epithermal vein deposits or porphyry copper deposits are currently exposed. Both occur in the Philippines.

9.2 PORPHYRY COPPER DEPOSITS

Porphyry copper is most important where the Pacific, West Philippine Sea, Celebes Sea and South China Sea basin lithosphere has subducted (Fig. 9.1), and least important when related to the Indian Ocean. Although disappointing with respect to porphyry copper, Sumatra and Java have been reasonably well mineralized in gold within the higher volcanic zones.

The volcanic arcs of west central Philippines, related to Tertiary subduction of Sulu Sea and South China Sea Basin lithosphere, are rich in porphyry copper. The older pre-Rhaetic Palaeo-Tethys basin of Thailand and Peninsular Malaysia (Fig. 9.1) has so far proved of only moderate value for base metal deposits.

9.2.1 Philippines

The geological details of the porphyry copper deposits have been well summarized by Divis (1983) and Sillitoe and Gappe (1984). They offer prime examples of the volcanic island arc type. They share many characteristics with western American porphyry deposits, but also possess several distinctive features (Taylor and Hutchison, 1979). Initial $^{87}Sr/^{86}Sr$ ratios for the Philippine porphyry copper plutons have a mean of 0.70367 with a standard deviation of only 0.00024 (Divis, 1983). These low values are

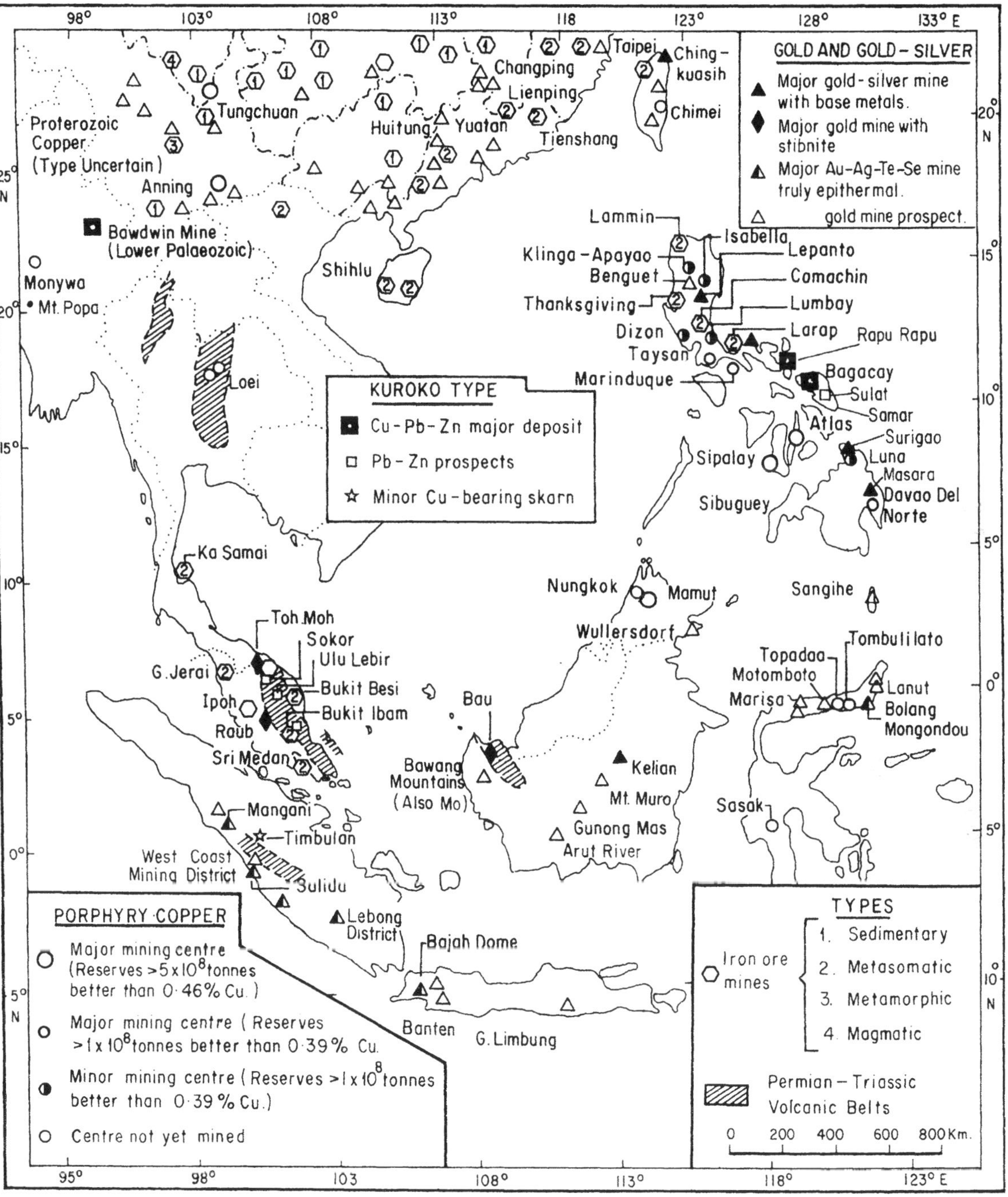

Fig. 9.1. Locality map of the most important volcanogenic and lead–zinc ore deposits of SE Asia (modified after Hutchison and Taylor, 1978).

characteristic of an island arc environment. There are three distinct areas of porphyry copper mineralization:

- The northern region, related to eastwards subduction of South China Sea lithosphere. The ages of mineralization group at 3 to 7 Ma (Pliocene) and possibly 15 to 17 Ma (Lower Miocene) in the Baguio area, and 15 to 20 Ma (Lower Miocene) in the Batangas–Marinduque area (Divis, 1983).
- The central zone, lying east of the Sulu Sea. Mineralized plutons date around 34 to 38 Ma (Lower Oligocene) in Negros (Burton, 1983) and 95 to 110 Ma (Albian) in the Cebu–Atlas area (Divis, 1983). All these volcanic episodes pre-date the SE Sulu Sea and their origin is obscure. Burton (1983) has pointed out that the SW Negros province is unique in the Philippines. The deposits contain greater amounts of molybdenum (average 0.015%) and lesser amounts of gold (average 0.06 g t^{-1}) than other Philippine deposits. He deduces that SW Negros represents a fault-bounded basement block, which has been subjected to extensive peneplanation, burial and later (Pliocene ?) uplift.
- The eastern region, related to West Philippine Sea basin subduction, dated Pliocene (Wolfe, 1972).

Over 40 potential deposits have been reported in the Philippines, and approximately nine have been in production. The central zone is the most spectacular, and contains the great Atlas Mine on Cebu, of more than 1000 Mt of 0.46 % Cu ore with 0.25 g t^{-1} Au. A close second is Sipalay mine on Negros Occidental with reserves of 662 Mt of 0.50 % Cu ore. Total known and inferred reserves from the Philippine porphyry copper deposits are approximately 3230 Mt of ore of average grade 0.46 % Cu and 0.30 g t^{-1} Au (Hutchison and Taylor, 1978). The main features of Philippine porphyry copper deposits may be summarized as follows:

- age of pluton and mineralization—Upper Cretaceous (Albian) to Pliocene;
- source pluton—quartz diorite or diorite;
- host rocks—Cretaceous to Mid-Tertiary volcanic or meta-volcanic rocks;
- ore body geometry—tabular;
- percentage of ore in source pluton as compared with host rocks—more than 50%;
- distribution of ore bodies—as clusters;
- average ore tonnage—133×10^6;
- average hypogene Cu grade—0.46%;
- ore zone alteration—potassic (**biotite, chlorite**, alkali feldspar);
- fringe area alteration—pyropylitic (**chlorite**, epidote, calcite, sericite, clay);
- zonation of alteration—overlapping, indistinct zones;
- hypogene sulphides—chalcopyrite, minor bornite;
- sulphide occurrence—as quartz-sulphide veinlets with lesser disseminations;
- weight percentage of sulphides in ore zone—2 to 6 % (average 3%);
- pyrite/chalcopyrite in ore zones—8/10 to 4/1 (average 2/1);
- supergene enrichment—negligible;
- oxidation and leaching—shallow partial oxidation to 100 m depth;
- associated gold—0.3 ppm;
- associated molybdenum—70 ppm;
- associated magnetite—abundant (average 3 %);
- anhydrite or gypsum—common as moderate amounts (often post-mineralization);
- intra to post-mineralization dykes—abundant and of variable thickness;
- post mineralization structures—abundant shatter fractures.

Hydrothermal breccias of intra- to post-mineralization age have been recognized as pipes, dykes and irregular bodies in 60% of the deposits, and a few of them carry copper ore. Zinc-rich mineralization commonly as veins is superimposed on or marginal to the ore zones, and constitutes the final event at some porphyry copper deposits (Sillitoe and Gappe, 1984).

9.2.1.1 The Atlas deposit

This deposit at Lutopan, Toledo City, Cebu Island, is the largest and the first to go into operation. It began operation in 1955. Up to 1982 the production was 1.704 Mt copper metal, 39 199 kg Au, 168 917 kg Ag, 70 t Mo and 814 504 t iron ore. The reserves are around 900 000 Mt, containing an average 0.45% Cu.

The country rocks are the Pandan Formation, an important part of which is of meta-basalt exhibiting pillow and flow structures, intercalated with lithic tuffs and greywackes.

The Pandan Formation is intruded by the late Lower Cretaceous Lutopan hornblende diorite porphyry and biotite-rich hornblende-diorite porphyry (Fig. 9.2). At Lutopan, the porphyry has a mushroom-shape, the surface outcrop measuring 200 m across and extending > 460 m down dip.

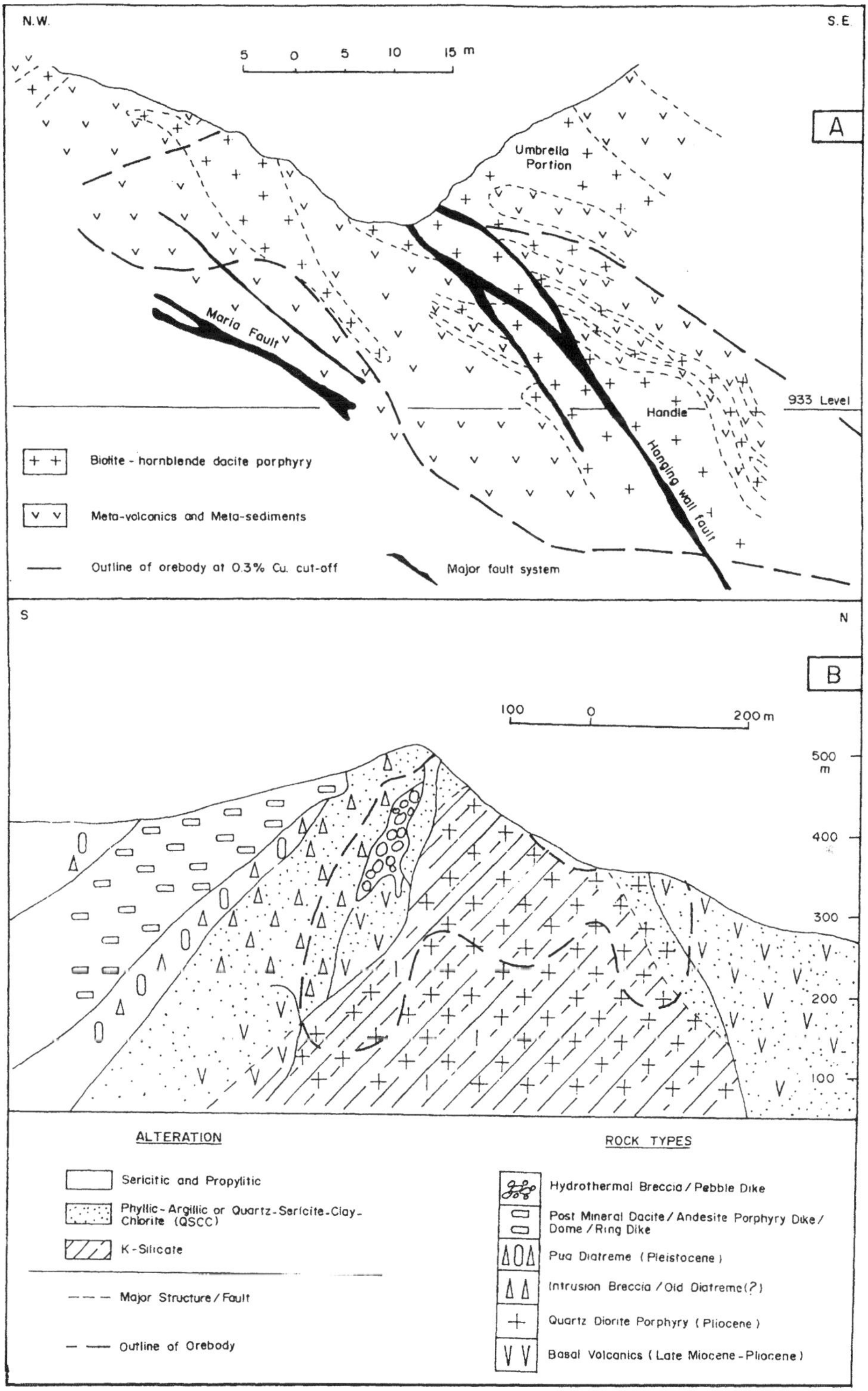

Fig. 9.2. Cross sections across two porphyry copper deposits of the Philippines. **A**: Atlas of Cebu Island (Lutopan orebody), after Bureau of Mines (1986). **B**. the gold-rich Dizon mine of the Pinatubo strato-volcano, west Luzon (after Malihan, 1987).

Faults have played an important role in mineralization location. The deposits are confined to a 3.6 km wide horst bounded by ENE trending right-lateral faults—North Barot and Cantabaco. The porphyry deposits were formed in NW and NE trending structures within the horst and related to its boundary faults.

At Biga–Bartot, breccia pipes have a diameter of 120 m and occur along the intersections of the faults. The 'pink-breccia' pipes are composed of pinkish orthoclase, and contain angular 1 to 3 cm long clasts of diorite and meta-volcanics. They have a strike length of 150 m, and can be traced down dip more than 3 km.

The mineralization-alteration includes chloritization, silicification, sericitization and biotitization. Epidote, montmorillonite, pyrophyllite and kaolinite occur near the orebody. Farther away is a pyropylitic zone of chlorite-epidote. Post mineralization gypsum is uniformly distributed throughout as vein fracture fillings (Bureau of Mines, 1986).

The ore minerals are chiefly pyrite and chalcopyrite, minor bornite, rare molybdenite, magnetite and traces of gold and silver. At Lutopan, the sulphides on the hanging-wall side constitute up to 10% of the rock, decreasing to 1.5% on the foot-wall. Secondary malachite, azurite, chrysocolla and cuprite occur as paints on the outcrop surfaces, with limonite and haematite as fracture coatings.

9.2.1.2 *The gold-rich Dizon porphyry*

The Dizon porphyry copper deposit has formed within the andesitic to dacitic Pinatubo strato-volcano, 80 km NW of Manila in the south Zambales Province of west Central Luzon. The gold-rich porphyries lie on a 300 km long west-facing arc.

The mine has yielded in 10 years about 182 000 t Cu and about 26 000 kg Au, and 48 000 kg Ag since it opened in 1979 (Malihan, 1987). The monthly production average is 2000 t Cu, 320 kg Au and 590 kg Ag. The ore body has 140 Mt of ore, associated with a small quartz diorite porphyry stock intruded into volcanic rocks. The mineralization is as stockworks, fracture fillings and disseminated sulphides, averaging 0.43% Cu, 0.93 g t^{-1} Au and 2.5 g t^{-1} Ag. Intense hydrothermal alteration developed in and around the stock, the sericite of which gave a K:Ar date of 2.7 Ma.

A post-mineralization diatreme broke through south of and adjacent to the orebody (Fig. 9.2), partly destroying it (Malihan, 1987). Later volcanic activity covered the orebody with a thick blanket of pyroclastic and lahar debris. It was subsequently exposed through erosion. The youngest mineralization is of narrow quartz veinlets cutting across the orebody. They contain mainly sphalerite and galena, with variable silver, tetrahedrite, cinnabar, pyrite and molybdenite (Bureau of Mines, 1986).

9.2.1.3 *The Thanksgiving Mine*

Many porphyry copper cupolas or minor intrusives have been emplaced wholly or partly into limestone country rocks. The hydrothermal-mineralization solutions have interacted with the limestone to produce copper or iron bearing skarns (Hutchison, 1983*a*). The best example in the Philippines is the Thanksgiving Deposit of the Baguio District (Callow, 1967).

The ore is unusual, but serves to illustrate that a variety of metals can be concentrated in limestone contact-hydrothermal zones. The mine comprises a number of small sulphide orebodies, extracted for their gold, silver, zinc, copper, lead, cadmium and sulphur contents. The ore is massive sulphide, mainly of sphalerite and pyrite, with minor chalcopyrite, galena, arsenopyrite, magnetite, haematite, chlorite, calcite, quartz and garnet. Tellurides and native gold are scattered throughout the ore.

The ore occurs as veins and pods, mainly along the contact between diorite porphyry dykes and limestone, and a few are entirely within the limestone. They vary from 3 cm to 150 cm in width, and are 30 m or more in length. The diorite porphyry is considered to be the source of the metals and the cause of the hydrothermal activity. The contacts are characterized by zones of garnet-clinozoisite and the deposition is considered to have been in the temperature range 350–400°C (Callow, 1967).

9.2.1.4 *Others*

Gold-silver veins are associated with a porphyry copper and pyrometasomatic deposit at Masara (Mercado *et al.*, 1987*b*). Fissure-filling gold-silver veins occur with a gangue of quartz and calcite containing gold and electrum. The gold values increase directly with the base metal sulphide contents. The mine reserves are estimated at 3.3 Mt averaging 8.6 g t^{-1} Au.

An unusual and interesting deposit occurs on Surigao Del Norte (Mercado *et al.*, 1987*a*). Gold occurs at Siana along the Philippine Fault Zone, as two modes: (a) fine free particles of electrum in the muddy limestone breccia and within the clasts, with grades of 3.5 to 4 g t^{-1}Au. The gold is associated with pyrite framboids; (b) as hydrothermal deposits within massive sulphide bodies in limestone breccia, assaying up to 641 g t^{-1} Au. The origin is thought to be partly placer, and partly hydrothermal, related to the emplacement of Miocene andesite. The mine had a checkered history. Underground mining began in 1938, was suspended during the war, resumed in 1946, closed again in 1960 because of frequent underground fires and insufficient ore. Newly determined reserves are 4 Mt averaging 6 g t^{-1} Au. It was re-opened in 1980.

9.2.2 Taiwan

The Chimei porphyry copper of eastern Taiwan has a similar setting to those on Luzon and the age of mineralization is considered to be 18 Ma or slightly older.

9.2.3 Sabah–Sulu Sea

The Mamut porphyry copper mine is unusual for two reasons: the source rock is adamellite porphyry and not diorite as in the Philippines, and the plutons are emplaced hypabyssally into Palaeogene turbiditic rocks (Trusmadi Formation) and ophiolite and not into coeval volcanic rocks, which are totally unknown in the region (Fig. 9.3). The average K:Ar age of the porphyry stocks is 9 Ma (Upper Miocene). The total reserves are 179 Mt of Cu grade 0.476% (Akiyama, 1984). Other metal contents are Pb 102 ppm, Zn 103 ppm, Ag 2.5 g t^{-1}, Au 0.66 g t^{-1}, Mo 7.5 ppm, S 2.28%. The ore body is only about one seventh the size of the Atlas Mine on Cebu.

The distribution of mineralization in this deposit is unique. About 50% of the mineralization is hosted by the source adamellite porphyry, 30% is hosted by serpentinized peridotite country rock, and the remaining 20% in the fine sandstones of the Trusmadi Formation (Kosaka and Wakita, 1978). There is a post-mineralization barren granodiorite porphyry which contains xenoliths of the earlier mineralized pluton. The hydrothermal alteration is interesting, for the serpentinite has been altered to an

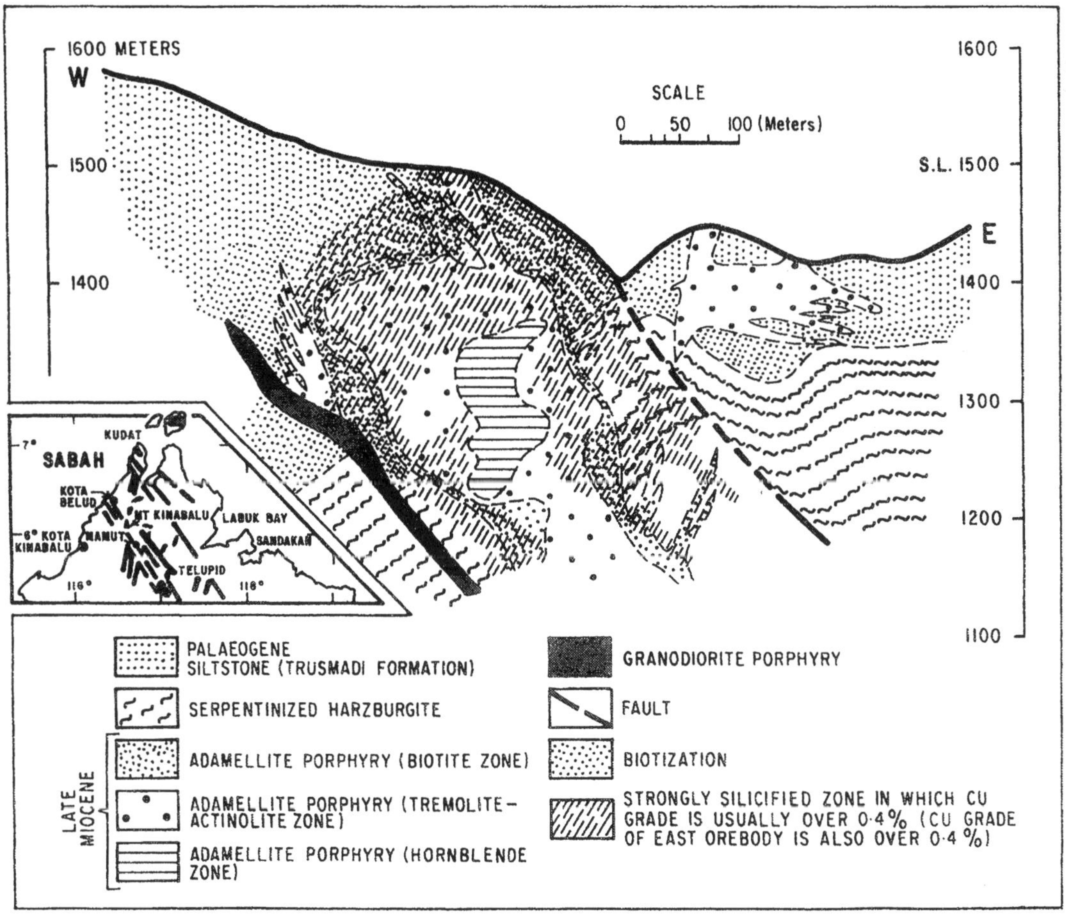

Fig. 9.3. Mamut porphyry copper mine of the Mount Kinabalu region of Sabah (after Kosaka and Wakita, 1978). The mineralized rocks are adamellite porphyry, siltstone and serpentinite.

assemblage of talc, antigorite, tremolite, chlorite and biotite with minor phlogopite. Silicification and deposition of biotite and sericite are characteristic in the other mineralized rocks.

The strongly silicified Mount Wullersdorf area of the Semporna Miocene–Pliocene volcanic arc contains chalcopyrite in association with galena and sphalerite in quartz veins. These occurrences are also associated with minor silver and gold, which has been panned from rivers draining the volcanic arc.

9.2.4 Indonesia

All the major Indonesian porphyry copper deposits are related to Miocene–Pliocene volcanic arcs resulting from subduction of marginal basin lithosphere on the Pacific side (Katili, 1974*b*). The main areas are at Tapadaa and Sasak (Fig. 9.1). Despite systematic geochemical exploration, Sumatra and Java have not revealed any economic porphyry copper deposits.

9.2.4.1 Sulawesi

The geology of the north arm of Sulawesi has been summarized by Carlile *et al.* (1990) and Kavalieris *et al.* (1992) The western Marisa region is ensialic, composed of rhyodacitic volcanics overlying a quartzo-feldspathic metamorphic basement. In contrast, the central Gorontalo region and the eastern Minahasa region comprise oceanic basaltic basement overlain by volcanic arc andesites. The centres of volcanism have migrated progressively eastwards from the early Miocene to the present (Carlile *et al.*, 1990). The Neogene volcanic rocks are comparatively well mineralized. Arc-related mineralization includes porphyry Cu-Au, porphyry Mo, volcanogenic massive Cu-Pb-Zn sulphide and magnetite Cu-Au skarn. Since the 1970s the region became the focus of extensive porphyry copper exploration leading to discovery of the presently known but unfortunately largely sub-economic deposits. There are three major prospects in the northern arm and one other prospect on the western arm.

Porphyry Cu-Au and porphyry Mo mineralizations have been K:Ar dated 2 to 4 Ma. Cu-Au mineralization in the north Sulawesi arc is associated with clusters of small hornblende diorite or quartz diorite stocks. Cu-Au mineralization includes skarn, polymetallic vein and high-sulphidation epithermal Au (+ enargite). Porphyry Mo is related to small stocks or roof apophyses of the Dondo Batholith. Cogenetic granites are exposed over 5000 km^2.

9.2.4.1.1 Tapadaa–Tombuililato (northern arm). Two clusters of porphyry-type deposits have been investigated and described by Lowdner and Dow (1977) and Taylor and Van Leeuwen (1980). At Tapadaa, the Bone Diorite, K:Ar dated 16 Ma, forms large high-level plutons intruded into Miocene andesite and tuff of the Bilungala Formation. The mineralization is wholly within the Bone diorite, and K:Ar dated at 2.5 to 5 Ma on biotite from the mineralized zone. Tombuililato is at a higher elevation and represents a higher tectonic level in which there is a swarm of small pipe-like bodies of quartz diorite porphyry intruded into the Bilungala volcanics. The mineralization affects the entire intrusion and extends a short distance into the wall rocks.

There is no classic zonation of mineralization at Tapadaa, which is largely confined to NW-trending structurally controlled blocks. The most widespread alteration is sericite-albite-magnetite associated with early-phase pyrite-chalcopyrite. Total sulphide content is low and biotite is rare. Ore grade is only locally present. There are centres of advanced argillic alteration containing higher pyrite contents. The alteration contains abundant quartz, together with alunite, kaolinite and pyrophyllite. Primary grades are in the range 0.15% to 0.35% Cu with less than 0.25 ppm Au. Low levels of Mo are widespread. Pb and Zn are low, supergene enrichment put grades up to 0.5 to 1.0% Cu in the chalcocite blanket.

Ore grades at Tombuililato occur with advanced argillic alteration containing quartz-diaspore-pyrophyllite. Gold in excess of 0.5 ppm is associated with hypogene copper grades exceeding 0.5% Cu. Outside of the mineralized centres, rhyolitic country rocks are regionally pervasively silicified and pyritized.

9.2.4.1.2 Malala A quartz monzonite and granodiorite batholith, K:Ar dated 4 Ma, intrudes into volcanic and sedimentary rocks of the Eocene Tinombo Formation. There are no coeval volcanics.

The mineralization consists principally of molybdenum with subsidiary copper and trace Pb and Zn (Taylor and Van Leeuwen, 1980). Total sulphide is low. The mineralization is fracture controlled and restricted to the roof of the quartz monzonite body and narrow contact zone with the Tinombo Formation.

Molybdenite occurs down to 200m depth, with grade decreasing downwards. It occurs as (a) coarse crystals within carbonate veins 1–10 cm thick, (b) hairline veinlets of pure fine molybdenite associated with quartz-sericite and clay, (c) medium grained flakes dispersed in a quartz-sericite gangue, accompanied by pyrite, and (d) as very fine flakes with fine pyrite disseminations in the quartz monzonite porphyry.

Pyrite is the most common sulphide but rarely exceeds 2% of the total rock volume. Others are chalcopyrite, galena and sphalerite. There is no well organized mineralization zonation. The impermeable nature of the wall-rocks has prevented the development of a hydrothermal system beyond the host intrusive (Taylor and Van Leeuwen, 1980).

Most of the molybdenite mineralization is concentrated in a 30 to 300 m wide zone, about 1.85 km long and extending downwards to 200 m. The zone trends parallel to the main 320° fault pattern. The reserves are estimated at 55 Mt of 0.24% MoS_2 grade.

Thick continental (sialic) crust is usually taken as a requirement for porphyry molybdenum systems (Hutchison, 1983*a*). Pre-Tertiary schists and gneisses outcrop about 8 km east of the deposit, suggesting a continental basement beneath the Tertiary volcanic arc. Furthermore, ongoing geochemical and petrological studies into western and northern Sulawesi demonstrate that underlying continental crust needs to be present to account for the high potassium contents and high Sr isotope ratios of the volcanic rocks (Coffield *et al.*, 1993). Accordingly a model may be suggested whereby Australian (?) continent, or a micro-continent like the Sula Spur, is underthrust westwards beneath this island arc, and has been involved in the magma genesis. Hence a molybdenum Climax-type porphyry system can be explained in this apparently unusual island arc setting.

9.2.4.1.3 Sassak (western arm) This deposit is of little economic significance but is the only porphyry copper deposit in SE Asia related to alkaline intrusives. The alkaline undersaturated rocks, K:Ar dated 10.6 Ma, consist of monzodiorite, monzonite and syenite, intruded into Upper Eocene shale, limestone and sandstone and Miocene andesitic lavas and breccias. The rocks are intensely altered to quartz-clay-sericite over an area 3 km^2. The andesites are propylitized. Drilled grades drop from 0.6% Cu at the surface to less than 0.1% at 40 m depth. The drop in grade is from chalcopyrite-bearing syenite to highly silicified and sericitized rock, so that the lower extension of the ore body appears to be faulted-out. Hypogene mineralization is both disseminated and as veinlets. The former is of magnetite, pyrite and chalcopyrite, with traces of electrum. The later veinlets contain pyrite, galena, sphalerite and chalcopyrite. Geochemical and petrological ongoing work (Coffield *et al.*, 1993), strongly suggests that these alkaline host rocks must be derived by anatexis of an underlying continental basement, which is not seen outcropping.

9.2.4.2 Sangihe

The Sangihe islands, forming the western N–S line of islands of the Molucca Sea, comprise mainly Recent and active volcanics (Fig. 9.1). Miocene andesites and intrusives are confined to the southern part of Sangihe island.

Porphyry copper mineralization is hosted within a diorite stock and andesite lavas, less than 1 km NE of Taware Ridge, not far from Gunung Kakiraeng and Gunung Awu, which last erupted in 1966. Alteration is generally weak and biotite-magnetite at depth appears to grade upwards to sericite alteration (Carlile *et al.*, 1990). Pyrite and chalcopyrite occur as disseminations and stringer zones, and quartz-carbonate veins carry pyrite, chalcopyrite, galena and sphalerite. The highest gold grades of up to 1 g t^{-1} appear to correlate with zones of high magnetite.

9.2.5 Myanmar

There are three separate and economically important porphyry copper deposits near Monywa in the Lower Chindwin district (Sabetaung, Kyisindaung and Letpadaung). The volcanic rocks are predominantly Pliocene–Quaternary, but unlike Mount Popa and Mount Taungthalon, the Monywa andesitic-dacitic volcanic formations have lost their volcanic geomorphology. The most important ore mineral is chalcocite, with minor pyrite and bornite. The supergene enrichment is over 300 m deep. Primary grades are only 0.05 to 0.1 % Cu, and the dominant sulphide is pyrite with lesser amounts of chalcopyrite and enargite. Neither molybdenum nor gold occur in recoverable amounts (Taylor and Van Leeuwen, 1980). A thick blanket of supergene enrichment contains chalcocite and represents substantial tonnages grading 0.8% Cu. Massive chalcocite-rich sulphide lenses are overlain by a leached zone containing haematite and some secondary copper oxides. Such a large build-up of chalcocite has given an enrichment of around 10:1 over protore. The deposits are therefore rather atypical of porphyry copper deposits, for the ore minerals are confined to the upper chalcocitized parts of the volcanic body, in which massive ore zones are very common (Goossens, 1978). A major factor in the secondary enrichment has been the hot dry climate with strongly seasonal rainfall.

9.2.6 Thailand

At least 12 base metal porphyry sulphides occur in the Loei district of north-eastern Thailand and 17 iron prospects have been identified (Jacobson *et al.*, 1969). They have not been mined except by primitive methods by local miners. Carboniferous sandstones and shales and Permian Saraburi Limestone are intruded by Triassic mainly andesitic volcanics. Granodiorite stocks, considered to be co-magmatic with the volcanics, have been K:Ar dated 230 Ma. The Loei volcanic rocks have been dated by Intasopa and Dunn (1994), and found to be considerably older than formerly held. The Loei rhyolite has given dates of 374 Ma (Devonian) and the basalt has given dates of 361 Ma (Devonian to Lower Carboniferous). The Loei andesite is Permo-Triassic.

Two porphyry copper deposits have been identified in northern Thailand and investigated, at Phu Thong Daeng (Fig. 10.4), 8 km S of Loei, and Phu Hin Lek Fai, 14 km E of Loei (Jacobson *et al.*, 1969). At Phu Hin Lek Fai, a diorite stock is mineralized by pyrite and chalcopyrite-bearing veinlets within a silicified zone. Total sulphide content is low and primary ore grade is 0.15 to 0.25% Cu. At Phu Thong Daeng, the primary mineralization occurs within porphyritic dacite host rocks. Copper grades are in the range 0.1 to 0.2%, overlain by an enriched chalcocite zone of grade 0.5 to 0.7% Cu. The surface is of limonitic gossan with pockets of malachite, which attracted ancient local mining. Neither molybdenum nor gold appear to be in significant quantity in the Thai deposits (Taylor and Van Leeuwen, 1980).

Shawe (1984) concluded, from limited drill data, that there is something of the order of 50 Mt of mineralized rock in the Loei area that averages 0.65% Cu. About half this amount is as near surface oxide zone, and half is of sulphide ore. Check assays for gold have been done (Davis *et al.*, 1987), from which it is reasonable to project at least 10 Mt of open pit ore in Phu Lon containing of the order of 2.0 g t^{-1} Au and 9 g t^{-1} Ag and up to 1.5% Cu.

According to Shawe (1984), another porphyry copper deposit is reported in NW Thailand at Wang Chao, about 50 km WNW of Tak, where porphyry intruded Upper Palaeozoic slate. He also reported one in the vicinity of Petchabun, but details are lacking.

9.2.7 Peninsular Malaysia

There is a zone of base-metal deposits in the central belt of the Peninsula similar to that around Loei (Fig. 9.4). They may be broadly interpreted as 'porphyry copper type' but some may be of submarine exhalative Kuroko type. The Tasek (Lake) Chini area of Pahang State, 20 km south of the Pahang River and 15 km west of the Kuantan–Segamat Highway, has been investigated for base-metal mineralization from 1977 to 1979 (Taylor and Toh, 1981). The area was selected because of arcuate outcrops of weathered skarn, and was investigated by stream geochemistry, geophysical surveys and finally by drilling (Fig. 9.4). The rocks are all deeply weathered following upon hydrothermal alteration to 27–103 m below surface.

The mineralization is contained within a Triassic sequence of volcanic and pyroclastic rocks, intruded by quartz porphyry. A few km to the east they are unconformably overlain by thick Cretaceous sandstones, which form prominent cuestas.

All drill holes cut either altered volcanics, with small local bodies of metasediments, or variably altered and mineralized quartz monzonite porphyry intrusives. Both volcanics and altered intrusives show a strong development of both ecto and endo garnet skarns (Fig. 9.4). The gently open-folded skarns have weathered to form manganiferous ironstone, resembling gossan. Magnetite skarn is of restricted occurrence. Copper sulphide mineralization may range up to 0.68% Cu with 0.3% Zn and 17 g t^{-1} Ag. Other drill-holes encountered 4.5% Pb, 1.5% Zn and 173 g t^{-1} Ag in garnet skarn. The mineralization is of four main types (Fig. 9.4):

- galena-sphalerite in garnet skarn;
- chalcopyrite with minor sphalerite in magnetite-pyrite skarn;
- chalcopyrite with lesser pyrite in pervasively fractured quartz porphyry;
- molybdenite with pyrite in quartz monzonite porphyry.

Less common types are molybdenite in altered garnet skarn within the volcanics (DH 12), primary chalcocite in garnet skarn (DH 7) and stibnite with galena/sphalerite (DH 13). Most rock lithologies are so deeply altered as to render petrographic description difficult. Argillic alteration further complicates the picture. The deeper drill cores penetrated the central potassic zone of the porphyry copper system. The prospect was relinquished because of the great depths of oxidization and weathering resulting in worthless gossan overlying the primary sulphide mineralization. The porphyry copper is also not of adequate grade but local galena/sphalerite skarns could be economic, but are difficult to locate satisfactorily by geophysical methods under the deep overburden

Surficial Fe-Mn weathered gossans were extensively exploited in the 1970s by indiscriminate shallow pitting and the excavated material trucked to the coast for export as untreated manganese ore. As soon as these ephemeral excavations hit solid bedrock or protore, the operations were abandoned, thus precluding a proper evaluation of the mineralization of the region and the development of any systematic hard-rock mining. The surficial manganese excavations extended from Tasek Chini northwards to the Bukit Sagor area, between Sungei Lembing and Kuantan. In one case, Bukit Bangkong, an excavation was abandoned when a hard magnetite skarn was encountered, because these small scale miners were not equipped for hard rock evaluation and mining.

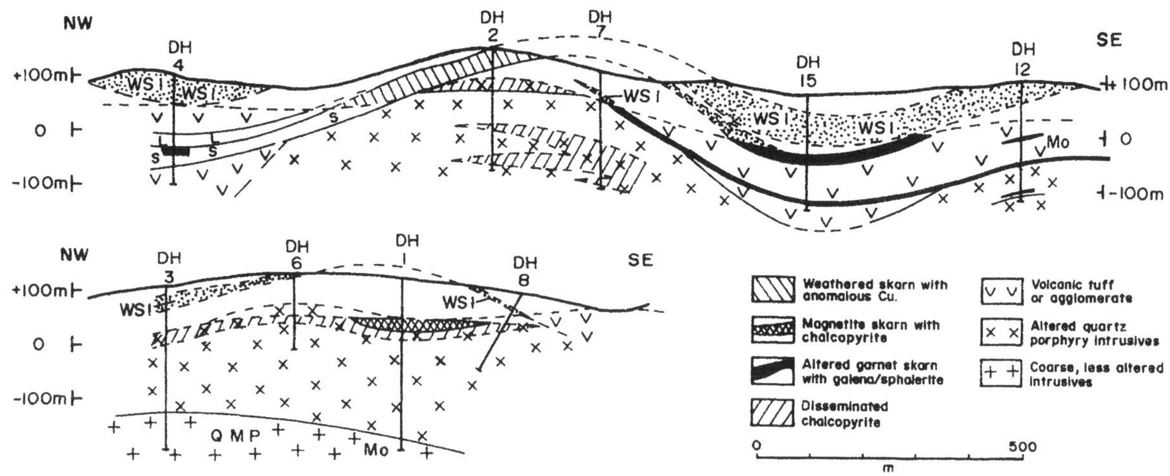

Fig. 9.4. Cross sections through the Chini Prospect, Pahang, Peninsular Malaysia (after Taylor and Toh, 1981). DH = drillhole; L = crystalline limestone; Mo = molybdenite occurrence; QMP = quartz monzonite porphyry; S = unmineralized skarn; WS1 = weathered skarn with anomalous Pb content. Elevation in metres.

9.3 KUROKO DEPOSITS

Kuroko deposits are strata-bound and genetically related to submarine acid volcanic activity. They represent a spectrum from proximal to a porphyry copper to distal from the volcanic centre (Hutchison, 1983*a*).

9.3.1 Philippines

The Cu-Fe-Zn Bagacay and Sulat mining areas of Samar Island are considered to be of Kuroko type. The Bagacay mine operated from 1957 to 1981, when the ore reserves were estimated at 1.425 Mt containing an average 2.4% Cu and 2.32% Zn. Until 1981, total production had been 133 933 t Cu, 1 126 kg Au and 75 318 kg Ag (Bureau of Mines, 1986). The deposit contains stratiform, lenticular gently dipping orebodies of pyrite, chalcopyrite, chalcocite and black copper oxides within Pliocene coaly beds contiguous with strongly argillized dacite and tuff. Chalcopyrite replaces the wood structures (Bryner, 1969). At Sulat, massive stratiform pyrite, chalcopyrite, bornite, sphalerite and galena occur as flat lenticular ore masses concordant with Pliocene sediments contiguous with Miocene volcanic and pyroclastic rocks (Hutchison and Taylor, 1978). The mineralization is restricted to dacite. The reserves are estimated at 32.5 Mt containing an average 0.61% Cu, 0.62 g t^{-1} Au and 21.0 g t^{-1} Ag.

9.3.2 Thailand–Peninsular Malaysia

Several minor Kuroko-type sulphide occurrences are found within the Permo-Triassic volcanic and pyroclastic rocks of central and eastern Peninsular Malaysia. The Galena and Manson's lodes of Ulu Sokor, Kelantan (Figs 9.1 and 10.4) are strongly stratabound. They contain up to 0.3% Sn and 20.3 g t^{-1} Ag, but all are too small to support mining. The Manson's Lode is a stratabound massive sulphide lens (350 × 50 × 10 m), which occurs conformably in a sequence of phyllite, limestone, rhyolite and tuff. It has been interpreted as a submarine-exhalative deposit of Permian age (Gan, 1980). Its mineralogy includes pyrrhotite, pyrite, sphalerite, galena, arsenopyrite and chalcopyrite. On the surface, it is characterized by limonitic and manganiferous gossan. Yeap (1993) considers these occurrences as submarine exhalative. The nearby Sungei Chekir anomaly is probably also exhalative.

The deposits at Sungei Lebir and Sungei Aring (Fig. 10.4) are also considered to be submarine exhalative (Yeap, 1993). Mengapur and Tasik Chini are also part of this belt. The Tasik Chini massive Cu, Zn, Pb, Fe sulphides, with barite and Fe-Mn ore, are best interpreted as submarine-exhalative, in keeping with their Ag:Au ratio of 200–300:1 (Yeap, 1993). The Mengapur deposit contains massive sulphide ore replacing skarn in the shape of a cylinder surrounding a dacite plug. A younger granodiorite has remobilized the mineralization, making classification difficult because of over-printing.

9.4 BESSHI-TYPE

Besshi-type deposits are stratabound massive cupriferous iron sulphides lacking any immediate volcanic association. The Hixbar mine on Rapu-Rapu, Philippines, has been classified as Besshi-type, but it may equally be considered as Kuroko-type. Reserves are estimated at 1.058 Mt averaging 1.17% Cu, 2.04 g t^{-1} Au, 14.16 g t^{-1} Ag, 30.38% Fe and 1.10% Zn. The massive main orebody is 140 m long, 60 m wide and 30 m thick, conformable with the schistosity of enclosing quartz-sericite schist, thought to be of meta-volcanic origin (Bryner, 1969). The mineralogy is mainly pyrite and chalcocite, with minor chalcopyrite, sphalerite, covellite and bornite. Localized gold values up to 46 g t^{-1} have been found in the overlying limonitic gossan (Bureau of Mines, 1986).

9.5 EPITHERMAL GOLD-SILVER MINERALIZATION

The major requirement for such mineralization appears to be the emplacement of minor andesitic to dacitic intrusions. They are probably the near surface expressions of plutons at depth and they provided the heat source for driving near-surface hydrothermal circulation. Most epithermal deposits formed within the temperature range 180 to 300° C and at depths rarely exceeding 1.5 km.

Basement antiforms are highly favourable environments for epithermal mineralization, but are not essential. Many of the large low-sulphidization epithermal gold vein deposits are situated on basement antiforms, which commonly include ophiolite (Mitchell and Carlile, 1994). By contrast, porphyry copper (-gold) deposits are mostly situated away from the antiform axis. Faults and fractures are clearly necessary for epithermal mineralization and they exerted a strong local control on mineralization. In most gold districts, mineralized structures are steeply dipping.

Epithermal deposits can be conveniently classified into two groups: a low-sulphur category in which gold and silver are the only economic metals, and a high-sulphur category which commonly includes the copper mineral enargite. They have come to be known as low-sulphidization and high-sulphidization respectively. The best examples in South-East Asia are to be found in the Philippines (Mitchell and Leach, 1991), where the upper levels of epithermal deposits are nowhere preserved and it is uncertain if the mineralizing fluids reached the surface. Accordingly a term such as 'hot spring deposits' cannot readily be applied.

9.5.1 Relationship to porphyry copper deposits

It has been widely held that epithermal vein-type mineralization develops high in the subaerial andesitic stratovolcanoes of magmatic arcs with porphyry deposits at deeper levels (Sillitoe, 1973). It has also been suggested that gold-rich porphyry systems form at a level intermediate between those of epithermal and gold-poor porphyry systems. This is known as the hydrothermal stacking hypothesis. In the Philippines, numerous deposits of both co-exist. In Baguio, the epithermal deposits are Pliocene, whereas the porphyry deposits are mid-Miocene. Overprinting of a porphyry deposit by epithermal veins is best explained by uplift and exposure of the porphyry system prior to epithermal mineralization (Mitchell and Leach, 1991). Throughout the Philippines, all outcropping porphyry systems are Middle Miocene or older whereas most epithermal deposits are younger than Middle Miocene. Regional uplift is therefore important in the scenario.

9.5.2 Philippines

The Philippines had an official 1989 gold production of 30 t, second only to South Africa in terms of unit land area. About 65% was from epithermal systems, the remainder mostly from gold-rich porphyries. An additional non-official 10 t or more gold is produced annually by small scale workers of epithermal veins and derived placers (Mitchell and Leach, 1991).

There are four main producing epithermal provinces: Baguio district of NW Luzon, Camarines Norte of SE Luzon, Masbate in central Philippines, and eastern Mindanao in the south-east, all strung out on a broad scale along the major left-lateral Philippine Fault system. Most deposits are of vein systems; stockworks are now being bulk-mined and placers are rare.

9.5.2.1 The Baguio District

The district has produced in excess of 800 t gold, more than any other gold field in the western Pacific. The epithermal deposits are confined to a 7 km wide N–S zone lying east of Baguio City (Fig. 9.5). The gold values are highest when the gangue is rich in rhodonite, and lowest when gypsum and anhydrite (Bryner, 1969). Most authors favour a Quaternary age for the epithermal mineralization. Low-sulphidization systems are the most common, and to the north occur the gold-enargite deposits of Lepanto, which may be considered the subaerial analogue of Kuroko deposits. They contain abundant copper and a suite of antimonial minerals and barite.

Mitchell and Balce (1990) have shown that the epithermal district occurs on or near the crest of an antiform of

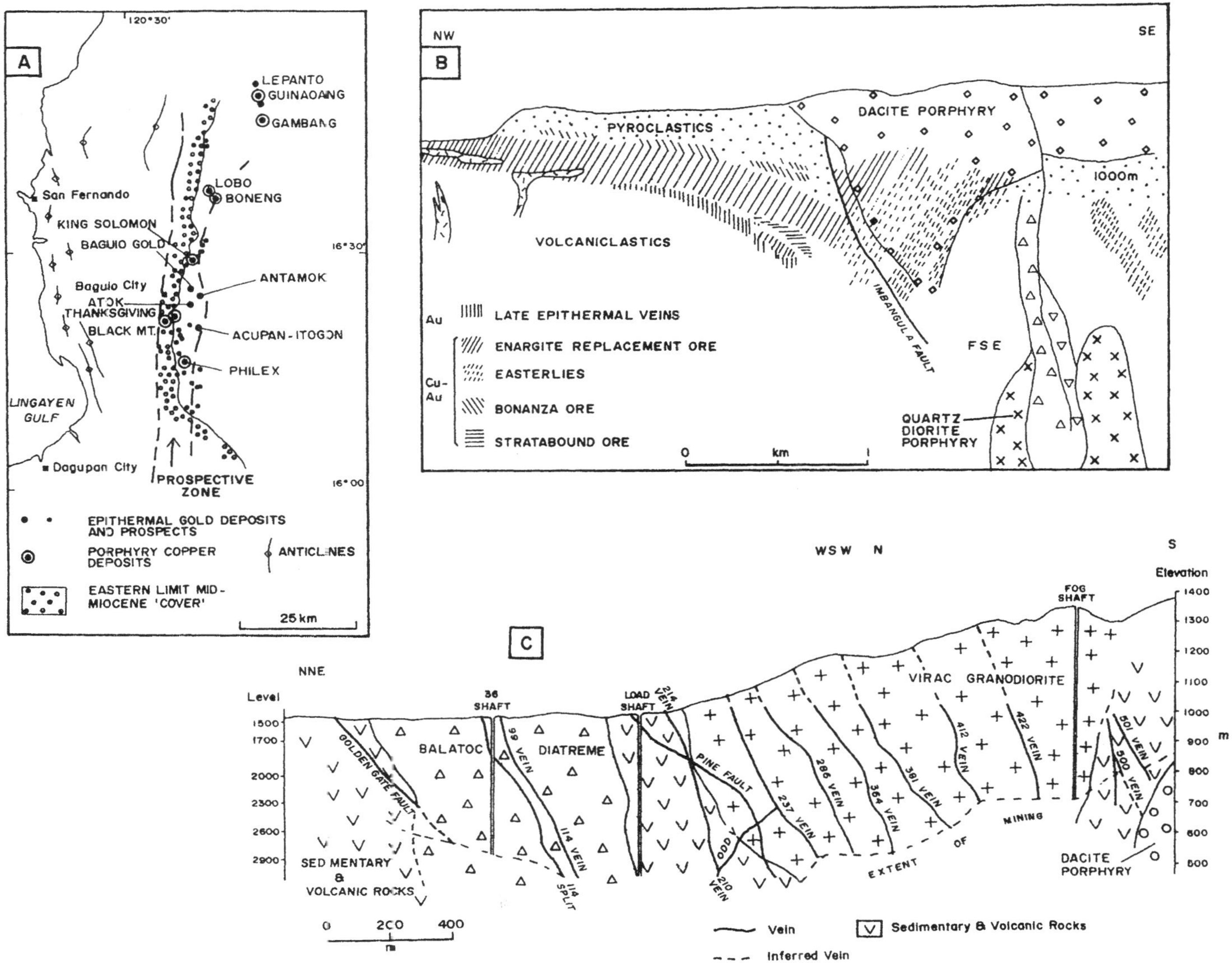

Fig. 9.5. **A:** Distribution of porphyry copper and epithermal gold-silver deposits in the Baguio district of Luzon. **B**: Cross section through the Lepanto Mine to the Far South-east (FSE) porphyry copper-gold orebody. **C:** Cross section through the Acupan epithermal mine, Baguio (after Mitchell and Leach, 1991).

the underlying Itogon ophiolite, which controlled the mineralization. The age of mineralization is inferred to be Pliocene or Upper Miocene, implying a magmatic heat source at depth, of which andesitic or lamprophyric dykes and breccias are the near surface manifestation. The district lies on line with the Quaternary–Pliocene shoshonitic volcanoes of Mt Cuyapo and Mt Arayat.

9.5.2.1.1 Hydrothermal alteration Alteration is widespread in the Baguio epithermal gold mining district. Propylitic alteration results in green-grey rocks which are slightly silicified and usually contain epidote and chlorite. The alteration affects most rocks in the district below elevations of 1200 to 1400 m and extends through the area that contains porphyry copper deposits.

Illite is widespread and more abundant adjacent to veins. At Acupan, Cooke and Bloom (1990) noted silicification zones up to 10 cm wide bordering some veins, within a sericite envelope up to 1 m wide, consisting of K-mica, quartz and pyrite, with or without illite.

Advanced argillic alteration of the Klondyke Formation has given a widespread clay blanket at elevations above 1200 to 1300m, clearly seen on satellite imagery. The altered rocks form spectacular cliffs and slopes of yellow to white weathering pyritic clay and silicic rocks, which are usually black and highly pyritic when unweathered.

9.5.2.1.2 Elevation The topographically highest workings are at 1300 m above sea level. From fluid inclusion studies, the hydrothermal water table during mineralization can be approximately estimated to have been at an elevation of 1300 to 1500 m relative to the present day sea level (Cooke and Bloom, 1990). This inferred surface of the boiling watertable lies somewhat above the projected base of the Klondyke Formation, suggesting an erosional palaeo-ground surface within the Klondyke Formation.

9.4.2.1.3 Antamok Mine This is the largest gold deposit in the Philippines. Underground mining began in 1917 and continues with 1500 t per day of rock of average grade 4 to 5 g t^{-1} Au. It has produced some 300 t Au and the gold/silver ratio is 11/10.

There are 3 major NE–SW veins—the 400, Keystone and Camote–Amigo—minor veins of the same trend, and NW–SE cross-cutting veins. Ore persists from the surface at 1100 m to below 450 m. Anhydrite veins dilute the ore at depth. The veins are tabular up to 6 m wide, largely composed of quartz. Ore grades are confined sharply to the veins. Keystone and 440 are nearly 3 km long and extend down dip over 600 m, increasing upwards in width. At the surface they form mineralized stockwork-breccias of tens of metres width of grade 2 to 3 g t^{-1}.

The veins are breccia fillings, cut by calcite, rhodonite and later anhydrite. Quartz veins predominate in the north, and calcite becomes important southwards. Base metals and tellurides accompany the gold.

The veins occur within the gently to steeply dipping Keratophyre Series of the Upper Eocene to Lower Miocene Zigzag Formation. There are few productive veins in the conformably emplaced hybrid diorite.

9.5.2.1.4 Acupan Mine This is the second largest mine in the Baguio district. It has seen a century of operation, with a 1988 production of 1100 t per day of rock of average grade 6.1 g t^{-1} Au at a cut-off of 4.2 g t^{-1}. Between 1929 and 1958 it produced almost 84 t of gold.

In the Acupan Mine, the main mineralization is pyrite within quartz veins. Pegmatite dykes contain molybdenite and wolframite. The high grade ore is mostly hosted by the Balatoc Diatreme, oval in plan and about 1 km maximum diameter, tapering downwards to a maximum known depth of 2 km. It is bounded on the N and S by sedimentary and volcanic rocks which are intruded by or faulted against the Virac Granodiorite (Fig. 9.5). The diatreme fill is of dacite boulders and smaller clasts of a variety of rocks in a fine grained matrix. Carbonized logs occur in it to depths of 650 m (Bryner, 1969). The age of the diatreme is probably indicated by 1 Ma fission track dates on the dacite (Mitchell and Leach, 1991).

The origin of the diatreme is open to debate. Support for a sedimentary or lahar origin is given by its similarity to sedimentary sequences outcropping west of the mining district. They comprise crudely bedded debris flow and interbedded dacite and quartz greywackes containing carbonized wood, assigned to the Zigzag Formation. On the other hand, the shape of the diatreme suggests an explosive origin (Wolfe, 1986).

The mineralization at Acupan occurs within veins and in the irregular 'GW' sub-vertical ore-bodies. Ore extends from surface at 1400 m elevation down to 520 m. NE–SW steeply dipping veins occupy a NE–SW zone which intersects the diatreme. Individual veins rarely exceed 1.5 m width. The mineralogy has been described by Cooke and Bloom (1990).

The earliest mineralization has the highest gold values, dominated by cryptocrystalline grey pyritic quartz and the development of hydrothermal breccias. Tellurides coexist with free gold. Later veins are of quartz, pyrite, adularia, K-mica, rhodochrosite and minor Mn-calcite. The pyrite, which occurs at shallow depth, is accompanied by base metals at deeper levels. Cinnabar and stibnite are rare. Late barren veins of anhydrite and quartz, calcite and pyrite, occur at depth and cut the mineralized veins.

The GW, or high-grade 'bonanza' ore-bodies, have plan areas up to 3000 m^2, are hosted by 'young breccias' and have grades up to 30 g t^{-1}. Gold occurs as veins and leaves in open spaces between andesite clasts. The GW orebodies may represent hydrothermal breccia pipes (Mitchell and Leach, 1991).

Fluid-inclusions indicate entrapment temperatures of 240 to 320° C and salinities less than 2.6 eq. wt% NaCl for the early grey quartz and 220 to 310° C with salinity less than 5.1 wt % for the later mineralizing fluids (Cooke and Bloom, 1990). Boiling occurred in both the early and late stage veins. Subsequent anhydrite deposition occurred between 214 and 299° C with very low salinities < 0.3 wt %.

9.5.2.1.5 Lepanto Lepanto has a history of mining extending back at least to the Ming dynasty. Commercial operations began in 1936 and production plus reserves are estimated at 0.9 Mt Cu and 122 t Au (Mitchell and Leach, 1991). The Lepanto deposit contains gold and silver in the form of tellurides, and silver in part within tennantite. The principal copper minerals are enargite and luzonite (Bryner, 1969). The ore also contains traces of tin which increase towards the volcanic centre.

The main ore is in the form of narrow enargite-gold veins and veinlets in which chalcedony is abundant. The ore is hosted by a fault controlled elongate pipe-like 10–50 m wide breccia body, in wedge shaped veinlets and clusters, and as replacement ore concentrated at the gently dipping unconformity between dacite pyroclastics and underlying volcaniclastics. However, most of the ore is in the volcaniclastic sediments (Fig. 9.5).

There are also 'easterlies' veinlet zones, within a funnel-shaped body of dacite porphyry. They are high grade massive sulphide replacement bodies which are 'stratiform' in the dacite pyroclastics and stratabound within the red and green beds of the meta-volcanics (Fig. 9.5). It is proposed that the mineralized solutions came from a volcanic centre and were trapped as a largely strata-bound deposit, and structurally controlled by major NW–SE faults. Auriferous quartz veins of low-sulphidation type also occur in the Nayak area of Lepanto. They are either over-printed on or immediately underlie the enargite-gold ore and considered to be younger than the enargite deposit.

The mineralization is confined to zones of intense silicification, comprising residual silica and alunite, surrounded by kaolinite, barite and other minerals resulting from reaction with acid sulphate fluids (Mitchell and Leach, 1991). The mineralization locally overlaps the advanced argillic blanket which has been assigned to the Far Southeast porphyry (FSE) system (Fig. 9.5). The latter contains 356 Mt of ore at 0.73% Cu and 1.24g t^{-1} gold.

The earliest hydrothermal system involved high-temperature fluids which formed the FSE porphyry deposit. The enargite-gold mineralization and associated acid alteration is probably younger and of Upper Miocene age, deposited in the range 200 to 250° C. The latest low-sulphidation system is possibly of Lower Quaternary age.

9.5.3 Taiwan

The most important hydrothermal vein gold deposit is at Chinkuashih, NE Taiwan. During its peak in 1935, it produced 2 to 2.5 t of gold, 6000 to 7000 t of copper, and silver increased to 2 t in 1961. The ore bodies are related to N–S faulting in Pleistocene dacite and sandstone (Huang, 1955). The ore minerals are pyrite, enargite, luzonite and native gold. The gangue is of quartz, barite, alunite, sphalerite, wurzite, galena, native mercury, cinnabar, native sulphur and kaolinite. Traces of famatinite in the luzonite-enargite ore have been mined for copper. The ore body is well zoned, with a Cu centre, grading outwards towards gold, and finally to mercury ore. Zinc and lead gangue minerals are found in the gold zone. Crustification and druzy textures are common. There is a marked outward zonation of alteration from silicification, through argillation, to chloritization. Where the ore grade is high, the pyrite occurs as octahedra, and where low as cubes; elsewhere it forms pyritohedra (Tan, 1976). The temperature of deposition was within the range 228 to 305° C (Folinsbee *et al.*, 1972). The metal contents are Cu 0.6% and Au 3 g t^{-1}, whereas the similar Luzon deposits have Cu 3.0% and Au 5 g t^{-1}.

9.5.4 Sumatra

Gold and silver hydrothermal quartz vein deposits have been mined at the Lebong and West Coast mining districts of Sumatra (Van Bemmelen, 1970). The gold and silver occur native as well as in selenium minerals with an Au:Ag ratio of 1:1 to 1:90. The gangue minerals are quartz, chalcedony, calcite and zeolite.

9.5.4.1 Lebong District

The Lebong district of Sumatra (Fig. 9.1), inland northwards from the historic coastal town of Benkulen, yielded 72% of the gold and 69% of the total silver produced in Indonesia this century. The mining district was the most famous in South-East Asia during the early years of this century. There were five mines, which together produced 89.15 t of gold and 843 t of silver up till 1940. However, one mine, Simau, was pre-eminent and between 1910 and 1940 it produced 37.87 t Au and 422 t Ag (Van Bemmelen, 1949). The ore occurred within strongly faulted andesite to dacite volcanic rocks with minor granodiorite. The mineralization is dated Miocene–Pliocene (Van Bemmelen, 1970).

At Simau the ore veins cut trachyte and 'Old Andesite' as well as adjacent sediments. The most important Tandaiberg vein yielded 57% of the total mine production. It strikes E–W and dips 85° to the north. The vein was intruded along and near to the contact between trachyte and andesite. The vein filling is of quartz, with calcite and zeolites. The most abundant ore minerals are pyrite, sphalerite, galena, chalcopyrite, argentite, and electrum. Selenium occurs as aguilarite. Native gold and silver occur in the supergene zone. The Au:Ag ratio averages 1:14, with a common range from 1:12 to 1:16.

Detailed assessment work is now being carried out in the Benkulen area of south Sumatra, which promises to once again become a major mining province of gold and associated silver.

9.5.4.2 West Coast Mining District

The mining district inland from Padang, is known as the West Coast Mining District (Van Bemmelen, 1949). It has continued to be important, while the Lebong District has declined.

9.5.4.2.1 Salida At the Salida and Gunung Arum mines, the gold and silver ore contains substantial amounts of silver tellurides, along with selenium and base metal sulphides. The Salida ores occur in two quartz veins within Miocene andesites. The Main Vein cuts the Leader Vein. They bifurcate upwards to become a vein swarm. The sulphide ore of the Main Vein has Au:Ag ratios variable from 1:40 to 1:20. The Leader Vein has ratios variable from 1:150 to 1:400 (Kieft and Oen, 1973). The ores show typical truly epithermal characteristics, including quartz crustification, colloform textures and diffusion banding of carbonate and sulphides. The main ore minerals are sphalerite, galena and tellurides with minor chalcopyrite, pyrite, enargite or luzonite and tennantite (Kieft and Oen, 1973).

9.5.4.2.2 The Mangani Mine This mine, lying on the equator north of Bukit Tinggi, is historically the most important gold producer in W and N Sumatra. It was discovered in 1907. The production total to 1980 was 5.66 t Au and 242 t silver. The mineralization is of silver-antimony sulpho-salts in a gangue of manganese minerals. Selenides and tellurides are absent.

The geology is dominated by a NW–SE graben formed by dextral splay faults from the main Semangko Fault. Quartz–rhodochrosite–rhodonite–Ag–Au veins are hosted by Tertiary pyroxene andesite and quartz phyllite conglomerate (Kavalieris *et al.*, 1987). There are also intrusive andesite dykes, pyroxene microdiorite and hornblende granodiorite. Quartz–epidote–pyrrhotite alteration is related to concealed intrusions.

The vein mineralization has an Ag/Au ratio > 25. The ore is banded with encrustation, infilling and brecciation textures. The veins rapidly terminate upwards. Average ore grade is 7 g t^{-1} Au and 279 g t^{-1} Ag. The veins are vertically zoned, becoming barren upwards into quartz-pyrite veinlets. The epithermal mineralization has resulted from boiling of fluids at depths between 200 to 600 m.

9.5.5 Java

Precious and base metals are abundant in west Java, associated with Miocene and Pliocene plutonic arc activity. The best known deposit is the Au-Ag one at South Banten (Van Bemmelen, 1970), associated with the 'Old Andesites'. By contrast the plutonic rocks are associated with Cu, Pb and Zn sulphides. Bayah Dome, West Java, is the largest low-sulphidization epithermal vein deposit in Indonesia. The host rocks are Middle Miocene sediments, ophiolite and volcanics and Gunung Ponker contains 104 t of Au (Mitchell and Carlile, 1994).

The prospect at Gunung Limbung, 40 km W of Bogor, West Java, has been described by Alderton and Sudharto (1987) and Alderton *et al.* (1994). The host rock is a Mid to Upper Miocene stock of quartz monzonite associated with vertical veins, a few cm to several metres in width, composed of quartz, sphalerite, pyrite, chalcopyrite and galena. The temperature of mineralization was mostly around 270° C. Wall-rock alteration gave K-feldspar in an inner zone and propylitization outwards.

9.5.6 Kalimantan

The epithermally mineralized region of east-central Kalimantan forms a prominent linear zone, approximately 450 km long, extending north-eastwards from the Arut and Sampit rivers (112° E, 2° S). From SW to NE, the mineralized centres are: **1**. Arut and Sampit rivers (100 km NE of Pangkalambun); **2**.Mirah; **3**. Gunung Mas; **4**. Masupa Ria; **5**. Mount Muro; **6**. Kelian; and **7**. Muyup. All of these regions support a well developed alluvial gold mining community. These systems are all related to Lower Miocene minor intrusives.

9.5.6.1 Arut and Sampit rivers

Sub-volcanic stocks of basalt, andesite and quartz porphyry cut through an Upper Mesozoic andesite-rhyolite volcanic sequence which contains interbedded sediments (Cuthbertson, 1988). The stocks are presumed to be of

Lower Miocene age. There is an important NE structure control to the distribution of the sub-volcanic intrusions.

Five styles of mineralization have been identified in this region (Cuthbertson, 1988), which presently supports large scale placer gold mining from the alluvium along the Monton and Arut rivers:

- Au + pyrite in siltstone-hosted quartz veins
- disseminated Pb, Zn sulphides + silver in siltstone
- massive Zn, Pb, Cu sulphides (+ Ag, Au) in siltstone
- Au + pyrite (+ Ag, Pb, Zn sulphides) in quartz veins and altered volcanics
- Au (+pyrite) in quartz veins and alteration associated with young igneous stocks.

The gold mineralization, which is associated with altered volcanics, seems to be spatially related to a quartz-feldspar porphyry intrusion. A different style of gold mineralization seems to be genetically related to young acid to intermediate stocks. A third style of mineralization is associated with thin quartz veins which trend 120° within folded siltstone. It is not yet established how the base metal mineralization is related to the gold mineralization.

9.5.6.2 Mount Muro

The Mt Muro prospect contains numerous steeply dipping gold-bearing quartz veins, centred around Gunung Baruh, Luit and Serujan, that formed epithermally during early Miocene calc-alkaline volcanism in central Kalimantan. The geology and mineralogy have been described by Simmons and Browne (1990). The near surface oxidized high grade zones of the veins are being hand-dug in trenches by more than 8000 people and production is estimated at 500 to 1500 kg annually. Hand-picked ore typically contains 30 ppm Au.

The region is largely comprised of Tertiary subaerial lavas, tuffs and volcanic breccias of intermediate calc-alkaline affinity. Dioritic stocks intrude the volcanic sequence and outcrop as cylindrical-shaped landforms. The last eruptive phase was of fine-grained basalt. The region is faulted in a conjugate NW and NE unmineralized pair. However the main trend of the mineralized veins is NNW.

The Kerikil veins at Gunung Baruh are characteristic of the mineralization style and represent one of the most intensively drilled targets. The veins trend NNW and extend discontinuously over 1000 m. The host rock is of porphyritic andesite flows, a hypabyssal basaltic andesite dyke and volcanic breccias.

Primary sulphide mineralization occurs in the Kerikil veins in four distinct stages related to hydrothermal brecciation and open-space filling. Stage I is represented by early hydraulically fractured and brecciated volcanic rock. Open space mineralization consists of microcrystalline quartz with subordinate adularia. Stage II is of chaotic, heterolithic, matrix-supported hydrothermal breccia. The gold values range from < 1 to 15 ppm. The breccia in-filling is predominantly microcrystalline light grey cloudy quartz, with subordinate adularia, carbonate, and fine grained sulphides (pyrite, sphalerite, galena, chalcopyrite, covellite, acanthite and electrum). Stage III is of quiescent mineral precipitation, leading to interlaminated fine grained quartz and pink rhodochrosite. Stage IV vein fill sporadically consists of microcrystalline quartz interlaminated with disseminated pyrite and other sulphides, with gold values ranging from 1 to 9 ppm (Simmons and Browne, 1990). The highest Au values are associated with a hydrothermal breccia (Stage II), and local occurrences of laminated quartz-sulphide veins (Stage IV). Quartz is the predominant gangue mineral throughout, with subordinate adularia, calcite, and rhodochrosite. Electrum and acanthite are fine grained (≤ 30 m) and associated with minor disseminated pyrite, sphalerite, galena, chalcopyrite, and covellite. Bleached zones of quartz-illite-pyrite envelop the mineralized veins, but where the host rock is less permeable, a weak alteration assemblage of epidote + chlorite + albite ± pyrite ± illite ± calcite formed instead. Fluid inclusions indicate that the mineralizing fluids were dilute and ranged between 207 and 253° C, indicating a deeply convecting hydrothermal system. The gold mineralization was restricted spatially and temporally to the hydrothermal brecciation. At Luit, the mineralization is similar, but the exposed veins were formed at greater depths and at higher temperatures of 238 to 262° C. The gold grades diminish rapidly with depth.

A late stage kaolinite over-print was produced on the deposits by shallow acid fluids during collapse of the geothermal system. These fluids also partially digested vein carbonate, adularia and rock fragments, forming enhanced permeability within the vein systems. The secondary permeability assisted in the development of supergene oxidation, which extends down to 100 m below the surface. Erratic supergene gold enrichment is mostly restricted to less than 25 m below the present land surface and is closely associated with Fe and Mn oxides.

9.5.6.3 Kelian

Kelian is the largest known gold deposit in Indonesia (Van Leeuwen *et al.*, 1990). It is located 250 km west of Samarinda, and named from the Kelian river, a tributary of the Mahakam. The deposit consists of two main orebodies, the West and East Prampus, and four smaller mineralized zones. They are spatially related to several small sub-volcanic andesitic bodies intruded into a pile of pyroclastic and sedimentary rocks. They have a combined resource potential of more than 75 Mt of ore at 1.8 g t^{-1} Au or 53 Mt at 2.0 g t^{-1} Au. Alluvial gold was discovered in the region in the early 1950s leading to a minor gold rush, with more than 100 to 300 kg/year production reported. After 1979, more than 10 000 people worked on the alluvium An estimated 20 t of placer gold have been produced since 1955.

The geology is dominated by a thick pile of sediments with local volcanic intercalations deposited in the western part of the Kutei Basin during Palaeocene–Oligocene times. An Upper Eocene silicic pyroclastic unit is at least 300 m thick. The sequence was folded and faulted with N and NE trends then intruded by a number of small porphyritic dacitic to andesitic intrusions, K:Ar dated at 23 Ma (Lower Miocene). Plio–Pleistocene basalts extensively cover the folded Eocene–Oligocene formations.

Some time after the emplacement of the andesite intrusives, a hydrothermal system was established, causing extensive alteration, mineralization and hydrothermal brecciation. The preserved part of this system is 1 km^2 in area and has a vertical extent of at least 600 m. Radiometric dating suggests the age of mineralization to be of Lower Miocene age. Four overlapping stages of mineralization-alteration are recognized (Van Leeuwen *et al.*, 1990):

- **I**: chlorite + carbonate + sericite + pyrite ± epidote alteration in the cores of the larger andesite bodies;
- **IIA**: localized fracturing and sericite quartz deposition;
- **IIB**: a major period of poly-phase fracturing and hydrothermal brecciation with associated boiling of hot (290 to 330°C) dilute and CO_2- rich fluids, and development of replacement/vein alteration of adularia + sericite + quartz + pyrite accompanied by subordinate As-Ag mineralization and possibly some gold mineralization. Progressive loss of CO_2 upon boiling caused a progressive increase in adularia and a corresponding decrease in deposition of sericite. Deeper within the system, a quartz-sericite-carbonate assemblage was formed without adularia;
- **III**: renewed magmatic activity, focused on East Prampus, released volatiles to the hydrothermal system, producing a hot (300 to 330°C) relatively saline CO_2- rich brine, and extensive fracturing and hydrothermal brecciation. This was accompanied by boiling, deposition of various carbonates ± quartz ± sericite and gold-base metal mineralization. Stable isotope data suggest the presence of a magmatic component. Mixing of the upwelling hot saline fluids with descending cooler steam-heated fluids has resulted in a complex zonation in carbonate mineralogy and a wide range of homogenization temperatures. Away from the centre of brecciation and boiling, gold and base metals were deposited in a more passive environment together with carbonate and/or quartz;
- **IV**: late stage influx of relatively cool ($<200°$ C) CO_2-rich steam-heated fluids from the top and margins of the system forming a kaolinite + Fe/Mn carbonate + pyrite ± haematite ± cinnabar assemblage.

The magmatic-hydrothermal event was followed by uplift and erosion of more than 600 m during the Middle–Upper Miocene and basaltic volcanism in the Plio–Pleistocene. A second cycle of erosion, which began in the Pleistocene, has removed most of the young volcanic cover exposing the presently known mineral deposit.

9.5.7 Sulawesi

The northern arm of Sulawesi contains examples of four major categories of gold mineralization (Carlile *et al.*, 1990): gold-copper porphyries within which gold is distinctly partitioned, gold and base-metal bearing breccias, epithermal Au of both high and low-sulphidization type and limestone-hosted Au of possible Carlin type (Kavalieris *et al.*, 1992). Between 1886 and 1931 several hard-rock gold mines produced a total of more than 19.5 t of gold and 28.8 t of silver. In the early 1980s the old mining districts were re-activated. It is estimated that local miners have produced at least 10 t of gold from Sulawesi.

Epithermal Au mineralization of low-sulphidation type is hosted mainly by Lower Miocene andesites. Typically it is of quartz-calcite veins ± adularia. A district-wide correlation between epithermal Au and porphyry Cu is not apparent. Due to erosion, the deposits are now exposed to deep levels. Most systems are characterized by propylitic chlorite-epidote wall-rock alteration (Kavalieris *et al.*, 1992). Most of the low-sulphidation systems are typified by quartz veins, which have limited grade and tonnage potential. All are sub-economic, and historic production nowhere exceeded 10 t Au.

9.5.7.1 Marisa, Gunung Pani

The Pani gold prospect at Marisa, in the centre of the northern arm of Sulawesi, has of the order of 30 Mt ore at 1.35 g t^{-1} Au. The mineralization has been summarized as epithermal, low-S and of quartz veinlets and fracture disseminations (Carlile *et al.*, 1990). Silver and base metals are very low throughout and the gold occurs as electrum with 20% silver. The mineralization is hosted in felsic volcanics (Fig. 9.6A). The rhyodacite centre is 3.5 km in diameter, consisting of porphyritic Miocene or younger lavas, breccias and fine grained pyroclastic rocks. Mineralization is related to the Baganite Dome. The ore is low grade, 0.8 ppm disseminated Au, with associated pyrite and acanthite occurring in minor quartz-haematite veins in silicified pyroclastics. The most important Au mineralization occurs at the Pari Ridge, associated with minor base-metal sulphides, quartz and adularia encrustations, but quartz veins are absent. It has been suggested that the low grade of mineralization and the observed fracture types are consistent with formation by degassing of a rhyodacite dome, but post-domal faulting is more likely to be the major control.

Chlorite + albite + anatase + pyrite ± quartz and adularia are ubiquitous in the rhyodacite. Chlorite was formed at 125 to 275° C and fluid inclusions in quartz indicate mineralization in the range 150 to 300° C (Kavalieris *et al.*, 1990). A complex hydrothermal history is indicated. Open system boiling, silicification and pyritization occurred in the upper parts of the Baganite Dome. This was followed by mixing with ground water causing the gold to deposit.

Alluvial gold derived from the nearby volcanogenic epithermal Au deposit has been mined for centuries and was dredged between 1910 and 1912 by Marisa Alluvial Mining, 10 km downstream from the Pani gold prospect, producing 18.5 kg Au (Kavalieris *et al.*, 1990; 1992). The average grade is 150 mg m^3.

9.5.7.2 Motomboto

This prospect occurs 30 km ESE of Gorontalo. It is hosted within Miocene andesitic lapilli tuffs, volcanic breccias, andesite and diorite. In low-lying areas, the mineralized rocks are unconformably overlain by flat-lying lake sediments and intercalated boulder beds (Fig. 9.6B). The prospect is described by Carlile *et al.* (1990) as epithermal, composed of high-sulphur stringers, veins and breccias. The ore mineralogy is of pyrite, enargite, luzonite and free gold. The mineralized zone extends over 5 km E–W and in places up to 300 m wide, and consists of vuggy silica bodies. The silica-alunite -pyrite alteration grades outwards through argillic to chlorite. The ore is controlled by breccia zones and faults.

9.5.7.3 Tombulilato

This Cu-Au porphyry district occurs 35 km ESE of Gorontalo. The Cabang Kiri East prospect (Fig. 9.6C) is approximately of 140 Mt grading 0.43% Cu and 0.58 g t^{-1} Au. The copper-gold mineralization is centred on quartz diorite stocks intruded into Miocene andesite. Most of the mineralization is within the intrusive rocks, and alteration-mineralization shows marked vertical zonation (Carlile *et al.*, 1990). A quartz-sericite-albite-chlorite-magnetite assemblage occurs at the deepest known level and passes upwards through chlorite-magnetite into argillic alteration that includes montmorillonite-sericite-chlorite-magnetite. A later upper acid over-print is developed in which kaolinite-diaspore-alunite is associated with illite. The mineralization is gold rich, close to 2 g t^{-1}. At the lower levels it is in the form of disseminations of chalcopyrite with lesser bornite. This style grades upwards into fracture and stockwork mineralization. The Au is associated with chalcopyrite and bornite and there is supergene chalcocite.

9.5.7.4 Lanut

Lanut lies 20 km SE of Kotamobagu in the eastern Minahasa region. Van Bemmelen (1970) records that between 1913 and 1931, the region produced 5 t of gold and 4 t of silver. Recent evaluation has indicated 19 Mt of ore of 1.8 g t^{-1} gold (Carlile *et al.*, 1990). Two distinct rock units occur in the mineralized area (Fig. 9.6D). The lower unit is of volcaniclastic sandstone and conglomerate, marine Lower Miocene mudstone, limestone and lahar deposits, with thin interlayered andesite flows. The upper unit is of andesitic lavas with some interbedded pyroclastic material. The volcanic section is at least 250 m thick and hosts most of the mineralization. The contact between the two rock units played an important role in localizing the mineralization, described as epithermal low-sulphur quartz veins as a stockwork and also as fracture disseminations and breccias. The early phase of mineralization comprises adularia on veinlet walls and in immediately adjacent wall-rock. This assemblage passes outwards through an illite-pyrite zone into regionally chloritic andesites. The upper part of the system is overprinted by later kaolinite-pyrite-marcasite. The sulphides are most strongly developed close to the mineralized zones.

The gold mineralization is associated with quartz veining. Much of it forms stockworks, but some have been shattered, resulting in mineralization as veinlet fragments within milled rock. The richest gold contents are in quartz veins up to 1.5 m wide. The veins are commonly brecciated and re-cemented.

9.5.8 Sangihe, Indonesia

Taware is located 2 km inland from the south coast. Panned gold was first discovered in 1986, leading to a minor gold rush from rich alluvium and quartz veins. It is

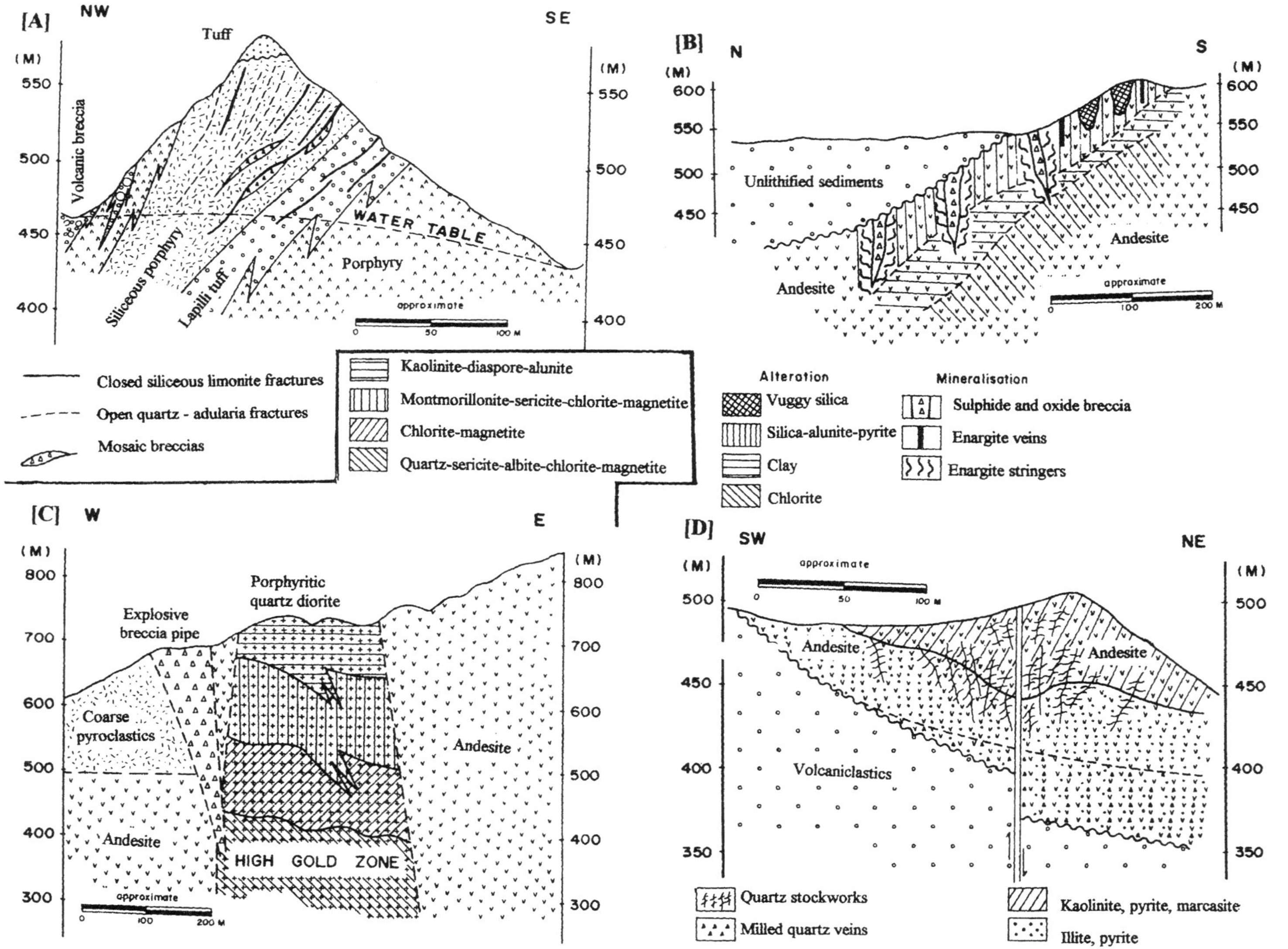

Fig. 9.6. Simplified geological cross sections through four gold porphyry deposits in the North Arm of Sulawesi, Indonesia (after Carlile *et al.*, 1990). **A**: Gunung Pani, **B**: Motomboto, **C**: Cabang Kiri East, Tombulilato, and **D**: Lanut.

estimated that local miners have produced at least 3 t of gold. Miocene andesitic volcanics, including lapilli tuffs, are intruded by dioritic stocks. Illite-pyrite alteration occurs in the andesites (Carlile *et al.*, 1990). The alteration extends outwards to twice the width of the quartz veins. Chlorite occurs laterally. Gold mineralization occurs in sheeted quartz veins and stockworks associated with minor pyrite, arsenopyrite, chalcopyrite, galena, sphalerite and sulpho-salts such as pyrargyrite. The NW-striking faults have controlled the mineralization.

At Binebase on the east coast of Sangihe, massive gossan and barite boulders were observed in 1988. The main country rocks are Miocene andesitic lapilli tuffs and dacite. Intense silica-pyrite alteration coincides with extensive brecciation zones (Carlile *et al.*, 1990). An illite-pyrite assemblage occurs lateral to the breccias and post-mineralization barite veining is strongly developed.

Gold and silver mineralization occurs in hydrothermal breccias associated with intense pyritization and silver sulpho-salts, including tetrahedrite. Galena and sphalerite are developed marginal to the main breccias, which are surface oxidized. NW-striking faults also control the mineralization.

9.5.9 Banda Arc, Indonesia

The Lerokis deposit on the active volcanic island of Damar, Banda Sea, is scheduled to produce nearly 2 t Au per year (Mitchell and Leach, 1991). A gold prospect has also been discovered on the extinct, uplifted and eroded island of Wetar (Fig. 1.3).

10
Non-volcanic epithermal deposits

Stibnite, stibnite-gold, and stibnite-gold-scheelite mineralization is confined to the Sundaland continental cratonic core area of SE Asia. This gold association is distinctly different from the gold-silver-telluride association of the Cainozoic volcanic arcs. By contrast it is confined to areas of continental crust characterized by high level granite, granodiorite and diorite plutons, which may have provided the heat source to mobilize the metals from the old continental crustal infrastructure. Volcanic activity is lacking. There is commonly an important mercury association (Figs 9.1 and 10.1). The major gold-bearing quartz vein mineralization style has no volcanic association and occurs notably in metapelite and carbonate country rocks. Although this mineralization association overlaps the tin belts on a regional scale, it is separate, unrelated to and younger than the tin-bearing granites. The extensive Cainozoic fracture system of Sundaland, combined with the high regional geothermal gradients, have facilitated the low temperature mobilization of the metals.

10.1 WESTERN BORNEO

The region is marked by important Miocene (8 to 23 Ma) epizonal intrusive activity. The calc-alkaline belt is marked by epithermal mineralization with gold, antimony and mercury in Sarawak and gold in west Kalimantan. Only gold mining persists in the Bau region. Much of the gold has been from placers, but the best primary production has been from Bau, near Kuching, Sarawak. The Bau mineralization is related to a line of small high-level Tertiary dacitic to granodioritic stocks, trending NNE perpendicular to the tectonic zones, emplaced into the Jurassic–Cretaceous Bau Limestone and Pedawan Formation. The epizonal plutons could have provided the heat source to drive the hydrothermal circulation responsible for the mineralization. The mineralization consists of complex arsenical ores containing native arsenic together with other arsenical species, stibnite, pyrite and free gold. The quartz rich veins and silicified limestone contain more free gold. Silver is subordinate. Fluid inclusion studies indicate the temperature of formation of the mineralization to be within the range 140 to 250° C (Hutchison, 1983*a*).

Sediment-hosted, or Carlin-type, Au deposits are conventionally thought to have been generated at shallow levels in a geothermal system and the Au scavenged from the host sedimentary rocks by meteoric hydrothermal fluids. However, Sillitoe and Bonham (1990) propose that the Au was contributed by magmatic hydrothermal fluids and deposited at the peripheries of base and precious metal districts, up to several kilometres from the progenitor intrusives. They classify the Bau District together with Carlin, Nevada, and Bingham, Utah, as having been formed in this way.

10.1.1 Bau district

10.1.1.1 Antimony

Antimony began to be produced from the Bau district in 1823 and since then more than 85 000 t have been produced (Wilford, 1955). Output declined rapidly and there is no present day production. Stibnite accompanies gold around Bau and mercury at Gading. The stibnite occurred as large granular masses and interlocking acicular crystals. The commonest gangue is quartz and calcite. Ore bodies are veins and tabular replacements in the Bau Limestone. Many mines worked eluvial ore in addition to the primary veins. The main deposits were at Paku, where stibnite-quartz ore occupied faults at the Jurassic–Cretaceous Bau limestone-shale contact. Mineralizing solutions appear to have migrated upwards through the limestone until they met the overlying impermeable shale of the Pedawan Formation.

Most of the ore mined from the Bau area was from eluvial and residual deposits, concentrated as masses in solution hollows in the limestone following weathering and erosion of the shale-limestone contact. Mining of the primary veins, which continue down into the limestone, was largely unprofitable because of rapid and irregular thinning of the veins and the expense of draining the mines in this region of high rainfall.

The limestone which forms Gunung Pangga has been folded into a dome-like structure, subsequently broken by normal faults, along some of which dacitic dykes have been intruded (Fig. 10.2). Hydrothermal solutions moved along the fractures to be impounded by the overlying impermeable shale of the Pedawan Formation (Wilford,

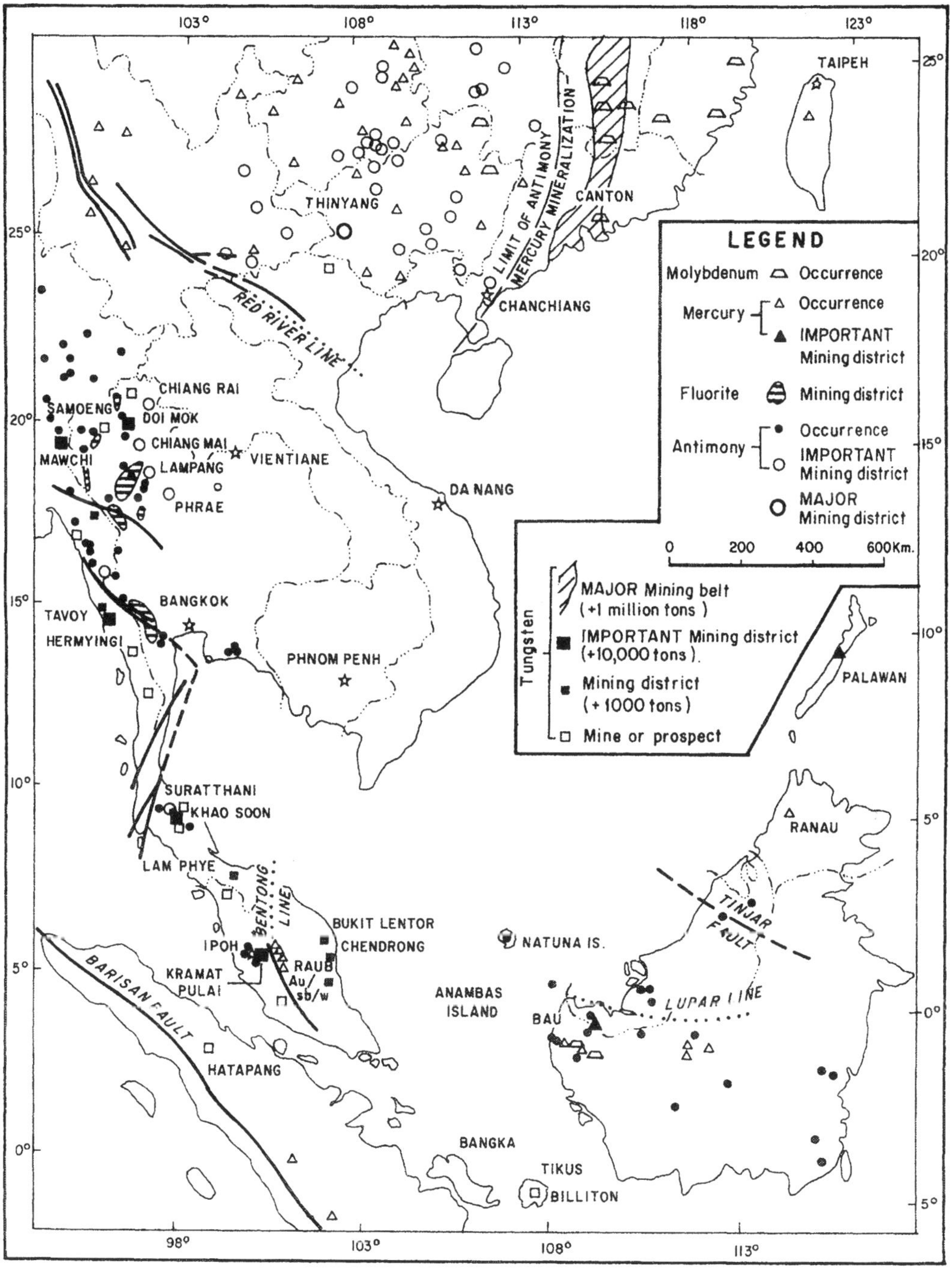

Fig. 10.1. Distribution of epithermal antimony and mercury in continental SE Asia. Outside the map area, there are unimportant mercury occurrences in Java and antimony in Sulawesi. Most productive lode gold mines in SE Asia at Bau (Sarawak) and Raub (Malay Peninsula) are associated with arsenic and scheelite, close to stibnite and mercury mineralization. Tungsten mines are commonly located in antimony-producing regions, but there is a zonal separation in China. (From Taylor and Hutchison, 1979.)

1955). The original ore bodies were probably veins of flat tabular replacement bodies in the limestone. Subsequent erosion removed the soft shale cover and produced the present rugged hill profile. During erosion, the high density antimony ore remained as residual deposits and they settled into solution cavities which developed in the limestone, especially along the fracture systems.

10.1.1.2 Gold

The Bau area has been known as a gold mining district since ancient times. Intensive mining by Chinese, who came from Sambas across the border, dates back to 1857 but flourished by 1882.

From 1864 to 1954 the Sarawak production, predominantly from Bau, was 37 t mainly extracted between 1900 and 1921 (Wilford, 1955); production peaked in 1905 and 1935, but continues until today. 1905 production was 2 t and 1934 was 0.9 t.

The ore deposits are mostly situated near the intersection of the ENE trending Bau anticline with a NNE line of Middle to Upper Miocene dacite porphyry and microgranodiorite stocks, sills or dykes (Wolfenden, 1965). The richest deposits are at the contact between the Bau Limestone and the overlying predominantly argillaceous Pedawan Formation. Others are in fault, fracture, and joint zones. Gold occurs in irregular quartz-calcite veins in silicified limestone zones. These veins also contained stibnite and native arsenic in the upper workings.

Coarse flakes and nuggets of gold occur in the alluvium of streams which drain the areas where dacite and microgranodiorite dykes and stocks are abundant, and much of the production was from residual and eluvial deposits associated with solution cavities in the limestone. Fine gold particles, rarely visible even under the microscope, characterize all the primary deposits of the Bau area. Alluvial gold and placer diamonds have also been derived from weathering of Tertiary conglomerate outcrops of the Silantek Formation and Plateau Sandstone (= Ketungau Basin of Kalimantan) (Wilford, 1955). Across the border in Kalimantan, Sukamto *et al.* (1988) have shown that gold is enriched in the basal Upper Eocene sandstones of the Ketungau Basin, which unconformably overlie the orogenic complex (= Rajang Group of Sarawak). These Upper Eocene conglomerates and sandstones became a secondary source for placer gold in contiguous Quaternary alluvium. The detrital diamonds probably also have their source in the same Eocene strata, but this has not been proved.

Mining in Sarawak was predominantly by panning and sluicing. Trace gold is widely disseminated in the acid stocks. The deposits at the limestone/shale contact are of auriferous silicified shale capping a mass of fine grained quartz ore that has replaced the underlying limestone. Stibnite is common in such deposits (Wolfenden, 1965).

The primary ore occurs in the limestone as: (i) quartz-calcite veins, (ii) quartz ore with calc-silicate minerals, and (iii) within the contact aureole of intrusions or in quartz veins within the intrusions. Residual deposits are of weathered ore in Au-bearing clay and as cave filling alluvium (Pimm, 1967). Recently, a large lens-like body of disseminated Au mineralization, associated with arsenopyrite and pyrite, was discovered in shale cut by minor dykes at Jugan. Present-day mining in the Bau area is by opencast pits.

Arsenical gold ore characterizes many deposits in the Krokong area of Bau. They contain native arsenic, arsenopyrite, realgar and orpiment (Pimm, 1967). A representative bulk analysis of such ore contains: Au 12.00 g t^{-1}, Ag 1.68 g t^{-1}, As 8.09%, Sb 1.91 %, Fe 1.33 % and Cu 0.00173%. Coarse gold is absent and gravity methods of recovery are unsuitable. Cyanidation was used to remove the undesirable sulphides. The old workings present an unsolved environmental problem.

Residual deposits are formed by weathering of the primary deposits. They have been the source of large amounts of gold and antimony. They occur mostly on the limestone flats. Alluvial ore occurs in sinkholes and channels and they have also been mined. Coarse flakes and nuggets occur in the alluvium of streams which drain the areas where dacite and microgranodiorite dykes are abundant. By contrast, fine particles, rarely visible even under the microscope, characterize all the primary deposits.

The sequence IA to IC (Fig. 10.2) shows formation of a secondary auriferous clay deposit resulting from weathering of a Carlin-type primary lode, as envisaged by Wilford (1955). **A** shows a typical primary ore body formed at the junction of the Bau Limestone and the overlying Pedawan Formation shale. **B** shows erosion starting to remove the overlying shale, and percolating water beginning to dissolve the upper part of the limestone, into the cavities of which the shale slumps. Finally at stage **C** erosion has removed the shale cover and weathered much of the ore to auriferous clay in which fragments of unoxidized ore remain (Wilford, 1955).

The sequence IIA to IIC shows weathering of an epithermal deposit where dacite penetrated as far as the limestone-shale contact. The dacite was weathered to clay, which slumped and covered the gold ore.

Alluvial gold and diamonds have also been derived from weathering of Tertiary Plateau sandstone and conglomerate. The diamonds were panned last century from the river gravels in the headwaters of the Sungai Sarawak Kiri. They and associated gold are interpreted to have been derived from the Tertiary conglomerate. They also

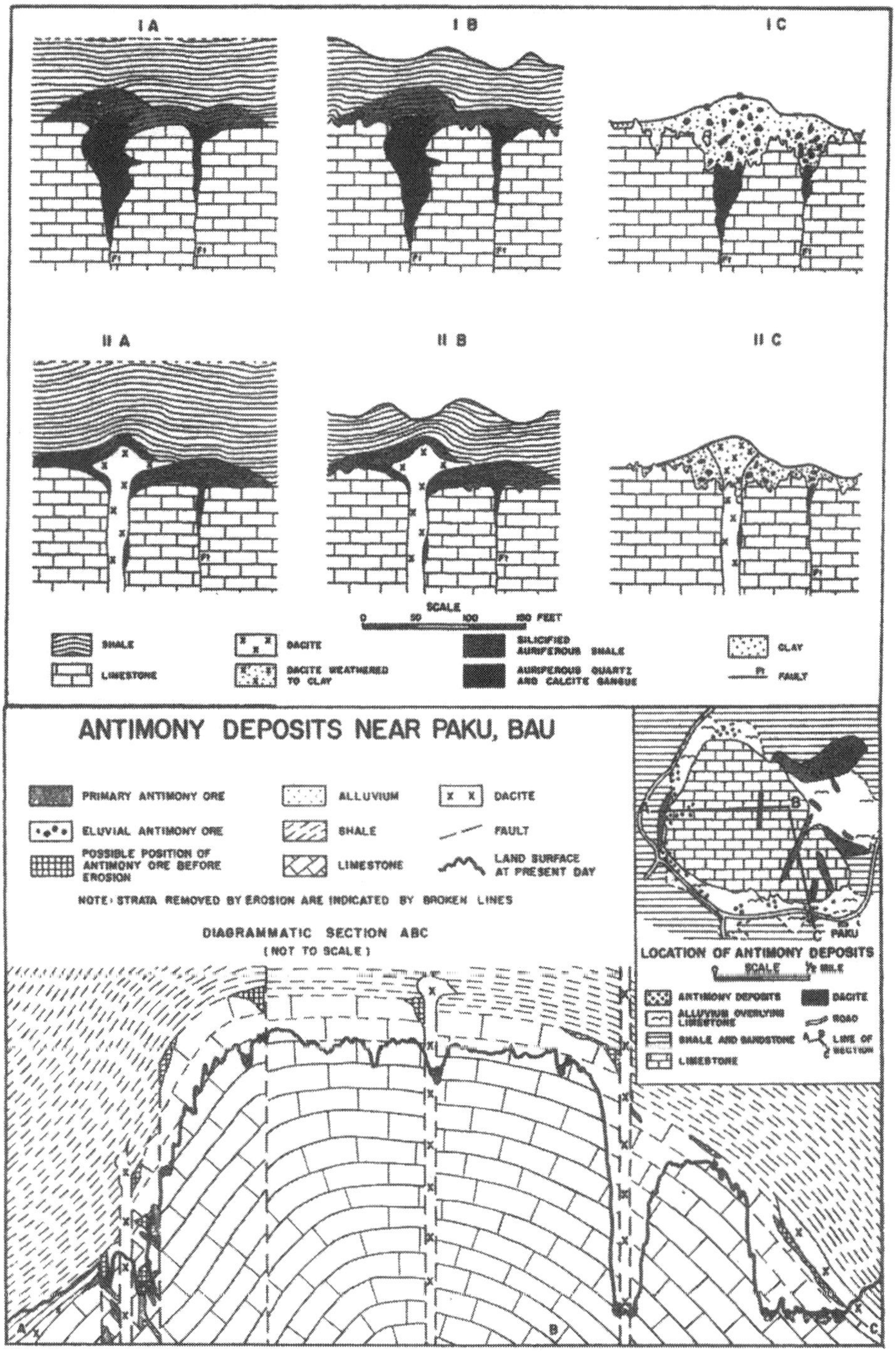

Fig. 10.2. Epithermal mineralization at Bau, Sarawak. **Top**: Diagrammatic sequence of events in formation of secondary gold deposits. **Bottom**: Antimony deposits at Paku near Bau. Both antimony and gold are genetically related to dacitic dykes and the ores concentrated near the Jurassic/Cretaceous limestone-shale contacts (after Wilford, 1955).

occur with pebble corundum and small quantities of gold (Wilford, 1955).

10.1.1.3 Mercury

Between 1868, when mercury was first exported, and 1949, the official production was 22 000 flasks (760 t) (Wilford, 1955). The rich Bau deposits occurred as cinnabar-quartz-pyrite-marcasite and realgar fillings of breccia in sandstone and shale, but the great production came from eluvial cinnabar (Wilford, 1955).

The main deposits were at Tegora and Gading, some 11 km south of Bau. The mercury mineralization is distinctly separate from the gold-antimony, occurring as cinnabar in breccias of sandstone or shale, cemented by quartz, pyrite and lesser barite, realgar and cinnabar. Production was mainly from eluvial ore and later from primary breccia.

The breccia localities occur close to dacite and micro-granodiorite stocks. The most abundant ore mineral is cinnabar, occurring mostly as crystalline encrustations, rarely massive and along the joints of silicified sandstone and shale. The main gangue is quartz, pyrite, calcite, stibnite, and talc. Barite and fluorite also occur and the breccia contains traces of gold.

The breccias are lens-shaped with overlying and contiguous eluvial and alluvial deposits. The dacite and microgranodiorite provided the geothermal heat engine for the mineralization.

10.1.2 West Kalimantan

Most of the gold production from Kalimantan has been from placer deposits, but coarse gold has been mined from quartz veins in hornfels at Gunung Mas. Mercury mining at Sambas, Kalimantan, dates back to 1760 (Van Bemmelen, 1970).

Calc-alkaline volcanic centres are aligned in a NNW trend across the Chinese Districts. The northernmost centre forming the Bawang Mountains has a granodioritic core containing scattered molybdenite in quartz veinlets with traces of chalcopyrite, scheelite and cassiterite. The volcanic rocks around the granodiorite have yielded alluvial gold and a cluster of mercury occurrences was reported (Van Bemmelen, 1970).

Gold occurs in five principal associations (Sukamto *et al.*, 1988): (a) Lower to Middle Cretaceous plutonic and volcanic rocks of the Schwaner Mountains; (b) sheared sedimentary and low-grade quartz-veined metamorphic rocks of the Cretaceous to Middle Eocene orogenic complex (= Rajang Group of Sarawak); (c) Upper Eocene basal conglomerates and sandstones of the Ketungau Basin; (d) Upper Eocene volcanic rocks; and (e) Upper Oligocene to Lower Miocene sub-volcanic plugs, especially near Sintang, comparable to Bau in Sarawak. Whatever the primary source, the gold is mined from alluvium, and in places placer diamonds, thought to have eroded from the same Eocene sandstones and conglomerates, have also been recovered.

10.2 PALAWAN

The large mercury mines of Palawan island (*Palawan Quicksilver*), Philippines, lie along the Ulugan faulted contact between an ophiolite mélange and the NE pre-Mesozoic micro-continent, a region characterized by various faults. Mitchell and Leach (1991) have pointed out that all the deposits—Palawan Quicksilver, Sulu Sea and Philippine Mercury—are all located on the NW margin of the Iratag tectonic window, near the Beaufort Thrust of Fernandez (1968). Within this zone, serpentinized harzburgite with a metamorphic sole, is overthrust over an ophiolite sequence of mudstone, chert and pillow basalt. This mining region accounted for 1.4% of the world production. From 1955 to 1975, the production was 2027 flasks (69.86 t) (Bureau of Mines, 1986). The Philippine peak of production was in 1970, at 173 flasks, and since then there has been a rapid decline to zero. Cinnabar is the chief ore mineral. Common gangue minerals are quartz, chalcedony, opalite, calcite, pyrite and marcasite (Fernandez, 1968). Yellow ochre results from cinnabar weathering. The occurrences are described as fissure veins, breccia fillings, replacements, cavity fillings and disseminations. Free metallic mercury occurs as globules in open fractures but is insignificant. Composite veins of quartz-calcite, in which cinnabar occupies 60% of the vein, may have widths up to 60 cm. Cinnabar veins range from stringers to as much as 15 cm thickness. It is very clear that the mercury is of very low temperature hydrothermal deposition from hot springs. Fernandez (1968) believes that the mineralization is subaerial and of Pleistocene to Recent age.

Some 40 km from the mercury deposits, stibnite occurs in quartz veins in the western flank of the Caramay Schist dome.

10.3 SULAWESI

High purity alluvial gold derived from metamorphic quartz veins is widespread in the neck of Sulawesi (connection between the northern and western arms of the island) (Kavalieris *et al.*, 1992). Quartz occurs as lenses along the foliation of schists as cross-cutting veins. The veins are generally less than 0.5 m wide. This kind of mineralization is similar to that of the Otago Schists of New Zealand. The relatively coarse gold is worked by local miners.

10.4 THE MALAY PENINSULA

The 'gold belt' extends northwards from Peninsular Malaysia through Thailand immediately to the east of and parallel to the Bentong–Raub Palaeo-Tethys Suture. Most of the mining has been from placer deposits, eroded from nearby lode deposits. However there have been two great lode mines, one in Peninsular Malaysia, the other in south peninsular Thailand.

10.4.1 Raub Australian gold mine

Gold, antimony and minor mercury mineralization occurs in a narrow zone within Permo-Triassic sedimentary and volcanic rocks east of the Bentong–Raub suture zone of Peninsular Malaysia (Figs 10.1; 10.3). Only gold has been of importance, with one mine, the Raub Australian gold mine, yielding over 80% of the recorded production.

From 1889 to 1938 the mine produced 21 t of gold. When it went into disuse around 1955, the gold-mining industry of the Malay Peninsula became insignificant, but a few small mines continue to operate on the overlying alluvium and the quartz lodes which outcrop beneath the alluvium. The ore bodies are hydrothermal replacements and fissure fillings located in zones of compressional shear faulting, known as the Eastern and Western lode channels. The minerals in the lode channels are quartz, calcite, gold, pyrite, arsenopyrite, stibnite, scheelite and traces of chalcopyrite and cerussite. Silver is generally absent. The stibnite occurs as large masses, and the gold frequently occurs within quartz, stibnite or arsenopyrite. The veins are largely conformable with the folded Permian calcareous shale. Quartz lodes also occur as discontinuous replacement bodies within the shale. There is no demonstrable relationship with igneous intrusions. However Richardson (1939) implied that the deposits

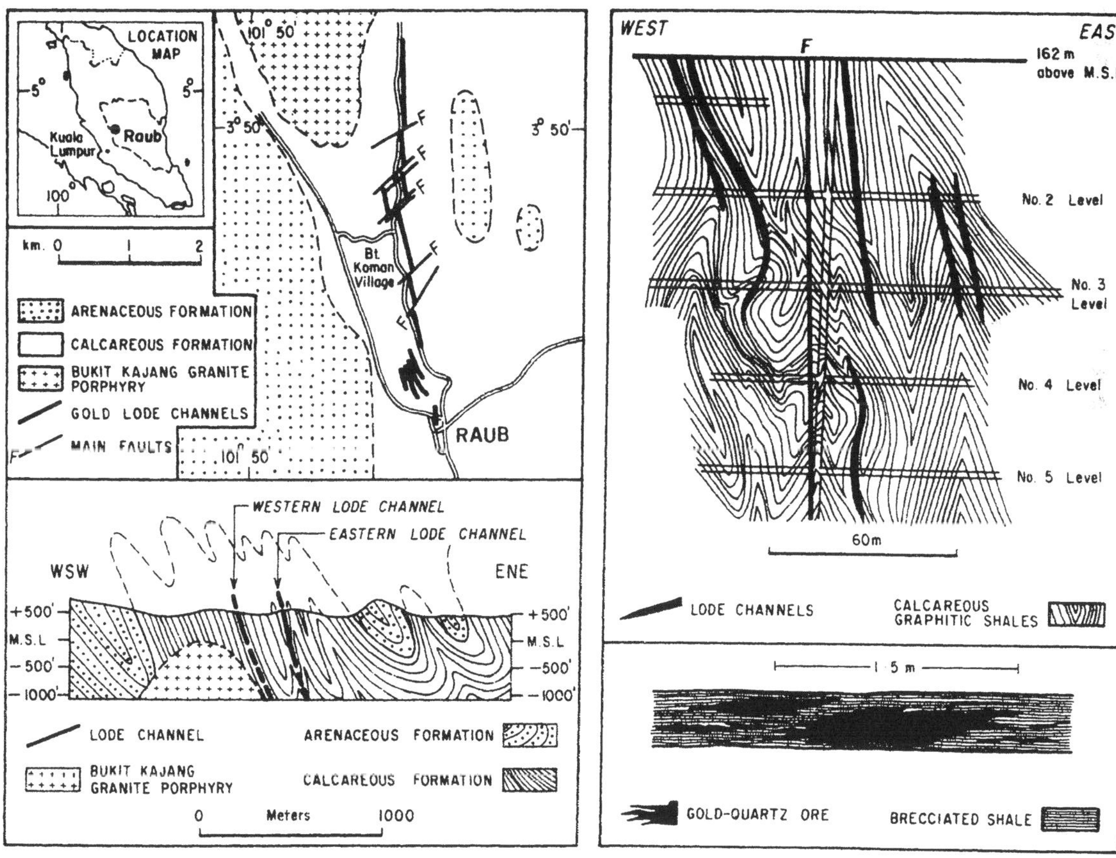

Fig. 10.3. The Raub Australian gold mine of Peninsular Malaysia (after Hutchison, 1983*a*; based on Richardson, 1939). The gold and stibnite occur in quartz veins (upper right) or in discontinuous replacements or 'sweat-outs' (lower right) in the Permo-Triassic sedimentary rocks. A genetic relationship to the quartz porphyry has been suggested.

were genetically related to the Bukit Kajang rhyolite porphyry, which is probably of Upper Triassic age. However most of the lodes resemble typical 'sweat-out' veins and pods of quartz that commonly occur in low grade metamorphic assemblages, such as the Schist Series not far to the west of the Raub mine. It is likely that the lodes and quartz veins resulted from mobilization of the economic elements from the country rocks by circulating meteoric hydrothermal solutions. If the Bukit Kajang rhyolite porphyry did play a role, it was only to raise the regional temperature to facilitate the meteoric water circulation.

10.4.2 Toh Moh

Farther north along the same Peninsular Malaysian tectonic line, a lode deposit was formerly mined at Toh Moh (Waeng) in south Thailand (Fig. 10.4). The geology of the mine is very similar to that of the Raub Australian gold mine (Brown *et al.*, 1951). The main quartz lode strikes N–S and dips 20° E, ranging from a few cm to 2 m thick. It occurs within slate and phyllite, but Brown *et al.* (1951) state that granite may also be the country rock. The average ore tenor was 10 g t^{-1}. A French company produced 1.95 t of gold between 1936 and 1940. Since then the deposit has been re-investigated, but considered uneconomic.

10.4.3 Others

The main gold belt of Peninsular Malaysia extends through Raub and follows the 345 to 360° - striking fabric of the Peninsula (Fig. 10.4), from south Thailand towards Malacca (Yeap, 1993). Throughout the length of the country, the gold occurs in quartz veins, reefs and lodes occupying steeply dipping shear zones generally in greenschist facies phyllitic to slaty meta-sediments. Small mines and river panning accounted for a significant gold production in the early years of this century. The primary occurrences are outlined in the following paragraph.

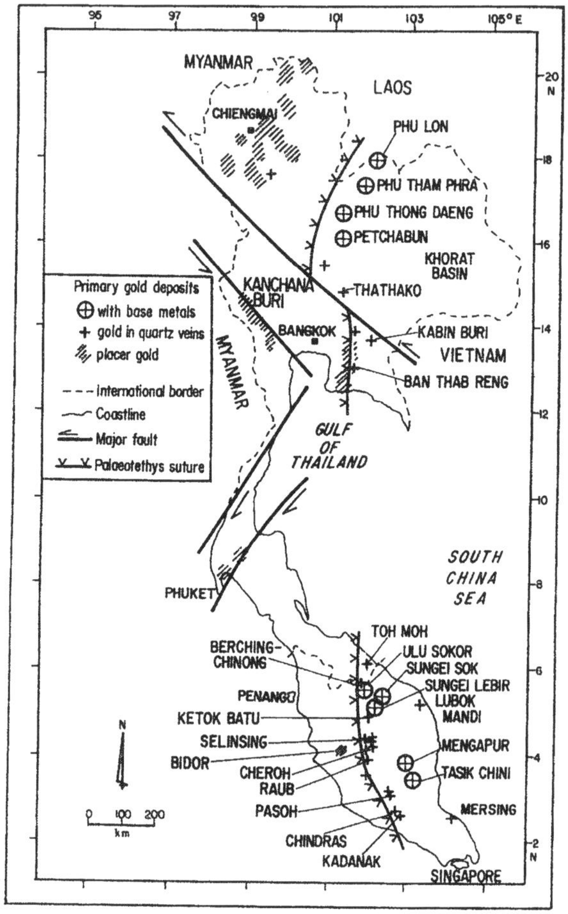

Fig. 10.4. Distribution of gold deposits in Thailand and Peninsular Malaysia, compiled from Davis *et al.* (1987), Shawe (1984) and Yeap (1993). All primary deposits have overlying or contiguous placer deposits resulting from their erosion. The placer deposits on the map have so far not led to properly assessed primary deposits. The gold-bearing volcanic related base-metal deposits are described in Chapter 9.

At Berching, Kelantan, the gold occurs in rich quartz veins hosted in quartz-mica schist and slate. At Kanan-Kerbau (Buffalo Reef), NW Pahang, sheeted veins and lenses of quartz-gold-stibnite occur within two well defined shear zones in slate, metamorphosed volcanics and in marble. At Selinsing, NW Pahang, the gold is in a lode, sheeted stockwork and disseminated in carbonaceous slates and metamorphosed volcanics (Yeap, 1993). At Tersand, NW Pahang, the gold-bearing veins occur in metamorphosed tuffaceous sandstone. At Cheroh, W Pahang, the gold occurs in silicified veins and stockwork in a 350°-striking felsite dyke. At Pasoh, Negeri Sembilan, there are two N–S zones of gold-bearing sheared and silicified granophyre. At Batu Bersawah, Negeri Sembilan, gold occurs in a pyritized 350°-striking lamprophyre dyke cutting calcareous slate. At Chindras, Negeri Sembilan, there is a gold-bearing stockwork vein related to a N–S felsite stock and dyke. The most southerly gold occurrence is at Kedanak, on the border between Malacca and Johore, where quartz veins occur in quartzitic and calc-silicate hornfels (Yeap, 1993).

Another important 345 to 355°-striking belt occurs close to the east coast of Trengganu, between Kampong Chenering and Marang (Fig. 10.4). A N–S striking belt of deformed phyllitic metasediments is cut by N–S striking gold-bearing quartz veins filling shear zones. There are other directions of narrow deformed 'sweat-out' quartz veins in the meta-sediments, but they are not mineralized. At Lubok Mandi, the 1 to 2 metre wide quartz dyke is being mined by open-cut at the Permint Mine. Higher grades in the quartz vein are manifested as visible gold flakes. Ore values are generally around 10 ppm, with local highs up to 40 ppm.

Erosion of the outcropping gold-bearing lode resulted in rich placer gold in nearby Quaternary valley alluvium immediately to the east near Marang. Placer mining supported a major 'gold rush' in the late 1980s, and mining continues on a small family-sized scale.

The Mersing Beds, low grade metasediments occurring south-wards along strike from Lubok Mandi, contain quartz veins. The veins are gold-bearing and erosion has given local placer deposits that have been mined on a small scale. However, no major lodes have been discovered.

Gold is also strongly associated with certain base-metal deposits in the Central Belt of Peninsular Malaysia. The Mengapur base-metal district [102°47′ E, 3°43′ N] has four types of mineralization largely associated with a large porphyry system (Wong and Lee, 1986). The system consists of a core of copper-molybdenum, surrounded by lead-zinc veins and an outer anulus of geochemical highs in Au-Ag-Sn. Iron-oxide replacement mineralization occurs within this system, predominantly along the contacts between granodiorite and metasediments as well as over high sulphide zones. The nearby Sungei Luit has produced placer gold since the 1820s, and a large mining town of alluvial gold miners was formerly sited here.

10.5 THAILAND

10.5.1 Gold

A N–S trending belt of gold deposits extends northwards from near the Sra Kaeo ophiolite locality (Fig. 10.4) in the Rayong District east of Bangkok (Davis *et al.*, 1987; Shawe, 1984). Gold mining from small placer deposits has a long history in Thailand. The most important are at Pa Ron in the Prachuap Khiri Khan Province. The placer gold ranges from extremely fine to the largest recorded nugget of 50 g. The gold province of Thailand is on the east, whereas the antimony province lies to the west.

The Tha Thako lode, 75 km N of Khok Samrong in Lop Buri Province, was discovered in 1890. Permian Saraburi limestone is intruded by quartz monzonite porphyry and quartz diorite dykes. The gold is in irregular quartz-calcite veins of tenor ranging from 2 to 90 g t^{-1}. The largest vein is 6 m wide and can be traced for 200 m (Shawe, 1984).

The Kabin Buri lode at Ban Bo Thong, Prachin Buri Province about 100 km E of Bangkok, was first mined in 1880. The country rocks are E–W striking Palaeozoic metasediments—quartzite and marble—intruded by diorite porphyry. The gold occurs mainly along the quartzite-marble contacts and the ore zone is strongly brecciated. Average grade is 5.5 ppm. Mining has been carried on near surface for over 100 m along strike (Shawe, 1984).

The Ban Bo Nang Ching Mine is 20 km SE of Sra Kaeo in Prachin Buri Province. The country rocks are Permian Saraburi Limestone and shale, intruded by diorite porphyry. The quartz veins are irregular. A 2 m wide vein strikes N–S. Most of the gold is mined from overlying placer deposits, derived from erosion of this vein system.

Both Shawe (1984) and Davis *et al.* (1987) deduced that the gold and other epithermal vein deposits of Thailand have resulted from hydrothermal mobilization along the extensive wrench fault system of the region, driven by the high regional heat flow, manifested by hot springs along the fault zones and high well temperatures in the offshore oil and gas basins. It is hard to avoid the conclusion that the mineralization must be Tertiary, but the ultimate source of the metals is obscure.

10.5.2 Antimony

Antimony occurs at many places in Thailand and Myanmar east of the Shan Scarp (Fig. 10.1). Annual production has been below 100 t in Myanmar with the annual Thai production increasing recently to over 7000 t. The main occurrences are around Phrae, Lampang, and along the Thai–Myanmar border region (Asnachina, 1981). There are four types of stibnite association: (1) stibnite-quartz with or without fluorite, (2) stibnite fluorite, (3) wolframite stibnite fluorite-quartz, and (4) scheelite-stibnite. The first three associations occur as small veins unrelated to igneous intrusions, but related to regional fault systems. The fourth association occurs as lenses and brecciated zones in Silurian-Devonian marble.

Antimony mineralization occurs in the Lampang–Chiangmai–Chiangrai district of NW Thailand in the form of stibnite-quartz, stibnite-fluorite, ferberite-stibnite-fluorite-quartz and scheelite-stibnite veins, pockets and replacement bodies (Asnachina, 1981). The most widespread is the stibnite-quartz association. The proposed genetic relationship of the antimony deposits with Triassic granites is unlikely because most of the deposits occur in fault zones that are of Tertiary age. The Sinoburmalaya terrain in which the deposits occur continues to be one of high heat flow, characterized by fault-related hot springs. It is therefore most likely that these epithermal antimony deposits are of Cainozoic age and have resulted from hydrothermal circulation along the deep fractures which characterize the region.

10.6 MYANMAR

Gold has been panned but no major deposits have been located in the country (Goossens, 1978). The Burma Gold Dredging Company carried out operations in the Myitkyina district of the Upper Irrawaddy River between 1903 and 1918, producing 1.76 t of recovered gold. Gold has been mined on a small scale from quartz veins, but generally hard rock mining is undeveloped. Gold mineralization at Kyauk Pahto (23° 46′ N; 95°56′ E) is confined to a narrow zone of tension fractures within the dextral wrench fault system (Wilson, 1987), part of the major Sagaing Fault of Myanmar. The mineralization is within silicified and localized breccia bodies, and has resulted from water movement within the seismically active fault zone. The ultimate source of the gold is obscure.

10.7 CHINA

China is considered to have greater reserves of antimony than any other country, and it produces about 20% of the world's output. The main deposits are in the south, predominantly in Hunan, which holds more than 60% of the nation's total. The Ksikuangshan district is the premier mining region. The ore is of stibnite in a gangue of quartz and the veins occur in sandstone. It is of low grade with 6 to 9% antimony on average. The district produces over 25 000 t of metal antimony annually (Ikonnikov, 1975). Other regions of Hunan are at I-yang, where quartz-stibnite veins cut schist and shale, and in the Meicheng district. In addition to the pure antimony ores of China, many of the tungsten and tin deposits also contain stibnite in the same hydrothermal vein systems.

As well as the main quartz-stibnite veins, there is also an important stibnite-galena-arsenopyrite replacement in carbonate rocks. The antimony mineralization usually occurs in regions of important shearing and brecciation.

11

Tungsten deposits

South-East Asia is a major tin-tungsten metallogenic province, and the distribution of the main tungsten deposits is shown in Fig. 10.1, redrawn from Taylor and Hutchison (1979). All tin provinces contain tungsten mineralization, but the converse is not true worldwide (Hutchison, 1982). All tin deposits are spatially and genetically related to S-type granites characterized by high initial Sr isotope ratios, indicative of a continental crustal multi-cyclic anatectic origin (Hutchison and Chakraborty, 1979). However, in some countries there are major tungsten provinces which are impoverished in tin, but South-East Asia is characterized by a close concordance of both metals, with SE China the prime example (Hutchison, 1982). Accordingly it may be concluded that South-East Asian tungsten mineralization is related to S-type granites, but this is not a worldwide generalization.

The mineralization is spatially and genetically related to three major arcuate granitic provinces of South-East Asia, referred to and described by Cobbing *et al.* (1992) as: (a) the Western, (b) the Main Range, and (c) the Eastern. The Western Belt is of mixed S and I type, the Main Range is overwhelmingly S type, and the Eastern Province has mixed I and S characteristics. The peak of granite magmatism in the Eastern and Main Range provinces occurred at 220 Ma (Upper Triassic). However, the Eastern province ranges to as old as Permian. The Main Range Province granites are characterized by initial $^{87}Sr/^{86}Sr$ ratios commonly between 0.720 and 0.725, whereas the Eastern Province has ratios close to 0.710. The Western Province of Peninsular Thailand and western Myanmar ranges from Jurassic to Tertiary, with a peak in the Upper Cretaceous. Those of Thailand are predominantly S-type with high initial Sr isotope ratios, while those of Myanmar are variable. The tungsten mineralization, for example at Hermyingyi mine, is related to S-type granite, Rb:Sr isochron dated 59 ± 2 Ma, characterized by an initial Sr isotope ratio of 0.7350 (Cobbing *et al.*, 1992).

The close concordance of tungsten and tin mineralization in south-east China is genetically related to Yenshanian (Jurassic–Cretaceous) high-level S-type granites (Xu and Zhu, 1988). The region is characterized by multi-cyclic granite magmatism, notably Caledonian, Hercynian, Indosinian and Yenshanian. Each of these granite-producing orogenic cycles has caused the magmas to become progressively more enriched in both tungsten and tin. It is therefore obvious that both these metals have been scavenged from the continental crust by these polycyclic granite-anatectic episodes.

China was the first country ever to produce tungsten, where the metal was discovered, is now the world's major producer, and has vast reserves in the extensive metallogenic province of the south-east (Fig. 11. 1). Thailand continues to be a producer but production from Myanmar and Malaysia, which in the early years of this century both had the largest mines in the world, has declined to insignificance.

11.1 CLASSIFICATION

The deposits of the region have been broadly classified into four types by Chong (1976) and Ikonnikov (1975):

- **1:** *Hydrothermal quartz veins*: this category is known as the South Jiangxi type, with subdivisions into: **1.a:** tungsten-tin in greisen, **1.b:** wolframite-quartz veins, **1.c:** scheelite-wolframite quartz veins, and **1.d:** tungsten-antimony-gold veins.

The most important examples are in the Nan Ling Range of South Jiangxi, China, but good examples were found at Mawchi and Hermyingyi in Myanmar and southward in Thailand and Peninsular Malaysia.

- **2:** *Scheelite skarn deposits*: they occur at granite-limestone contacts. Examples occur in SE Hunan, at Doi Mok in Thailand, and at the now exhausted Kramat Pulai mine in the Kinta Valley of Malaysia.

- **3:** *Wolframite-scheelite-sulphide veins*: They consist of fine grained wolframite, together with scheelite and base metal sulphides, with examples in south-east China.

- **4:** *Placer deposits*: They are either residual, eluvial or marginally alluvial, occurring less than 5 or 6 km from the primary source. Tungsten content dies off more rapidly than the associated tin. They are of minor importance and occur mainly in SE China.

Cassiterite is a frequent but by no means constant companion of tungsten in China and throughout SE Asia. The major deposits are shown in Fig. 10.1. The most important groups, from a production viewpoint for the whole region are, in declining order: 1b, 2, and 1a. Types 1b and 2 commonly have little or no associated cassiterite, but the vein type in greisen (1a) contains significant cassiterite. The closest association of tin and tungsten is to be found in the Xihuashan mining province of south Jiangxi, the Tavoy and Shan states of Myanmar and the immediately adjacent parts of Thailand. The Indonesian parts of the tin-belt are markedly deficient in tungsten, with only one small producer known.

11.2 CHINA

Xu *et al.* (1982) consider that SE China has become so enriched in both tungsten and tin deposits as a result of multi-cyclic anatectic events within the continental crust. The granites are related to the Caledonian, Hercynian, Indosinian and especially to the Yenshanian orogenies, all of which resulted in widespread S-type granites, with which the ore deposits are spatially and genetically related. South-east China, and particularly Jiangxi Province west of Ganzhou (Fig. 11.1), is the world's foremost tungsten metallogenic province, as a result of which China produces approximately 50% of the world's supply. The major reserves are in two forms: (1) hydrothermal wolframite-quartz veins, and (2) pyrometasomatic scheelite deposits. Next in importance comes (3) stratiform or stratabound scheelite-stibnite-gold deposits, and of very minor significance are (4) tungsten-bearing haematite beds, (5) porphyry tungsten deposits, and (6) placer deposits.

The Nan Ling 'Caledonian' foldbelt is intruded by numerous high-level Yenshanian (190 to 130 Ma)

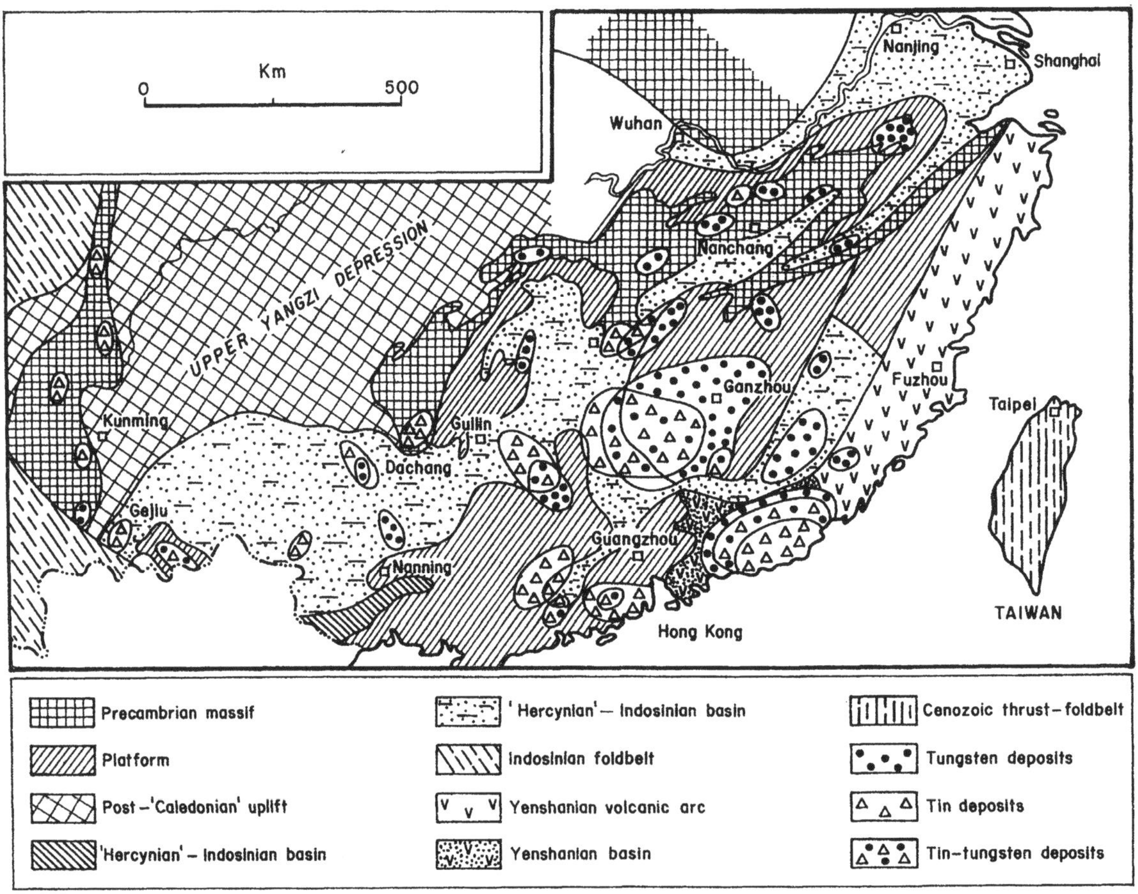

Fig. 11.1. Distribution of the major tin and tungsten deposits of South China showing their spatial relationships (after Xu and Zhu, 1988).

granites, which are surrounded by andalusite-bearing thermal aureoles. The fissure vein systems, which centre around granite cupolas, have been the focus of the tungsten mineralization. The tungsten deposits of south China have been subdivided by Li and Xu (1982) into the following types:

11.2.1 Veinlet disseminations

This type occurs at the intersections of regional fractures in the 'Caledonian foldbelt'. Many are of apophyses of fine grained albitized and greisenized muscovite granite. Fine grained wolframite, associated with Nb and Ta minerals and molybdenite, is disseminated in dense stockworks. The Dajishan and Xiatongling mines are good examples.

11.2.2 Skarn types

In the borderlands between Hunan and Jiangxi provinces, the Lower Yenshanian (Jurassic) granites intrude Upper Devonian to Upper Carboniferous limestones. Scheelite deposits have been formed along the contact zones. The ore body geometry is irregular and controlled by the nature of the contact. The major ore minerals are scheelite, galena, sphalerite, chalcopyrite and pyrrhotite, and the skarn consists of garnet, diopside, wollastonite, epidote, vesuvianite, scapolite, plagioclase and chlorite. Scheelite is mainly granular and disseminated within massive sulphide ore. An example is the Baoshan Mine.

11.2.3 Pegmatite ore

Pegmatite veins occur locally and may extend into the quartz vein deposits from the granite cupolas. They contain wolframite and Nb and Ta minerals.

11.2.4 Quartz-(feldspar) veins

This is the most important ore type of the Nan Ling Ranges of Jiangxi Province with the foremost examples in the Xihuashan tungsten ore-field SW of Gangzhou (Fig. 11. 1).

The vein systems emanate from high level cupolas of Yenshanian biotite and two-mica granite batholiths intruded mainly into folded Lower Palaeozoic strata of the Caledonian foldbelt (Fig. 11.2), but also including older and younger strata. The quartz veins are mineralized within the outer zones of the plutons as greisen-bordered veins, but much of the mineralization is exo-contact within the thermal aureole rocks. The vein systems vary from place to place and form a variety of patterns in response to cupola geometry and regional structures. The quartz veins are exceedingly rich in coarse grained wolframite on a spectacular scale and are characterized by a great variety of associated minerals, many of which are euhedral.

More than 50 minerals have been identified (Li and Xu, 1982). Some quartz veins carry mainly W-Sn mineralization, others W-Bi, W-Mo and W-Cu. The common gangue minerals are quartz, feldspar, muscovite, zinnwaldite and topaz, as well as fluorite and calcite. Wall-rock alteration is highly characteristic, with greisenization within the granite, and silicification without.

Liu (1982) has summarized the salient features of the wolframite-quartz veins of south Jiangxi, typified by the Xihuasan mining district of the Nan Ling Caledonian foldbelt (Fig. 11.2). There is always a spatial association with shallow or concealed high level small Yenshanian cupolas. All deposits are genetically related to strong hydrothermal alteration of the granitic cupolas, the upper parts of which have been greisenized. Biotite granite gives way upwards to two-mica granite in a zone from which the tungsten-bearing quartz veins emanate upwards into the thermally metamorphosed Cambrian strata. In some cases the veins extend beyond the contact aureole and mining is facilitated by their outcrop geometry.

The commercial ore zones may be within the greisenized granite as at Xihuashan, but only 12 km to the NE the commercial ore is entirely exo-granite within the thermally metamorphosed Cambrian strata (Yang and Lu, 1982), as at Piaotang (Fig. 11.2).

The following mineralization stages can be discerned (Yan *et al.*, 1980):

- The early Silicate Stage is characterized by skarn formation, alkali feldspar growth and greisenization. Mineralization may be disseminated or may form commercial tungsten-tin deposits. The temperature range was 500 to 400° C.
- The Oxide Stage is the main ore-forming stage for tungsten. The mineral paragenesis is: molybdenite, followed by (molybdenum-bearing scheelite), wolframite-quartz, lithium-containing muscovite-beryl-cassiterite-wolframite-(feldspar)-quartz and zinnwaldite-cassiterite-wolframite-sulphides-quartz. The temperature range was 400 to 330° C.
- The Sulphide Stage represents declining intensity of tungsten-tin mineralization. The main minerals are (cassiterite-wolframite)-scheelite-sulphides-quartz. The temperature range was 330 to 185° C.
- The Carbonate Stage is characterized by weak tungsten-tin mineralization. The paragenesis is fluorite, scheelite, pyrite, quartz and carbonates. The temperature range was 185 to 129° C, as determined by fluid inclusion studies (Yan *et al.*, 1980).

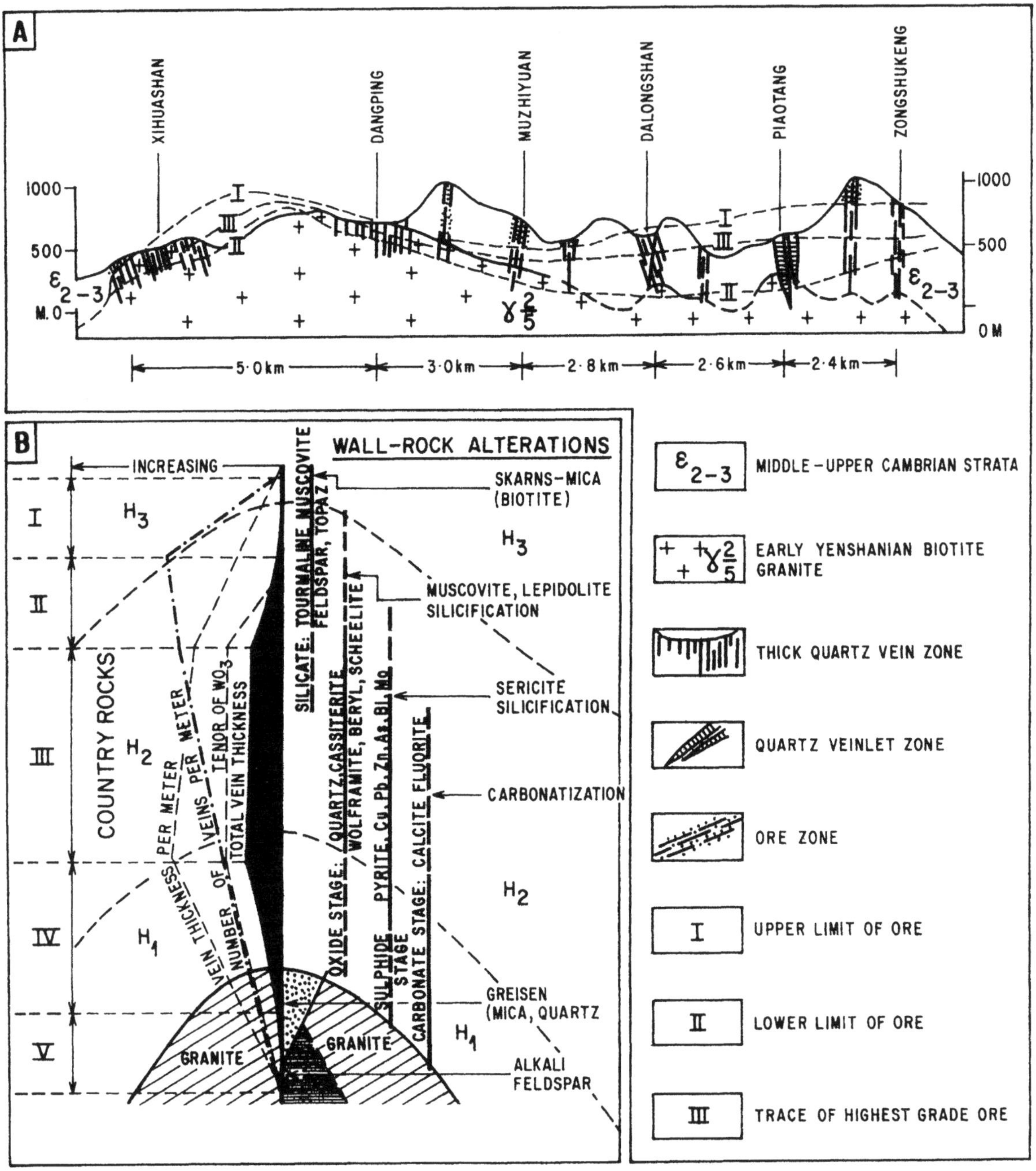

Fig. 11.2. The wolframite-quartz veins of the Xihuashan district of south Jiangxi, southern China. **A**: Cross section to show the relationship to granite cupolas. In the south the ore is within the granite, and exo-granite in the north (after Yang and Lu, 1982). **B**: Vertical zoning pattern for the veins (after Hutchison, 1983*a*). Zone I = Mica-quartz veinlets, Zone II = Densely distributed veinlets (poor in mineralization), III = Veinlets and vein zone (principal commercial ore), IV = Vein zone (commercial ore), V = Sparsely distributed vein zone (barren). H = Thermal hornfels aureole (1 = highest grade, 2 = hornfels, 3 = spotted country rocks).

Wall-rock alteration changes with style of mineralization (Fig. 11.2). At the silicate stage, skarn formation and feldspar development predominate. At the early wolframite-molybdenite stage, biotite is characteristic. At the main wolframite stage, greisenization and silicification are important with extensive development of tourmaline, topaz, muscovite and fluorite. At the sulphide stage, sericitization, chloritization and silicification are important, while the carbonate stage is characterized by carbonatization.

Reversed vertical zoning is commonly observed in the veins (Yan *et al.*, 1980), summarized in Fig. 11.2. In the top part (I), the main minerals are muscovite and quartz. Sometimes topaz, zinnwaldite and tourmaline may be important, with or without molybdenite, cassiterite and wolframite. The minerals are usually fine grained. Wolframite forms tabular or acicular crystals. Comb-structured quartz veinlets and planar mica veinlets are developed, and they usually indicate underlying mineralized zones at greater depth.

In the upper part (II), oxides and silicates are important, such as zinnwaldite, muscovite, feldspar, tourmaline and topaz, as well as beryl, cassiterite, wolframite and quartz. Sulphides such as arsenopyrite, molybdenite, bismuthinite and pyrrhotite may also occur. The ores are commonly banded. Wolframite occurs inside an outer zinnwaldite band, and it is associated with cassiterite, beryl and topaz. Wolframite crystals are tabular. The ore bodies are composed of small veins and veinlets or zones rich in veinlets ('streaky bacon ore').

In the middle part (III), wolframite, cassiterite and other oxides predominate. Sulphides increase in quantity and variety. The gangue is often clearly banded. Miarolitic cavities and brecciated structures frequently occur. The wolframite is of large tabular crystals unevenly distributed. This is the richest part of any vein system (Fig. 11.2).

In the lower part (IV) the economic mineral content diminishes, but sometimes scheelite, stannite and helvite appear. In the bottom part (V), the tin and tungsten minerals disappear almost completely. However fluorite, ankerite, siderite, rhodochrosite, calcite and other carbonates increase in amount. The barren veins disappear with depth into the granite.

The southern limit of the Nan Ling Range is in Hong Kong, where the Upper Yenshanian uppermost Jurassic 140 Ma old Needle Hill granite contains wolframite-bearing quartz veins.

11.2.5 Porphyry tungsten

The Yangchuling porphyry tungsten deposit occurs in a district of porphyry copper deposits. A minor quartz diorite-granodiorite porphyry (142 to 145 Ma) intrudes into Proterozoic slates (Li and Xu, 1982). It is accompanied by an explosive breccia pipe. The porphyry is rich in tungsten (40 to 150 ppm) and molybdenum (4 to 300 ppm). Hydrothermal alteration is characteristic, resulting in biotite, sericite, chlorite, carbonate and quartz deposition. Within the alteration zones, scheelite and molybdenite mineralization is developed, in association with bismuthinite, chalcopyrite and pyrrhotite, mainly in the form of veinlets and disseminations.

11.2.6 Breccia pipes

This type occurs in close association with porphyry deposits. They occur within important fracture zones. A common plan area is less than 1 km^2. In cross section it is funnel shaped containing shattered and collapsed angular to sub-rounded country rock breccia clasts, cemented by hydrothermal quartz and fluorine and boron-bearing minerals. The breccia pipe connects with a granite at depth. The major commercial ore minerals are scheelite and molybdenite and the ore exhibits a veinlet, disseminated and brecciated structure.

11.2.7 Stratabound deposits in limestone

These stratabound deposits are well developed in Cambrian and Devonian limestones, recrystallized and skarnized by granites.

The Jiaoli mine is a good example, situated in the N–S trending Caledonian foldbelt. A biotite granite stock (202 Ma) discordantly intrudes a sequence of Upper Cambrian sandstone, limestone and slate. Several concordant ore beds occur far from the granite contact and they have been folded together with the strata. The mineralized layers include calcareous sandstone and limestone. The ore minerals are scheelite, sphalerite, galena, chalcopyrite and molybdenite in a skarn gangue of garnet, pyroxene, wollastonite and tremolite. The scheelite is very fine grained and disseminated and the ore may be banded, massive or in veinlet form.

One interpretation is that the hydrothermal fluids readily penetrated parallel to the bedding from the granite plutons. The alternative view is that at least some of the mineralization was in the strata before the granite intrusion and was partly mobilized by it. Some deposits have composite characteristics between contact skarn and stratabound geometry.

11.2.8 Stratabound deposits in volcano-sedimentary formations

Proterozoic tuffaceous slate formations in the Jiangnan uplift have been converted to skarn ore. Fine grained diopside skarn contains disseminated scheelite. Some deposits have no obvious association with granite. In the Woxi Mine, stratiform and lenticular ore beds are conformable with the volcano-sedimentary strata. The commercial minerals are scheelite, stibnite and native gold with minor siderite, galena, sphalerite, arsenopyrite and wolframite. The wall-rocks are strongly silicified.

11.2.9 Tungsten deposits in terrigenous clastic rocks

Stratabound scheelite, showing a veinlet-disseminated structure, occurs in biotite-quartz hornfelsed sandstone of Devonian age in the Yaogangxian Mine.

In the Shuikou Mine, the tungsten-bearing layers are of Carboniferous siltstone. The ore mineral is wolframite in a gangue of quartz, tourmaline and topaz, both disseminated and in inter-stratified fractures.

11.2.10 Tungsten-bearing sedimentary iron formations

The Dabaoshan deposit is an example of a tungsten deposit occurring in a Devonian siderite (limonite) iron formation. The ore bed occurs in limestones and the tungsten mineralization is related to copper-bearing pyrrhotite and siderite.

Tungsten-bearing haematite beds occur in Carboniferous strata in Fenglin. The ore is mainly of haematite with minor amounts of wolframite and veinlets of scheelite.

11.2.11 Placer deposits

Wolframite-cassiterite alluvial placers occur in intermontane basins, with examples at Fentian and Yangmeisi. However placer deposits are not of great importance in China, and more than 95% of its tungsten and tin are won by hard-rock mining, mainly underground.

11.3 THAILAND

The tungsten deposits of Thailand occur on the Sinoburmalaya Block, west of the Uttaradit–Luang Prabang suture zone, with extension southwards into Peninsular Thailand. They are of the three following types (Charoensri, 1982).

11.3.1 Stockworks and breccia-filling deposits

Only two localities are known, but the one at Khao Soon (8° 30′ N; 99° 29′ E), in Si Thammarat Province of Peninsular Thailand, has contributed more than half the total Thai tungsten production of the past decade. It was discovered in 1970 and most of the production is by unlicensed miners that honeycomb the hillsides with small diggings and adits. Production has now declined because of insurgency in the region and a lack of proper reserve assessment by drilling. It had a monthly production of 180 t. The ore deposits have formed as veins along N-striking brecciated zones or as pipes in irregular cylinders of breccia rocks. There are eight major breccia pipes and two N–S striking veins (Shawe, 1984). The main funnel shaped breccia pipe, as much as 10 m diameter, cuts altered pervasively silicified meta-sedimentary rocks of the Upper Palaeozoic Kanchanaburi Series, which underlie the Permo-Triassic Ratburi Limestone. The rock sequence includes quartzite, tuffaceous greywacke and phyllite. The 300 m deep mine occurs near both the axis of a N–S-trending anticline and a N–S fault (Arrykul *et al.*, 1978; Shawe, 1984). The single 400 m × 800 m ore body exhibits lenses, pockets and pipe-like shapes. The ore is of fragmented angular altered country rocks cemented by a matrix mat of ferberite, some quartz and rarely pyrite. The brecciated fragments mostly range between a few mm to 10 cm diameter and are angular. The ore mined ranges between 5 and 12% W. Stibnite and marcasite occur as accessories in the upper parts. Silicification and pyritization are common in the rock fragments and wall-rocks. Ferberite-quartz veins occur in the hornfelsed quartzite and phyllite. Ferberite and stibnite may also be found in vugs in the mudstone country rocks. The overlapping of the higher temperature wolframite by later stibnite indicates a xenothermal classification. The ferberite contains on average 5% manganese tungstate and hubnerite occurs, but only very locally in the deposit. It is therefore likely that the deposit formed in a near surface environment and meteoric water may have played an important role. It may be significant that hot springs are characteristic of the region, which is dominated by N–S trending faults (Arrykul *et al.*, 1978). Khao Soon is, however, unique in the region, for no igneous rock is known nearby.

The Doi Ngom deposits of northern Thailand are confined to silicified breccia zones in Permian metasediments. The breccia developed as a result of the interaction between two major reverse faults, and mineralization is genetically related to hydrothermal activity from the neighbouring granite. The main ore is of ferberite, with fluorite and stibnite, cementing the breccia blocks and also occurring as veinlets in silicified country rocks. Traces of scheelite are also present in the volcanic rocks.

11.3.2 Tungsten-bearing quartz veins

They usually occupy the exo-contact zones around granite batholiths, and are common throughout the country. The major producing areas are at Mae Liang, Mae Lama, Pilok, Wang Pha and Lam Phaya. The veins extend from granite cupolas into contact aureole rocks. The veins are predominantly of quartz, containing wolframite. However there is a variety of accompanying minerals, such as scheelite, cassiterite, pyrite and arsenopyrite. Tungsten also occurs in pegmatites occurring in the outer parts of the granite cupolas.

Tan (1989) has described an interesting set of tungsten ore bodies at Wang Phar, 32 km WSW of HatYai in south Thailand. The wolframite-bearing quartz-muscovite greisen veins have been structurally impounded and localized within the Triassic (Main Range) granite beneath an impervious roof of meta-sediments.

11.3.3 Contact metasomatic skarn deposits

These are now the main source of tungsten in Thailand. The Doi Mok of northern Thailand is the most important in the country, discovered in 1969. The ore body is a skarn zone between biotite granite and Silurian–Devonian rocks especially in limestone lenses. Scheelite occurs as lenticular pockets in the contact zone, disseminated in the marble and granite, within scheelite-quartz veins in the granite, tuffaceous sandstone and quartz mica schist. Scheelite is the only mineral produced from the district. The gangue minerals are pyrite, arsenopyrite, chalcopyrite, pyrrhotite, galena, tourmaline, fluorite, calcite, quartz and skarn minerals (Charoensri, 1982).

Scheelite-cassiterite ore bodies at Samoeng district are of contact metasomatic and vein type. The former developed along the contact between leucogranite and Palaeozoic sediments, hornfelsed to assemblages of diopside with small amounts of feldspar, biotite, quartz, together with fine grained disseminated scheelite and euhedral coarse grained cassiterite.

11.4 MYANMAR

Over 120 Sn and W occurrences are known in Myanmar along the westernmost part of the Shan States and especially in Tenasserim in the area east of Tavoy (Bender, 1983). Tungsten seems to become progressively more important northwards in reciprocal relationship to tin (Goossens, 1978), and the southern end of this mineralized belt is at Phuket, Thailand, where tungsten is a very minor commodity. The deposits of Thailand lie along a belt, which extends northwards from Phuket, Thailand, along the Mergui–Tavoy coastal region of Myanmar—the Central Granitoid belt of Khin Zaw (1990) or Western Granite Province of Cobbing *et al.* (1992), thence northwards along the western margins of the Shan Highlands to the vicinity of Mandalay. The granites of this belt have been described by Cobbing *et al.* (1992) and shown to be of mixed I and S-type and predominantly of Upper Cretaceous to Eocene age. In these deposits both cassiterite and wolframite occur in the same lode, as they do in southern China. However in Thailand, Malaysia and Indonesia they occur in separate lode systems, though they may be in geographical proximity.

11.4.1 Mawchi Mine, Kayah State

The Mawchi deposit (18° 49′ N; 97°10′ N) in Karen State (Fig. 10.1) has the greatest known reserves of the country, estimated by Goossens (1978) at 712 540 t of ore at 3.02% Sn +W. The 1938 production was 4646 t of mixed concentrates yielding 1784 t metallic Sn and 1494 t W. From 1939 to 1940, this mining region produced 60% of Myanmar's total, accounting for 17.4% of the world's output of tungsten.

Porphyritic megacrystic and medium grained biotite granite intrudes into the Carboniferous Mawchi Series (= Mergui Series) and has imposed a low grade thermal aureole (Khin Zaw and Daw Khin, 1983). Cobbing *et al.* (1992) did not actually date the Mawchi granite, but the Gua Chaung granite, a short distance to the north, has provided a Rb:Sr isochron of 22 ± 7 Ma, with an initial $^{87}Sr/^{86}Sr$ ratio of 0.7157, indicating a sialic crustal origin. Other granites slightly to the north have given ages of 80 ± 82 and 139 ± 21 Ma.

The steeply dipping N–S striking quartz veins cut the granite cupola and extend into the contact aureole composed of slate, sandstone, calcareous shale and limestone of the Mawchi Series. There are 60 major veins and in places they form a stockwork. They are of thickness up to 2.5 m, with a 1 m average. Mineralization has been explored down to 300 m below surface. The veins are predominantly of quartz, and contain cassiterite, wolframite, scheelite, and accessory amounts of arsenopyrite, pyrite, chalcopyrite, bismuthinite, molybdenite, and galena. The gangue is of quartz, with feldspar, mica, calcite, magnetite, fluorite, tourmaline and minor topaz and beryl (Khin Zaw and Daw Khin, 1983). The ore usually contains 1.54% Sn and 0.75% WO_3. The concentrates usually contain 32% Sn and 34% WO_3.

Fluid inclusions indicate a temperature range of 120° to 270° C, similar to Hermyingyi. In scheelite the range is 170 to 270° and in fluorite from 120 to 220° C.

11.4.2 Hermyingyi, Tavoy

The Hermyingyi deposit occurs 40 km NE of Tavoy, Tenasserim (Fig. 10.1). It was at one time the world's

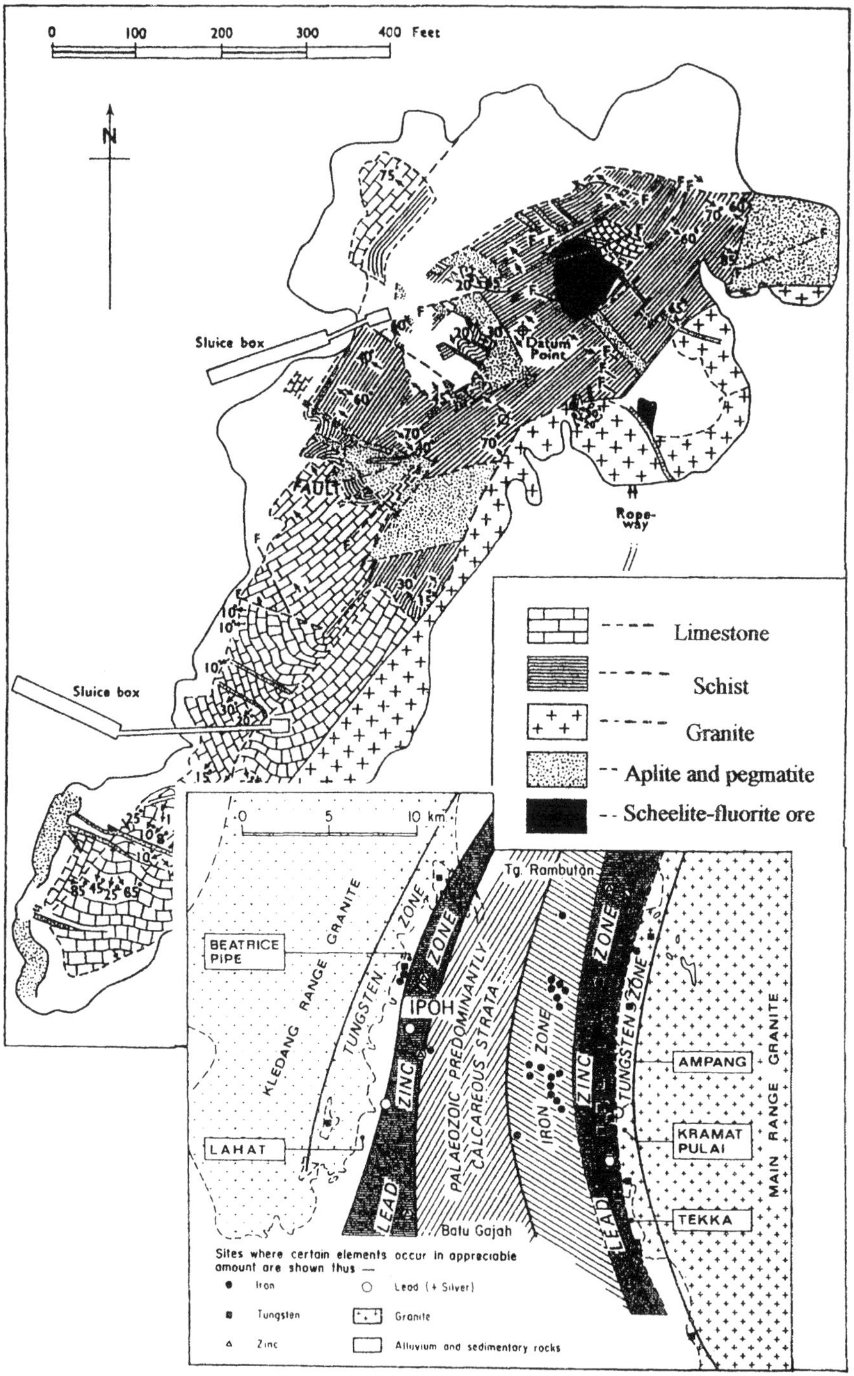

Fig. 11.3. Upper: The Kramat Pulai scheelite-fluorite mine (after Willbourn and Ingham, 1933). **Lower**: Metal zonation and locality map of the Kinta Valley of Peninsular Malaysia (From Hutchison, 1983*a*, after Hosking, 1973).

greatest producer of tungsten. The 1917 production was 1051 t of concentrate. The 1939 production was 127 t of Sn and 345 t of W concentrates. The ore reserve was estimated by Goossens (1978) at 90 079 t of ore at 1.04% Sn + W concentrates.

The megacrystic Hermyingyi biotite granite has been precisely dated by Cobbing *et al.* (1992). A Rb:Sr isochron of nine well separated points results in an age of 59 ± 2 Ma with an extremely high initial $^{87}Sr/^{86}Sr$ ratio of 0.7350. These Palaeocene rocks are highly fractionated leucogranites with 75 to 78% SiO_2. All other granites of this region have I-type chemistry, but the very high Sr isotope ratio at Hermyingyi indicates an extremely strong continental crustal signature (Cobbing *et al.*, 1992). The granite intrudes and contact metamorphoses the predominantly argillaceous Mergui Series country rocks.

Mineralization occurs in N–S striking near vertical quartz veins, which fill fissures in a cupola of the granite and the overlying contact aureole meta-sediments. There are 300 major branching veins, of thickness ranging from < 3 cm up to 2 m. The veins contain wolframite, cassiterite, molybdenite, pyrite, sphalerite, chalcopyrite, galena, bismuthinite and native bismuth (Khin Zaw, 1978; Bender, 1983). More than 40 veins have been worked, with an average WO_3 content of 0.3% and Sn of 0.05%. Fluid inclusions indicate a temperature range of 186° C to 250° C for the quartz (Khin Zaw, 1978).

11.5 PENINSULAR MALAYSIA

The country was a major producer of tungsten in the early decades of this century, but production has dropped to zero. The deposits have been described by Chong (1976) and summarized by Rajah (1982). Hydrothermal quartz vein deposits occur both east and west of the Bentong–Raub suture.

11.5.1 Kramat Pulai

At the peak of its production in 1936, the Kramat Pulai skarn deposit mine had an annual output of more than 1400 t, the largest in the world, making Malaysia (then Malaya) the third largest producer after Myanmar (then Burma) and China.

The main ore body lay under 7 m of alluvium at a distance of 30 m from the outcropping Triassic Main Range Granite (Fig. 11.3). It was an irregular body in Palaeozoic schist interbedded with and faulted against limestone (Ingham and Bradford, 1960). The rocks are traversed by dykes and veins of aplite and pegmatite. They vary in width from 60 cm to 3.8 m and dip steeply with a strike perpendicular to the granite contact, bearing 90 to 115°, and are often deeply weathered. The ore is entirely of fluorite and scheelite of a coarse pegmatitic texture. Gangue amounts to less than 1% and is chiefly quartz. The country rocks contain garnet, vesuvianite, axinite, tremolite and actinolite. Tourmaline and cassiterite occur in association with the pegmatites. The ore body formed anticlinal folds plunging 8° towards 240°. The scheelite content of the main ore body ranged from 3 to 15%, and it occupied the centre of a limestone anticline under schist, which is cut by an aplite dyke. The schist had acted to impound the hydrothermal fluids, but some leak-away fractures were healed by scheelite-fluorite ore. In addition to scheelite, Kramat Pulai also produced more than 8000 t of cassiterite. Other by-products were corundum, diaspore, fluorite, galena, yttrotungstite, ferberite, wolframite, stolzite, tungstite, stibnite, sphalerite, arsenopyrite, pyrite, pyrrhotite, tremolite, beryl and grossularite (Ingham and Bradford, 1960). The corundum was usually found as pebbles and boulders in clay at the limestone-granite contact. The ferberite occurred along with yttrotungstite in a fault fissure in weathered granite.

11.5.2 Others

Elsewhere in Peninsular Malaysia, scheelite may occur in deposits where there is no wolframite. But the two occur at Chenderong, where there is also a cassiterite association. At Ulu Langat, scheelite locally cements fractured cassiterite in greisen-bordered veins (Hosking, 1973).

Wolframite-bearing quartz hydrothermal veins occur in many regions of Peninsular Malaysia, from Sintok in Kedah, west of the Bentong–Raub suture, to Bukit Lentor and Chendrong in Trengganu, east of the suture (Rajah, 1982). The host rocks of the vein system usually are thermally metamorphosed meta-sediments within the aureoles of granites. The accompanying minerals are pyrrhotite, pyrite, chalcopyrite, sphalerite, galena and bismuthinite.

Placer wolframite and scheelite was commonly recovered from alluvial tin mining in a number of localities.

11.6 INDONESIA

Wolframite was produced commercially from Tikus Mine on Billiton, but it closed prematurely in 1921 (Suryono and Clarke, 1982). Cassiterite and wolframite were mined from a sub-vertical brecciated pipe within granite. The core of brecciated quartz is surrounded by weakly mineralized greisen.

Wolframite-bearing quartz veins, associated with greisen, are also known from the other tin islands of Singkep, Kundur and Bangka, but there has been no production since 1941.

Geochemical exploration has led to the discovery of several tungsten occurrences in northern Sumatra, notably at Hatapang (Fig. 10.1), where quartz dykes cut the granite and extend into the country rocks (Hamidsyah and Clarke, 1982).

12
Tin deposits

The cratonic core of peninsular Sundaland was the source of more than 70% of the tin mined this century in the whole world. The major producing centres lie in a broad arc 3400 km long and up to 800 km wide. The distribution of tin is strongly heterogeneous, with six major centres accounting for over 75% of the mined output. The balance has come from a further fourteen minor fields (Fig. 12.1).

The tinfields fall into two distinct types, those where virtually all the production is derived from mining placers, and those with a significant deep mined production. The former are by far the larger and the first four fields (Fig. 12.1) fall into this group. They are the unique feature of the region and the reason for its former dominance in world production. The great tin girdle dies out abruptly at Billiton (Belitung) and cannot be traced eastwards into non-cratonic Borneo.

12.1 ALLUVIAL PLACER FIELDS

Over 95% of the present tin output of Malaysia, Thailand and Indonesia is from Upper Cainozoic alluvial placers. The major tin sources are in braided streams, piedmont fans and residual elutriation placers together with rich eluvial and colluvial placers. They are spatially restricted mainly to the flanks of geomorphologically positive granite features and their contact zones with the country rocks.

The stratigraphic relationships of the Upper Cainozoic alluvial deposits have been summarized by Batchelor (1979). The Old Alluvium, or 'Older Sedimentary Cover', is a complex of unconsolidated sediments that overlies the scattered occurrences of Lower Pliocene boulder beds and bedrock. It ranges in age from Upper Pliocene to Lower Pleistocene (Fig. 12.2). The sequences are often complicated, slumped, extremely lenticular, and often difficult to decipher (Stauffer, 1973*b*). It has been conclusively shown from sedimentological studies of the great alluvial tin field of the Kinta Valley (Sivam, 1969) that the Old Alluvium is of fluvial origin. Its extreme lateral variability is a typical feature of fluvial braided stream sedimentation. The fluvial environment is also indicated by peat, peaty clay and occasional piles of partly lignitized branches and trunks of trees occurring as discontinuous beds and lenses. The abundance of fossil trees and large blocks of amber (local name *damar*) indicate that a good forest occupied the hilly areas outside of the braided streams. The existence of a deeply weathered regolith bedrock is illustrated by the decomposed nature of the clasts in the Old Alluvium.

The Middle to Upper Pleistocene Younger Alluvium is unconformable on the Old. In contrast its clasts are less weathered, it is less slumped and less consolidated than the Old Alluvium. It is also fluvial and its occurrences are limited to positions adjacent to the present-day river system (Stauffer, 1973*b*). Braided streams also characterized the Younger Alluvium, at least in its earlier episodes.

Evolution of the alluvium and its associated placers may be summarized as follows, largely after Batchelor (1979): the extremely low Upper Miocene to Lower Pliocene sea levels resulted in the extensive Sundaland continent. Conditions were ideal for the formation of a deep lateritized regolith over the landmass. Greisen-bordered vein-swarms in the border zones of predominantly Triassic granite (Upper Cretaceous–Eocene around Phuket) have been the main source of cassiterite to the placers of the western tin belt (Fig. 12.1). The eroding hydrothermal veins contributed large amounts of vein quartz to the regolith and hence to the alluvium. Placer formation generally requires deep lateritic weathering of the primary sources, producing quartz and resistate grains such as cassiterite and ilmenite in a sectile kaolinitic matrix. Sinking of the freed cassiterite grains in the soil to a position above the weathered bedrock sometimes allowed sufficient concentration to form eluvial placers. Down-slope creep of the regolith, assisted by seasonal heavy rainfall, led to the formation of colluvial placers, generally found as superficial cover on stream divides, at less than a few hundred metres from the primary source. Iron-cemented tin-bearing conglomerates are also formed by this process.

The most useful classification of tin placers is that adopted in Indonesia (Taylor, 1986). *Kulit* deposits result from *in-situ* weathering, with an important locality at Gopeng in the Kinta Valley. Cassiterite in shale interbeds in the thick limestone sequence was concentrated into the lateritic clay left behind by karstic weathering of the limestone.

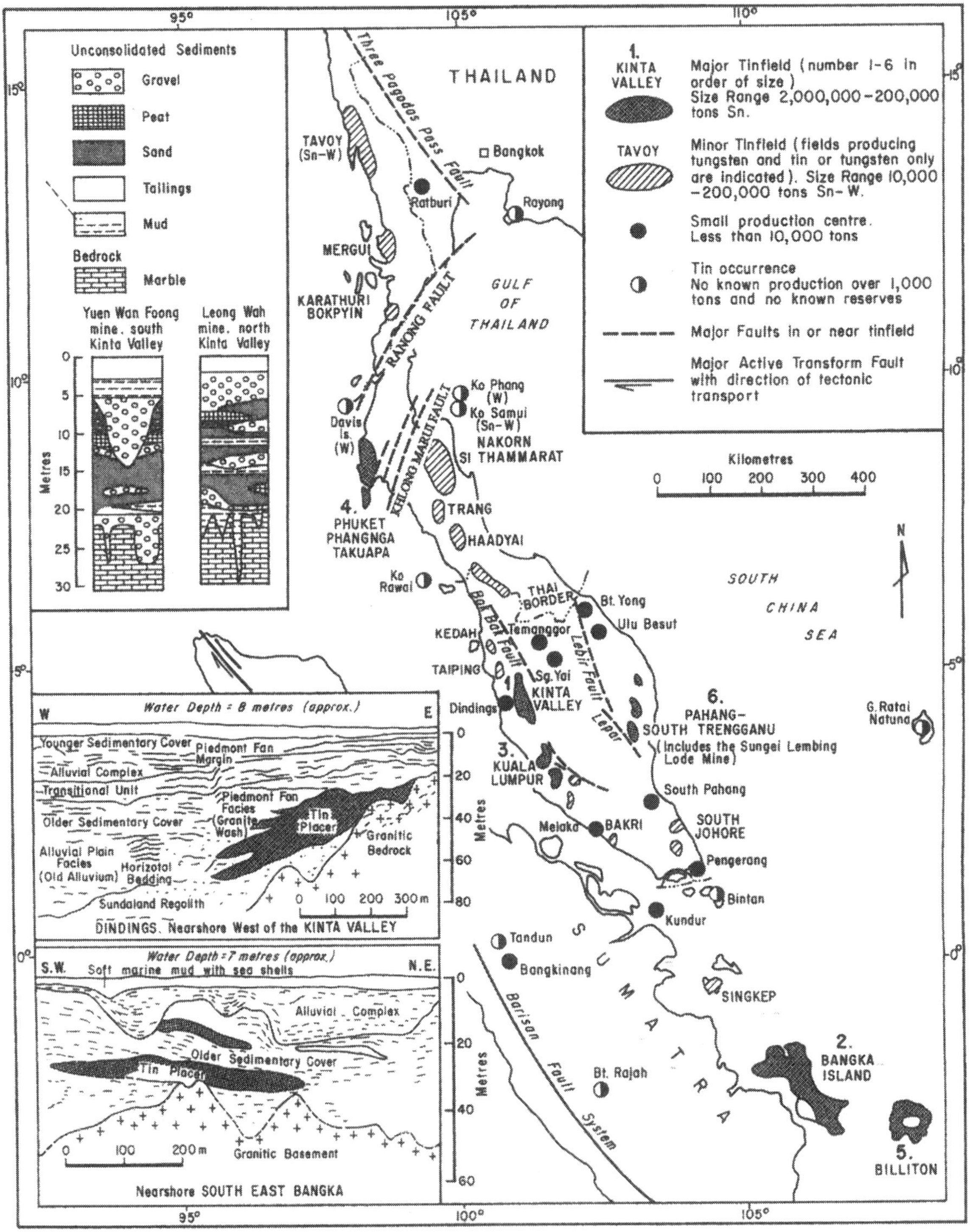

Fig. 12.1. Distribution of South-East Asian tin-fields, which are predominantly placer-type. After Hutchison (1983*a*). Inset (**upper**): mapped details of two mines; (**lower**): seismic sections of two areas.

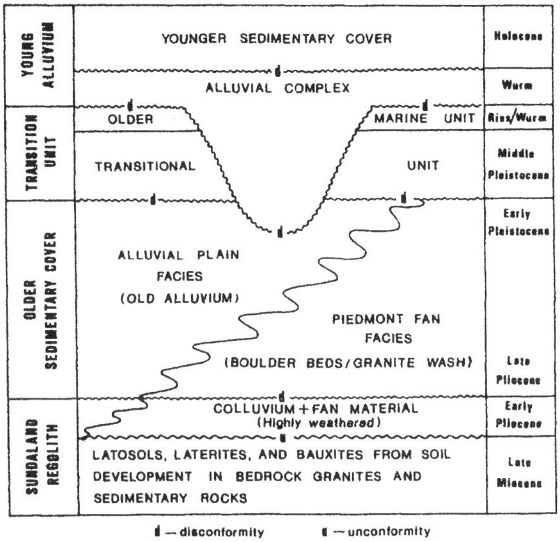

Fig. 12.2. Regional Upper Cainozoic stratigraphic relationships in Sundaland, South-East Asia, used for the classification of the important cassiterite placers (after Batchelor, 1979).

As sea level continued to rise and the size of the Sundaland continental land-mass decreased, increasing rainfall led to gullying, especially along more easily erodable fault and mineralized zones in the bedrock. Major piedmont fan placers were thus formed in an oxidizing environment. Scrivenor (1928) called these 'torrential deposits', and they are sometimes called 'granite wash', characterized by extremely poor sorting. This granite wash contains some of the highest heavy-mineral contents (Sivam, 1969). The piedmont fan placers form massive discontinuous wedges along valley sides, thickening and dipping away from the bedrock at angles from 5 to 30°. They have the greatest economic importance and include the Western Boulder Clays of the Kinta Valley, which lie adjacent to the granite contacts. They are extremely rich in cassiterite, but are of patchy occurrence and lie less than 1 km from their primary provenance. The cassiterite is coarse grained, as it is also in the eluvial and colluvial placers, and the crystals are angular with some faces preserved. There is a tendency for higher heavy-mineral contents to be associated with the sediments of coarser grain size.

Excellent examples of these mass flow boulder bed deposits were mined in the Gopeng area of the Kinta Valley and at Ranong in South Thailand. Another occurred in the Khao Yai area near Sichon, south Thailand (Taylor, 1986).

Braided streams began to extend across the coalescing fans. This resulted in enrichment of the piedmont placers by removal of the clay and sand that accumulated down-slope in broad basement troughs. Increased precipitation near the end of the Lower Pleistocene resulted in entrenchment of perennial braided streams. Their head-wards erosion into the bedrock opened up new sources of primary deposits for erosion. During down-working of the streams, the stanniferous gravels remained *in situ* but were subjected to further concentration by elutriation to form the exceedingly rich *kaksa* (Indonesian) or *karang* (Malaysian) placers, which lie as a thin layer on bedrock piedmont fans (Aleva, 1973). These also contain angular coarse-grained cassiterite. Gradients of 1 in 200 allowed cassiterite dispersion but gradients as low as 1 in 400 allowed concentration by selective fluviatile transportation to form braided stream placers, with transportation up to 5 km from the source, especially where there was a smooth bottom to the streams. They show a marked decrease of cassiterite grade and grain size downstream, but they have greater thicknesses and a wider and more uniform distribution than the piedmont fan deposits.

Following planation and lateritization of the Old Alluvium surface, late Middle Pleistocene sea levels rose to form a great expanse of shallow seas across the Sunda Shelf (Fig. 12.2). Increased precipitation gave rise to rivers with high discharge volumes, forming a second major phase of braided stream development. The material washed away includes the finer cassiterite, which may be trapped downstream to form *mintjan* or large *padang* placers of gigantic width and extent up to 35 km from the primary source (Taylor, 1986). It is likely that many, if not all of the large high grade deposits have been subjected to several cycles of reworking as a result of rising sea levels, but in most cases the latest cycle has effectively destroyed the earlier history.

Slumping of the old placers into solution hollows in limestone bedrock gave rise to some cave placers. Streams flowing through underground karst river systems filled some large caves with coarse current-bedded alluvium containing rich tin placers, exemplified by Kaki Bukit in north-west Peninsular Malaysia (Stauffer, 1973*b*).

Two major marine transgressions 120 000 and 10 000 years ago resulted in elutriation and enrichment by wave action and tides, and long-shore currents redistributed the coastal placers. Rich beach placers were formed along relatively stationary strand lines. The Thai deposits of Tongkah Harbour, Phuket, may be of this type.

Economic meander stream placers in the Younger Alluvium are rare, except on Bangka island (Fig. 12.1). They formed as channel deposits, or point bar deposits on the inside of meanders, especially after the minimum sea level was attained 17 000 years ago (Batchelor, 1979).

Taylor (1986) summarized the grain size characteristics of cassiterite in placer deposits. Cassiterite grain sizes in excess of 300 μm resist transport and accumulate as a residuum with angular gravels to form *kaksa* deposits. Such grains do not travel more than 6 km from the primary provenance. Cassiterite of grain size in the range 300 μm to 75 μm travels with the sand fraction to be deposited as *mintjan* placers, with known transport distances from 10 to 35 km from the primary provenance. Cassiterite finer than 75 μm is lost from the fluvial system.

Placers of the three tin belts vary in a way which reflects the placer generating potential of the primary mineralization and the topography. The western Thai fields of Phuket and Renong have yielded very rich *kaksa* and *kulit* deposits in the immediate vicinity of the granites. In the Main Range belt of Peninsular Malaysia, both *kaksa* and *mintjan* placers have been important producers. More than 40% of the total alluvial tin, produced or remaining, in Selangor was contained in *mintjan* placers located more than 10 km from the primary source (Taylor, 1986). Alluvial tin production from the Eastern Belt has been much lower than from the Main Range fields or from Bangka. This is because of the finer grain size of the largely exo-granitic primary mineralization of the eastern belt. An important exception was the Gambang tin field, where rich *kaksa* placers were formed on and near the mineralized granites. The lower tin-bearing alluvial levels are dated by their contents of tektites which showered in 0.7 Ma ago (Stauffer, 1978).

12.2 PRIMARY TIN FIELDS

Much has been written about tin metallogenic belts and the maps presented often did not pay attention to the regional tectonic fabric. Hosking (1970*b*) made a suggestion that the main SE Asian tin-belt from Billiton, through the Malay Peninsula, to Mawchi in Myanmar, may bifurcate with a branch through the Gulf of Thailand towards Rayong extending into Indochina and SE China. Hutchison (1988*a*) showed that such belts should not be drawn on maps, and that the main belt through Billiton and Mawchi is totally independent from that of SE China.

The two major fields in the tin-belt proper with significant deep mined production, Billiton island and Pahang–Trengganu, show a number of common features and each has one major underground mining area. The mines are at Sungei Lembing near Kuantan, and Kelapa Kampit in Billiton. The lodes are developed in contact metamorphosed Carbo-Permian argillaceous sedimentary and intercalated andesitic tuffaceous rocks within the aureoles around high level Permian to Triassic granitoids.

In both Billiton and the East Coast Belt there occurs mineralization of magnetite, pyrite/pyrrhotite and cassiterite, which appears to be interbedded in the clastic sedimentary sequence. The so-called 'bedded lodes' of Kelapa Kampit (Adam, 1960) are of this type, as are the massive 'skarn' ores of Bukit Besi, Trengganu. While these ores have been traditionally regarded as replacement deposits related to the granite, they have been suggested to be volcanogenic-exhalative (Hutchinson, 1981, 1986). Similar iron-rich skarn beds are also present at Pelapah Kanan, but they have been later intruded by granite and cut by a network of cassiterite-feldspar-quartz veins.

A wide variety of primary lode deposits is known from the other major tin fields (Hosking, 1970*a*, 1977, 1979) but, with few exceptions, production has been insignificant. Even where substantial output has been obtained, as at Mawchi in Myanmar or Pemali in Bangka, the deposits have very limited depth extension with no through-going vein systems into the country rocks. Mineralization is usually closely confined to the granite-sediment contact zones with little continuation in depth into the granite. Stockworks of minor veins and stringers are frequently developed in argillaceous rocks, but bulk grades are usually so low that only the weathered portions can be worked by hydraulic methods. This is normally done as an extension of *kaksa* placer mining, and the production of such mines is traditionally recorded as wholly alluvial, as at the Tekka Mine, Kinta Valley, near Ipoh.

Though individually small, the aggregate amount of tin in the vein swarms and stockworks of the contact zones is very large, and Quaternary erosion has concentrated the cassiterite into rich alluvial tinfields.

12.2.1 Mineralogical associations

Hosking (1979) set out the different mineral and metal associations and showed that certain associations have limited regional distribution:

1. The association of cassiterite with pyrrhotite and magnetite (Sn-Fe) occurs in bedded or skarn ore, with examples at Bukit Besi in Trengganu and at Batu Besi in Billiton.
2. The association of cassiterite with chlorite (Sn-Cu) occurs in the major vein deposit of Sungei Lembing in Pahang. Associations 1 and 2 characterize the Malaysian Eastern Belt and Billiton.
3. An association of cassiterite with malayaite (and varlamoffite) (Sn-Ca) forms skarn deposits at Pinyok on the Thai Border and at Chenderiang in the Kinta

Valley. Malayaite is of very restricted occurrence in South-East Asia.

4. An association of cassiterite with stannite and complex sulphides (Sn-Cu-Pb-Sb) forms xenothermal veins at Lahat–Menglembu in the Kinta Valley.
5. The association of cassiterite with lepidolite (Sn-Li-Ta) occurs in pegmatites, well known at Reung Kiet in Phangnga, south Thailand. Lepidolite pegmatites are scarce elsewhere.
6. The association of cassiterite with wolframite (Sn-W) forms vein swarms, with an excellent example at Mawchi in Myanmar and Tikus in Billiton. The general association of tin and tungsten is widespread, but the closeness of the association varies widely. Generally on a local scale the tungsten and tin mineralizations exist separately. The closest association is at Mawchi, where both wolframite and cassiterite exist in the same quartz veins. This phenomenon is also characteristic of the rich wolframite-quartz veins of the Xihuashan region of the Nan Ling Ranges of southern China (Liu,1982). However in the rest of South-East Asia, such a close association is remarkably rare.
7. The association of cassiterite with columbite (Sn-Nb-Ta) is found in pegmatites in the Gunung Jerai area of Kedah and the Bakri area of Johore. Cassiterite from these fields is extremely pleochroic (Hosking, 1973). Significant contents of columbite-tantalite are also reported from Phuket and Phangnga (Garson *et al.*, 1975). Elsewhere Nb-Ta minerals are rare.
8. The association of cassiterite with arsenopyrite and pyrite (Sn-As-Fe) occurs in stockworks at Labu on the Thai border.
9. The association of cassiterite with complex polymetallic sulphides and sulpho-salts (Sn, Cu, Pb, Fe, Zn, W, S, with minor Mo, Bi, Be, In, Ag, As) occurs as skarn deposits and associated vein systems in the south China province (Moh, 1986; Chen and Wang, 1988). The two giant hard-rock mining districts are at Gejiu and Dachang, and deposits of this kind are generally unknown elsewhere in South-East Asia. A few small occurrences of similar mineralogy are worked on a small scale in the Yala district of South Thailand, where massive sulphides with Pb, Zn and minor Cu and Sn, capped by iron-rich gossans, occur in the contact zone between granite and limestone.
10. The association of cassiterite with ilmenite, zircon and monazite (Sn-Fe-Ti-Zr-Ce-Yt) is represented by the alluvial placer deposits of South-East Asia with the best examples in the Kinta Valley and Kuala Lumpur tinfields (Fig. 12.1). After the cassiterite has been separated from the placer material, the heavy mineral concentrate (*amang*) is separated into its various components by a number of *amang* factories. Economic minerals recovered are ilmenite, zircon, xenotime, monazite and struverite. The predominant heavy mineral is ilmenite, indicating the universal association of cassiterite with ilmenite-series S-type granites (Ishihara, 1977).

12.2.2 Major faulting in and around the tin fields

A number of the more important mining fields are bounded or transected by faults, some of which are of regional significance (Fig. 12.1). The south China tin metallogenic province is abruptly terminated by the right-lateral strike-slip Red River Fault, only a short distance south of the great Gejiu tin field (Fig. 12.1).

The Phuket–Phangnga field terminates abruptly to the east against the powerful Khlong Marui Fault (Garson *et al.*, 1975). The sub-parallel Phangnga Fault cuts through some of the stanniferous granites and is itself mineralized by lepidolite pegmatites indicating that some of the faulting is pre-mineralization (Fig. 12.1).

The great tin field of the Kinta Valley terminates abruptly on both the east and west sides along roughly N–S fracture zones which bisect the bounding granites. The Kuala Lumpur tin field is transected by a complex suite of sinistral wrench-rotational faults. These are thought to be young and to post-date the mineralization, and massive poly-phase quartz dykes intruded into the granites along these faults, for example the Klang Gates quartz dyke, are barren of tin mineralization.

12.3 TIN DEPOSITS OF SOUTH CHINA

South China is not only the world's greatest tungsten metallogenic province, but is also an outstanding region for tin (Fig. 11.1). The distributions may be spatially separate but commonly overlap in south Jiangxi (Xu and Zhu, 1988). Although there is a tin association with Proterozoic, 'Caledonian' and 'Hercynian' magmatic rocks, the paramount association is with Yenshanian crustal anatectic (S-type) granites. The tin deposits of SE China have been described by a number of authors in Hutchison (1988*b*).

The famous wolframite-bearing quartz veins of the Xihuashan district of Jiangxi (Liu, 1982) commonly contain economic amounts of euhedral cassiterite. The formation of these quartz lodes is clearly related to Yenshanian granites (predominantly Cretaceous). Much more economically significant for tin and more difficult to decipher are the tin-polymetallic sulphide deposits, exemplified by Gejiu[1] (Peng and Cheng, 1988) and Dachang (Chen *et al.*, 1988), which are respectively the first and second largest hard-rock deposits of cassiterite in the world. The proven reserves of Dachang are estimated at nearly 100 Mt of ore of grade between 0.5 and 1.0 % Sn, whereas the ancient mining district of Gejiu has already produced 1 Mt of Sn metal. Both these mining districts have similarities to the hard-rock deposits of Tasmania in their stratabound and mineralogical characteristics, but remarkably no similar deposits have been found elsewhere in South-East Asia, despite the fact that the great alluvial tinfields of the Kinta Valley and Kuala Lumpur areas occur in similar limestone country.

12.3.1 Gejiu[1]

A 3 km thick sequence of Middle Triassic limestone (Gejiu Formation) has been folded into a NNE-trending Laochang anticline and intruded by Middle to Upper Yenshanian porphyritic biotite granite (115 to 84 Ma by K:Ar) and equigranular biotite granite (80 to 64 Ma) as described by Wang (1983) and Peng and Cheng (1988). The granite lies between 200 m to over 1 km beneath the mining surface. The granites are S-type, rich in monazite and xenotime. About 10 giant to medium sized deposits have been found, of which the Laochang (Fig. 12.3) is the largest. The deposits occupy the anticlinal axis and are spread out in a NNE direction over a distance of 20 km.

The mineralization may be subdivided into four types (Fig. 12.3): contact zone, intra-formational concordant, intra-formational discordant vein type, and placer. Silicate mineralization formed at temperatures of 425° C, oxide at 238 to 358°, sulphide at 277 to 386°, and carbonate mineralization at 290 to 305° (Wang, 1983).

The major tin mineral is cassiterite, together with minor stannite. The cassiterite is accompanied by a large variety of sulphides and sulpho-salts, notably of Pb, Zn, Cu, Mo, Fe, and there is also an association with W, Bi, Be, In, Ag and F. The four main mineral assemblages are:

a. Li-Be-Nb-Ta-REE-Sn occurring in pegmatite-type deposits;
b. W-Sn occurring in quartz veins;
c. Fe-Sn occurring in tin skarns;
d. Sn-Cu-Pb-Zn-Sb sulphide assemblages.

In the first three, tin is subordinate, but it is the chief element in the sulphide deposits. There are also tourmaline-cassiterite veinlet zones and cassiterite greisens.

12.3.2 Dachang

In the Dachang ore-field, a thick sequence of shallow marine Devonian to Carboniferous platform limestone, with thin shale and important stromatolite horizons, was folded in late Triassic time (Indosinian Orogeny) and intruded by the Middle to Upper Yenshanian S-type Longxianggei biotite granite (Fig. 12.3), dated 77 to 138 Ma by K:Ar and 149 Ma by Rb:Sr. The younger range may represent hydrothermal alteration, but the mining area clearly contains poly-phase granites with mutually intrusive relationships (Chen *et al.*, 1988). The Dachang ore-field contains approximately 110 Mt of ore at 1% Sn, 3 to 5% combined Cu, Pb, Zn and Sb, and 100 to 300 ppm Ag. It was discovered in the 1950s, but mining of the outcropping fissure epithermal veins goes back at least 300 years.

The mineralization is subdivided into the three distinctly different types described below.

12.3.2.1 Cassiterite-poly-metallic sulphides

This kind of deposit forms very large, massive, often strata-bound and sub-horizontal ore bodies, at least partially of the carbonate replacement type. They appear to be easily mineable by modern bulk-mining methods. The metallurgical problems have been solved. Practically all the Sn produced from Dachang comes from these sources. The main ore minerals are: fine-grained cassiterite, pyrrhotite, marmatite, arsenopyrite, jamesonite, franckeite, chalcopyrite, galena, stibnite, fluorite and marcasite. The associations may be further subdivided into five: cassiterite-sulphides, tin-bearing sulphide-sulphosalt-carbonate, stibnite-quartz-calcite, realgar-orpiment-calcite, and cinnabar-pyrite-quartz. These associations represent a range of deposition from 450 to 250° C (Chen *et al.*, 1988).

12.3.2.2 Skarn-type

The proximal skarn ore bodies are of intermediate size, all within the granite metamorphic aureoles. They contain garnet and pyroxene. Their economic significance seems to be subordinate to the former category. Their exploration is difficult because of irregular geometry. The major ore minerals are cassiterite, marmatite, chalcopyrite, arsenopyrite, scheelite, molybdenite, native bismuth, fluorite, calcite, wollastonite and andradite, whose Sn content attains several parts per thousand. Fluid inclusion studies show that the temperature of formation ranged from 600 to 450° C (Fu *et al.*, 1993).

[1] Readers unfamiliar with the new preferred English transliteration of Chinese names may know this as Kochiu

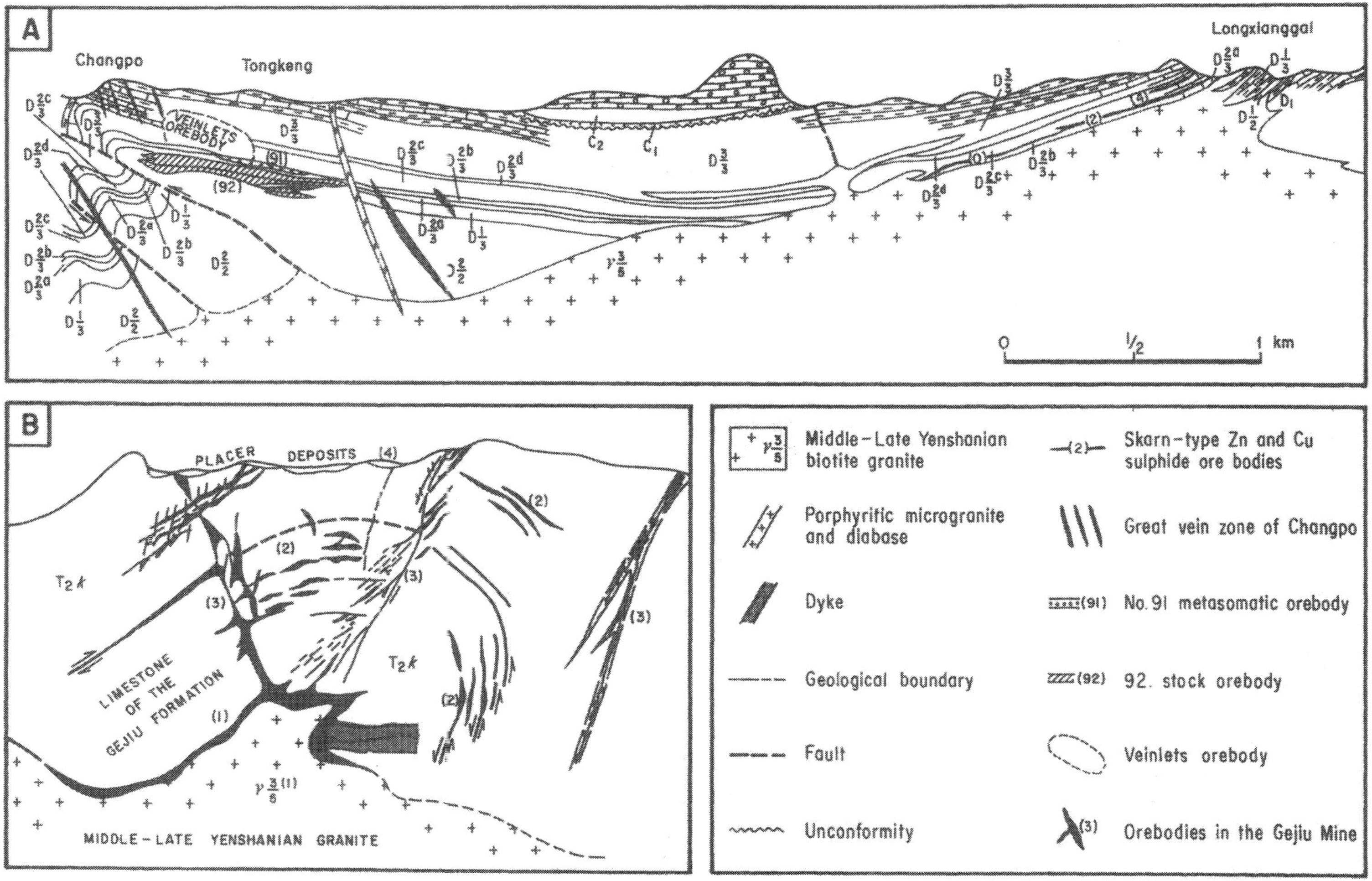

Fig. 12.3. The two great hard-rock tin-fields of southern China. **A:** Cross section through the Dachang field. C_1 = Carboniferous limestone, C_2 = shale and sandstone, $D_{2\backslash3}$ = Devonian limestone-shale sequence, $D_{2\backslash3d}$ = Mega-lenticular limestone, $D_{2\backslash3c}$ = Mini-lenticular limestone, $D_{2\backslash3b}$ = Fine banded stromatolitic limestone, $D_{2\backslash3a}$ = Thick banded limestone, $D_{1\backslash3}$ = Siliceous shale, $D_{2\backslash2}$ = Marl, $D_{1\backslash2}$ = Calcareous shale, D_1 = Argillaceous layer (after Chen *et al.*, 1988). **B:** Cross section through the Gejiu field. T_{2k} = Triassic Gejiu Formation limestone. Ore types are 1 = Contact zone type, 2 = Intra-formational concordant type, 3 = Intra-formational discordant vein type, 4 = placer type (after Peng and Cheng, 1988).

12.3.2.3 Tungsten–antimony type

Vein deposits are predominant and the mines are comparatively small. The veins have sub-vertical geometry. These are truly epithermal veins, occurring farthest from the granite of all the Dachang deposits (Fig. 12.3). The lodes are characterized by banded chalcedony and fluorite, which indicate a depositional temperature range of 180 to 150° C. The main mineralogy of the quartz veins is scheelite, wolframite, stibnite, berthierite, marmatite, galena, chalcopyrite, pyrite, and fluorite.

12.3.2.4 The Changpo–Tongkeng deposits

These form the largest and most representative of the Dachang field (Fig. 12.3). There are six mineralization zones:

1. The deep ore is represented by veinlet and disseminated sulphides in strongly skarnized and feldspathized Middle Devonian strata. It yields pyrrhotite, marmatite and arsenopyrite, superimposed by veinlets of jamesonite.
2. Interstratified stockworks and layers of veinlets of the No. 92 orebody (Fig. 12.3) occur within siliceous shale at the base of the Upper Devonian. The stockwork and veinlets are of the cassiterite-quartz-sulphide assemblage and in addition veinlets of late stage cassiterite-sulpho-salt-sulphide cut across the bedding plane lodes.
3. Stratiform No. 91 orebody in Upper Devonian striped limestone consists of cassiterite-arsenopyrite-marmatite-pyrite-quartz-potash feldspar-tourmaline and a later pyrrhotite-marmatite-muscovite-tourmaline-daphnite assemblage. The stockwork and veinlets are of the same composition. Also various Bi-Ag and Cu-rich late stage sulphide-sulphosalt veinlets are commonly superimposed on stibnite-quartz-calcite veins.
4. Veinlet orebodies occur above the stratiform orebody. They are of different composition. On both flanks of the veinlets are cassiterite and sulphide disseminations in the tourmalinized country rocks.
5. Large lodes occur in Upper Devonian sequences, controlled by NE-trending tensional features in the transverse anticline. The lodes are of cassiterite, marmatite, pyrite, pyrrhotite, arsenopyrite, jamesonite and manganocalcite. The country rocks are strongly silicified and tourmalinized. Later veins of stibnite and franckeite cut the larger lodes.
6. At the top of the deposit, in Upper Devonian strata, are minor fissure epithermal veins, with an important tungsten association.

The hydrothermal mineralization-alteration is broadly zoned outwards and upwards from the granite, and fluid inclusion studies indicate a progressively decreasing temperature of formation away from the granite, consistent with the S-type granite being the source of the mineralization. The single high-angled fault played an important role in channeling the mineralizing fluids (Fu *et al.*, 1993), which were deposited as follows:

- proximal ore skarn zone containing andradite garnet and pyroxene, deposition within the range 450–600° C. The garnet incorporates Sn in its structure;
- tourmaline silicified zone, deposition within the range 350–450° C;
- sericite ± cassiterite zone, deposition within the range 250–350° C;
- carbonate (sulpho-salt) vein zone, deposition at a temperature < 250° C.

Various isotope studies indicate an Upper Mesozoic age for the ore, proving its genetic relationship to the Yenshanian biotite granite. However, the model age of lead isotopes in the jamesonite is Devonian. Both the presence of methane in fluid inclusions and organic relicts in the ore are indicative of a syn-sedimentary origin. The common occurrence of stromatolitic layers in the Middle Devonian limestones and other sedimentological features indicate extremely shallow water conditions of deposition, and it is therefore most likely that the sulphur and perhaps iron were enriched in the Devonian strata. Continued research is required to consolidate the theory of a syngenetic origin of the limestone-hosted stratabound-type of sulphide-cassiterite mineralization at Dachang. However, the present configuration and mineralogy of the ore-bodies has been determined by the Yenshanian biotite granite and its hydrothermal system. It is reasonable at this stage to conclude that some of the ore components date back to the Devonian, while others were contributed by the Cretaceous granites. However the richness in sphalerite may also suggest a Mississippi Valley-type of mineralization, which could have been localized any time between the Devonian and the Cretaceous.

12.4 THAILAND

The various tin-granitoid belts continue northwards from Peninsular Malaysia into Thailand. The Eastern Belt continues north of the Gulf of Thailand in the Chanthaburi province near the Cambodian border. However, this eastern belt disappears northwards within a short distance beneath the Khorat Basin. Granitoids in this belt have been dated Upper Cretaceous (Cobbing *et al.*, 1992). Cassiterite occurs in streams and beach placers (Tantisukrit, 1978).

The Main Range Belt of Peninsular Malaysia continues northwards through peninsular Thailand, where the granites, as in Malaysia, have been dated Upper Triassic (Cobbing *et al.*, 1992). The belt extends north of the Gulf of Thailand to the Rayong district, SE of Bangkok. A large number of greisen-bordered hydrothermal quartz vein systems have been mined throughout the province (Tantisukrit, 1978).

The Triassic granite belt continues northwards to the Lampang, Chiengmai and Chiengrai district of NW Thailand, where hydrothermal vein and skarn deposits occur at the granite contact zones (Tantisukrit, 1978). The size of these deposits, however, is much smaller than in peninsular Thailand, which is the main tin-mining province of the country. It has been demonstrated by Lehmann and Mahawat (1989) that the tin deposits of west central Thailand are genetically related to the most highly differentiated granites. However, hydrothermal fluid-rock interaction in the highly fractionated low-oxygen-fugacity granitic rocks may produce a highly efficient removal of tin, leading to anomalously low Sn background values in the host granites. Therefore the granites that have been affected by large-scale hydrothermal tin depletion may be regarded as good indicators when searching for nearby tin deposits.

The unique feature of Thailand, not found in Peninsular Malaysia, is the Upper Cretaceous Western tin-granite belt through Phuket and Ranong, described below. Granites of this belt are of mixed I and S type and have been dated by the $^{40}Ar/^{39}Ar$ method within the range 50 to 88 Ma (Charusiri *et al.*, 1993). This Upper Cretaceous granite belt is separated from the Upper Triassic Main Range Belt by the major NE-trending Ban Kram fault zone.

12.4.1 Phuket

The island of Phuket represents the southern extremity of an important mineralized belt which extends southwards from Mandalay, along the western margin of the Shan Highlands and along the Tenasserim coast of Myanmar. The belt is characterized by epizonal Upper Cretaceous to Eocene granites of mixed I and S-type affinities (Cobbing *et al.*, 1992). Whereas in Myanmar, to the north, tungsten is very important to dominant, it is of very minor importance in Phuket, which represents one of the major tin mining districts of SE Asia (Fig. 12.1).

The granites of Phuket island have been dated by Putthapiban and Gray (1983) by Rb:Sr isochrons. The Kata Beach granite gives an age of 98 ± 7 Ma with an initial $^{87}Sr/^{86}Sr$ ratio of 0.7193. The Ao Nai Thon granite was dated 94 ± 12 Ma with an initial ratio of 0.7277. The granite at Khao Tosar gave an age of 84 ± 1 Ma with a ratio of 0.7406 and the Khao Prathiu granite gave an age of 82 ± 4 Ma with an initial Sr isotope ratio of 0.7144. All of these Upper Cretaceous granites are therefore derived from continental crust, as also shown by their high ε Sr and high (negative) ε Nd values (Cobbing *et al.*, 1992).

Most of the tin production in the three mining districts of Phuket, Phangnga and Takua Pa is derived from Quaternary alluvial deposits. The peak of production was reached in 1968. In terms of total production from 1871 to 1970, Phuket's tin-in-concentrates production amounted to 166 700 t, Takua Pa 111 175 and Phangnga 75 776. For this period, the total Thai production was 638 326 t. In the early years the tin was locally smelted, then later sent to Penang, but from 1965 onwards it was all processed at Phuket. Associated minerals are monazite, xenotime, and columbite-tantalite concentrates, with 1971–1990 production respectively 3 486 275 and 1272 long tons (Nakapadungrat and Maneenai, 1993).

The Tongkah Harbour and Aokam Tin sea-going dredges worked rich alluvial deposits in the bays around Phuket island (see also Chapter 15—diamonds). The alluvium is covered by a thin layer of marine mud, indicating that the region was all formerly above sea level and has been drowned. This is obvious by the great number of karstic Ratburi Limestone hills which rise from the sea at Phangnga. The alluvium dredged consists of clay, sand and coarse gravel, carrying boulders of granite. Onland, mining of the alluvium was carried out mostly by gravel pumping and hydraulic mining. Associated with the cassiterite in the alluvium are other heavy minerals of commercial value: ilmenite, niobium/tantalum minerals, monazite and xenotime. Tantalum was formerly smelted at Phuket until the plant was closed because of labour trouble.

It is possible that most of the cassiterite worked from the alluvium was originally derived from erosion of stockwork zones at the margins between the Phuket Group and the Upper Cretaceous biotite granite plutons. These zones extend up to 2 km from the granite contact and consist of a hornfels zone containing a network of narrow veins, dykes and sheets of quartz, aplite, pegmatite and fine-grained granite (Garson *et al.*, 1975). The stockwork zones are also mined in pits by gravel pumping where the contact zone has been deeply weathered.

Pegmatites have been mined. Cassiterite occurs in mica-tourmaline pegmatites and in lepidolite pegmatites. The grade of the latter averages about 0.1 to 0.2% Sn (Garson *et al.*, 1975). The cassiterite is associated with Li and Ta-Nb minerals (Nakapadungrat and Maneenai, 1993). The pegmatite mines occur in Takua Pa (Khao

Lam Ru, Khao Kata Khwam and Khao Khanim), Phangnga (Khao Po) and Phuket (Kathu Valley). They intrude both country rocks and the granite. The pegmatites occur as swarms of dykes of width variable from 1 to 15 m. Except for Khao Kata Khwam, where the pegmatite dykes are oriented NW–SE, all others are oriented parallel to the major NE–SW Khlong Marui and Lam Ru faults (Nakapadungrat and Maneenai, 1993).

Another type of mineralization is as argillic disseminations, found in the Nok Hook mining district. The Nok Hook tourmaline-muscovite granite hosts the mineralization within its marginal zones. Cassiterite and Ta-Nb minerals occupy the muscovite rich (greisenized) marginal zones of specialized granites.

12.5 MALAYSIA

Malaysia used to dominate the world production of tin and the kinds of its deposits have been summarized by Hosking (1973, 1988). The industry is now decimated and tin dredges have vanished from the landscape of Kuala Lumpur and the Kinta Valley. The primary deposits are spatially and genetically related to S-type ilmenite-series granites, and divided by Scrivenor (1928) into the Western and Eastern belts (Fig. 12.4). To avoid confusion with Thailand, the Malaysian 'Western Belt' should be referred to as the 'Main Range Belt'.

The Main Range belt, which is composed of the Triassic Main Range Granite Batholith (Cobbing *et al.*, 1992), is overwhelmingly the most important in terms of the number and importance of mines and contiguous placer fields. The primary deposits may be subdivided into the following:

12.5.1 Hydrothermal veins and lodes

This class is the most significant source of primary tin, and by erosion, of placer tin. Wall-rock alteration in the form of greisenization, tourmalinization, haematitization, kaolinitization and silicification is common.

12.5.1.1 Cornish-type lodes (Sungei Lembing)

These extensive lode systems are rare, and only the Sungei Lembing area of the Eastern Belt provided a large underground mine, which operated successfully from the 1880s until 1987.

The mine is situated about 20 km west of Kuantan, and its geology has been described by Fitch (1952). A thick monotonous sequence of Viséan to basal Namurian shale with subordinate siltstone and sandstone generally dips 30° to the ENE. The strata are fossiliferous and ripple-marked. At the mine location, the rock sequence is intruded by a ridge of Permian granite (Hutchison, 1983*a*). The thermal aureole around the granite contains chiastolite (Fitch, 1952). At its peak in the 1970s the mine had workings to a depth of 340 m. The recent slump in the tin price caused its closure in 1987, but it has been revitalized on a restricted scale.

The tin lodes are steeply dipping mineralized faults that have been repeatedly re-opened. The principal lodes strike due E and also ESE and the common dips are 70 to 90°. However, a few have dips as shallow as 20°. Most lodes lie about 600 m from the top of the granite (Fig. 12.5). As in Cornwall, SW England, the mineralization does not extend down into the granite, although the faults and quartz veins continue downwards. The zone of mineralization occupies a band about 600 m wide that lies parallel to the granite contact (Fig. 12.5). There is a sharp distinction between this Eastern Belt style of mineralization and that of the Main Range, in which the lodes occur impounded within the contact zone of the granite cupolas.

Much the same type of mineralization occurs down-dip for at least 600 m, indicating that the temperature gradient was very gentle and that meteoric water circulating through the country rocks played a dominant role. The lack of obvious zoning in individual lodes is a feature of all South-East Asian deposits (Hosking, 1970*a*) and in this they differ markedly from those of SW England. As in Cornwall, cassiterite of several generations can be recognized (Hosking, 1973). The broad paragenesis appears to be: cassiterite, followed by arsenopyrite, pyrite, pyrrhotite, sphalerite (with exsolved chalcopyrite), chalcopyrite, sphalerite (without exsolution) and galena. Quartz was introduced at various times. A large amount of iron-rich chlorite was formed immediately after the cassiterite. The frequent re-opening of the lodes led to a commonly brecciated texture. Further complications resulted from pre- and post-mineralization faulting, but the structural history of the Sungei Lembing mine was never satisfactorily resolved. In several cases, the lodes follow propylitized dacitic dykes along part of their extent, and these dykes have played a role in locating the lodes, yet the dykes themselves do not appear to be mineralized, presumably because they remained unfractured. Outcropping lodes have been oxidized and weathered down to as much as 300 m from the surface.

Fluid-inclusion studies (Foo, 1977) indicate that repeated opening of the lodes resulted in fluid boiling. The quartz deposition occurred in various stages, at 272° C, 180° C, 148° C and 70° C.

Wolframite has been produced nearby in the past at Kuala Sungei Perong (Fig. 12.5) and also from the

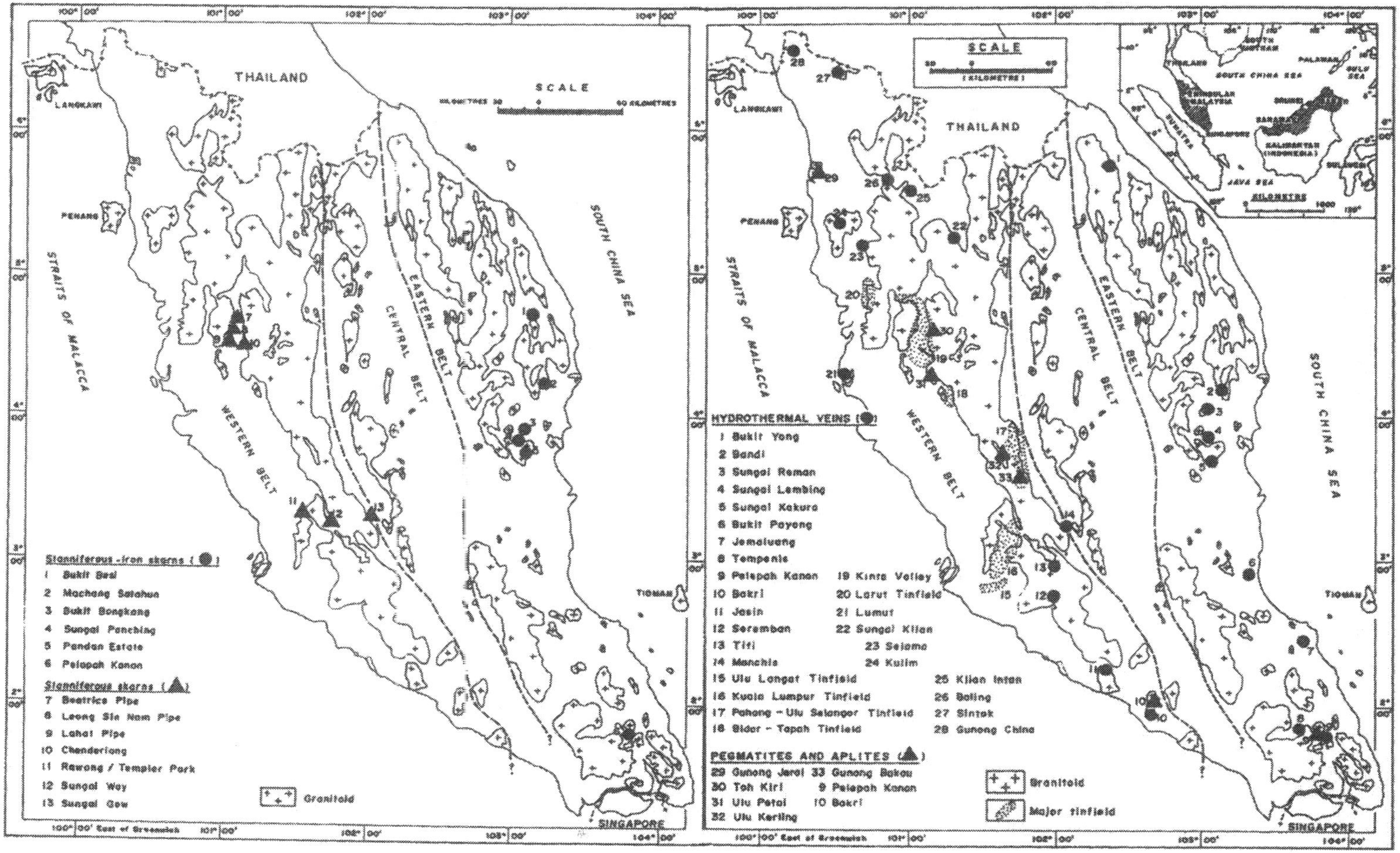

Fig. 12.4. Distribution of the most important primary tin deposits, which have been mined in Peninsular Malaysia (after Chu *et al.*, 1988). **Left**: tin-bearing skarns, **Right**: hydrothermal veins, pegmatites and aplites.

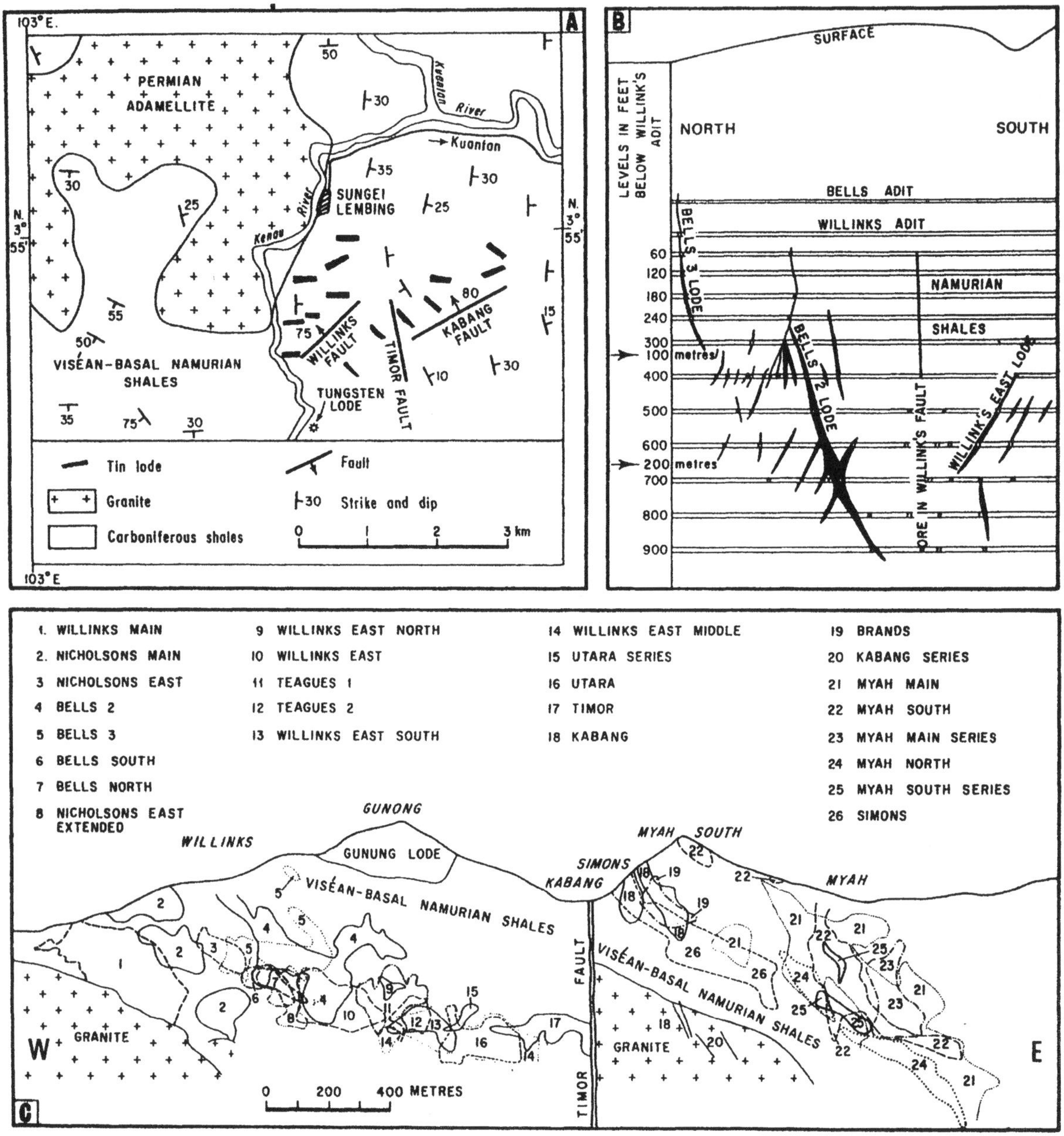

Fig. 12.5. The Sungei Lembing underground lode tin mine, Pahang, Malaysia (after Hutchison, 1983, based on Fitch, 1952 and Hosking, 1973). **A**: simplified map, **B**: vertical N–S section showing the Cornish-type lodes in relation to the Willinks Fault. **C**: E–W section with the lodes projected on to the plane of the section showing that the lodes lie parallel to and away from the granite contact.

granitic area at Ulu Sungei Reman, where it occurred in vein quartz along with tourmaline and chlorite (Fitch, 1952). The amount was always small compared with the cassiterite, and although occurring in the same general district, wolframite and cassiterite do not occur together in the same lode. This is in strong contrast to south China and Myanmar, where wolframite and cassiterite co-exist in the same lodes.

12.5.1.2 Veins, stringers and stockworks

These are more prolific and especially characteristic of the Main Range Belt (Fig. 12.4). The hydrothermal system was impounded close to the granite contact zone. The mineralization is commonly accompanied by greisenization in the form of muscovite-tourmaline and quartz-topaz assemblages (Chu *et al.*, 1988). Some of the best documented examples have been described from the renowned tin-fields of the Kinta Valley, Klian Intan, Kuala Lumpur, Ulu Langat and Ulu Selangor–Pahang Border areas of the Main Range (Hosking, 1973). Most of the cassiterite-bearing veins are small, generally less than 0.5 m in width, but in places as wide as 4 m.

The Kinta Valley, which was the world's foremost producer of tin, showed the greatest development of hydrothermal veining. A central zone, comprised of Lower Palaeozoic schist, phyllite and limestone, is bordered on the east by the Main Range Batholith, and on the west by granitoids of the Kledang Range (Fig. 11.3). Its mineralization has been summarized by Rajah (1979).

Several of the vein deposits of the Kinta Valley were mined by open cast methods. The Tekka deposit (Fig. 12.4) is the best known. The intra-granite cassiterite-bearing veins at Tekka occur in two vein swarms, separated by an elongated schist roof pendant. The N-dipping veins also extend into the schist. The veins have a mineralogical complexity. They demonstrate telescoping of both high and low temperature minerals (Teh, 1981). Associated with the cassiterite are scheelite, wolframite, stannite, varlamoffite, columbite-tantalite, galena, stibnite, haematite and fluorite.

Primary vein deposits around the Bujang Melaka Pluton of the Kinta Valley have been studied by Schwartz and Askury (1989). A low-salinity acid Sn-bearing aqueous fluid, partly mixed with CO_2, separated from the crystallizing granitic magma. The residual fluids interacted with the solidified granite, cooled, and deposited cassiterite in the 440° to 240° C range at an approximate pressure of 1 kbar. The crystallization of cassiterite was followed by arsenopyrite, pyrite, chalcopyrite, sphalerite, galena, Bi sulpho-salt, and native Bi. The style of mineralization is of greisen type in the muscovite granite within the pluton, and vein type along the contact zone with the country rocks.

At Klian Intan, the stockwork was mined by open cast methods. The E–W trending cassiterite-bearing veins were emplaced into phyllite. They vary from a few mm to more than 15 cm in width. The stockwork contains tourmaline, pyrite and minor galena, sphalerite, arsenopyrite, pyromorphite and native bismuth. Haematitization is widespread (Chu *et al.*, 1988).

Such low grade large tonnage deposits were amenable to open-cast mining. Because of the high degree of weathering, some vein, stringer and stockwork systems were mined by gravel pump methods, so the mines were classified as 'placer' in production records, rather than hard-rock.

12.5.2 Skarn mineralization

This style of mineralization occurs were the hydrothermal fluids have inter-reacted with limestone country rocks (Fig. 12.4). It has been subdivided by Hosking (1973) into 'stanniferous skarns', described here, and 'stanniferous-iron skarns', described in Chapter 14.

Stanniferous skarns occur mainly within the Main Range Belt (Fig. 12.4). Mineralization exists as stanniferous veins and pipe-like bodies in limestone which has been intruded by granite. The best known example was the Beatrice Pipe. It was a pipe-like body extending 150 m horizontally into limestone, then 90 m vertically, before being faulted-off (Ingham and Bradford, 1960). The deposit contained cassiterite, stannite, arsenopyrite, pyrite, pyrrhotite, molybdenite, bornite, tremolite, fluorite and fluoborite. The cassiterite occurred in bands parallel to the border of the pipe, and interstitially to arsenopyrite in the main body.

Another noteworthy example of skarn was the Leong Sin Nam Pipe, near Menglembu. Localized in dolomite, it consisted of cassiterite, arsenopyrite, pyrite, stannite, jamesonite, tennantite, tremolite and calcite. The nearby Lahat Pipe was similar. The mineral malayaite was not recorded in these pipes, but occurred in skarns at Rawang, Templer Park, Sungei Way and Sungei Gow (Alexander and Flinter, 1965).

12.5.3 Pegmatites and aplites

Cassiterite-bearing pegmatites and aplites are small and unzoned. A few have been mined for cassiterite. The best known are near Gunung Jerai, Kedah, and Bakri, Johore. At both localities, columbite-tantalite is associated with the cassiterite. Bradford (1972) records that the Gunung Jerai cassiterite contains a higher content of Ta than Nb.

At Gunung Jerai (Kedah Peak), the granite domed up and intruded Lower Palaeozoic meta-sediments. Pegmatites cut both the granite and the meta-sediments (Bradford, 1972) mostly in the southern margin. Quartz veins are more common to the north. Both pegmatites and quartz veins are tin-bearing, but columbite is more common in the pegmatites. The contiguous placer deposits contain cassiterite, columbite-tantalite, tourmaline, almandine garnet, muscovite, monazite, ferroan gahnite, topaz, betafite and complex Nb-Ta-Ti oxides (Bradford, 1972).

Farther south, stanniferous pegmatites and quartz-cassiterite veins cut the sedimentary rocks at Bakri (Fig. 12.4). The contiguous placers have been mined and columbite and monazite were important by-products.

12.6 INDONESIA

The South-East Asian tin province extends southwards from the Malay Peninsula, through the so-called 'tin islands' of Indonesia, to abruptly terminate at the island of Billiton (Fig. 12.1). But the mineralization on Billiton Island is great, varied, and unique.

12.6.1 Billiton

The island, also known as Belitung, is a topographically low peneplain with residual hills. Its tin production has been predominantly from placers, mined onland by gravel-pumps and offshore by sea-going dredges. In recent years its production has been around 5000 t/year. About 3/4 of the island is composed of Permo-Triassic sedimentary sequences with minor volcanic intercalations. The marine strata are of sandstone interbedded with claystone and siltstone, radiolarian chert, spilitic lavas and tuffs, but fine grained sandstone is predominant. Regional metamorphism is low to absent (Aleva, 1973; Adam, 1960).

These sedimentary rocks are intruded by Upper Triassic granite and minor gabbro. The largest granite batholith of Tanjung Pandan occupies the NW corner of the island and has been dated 215 ± 3 Ma with an initial $^{87}Sr/^{86}Sr$ ratio of 0.7140. Despite the richness in tin deposits, the main highly evolved ilmenite-series Tanjung Pandan granite contains only 2 to 4 ppm Sn. Lehmann and Harmanto (1990) deduced that this granite had been the source of much of Billiton's hydrothermal tin deposits, and has in the process become extensively depleted. There are several hard-rock primary tin deposits on Billiton, but only two have been selected because of their importance and contrast.

12.6.1.1 Tikus Mine

This mining area occurs in the centre of the Tanjung Pandan granite outcrop. Mining of the alluvial and eluvial deposits began in 1899. In 1907 a quartz body with visible cassiterite and wolframite was exposed in the bottom of a pit, and shaft sinking commenced in 1908, initially to mine greisen of 2.4% Sn. By 1915 the mine was well developed with reserves estimated at 600 000 t grading 0.45% Sn. Later the greisen was left in favour of the quartz body and in its last 3 years 583 t of Sn and W concentrates were produced. Flooding and severe loss of life led to closure in 1920. At the time of closure, the mine was estimated to have reserves of more than 54 000 t at 0.4% Sn and 0.5% W. Only the quartz was considered ore, and the greisen was no longer regarded economic. The geology is described by Wing-Easton (1937), and Suryono and Clarke (1982). Generally throughout South-East Asia, Sn and W mineralization occur closely in the same district but in separate mines. Tikus is an example where the two occur together in the same lode.

The mine deposit is in the form of a sub-vertical pipe of diameter 80 m wholly within the Tanjung Pandan biotite granite. The pipe core is of brecciated quartz, surrounded by a variable width of greisen. Both the quartz and greisen bodies contain numerous blocks and fragments of granite and aplite wall-rock, as well as of silicified and veined sandstone, probably resulting from stoping of roof rocks. Formation of the pipe appears to have involved several episodes of fracturing and mineralization. The quartz shows brecciation and multi-stage quartz growth. The greisen is a medium-grained siliceous rock containing 10 to 50% topaz, 1 to 5% of various metallic minerals and muscovite in a quartz matrix.

Cassiterite and wolframite occur as single crystals, random groups or large aggregrates in the quartz body. The cassiterite is of short prismatic habit, up to 3 cm long, and the wolframite is as aggregates up to 6 cm. The ore content of the greisen is lower. A suite of sulphides including pyrite, arsenopyrite, galena, and sphalerite is common in both the quartz and the greisen. These occur accompanying the later generation of quartz, which may cut through the earlier larger crystals of cassiterite and wolframite (Wing-Easton, 1937). It appears that the formation of the Tikus Pipe was controlled at the intersection of several fractures in the granite.

12.6.1.2 Kelapa Kampit

This is the most important primary deposit in Indonesia, for it has produced 25 000 t of Sn since 1906. Present day poduction comes from an open pit in the Nam Salu bedding-plane lode (Schwartz and Surjono, 1990). Formerly mining was underground (Adam, 1960).

The mine lies in NE Billiton (Belitung). It lies within the Permo-Triassic sedimentary formations, which are intruded by Upper Triassic biotite granite, granodiorite and gabbro. Meta-sandstone, meta-siltstone, phyllite and chert are the main country rocks The sequence at the mine strikes 110° to 130° and the strata dip nearly vertically. There are some dacite dykes in the mine area, but no granite outcrop is known in the vicinity.

Cassiterite lodes occur over an area of 3 × 1.5 km and were mined to a depth of 400 m below surface. They are of two distinct types: (a) bedding-plane lodes parallel to the sub-vertical bedding of the country rocks, and (b) discordant, cross-cutting fissure veins (Adam, 1960). The average thickness of the bedding-plane lodes varies from 0.1 m to 35 m, whereas the fissure veins usually have a thickness of 0.1 to 1 m. The cassiterite is finer grained in the bedding-plane lodes and magnetite is much more abundant in them than in the fissure veins. There are also some iron-rich tin deposits outside of the mine area which may be similar to Kelapa Kampit. Most of the fissure veins are related to steeply dipping faults, which strike NE, and intersect the bedding-plane lodes. The bedding-plane lodes occur within phyllite, containing tuff horizons, and at the contact between iron-rich phyllite and more competent meta-sandstone or chert. Often there is evidence of fault movement or brecciation along these contacts (Schwartz and Surjono, 1990).

The Nam Salu horizon is composed of steeply dipping iron-rich phyllite with minor calc-silicate hornfels containing concentrations of 500 ppm Sn average in the weathered phyllite outcrops and 1 to 2% Sn in fresh material in the mine pit. The Nam Salu phyllite is an iron formation with tuff layers which have been altered to stilpnomelane-biotite-chlorite assemblages. The phyllite is banded, defined by variations in the contents of magnetite (average 4 vol. %), pyrrhotite (2%), pyrite (1%), ilmenite (0.5%), detrital quartz (0.5%), siderite (< 0.5%) and rutile (< 0.5%). These are enclosed in a finely crystalline felted phyllosilicate matrix. Where the iron content is low, the phyllosilicate is of very fine grained intergrown biotite, stilpnomelane and chlorite, with subordinate sericite. The cassiterite occurs as dipyramidal short prismatic or anhedral equidimensional grains randomly disseminated in the phyllosilicate matrix, grain size varying between 2 and 200 μm with an average 30 μm. It has a tendency to be concentrated in bands which are rich in pyrrhotite and magnetite.

The banded ore is intersected by thin veinlets of average width 0.1 mm, They are mostly filled with sulphides and phyllosilicates. Apart from the ore-grade Sn concentrations (1–2%), there are a number of elements in the phyllite whose concentration suggests that granite has been the cause of the hydrothermal mineralization. They are K_2O (average 3.98%), Rb (1270 ppm) and Cs (325 ppm). Schwartz and Surjono (1990) conclude that only residual (hydrothermal) fluids, separating from a granite magma, are capable of producing such high concentrations in the Kelapa Kampit phyllite. The most likely source of the Sn-bearing fluids is the Tanjung Pandan granite pluton, which is known to have been depleted in Sn (Lehmann and Harmanto, 1990), and probably underlies the Kelapa Kampit deposit at depth. Hutchison and Taylor (1978) suggested that there may be some similarity between the Kelapa Kamput bedded-lodes and the tin occurrences at Bukit Besi, Trengganu, Malaysia. However, the submarine exhalative origin proposed by Hutchinson (1981) is most unlikely.

13

Mississippi Valley-type epigenetic deposits

13.1 DEFINITIONS

Mississippi Valley-type deposits are stratiform concentrations predominantly of galena and sphalerite, most commonly but not exclusively, in carbonate host rocks. Fluorite and/or barite, pyrite and chalcopyrite, are usually also present. The host rocks are exclusively shallow water, marine carbonate-dominated sequences, and they always occur on the rim of a large sedimentary basin or over basement highs at or near limestone/dolomite or limestone/shale interfaces, and at any depth (Hutchison, 1983*a*). The deposits are truly epigenetic, that is not formed contemporaneously with the host rocks. The low temperature hydrothermal solutions are considered to have moved laterally out of a major sedimentary basin towards its margins, where the change in salinity caused the deposition. Mississippi Valley-type deposits are therefore considered to be laterally secreted epigenetic deposits. The fluorite in such deposits often contains hydrocarbon globules, suggesting that large scale brine migration up-dip from the basin may also have carried the metals. These then became concentrated on the basin margins where the saline solutions encountered fresh or less saline water and sedimentary facies changes, for example to carbonates, which provided suitable host rocks for mineral deposition.

The continental regions of SE Asia, which are dominated by Palaeozoic platform carbonates, seem to offer ideal settings for Mississippi Valley-type epigenetic deposits, and some have been identified. The regional distribution of lead and zinc deposits is shown on Fig. 9.1 and that of fluorite on Fig. 10.1.

13.2 MYANMAR

13.2.1 Bawdwin lead-zinc mine

This famous deposit occurs 700 km NNE of Rangoon at 23°07' N, 97°18' E in the north Shan State (Fig. 9.1). Mining can be traced back to 1412. Peak production in 1930 achieved 530×10^3 t of ore (78×10^3 t Pb, 57×10^3 t Zn concentrate, 198×10^3 kg Ag). High-grade ore is exhausted and production is now less than 160×10^3 t per year. Underground reserves are estimated at 5×10^6 t at 8% Pb, 5.6% Zn, 0.6% Cu and 112 to 140 g t^{-1} Ag (Robertson Research, 1975). Because of the close association with rhyolite, this deposit has traditionally been classified as Ordovician 'Kuroko-type'. It is more appropriately considered epigenetic.

The mineralization is predominantly of silver-bearing galena and sphalerite. The ore is concentrated in a narrow 4 km long zone extending NW–SE. The high-grade massive Pb-Zn lodes are surrounded by a low grade halo of Cu or Cu-Pb mineralization, and numerous small lodes are linked by barite dyke-like veins (Brinckmann and Hinze, 1981). Every grade from pure sulphide to barren meta-sediment can be found. The ore is of fine grained galena, sphalerite and chalcopyrite, frequently banded, and of sphalerite-galena veinlets in the lower grades (Robertson Research, 1975). The high-grade ore has a small amount of quartz and calcite gangue. The ore commonly replaces the cores of altered feldspar and shale inclusions. The Chinaman Lode outcrops and is gossanized in the Quarry Mine (Fig. 13.1). A younger set of Cu-Ni-Co ore veins sharply cuts across the main Pb-Zn-Ag orebody. On the whole, the ores are fissure-filling vein-type and form steeply dipping lodes. The main country rock is the Bawdwin intrusive rhyolite, which is strongly sericitized and silicified, with SiO_2 wt % commonly > 70%, K_2O > 8% and Na_2O generally < 0.1% (Muller and Weiser, 1981).

The rhyolite (also known as the Loi Mi quartz porphyry) discordantly intrudes a sedimentary sequence, extending from the Precambrian Mogok Series, through the Upper Precambrian to Cambrian Chaung Magyi Group, the Cambrian to Lower Ordovician Bawdwin Volcanic Formation, to Ordovician meta-arkose of the Pangyun Formation (Fig. 13.1). The strata have been intensely brecciated by hydrothermal activity associated with the porphyry. The strata generally dip E but are deformed into a series of SE plunging anticlinal folds, which are cut by NW–SE thrust faults to which the lodes are related (Fig. 13.1). Further east the stratigraphy continues up to the Devonian to Triassic Shan Dolomite Group.

The rhyolitic and sericitized/silicified similarity between the Bawdwin Volcanic Formation, tuffs and agglomerates, has led almost all workers to conclude that the intrusive Bawdwin Rhyolite is also Ordovician.

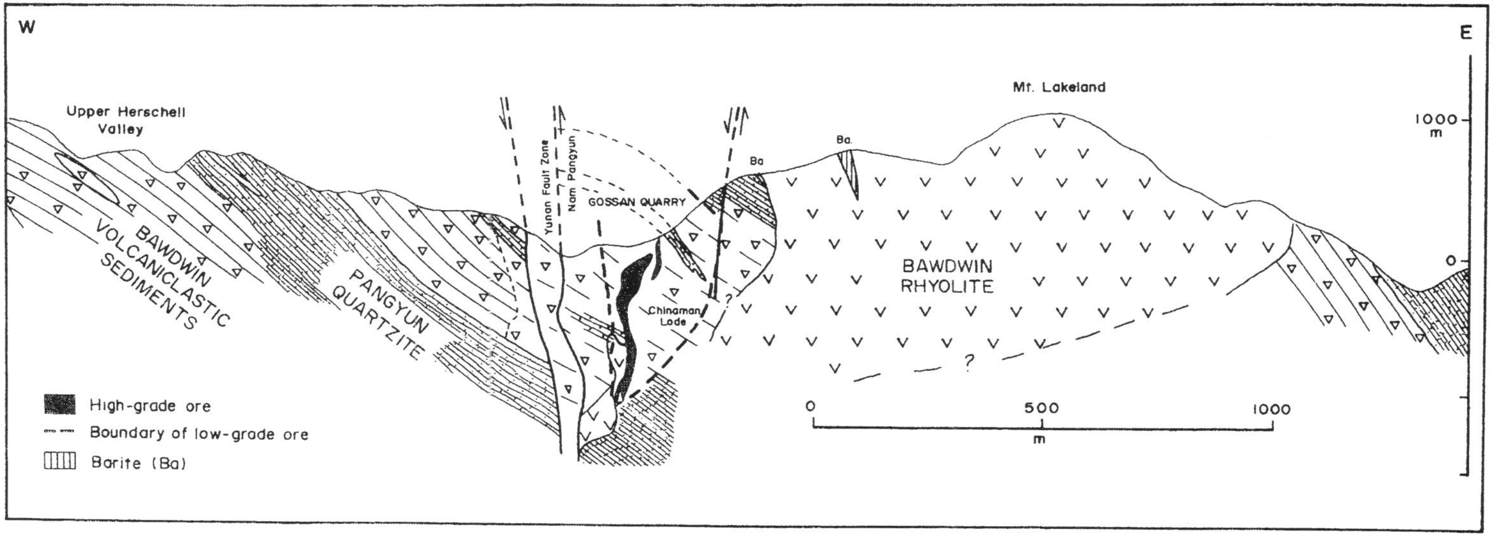

Fig. 13.1. Cross section through the gossan quarry and Chinaman lode of the Bawdwin lead-zinc-silver lode mine, Northern Shan State, Myanmar (after Brinckmann and Hinze, 1981). The Lower Palaeozoic strata are intruded by the Bawdwin Rhyolite (also known as the Loi Mi quartz porphyry). Traditionally this historically famous deposit has been classified as volcanic-related, perhaps of Kuroko type.

However, using fresh specimens from outside the mineralized fault zone, Lenz and Muller (1981) obtained a good Rb:Sr isochron giving an age of 211 Ma, with an initial $^{87}Sr/^{86}Sr$ ratio of 0.75553. Because of the high ratio, and because the isochron includes several Ordovician volcanogenic sediments, the authors conclude that the rhyolite cannot be Triassic. Further west of the mining area, a Rb/Sr age of 339 Ma was obtained for the Tawn Peng granite, which is assumed to be related to the Loi Mi quartz porphyry (Bawdwin intrusive rhyolite) (Mitchell *et al.*, 1977).

If we assume that the 211 Ma isochron represents the age of sericitization of the Bawdwin Volcanic Formation, then what has been dated is the alteration halo related to the clearly epigenetic ore lodes, which therefore appear to be of Triassic age and epigenetically emplaced into NW–SE striking structures related to the Indosinian Orogeny.

If the Bawdwin intrusive rhyolite is indeed consanguineous with the Ordovician volcanogenic sediments, then the mineralization cannot be rhyolite related and a Kuroko-type of mineralization is unrealistic. The most satisfactory classification is Mississippi Valley type, filling the N–S fissures epigenetically during the Upper Triassic stage of the Indosinian Orogeny. Mississippi Valley type deposits do not need to occupy fissures in limestones, as they do elsewhere in northern Thailand.

13.2.2 Others

The **Theingon** Pb-Zn deposit is located within Lower to Middle Ordovician Wunbye Formation limestone near Bawsaing (20° 57' N, 96° 48' E), about 70 km SSE of Mandalay, in the Southern Shan State (Khin Zaw *et al.*, 1984). The orebody varies from 15 to 61 m in thickness and has a maximum width of 21 m and a length exceeding 305 m, extending farther and deeper towards the south. The ore zone is a fissure-controlled solution channel, trending 340°, in the limestone host-rock, filled with yellow soft plastic clay in which galena is the main ore mineral, irregularly distributed as disseminations, stringers, nodules and blocks embedded in a silty and sandy clay matrix. The principal ore mineral is galena. There are minor amounts of sphalerite, tennantite, digenite, chalcopyrite, covellite, argentite and pyrite. The gangue is of calcite and quartz. The mine has long been a source of silver-bearing ore, dating back to the fourteenth century. Annual production is about 600 t of lead concentrates averaging 60% Pb. The estimated reserves are 185 400 t averaging more than 7% Pb.

The **Phaungdaw** lead deposit lies about 100 km SSE of Mandalay, where mineralization occurs within marble of the metamorphosed Ordovician Phaungdaw (= Wunbye) Formation (Khin Zaw and Goossens, 1989). From 1904 to 1914 the mine produced 2200 t of lead ore. Subsequent post-war production brought the total to 43 600 t. The mineralization forms stratiform lenses of 1 to 2 m thickness conformable to the bedding. The ore is predominantly of galena, with minor sphalerite, pyrite, chalcopyrite, and tetrahedrite-tennantite, in a gangue of calcite and quartz

Goossens (1978) lists several other smaller galena-barite-(sphalerite) deposits which have Mississippi-Valley characteristics, mostly occurring within Ordovician limestones in the Shan State province, which continue southwards into Thailand.

Barite is widely distributed in the Ordovician limestones. Near Bawsaing the barite deposits are stratabound and occupy a definite horizon in the upper part of the Ordovician Wunbye Formation, beneath the Nan-On Formation. The oil company requirements are mined from the Bawetaung deposit, near Bawsaing, and formerly from Maymio east of Mandalay. The barite reserves are estimated at over 1 Mt of good quality ore. It is always associated with minor galena and contains altered fragments of limestone.

It is surprising that Myanmar is deficient in fluorite deposits since adjacent Thailand is so rich. Only three occurrences are known (Goossens, 1978). Two are associated as a minor mineral with tungsten and tin ores, and only the Bawhningon deposit (20° 31'N; 96° 37'E) has been mined for fluorite. The fluorite occurs as veins in Upper Palaeozoic limestone.

13.3 INDOCHINA

All barite occurrences of the North Trungson region occur within the Samnua rift zone, which surrounds the Phu Hoat uplift. The barite bodies occupy shear zones trending NW–SE and N–S, cutting Triassic volcano-sedimentary formations (Tran and Nguyen, 1986). The barite is usually accompanied by quartz, but some deposits also contain galena and mercury.

13.4 THAILAND

13.4.1 Lead-zinc

Several lead-zinc orebodies occur in the Upper Triassic to Lower Jurassic Kamawkale Limestone in Tak Province, NW Thailand. The primary ore is epigenetic stratabound within dark grey fossiliferous dolomite. The ore is of sphalerite, galena and pyrite with a gangue of calcite, dolomite and quartz (Tantisukrit, 1978). The primary sulphides have been oxidized to form a secondary overlying zone rich in hemimorphite and smithsonite. Proven reserves at 26% Zn are 3.8 Mt. The deposit ceased production in 1976.

Pa-Daeng (16° 39′ N, 98° 40′ E) is the largest known zinc deposit of Thailand (Rung Songsiri, 1978). The deposit lies 12 km ESE of Mae Sod and 60 km SW of Tak, near the Myanmar border. It was discovered in 1956 and is readily seen from the air because of a metal-rich gossan which does not support forest growth, hence the name, meaning 'red cliff' (Shawe, 1984). The country rocks are the Triassic–Jurassic Kamawkale Limestone formation, here composed of an interbedded limestone-sandstone sequence regularly dipping to the NW (Rung Songsiri, 1978). The main body is an irregular pod 500 m × 250 m and there is a NW-trending line of smaller bodies, indicating some structural control. The deposits preferentially occur within the limestone beds, but also occur within sandstone. The ore consists of large replacement zones, and lamination and brecciation are common. The deposit is remarkable for its high grade, undoubtedly due to supergene enrichment, as indicated by the unusual mineralogy: hemimorphite and smithsonite, with local concentrations of hydrozincite and loseyite. The supergene oxidation has resulted in a vertical zonation, and the surface outcrops are of a very high grade (> 40% Zn) duri-crust of up to 2 m thickness. Associated gangue minerals are quartz, clay minerals, calcite, dolomite and iron minerals. There are no sulphides. The ore has a cadmium content ranging from 0.1 to 1.0%. After drilling, the reserves were calculated at 3 777 400 t averaging 26.6% Zn at a 10% Zn cut-off.

A lead-zinc mining district occurs in Kanchanaburi Province. The ore is a replacement or fissure-filling in Ordovician limestone, and is of fine grained galena and sphalerite accompanied by minor amounts of tetrahedrite and boulangerite. It occurs in layers, lenses, pockets and shear zones. The ore has low temperature Mississippi Valley characteristics.

13.4.2 Barite

The barite deposits of Thailand are stratabound as in Myanmar, but also occur filling major fissures (Asnachina and Chantaramee, 1986). At Mae Hong Son, barite veins occur cross-cutting Permian clastic strata. However the deposits at Lamphun and Phrae are stratiform within Ordovician Hod Limestone and Permo-Triassic volcanic group respectively. The largest deposit of northern Thailand is the stratiform deposit in the Ordovician Hod Limestone at Phu Mai Tong, about 80 km S of Lamphun. In south Thailand, fissure-filling barite occurs in Cambrian Tarutao Group strata in the Surat Thani Province. At Ti Mong Ta (NE Kancharaburi province), Ordovician limestone and shale contain a layer of interlaminated galena and white barite up to 2 m thick (Pitrakul and Tantisukrit, 1973).

13.4.3 Fluorite

Thailand continues to be a major producer of fluorite, with a total production from 1962 to 1972 of more than 2×10^6 t, and a peak production in 1971 of 429 493 t. There are three main mineralized provinces in the most intensely faulted districts of north-west Thailand, centred on Lamphun and Chieng Mai; West-Central district of Petchaburi and Kanchanaburi, and the Peninsular Thailand Surat Thani province (Shawe, 1984). In 1971, the NW highland Lampun district accounted for 159 920 t. The West-central district accounted for 77 790 t, increasing to 112 587 t in 1972. The peninsular province is relatively unimportant.

In the mountain ranges of western Thailand extending into Myanmar, fluorite occurs as epithermal veins which have cemented brecciated fault zones in granites and Palaeozoic and Mesozoic strata. The veins consist predominantly of fluorite, which is characteristically botryoidal but also massive, and silica (Asnachina and Pitragool, 1978). The lenticular ore bodies swell and pinch along strike. Locally fluorite constitutes up to 80% of the veins, or may be accompanied by more abundant chalcedony and milky quartz. Reserves are estimated at 200 000 t at Yang Hak in Ratburi Province, 500 000 t at Doi Tao in Chieng Mai Province, and 3 600 000 t at Ban Hong in Lamphun Province. Other workers estimate the reserves somewhat less. However most Thai deposits have not been properly assessed by drilling, and reserves are estimated as the opencast mining progresses.

The ore body at Huai San, 8 km NW of Fang in NW Thailand, is a pipe-like lens which is strongly internally brecciated. The country rocks are Carboniferous shale and limestone. A prominent NW-trending fault separates Tertiary sedimentary rocks of the Fang oil-bearing basin from the Carboniferous shale-limestone sequence (Shawe, 1984). Horizontal slickensides indicate some strike-slip movement. There is a broad zone of shearing along the fault, wherein the fluorite was localized. The ore body is mined from a zone 300 m long by 50 m wide. The deposit contains more than 100 000 t of light coloured fluorite which is massive and botryoidal. The Fang hot spring occurs within gneissic granite only 1 km north of the Huai San deposit. The age of fluorite mineralization accordingly is thought to be Tertiary.

At Doi Pha Lad, 10 km SW of Fang, there is a sequence of E–W striking veins cutting the granite, of thickness variable between 1 cm and 5 m. Light green fluorite is mined from a gangue of chalcedony, and there are traces of galena. The production was 600 to 700 t per month, with fluorite reserves of about 50 000 t. At Huai Ngok-Ngak, 60 km SW of Chiang Rai in the Ban Pon Pu Fueng mining district, massive and laminated fluorite, brecciated in places, occurs in NW trending veins which pinch and swell. Ore reserves are estimated at 13 000 t.

Ban Hong, 50 km S of Lamphun, is the largest deposit, which began mining in 1962. There are three long thick intersecting veins: No. 1 trends 320°, is 800m long and up to 30m wide; No. 2 is parallel to and 200m NE of vein 1. It is 420m long and 5–10 m wide; No. 3 branches north from vein No. 1 and curves ESE. In 1965, about 3000 t per month were shipped, and by 1964 at least 200 000 t had been extracted (Shawe, 1984). Reserves are variously estimated at 150 000 to 3 550 000 t. The country rocks are Silurian to Carboniferous greywacke and limestone, which strike NW–SE parallel to the veins. The fluorite is coarsely crystalline and massive, green, black and yellowish. Associated minerals are chalcedony and opal forming the cement in breccia zones.

At Pai, 30 km E of Mae Hong Son, fissure veins occur in granite and limestone. The mineralogy of the veins is fluorite, quartz, chalcedony, stibnite, pyrite and haematite. The orientation of the veins is NW–SE, but they may also lie N–S. At Muang Maha Lana, the largest mined vein is 6 m wide and 700 m long, striking SW, and containing botryoidal vari-coloured fluorite and chalcedony. At Doi Pha Dang, 65 km S of Mae Hong Song, fluorite occurs in fissure veins along the contact between granite and Lower Palaeozoic shale and limestone. At Muang Toho Sang Yo, NW-striking veins of thickness 0.75 to 2.0 m, composed of chalcedony, fluorite and pyrite, occur within the granite. At Muang Thai Prasit, also in the Pai district, N-S breccia veins cut the granite. They are up to 2 m wide, and are composed of fluorite, chalcedony and pyrite.

At Doi Chang, 4 km NE of Mae Sariang, veins of fluorite, quartz and chalcedony cut Ordovician limestone. They trend both NW and E, and the thickest is 4 m. At Sahachart, 2 km E of Mae Sariang, two fluorite veins occur in Lower Palaeozoic slate along a thrust, and also in Triassic shale and limestone.

NW-striking veins of high grade fluorite and chalcedony cut shale, sandstone and limestone at Huai Poy, 15 km south of Hod. At Doi Pong Kham, 20 km SE of Hod, fluorite veins occur within N–S trending quartz veins.

Mining at Salak Phra, Kanchanaburi, began in 1971. An estimated 200 000 t of fluorite have been mined. The deposit is of banded fluorite, locally brecciated, along with calcite and chalcedony. The veins, of average width 3–5 m, trend NNW, N and NNE and there is also a pipe containing about 1 000 000 t of fluorite within Ordovician limestone at Khao Ka Ta Yai, south of Salak Phra.

Most of the Thai fluorite deposits occur as fissure fillings which have been periodically reactivated and sometimes brecciated, and some occur as cavity fillings. The deposits are epigenetic and unrelated to host rock lithology and age. Veins occur in host rocks ranging in age from Cambrian to Mesozoic. In the major NW Thailand province, 35% of the fluorite veins occur within granite, 23% within Ordovician limestone, 23% within Silurian–Devonian clastic rocks, 5% within Carboniferous clastic rocks, and 14% within Triassic clastic rocks of the Hong Hoi Formation of the Lampang Group (Asnachina and Chantaramee, 1986). The most favourable host rock for the fissure vein deposits is limestone but fluorite is mined from veins in granite, sandstone, shale and slate. Chalcedony is the common associated mineral, but calcite is abundant when the veins occur in limestone. Barite occurs at Khlong Thom in Peninsular Thailand. Stibnite occurs with fluorite in NW Thailand. Minor amounts of galena are commonly associated with the fluorite.

The fluorite deposits occur in clusters where the country rocks are intensely fractured, especially by Cainozoic strike-slip faulting. That the deposits are overwhelmingly of vein type was demonstrated by Asnachina and Chantaramee (1986). The fluorite occurrences coincide exactly with major lineaments seen on air-photos. In NW Thailand, these lineaments are predominantly of a 320° trend, with 30° also quite common. The fluorite deposits occur in regions characterized by present day hot springs and may be genetically related to them. The deposits have resulted from wide hydrothermal crustal circulations resulting from high geothermal gradients, focused along the abundant trans-extensional strike-slip faults. The intensely faulted zones occur along the margins of the Cainozoic sedimentary basins. The age of the fissure filling is therefore from Upper Tertiary to Recent (Shawe, 1984), although Asnachina and Chantaramee (1986) suggest that a Mesozoic age is possible. The ultimate source of the fluorine to form fluorite is presumed by many people to be the granites of all ages, but clearly the solutions have travelled far and such a hypothesis is generally contradicted by the absence of economic fluorite deposits in Malaysia, which has more abundant granites of similar kind (Cobbing *et al.*, 1992). The fluorite deposits of Thailand are typical of other worldwide occurrences (Hutchison, 1983*a*). The commonest world occurrence is as fissure veins notably in carbonates and granites. The main ore bodies in the Pennines of England, known as *rakes*, are the filled, generally vertical, wrench-fault fractures. Most *rakes* are horizontally slickensided. There are comparable examples in Thailand. In the Pennines, *scrins* are the open joint fillings. There are also *pipes*, representing filling of solution voids in limestone, and *flats* are veins lying parallel to the bedding (Hutchison, 1983*a*). All of these variants can be found in Thailand. A comparable theory of genesis is therefore applicable, and there is a genetic relationship to barite and base metals, notably galena. The ores were deposited from low-temperature brines migrating up-dip towards the margins of sedimentary basins. A severe problem arises from timing of brine source and migration. The most conveniently available would have been from the Khorat Basin, which was very saline by Cretaceous time, but migration in Tertiary time from the Malay Basin is also probable. The problem of accounting for the fluorine is a universal one, commonly attributed to granitic or alkaline magmatism emplaced beneath the marine basin from which the hydrothermal circulations were initiated. Once the solutions passed through the fractures and deposited their fluorite and other Mississippi Valley type minerals, they must have reached the surface as hot springs. The dating and relationship to the present-day hot springs of Thailand remains to be isotopically analysed.

13.5 MALAYSIA

It is surprising that despite similar stratigraphy with Thailand and Myanmar, the Sinoburmalaya part of Peninsular Malaysia has never yielded Mississippi Valley-type Pb-Zn deposits. Perhaps they have never been recognized in a country which was traditionally obsessed with exploration for tin deposits. However the Mengapur and Sok prospects in Pahang and Kelantan in the Central and Eastern Belts may have epigenetic characteristics, though it is more likely that the localization of the latter deposit is granite-related. A mineral zonation in the Kinta Valley (Hosking, 1973) indicates that a lead and zinc zone occurs towards the valley centre, away from the surrounding granite (Fig. 11.3). This may indicate the possibility of Mississippi Valley-type Pb-Zn deposits in the limestones west of the Bentong–Raub suture, though any such deposit, should it be found, would have been largely remobilized by the abundant Upper Triassic granites.

Only three barite deposits have been mined in Malaysia, though several residual boulder outcrops have been noted. All occurred east of the Bentong–Raub suture within the Central Belt. The deposits were briefly reviewed by Teh (1994). The Sungei Trengan area (Trengganu) exhibits syn-sedimentary stratiform mineralization. The barite beds occur within a siltstone-shale sequence. Individual barite beds vary up to 3 m thickness. Trengan Barite was mined until 1983 near the eastern slopes of the Gunung Gagau plateau. The barite occurred in fissures and also stratabound within Carbo-Permian (?) strata (Rajah, 1987). Mining was abandoned when the area was flooded by the Kenyir dam.

At Tasek Chini (Pahang), barite filled a steeply dipping fault zone 3.5 m wide at Bukit Botol. The barite was black due to its high iron content. The barite occurrences had a spatial relationship to residual manganese deposits (Hosking, 1973). There is a continuing barite production from the Tasek Chini area, not from a vein deposit, but from a stratiform deposit, associated with base metal. The massive sulphides have high gold and silver contents.

Barite continues to be mined on a small scale from near the base metal deposit of Mengapur in west Pahang. The Jenderak deposit is stratiform, interbedded with carbonaceous mudstone, sandstone and tuff. The barite horizon has a maximum thickness of 5 m and is offset by several strike-slip faults.

Fluorite has never been found in economic concentrations in Peninsular Malaysia, but it is widespread as a joint and fault filler in granite terrains, and was the main gangue of the Kramat Pulai scheelite deposit (Fig. 11.3).

In summary, it appears that the Mississippi Valley-type epigenetic terrain dies off in importance southwards from the Shan Highlands of Myanmar and the mountain terrain of NW Thailand, and that Peninsular Malaysia occupies the uneconomic outer fringes of this province.

14
Iron ore deposits

Iron ore deposits are widespread throughout the Circum-Pacific (Park, 1972). However significant production has been confined to three areas: the south-east China province from Hainan island through Kwangtung and Fukien, the Malay Peninsula and the Philippines (Fig. 9.1). They are predominantly of contact pyrometasomatic type with associated sedimentary deposits occurring in China.

The contact deposits are associated with plutons which are emplaced at a range of depths, from deep seated batholiths to minor sub-volcanic stocks. Volcano-plutonic arcs, such as the Philippines, are favoured and diorite is the most appropriate pluton to give rise to iron concentrations at its contacts. Limestone represents the most favourable country rock, in which case the deposit is associated with skarn and called contact pyrometasomatic. Tholeiitic magmas, which are emplaced at or near the volcanic front, are more likely to evolve to iron-enrichment, as compared with the calc-alkaline magmas, which are emplaced behind the volcanic front.

The model proposed for the Cornwall-type contact deposits of Pennsylvania can be widely applied throughout volcano-plutonic arcs, especially in SE Asian arcs, which are rich in suitable iron-rich diorite and/or andesite. Readers unfamiliar with this well described deposit are referred to Lapham (1968), Eugster and Chou (1979), and a summary in Hutchison (1983*a*).

14.1 CHINA

There is an impressive belt of contact metasomatic deposits throughout the Kwangtung–Fukien 'Caledonian' foldbelt, genetically related to the abundant Yenshanian cycle of Upper Jurassic granites. The Shihlu-type contact iron ores occur in Hainan. Tayeh-type ores occur in Kwangtung. Oolitic sedimentary ores range in age from Devonian to Jurassic (Ikonnikov, 1975).

14.2 PENINSULAR MALAYSIA

Peninsular Malaysia was a major producer of iron ore with a peak exceeding 7 Mt in 1964. All major deposits have been worked out. The ore deposits are of Permian to Triassic age (Bean, 1969). Most authors postulate a hydrothermal replacement origin related to subjacent or nearby plutons, but Taylor (1971) considered the Bukit Ibam deposit (Fig. 14.1) to be of volcanogenic origin. A common feature of all the ore bodies is the strong supergene enrichment by tropical weathering processes, as at Pelepah Kanan in the Kota Tinggi district of Johore, where magnetite ore is capped by a thick crystalline martite deposit. Total iron production from Malaysia was around 100 Mt. No major ore bodies remain. Tin associated with magnetite occurs at Bukit Besi and in a number of smaller occurrences within the Eastern Belt of Peninsular Malaysia, particularly at Pelepah Kanan and also on Billiton Island, Indonesia. Most of the Eastern Belt iron ore mines ceased production because of increased tin contents. The Eastern Belt iron deposits were classified as 'stanniferous iron skarns' by Hosking (1973), a classification subsequently followed by Chu *et al.* (1988). The exceptional purity of the Ipoh haematite deposits, by contrast, is remarkable since they occurred in the centre of the great Kinta Valley tin field, but a magnetite-cassiterite association does occur.

14.2.1 Bukit Besi

This mine, 24 km west of Dungun, was the largest open cast iron mine in SE Asia (Fig. 9.1). The history of mining went back to 1927 and a peak production of 3 Mt was attained in 1964. Thereafter, production rapidly declined and large-scale operation ceased. Total production was of the order of 40 Mt.

The Carbo-Permian shales locally contain minor sandstone and limestone. The sequence has been folded and subjected to low-grade metamorphism, then further deformed and contact metamorphosed by Permian granitoids (Bean, 1969). The marble and calcareous shale horizons were converted to skarn containing garnet, pyroxene and amphibole assemblages. Near the granite contact, calcareous bodies were converted to amphibolite containing magnetite and pyrrhotite.

Irregularly shaped iron ore bodies all lie close to the granite. The ore has formed by replacement of shale, composed of magnetite, secondary martite and haematite enclosing relicts of unreplaced country rock. The major ore bodies, now mined out, were tabular and distinctly

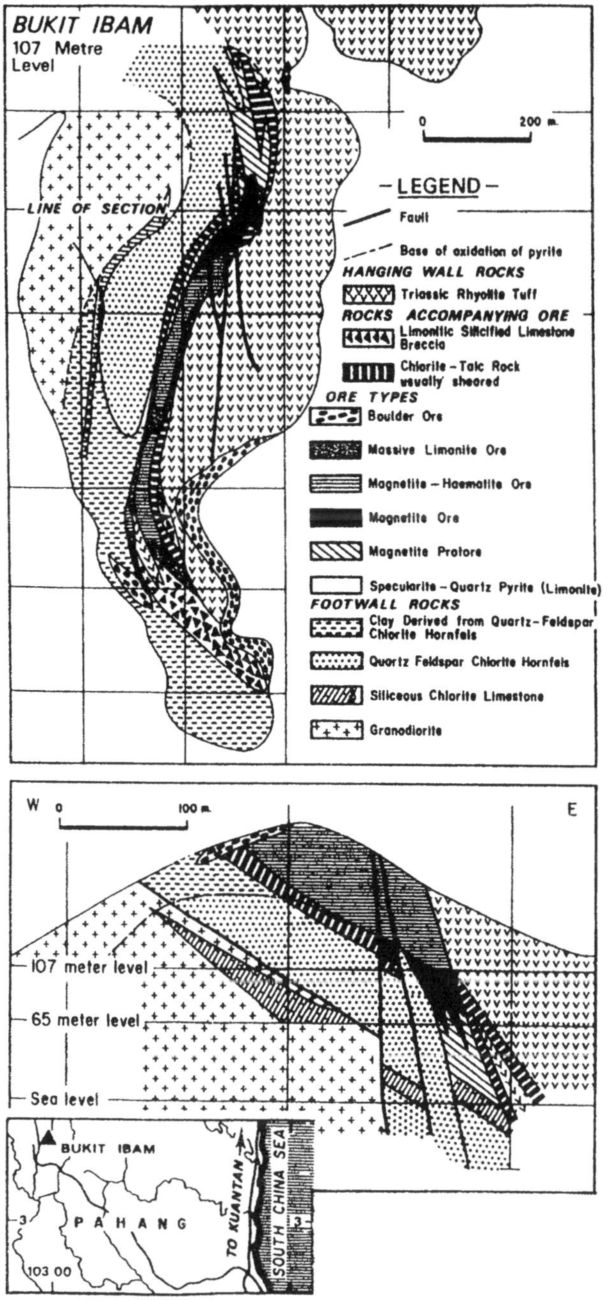

Fig. 14.1. The Bukit Ibam iron deposit of Pahang, Peninsular Malaysia, is a strata-bound structurally controlled replacement magnetite ore body, possibly of Triassic age (after Taylor, 1971 and Hutchison, 1983*a*). **Upper**: plan of the deposit at level 107 m. **Lower**: E–W cross section.

stratabound. They contained significant pyrite. They also occur as replacements of roof pendants within the main granitoid outcrops. The skarn ore is of massive magnetite in a gangue of garnet and diopside and it forms tabular bodies (Bean, 1969).

Ore bodies were preferentially formed where the sedimentary rocks dip towards the granite contact, facilitating the ingress of hydrothermal fluids. Locally there occurred marked concentrations of cassiterite owing to replacement of the host rocks around fracture zones. At Sri Bangun, the fine grained cassiterite is associated with iron oxides in the shales. At Batu Tiga, the ore consists of early magnetite, followed by cassiterite and scheelite, chlorite, pyrrhotite, chalcopyrite, and late pyrite (Hosking, 1973). These ores developed in a calc-silicate host and the deposition was controlled by fault zones, and impounded by overlying meta-quartzite.

Apart from these tin concentrations, a certain amount of tin was present throughout the iron ore, of average content 0.07% (Bean, 1969). This tin is structurally held within the magnetite.

Hosking (1973) defined the deposit as hydrothermal and metasomatic, because the cassiterite mineralization was structurally controlled. Such a fine distinction is not necessary, and contact pyrometasomatic skarn deposits are both hydrothermal and metasomatic (Hutchison, 1983*a*).

It is unrealistic to imagine that the same granitoid that provided the tin also gave the iron. The volcano-plutonic belt contains abundant diorites as well as granite. Either one, acting as a heat source, could have caused circulation of meteoric water throughout the granite as well as the diorites and sedimentary rocks to mobilize both iron and tin and deposit them around the granitic heat source. It is reasonable to assume that an ilmenite-series granite is the source of tin, while a magnetite-series diorite the source of the iron, but the later granite may have provided the heat engine. The possibility that the sedimentary rocks were iron enriched also cannot be ruled out, since there was contemporaneous volcanism during their deposition.

14.2.2 Bukit Ibam

At the peak of its production in 1965, this mine in SE Pahang State had an annual output of more than 2 Mt of grade 60 to 64% Fe. The ore-body formed a single lens of maximum dimensions of 930m × 240m × 60m. It is in part concordant with and in part discordant to the surrounding rocks and is cut by faulting sub-parallel to the ore lens (Fig. 14.1). The ore is mainly strata-bound between an overlying sequence of Triassic rhyolite tuff and ignimbrite of thickness about 300 m, and underlain by a sequence of siliceous hornfels, impure limestone and siliceous schist, which has been intruded by a hornblende diorite. An interesting feature of the diorite is its absence of magnetite, in contrast to other plutonic bodies exposed in the neighbourhood (Taylor, 1971). The ore occurs within the metamorphic aureole of the diorite.

The primary ore can be divided into an enriched upper oxidized zone of haematite-magnetite, and an underlying protore containing well crystallized single crystals or aggregates of magnetite (up to 10 cm diameter) in a green matrix composed of actinolite, anthophyllite, talc and serpentinite. Copper, zinc and bismuth were present in minor amounts, copper (0.08%) being mainly in the limonitic ore along the hanging-wall in a zone of secondary sulphide mineralization. Zinc (0.07%) was also mainly in the limonitic ore, and also in chloritic rocks along the hanging-wall. Bismuth (average 0.06%) was in the friable haematite-magnetite ore. The ore is confined to a zone within a sheath of sheared and chloritized rocks. The diorite is not in contact with the ore, but occurs at a depth of less than 60 m below it. The major fault that cuts the ore body is itself mineralized and the overlying altered rhyolitic rocks contain irregular patches of specularite-pyrite mineralization.

Although Taylor (1971) considered that the ore body was contemporaneous with the rhyolitic tuffs, the Cornwall-type mineralization holds many similarities, including the contact metamorphic mineral assemblage and the subsequent hydrothermal metasomatic changes. It seems therefore appropriate to apply the model of Eugster and Chou (1979) to this deposit, and to suggest that the iron has been removed from the underlying diorite and deposited in a favourable stratigraphic setting beneath a thick rhyolite tuff formation by a hydrothermal circulation cell. The succession underlying the rhyolite tuff contains limestone horizons, and the ore could have replaced a favourable limestone bed. Indeed, many of the smaller ore deposits in the Bukit Ibam vicinity are related to limestone beds and skarns. The serpentinization of the protore matrix favours the interpretation of hydrothermal alteration of a previous metamorphic assemblage within the aureole of the diorite, which could have included pyroxenes.

The mine is now abandoned, but for some time thereafter the serpentinite was used for a local carving industry under the term 'Dara Jade', also since abandoned.

14.2.3 Pelepah Kanan

This deposit is an excellent example of a contact pyrometasomatic iron deposit which has been rendered commercially useless by high tin contents (Fig. 12.4). Cassiterite was produced where it could be successfully separated from the iron ore. The mine, near the Kota

Tinggi Waterfall in eastern Johore State, has a history of underground mining as far back as 1939 and was operated by the Japanese Occupation forces as an open-cast mine. It continued under private ownership until recently.

The opencast hillside mine is composed of massive magnetite, largely altered to martite, resting on bedded meta-quartzite and calc-silicate hornfels, intruded by Triassic porphyritic biotite granite. The calc-silicate rocks have resulted from metamorphism of a sequence of calcareous shale, siltstone and mudstone now containing the following mineral assemblages: quartz-epidote-sericite; quartz-diopside-epidote-sphene; and quartz-diopside. Near the granite contact, there is a garnetiferous hornfels and some rocks contain cordierite porphyroblasts (Hosking, 1973).

Mineralized veins, of thickness from 30 cm to less than 2 cm, intrude the meta-sediments usually concordantly, but some are vertical and form a stockwork. The veins are of variable mineralogy, containing the following paragenesis: alkali feldspar, followed by biotite-chlorite, coarse cassiterite, scheelite (?), lollingite, acicular cassiterite, ankerite, siderite, haematite and fluorite. Quartz is abundant and overlaps the above sequence. Although feldspathic, the veins are not pegmatite and there are two generations of cassiterite (Bean, 1969). The earlier is characterized by intense colour zoning, which is most striking; later needles are never zoned. The cassiterite is ferromagnetic and its magnetism is destroyed by heating to 830° C for 15 minutes (Flinter, 1960). The darker the cassiterite zone, the greater its magnetic susceptibility. Generally the magnetization appears to be due to hydrated ferrous stannate. However, the magnetic character is also due in part to magnetite inclusions within the cassiterite (Grubb and Hannaford, 1966).

It is generally held that this contact deposit is a hydrothermal iron-tin replacement body developed above the granite contact beneath an impermeable cap rock that impounded the fluids (Hosking, 1973). However, one must conclude that the sediments were enriched in iron before the contact metamorphism, for their mineralogy includes the iron-rich amphibole grunerite. Hydrothermal introduction of tin mineralization from the biotite granite is therefore imprinted upon and mixed with an earlier iron-enrichment, perhaps hydrothermally leached from a diorite source by fluids circulating and being driven by the granite heat source.

This large iron ore deposit has proved of no commercial value because the tin content is too high for smelting. Up to 1% Sn has been reported from random samples of the iron ore (Chu *et al.*, 1988). The mine has been restricted to produce cassiterite from the veins near the granite contact. Cassiterite has also been produced from the weathered skin of the iron ore, but the recovery is poor.

14.2.4 Others

Other skarn type iron deposits (Fig. 12.4), in which tin forms an important constituent, include Machang Satahun (Trengganu), Bukit Bangkong, Pandan Estate and Sungei Panching (Pahang). Hutchinson (1984) regarded the Sn-Fe skarns in general to have a volcanic-exhalative origin. However a pyrometasomatic skarn origin, related to granitoids, is much more in keeping with the field evidence.

14.3 INDOCHINA

Two types of contact iron deposit have been described from North Central Vietnam within the Trungson region, bounded by the Song Ma Fault in the north and the Tamky-Phuocson Fault in the south (Tran and Tran, 1986). The first type is at the contact between Upper Palaeozoic limestone and gabbro, in the Langsam area. The contact zone contains skarn minerals and magnetite, with grades up to 65% Fe, but the reserves are limited. The second type at Thachkhe is of ore bodies within the contact zone between limestone and granite. It has magnetite ore grades up to 62% Fe and the reserves are good. The Thachkhe iron ores were formed along repeatedly opened fault zones.

14.4 PHILIPPINES

Pyrometasomatic iron ore deposits are closely associated with dioritic rocks of Tertiary age (Bryner, 1969). An interesting feature is the presence of copper and zinc, and most porphyry copper deposits are rich in magnetite. Peak Philippine production of iron of 2 254 556 t was reached in 1973 (Bureau of Mines, 1986). Since then, production has dramatically declined.

In addition to contact pyrometasomatic deposits, iron ore occurs in the form of beach and alluvial placers in many localities on Luzon and Leyte. The nickel laterites, especially of SW Palawan, are important sources of residual iron.

14.4.1 The Larap deposit

This is the largest deposit in the Philippines, occurring in Camarines Norte at 14° 15′ N, 122° 39′ E. It was first mined in 1918 and production continued until 1974. The ore reserves were estimated in 1982 to be 12.5 Mt of ore of grade 10% magnetic iron, 0.29% Cu, 0.04% Mo and 3.00% S (Bureau of Mines, 1986).

In the mine area, the Larap Volcanics and associated sedimentary rocks are intruded by Middle Miocene diorite and the ore bodies occur within skarn zones following the structure of the bedded Universal Formation rocks. The primary ore is of magnetite, converted to haematite in the weathered zone. High grade ore is of magnetite with little pyrite and other sulphide minerals. Magnetite:pyrite ratios are commonly 9:1. The ore commonly contains 40 to 35% Fe and 2 to 6% sulphur. The ore also contains traces of gold, uranium and cobalt. The gangue is of garnet, pyroxene, epidote, amphibole, calcite, chlorite and apatite.

The Larap Deposit grades downwards into a Mo-rich porphyry copper deposit in the lower levels.

15
Other deposits

15.1 BAUXITE

The broad SE Asian region contains examples of all the major types of bauxite deposits. Australia is the world's major producer of bauxite (more than 33%), with major producing centres on the southern margin of SE Asia at Weipa on the Cape York Peninsula and in the Gove Peninsula of the Northern Territory (Hutchison, 1983*a*). The bauxite is in the form of lateritic residual caps upon Cretaceous and Tertiary sedimentary formations.

The southern extremity of the eastern Malay Peninsula, extending to the Riau Archipelago of Indonesia, has been favoured for bauxite weathering by its long tectonic stability since early Jurassic times and the low present day elevation of the coastal regions of southern Sundaland. Traditional mining areas are the island of Bintan south of Singapore (Van Bemmelen, 1970) and the Pengerang SE tip of Peninsular Malaysia, where the bauxite is developed as a residual lateritic capping on the Middle to Upper Triassic ignimbrite and sub-volcanic granophyre (Grubb, 1968). There was former production in west Sarawak and there is future potential in the northern part of west Kalimantan.

Bauxite deposits on Samar island, Philippines, overlie Lower Miocene karstic limestone. Santos-Ynigo and Pacis (1981) have shown that these karst bauxites resulted from *in-situ* weathering of Mid-Pleistocene ash falls over the karstic topography from the nearby volcanic arc. They therefore have an origin similar to the well known Jamaican deposits (Hutchison, 1983*a*).

SE China has important bedded bauxite deposits of Carboniferous and Permian age (Ikonnikov,1975), and north Vietnam has bedded Upper Permian bauxite.

15.2 GEMSTONES

The famous sapphire, ruby, zircon and spinel gemstones are worked in several localities from alluvial and eluvial gravels derived from nearby outcrops of the extensive SE Asian Pliocene–Pleistocene high-alkali basalt fields. The basalts are usually basanite and nephelinite, dated by K:Ar at 0.44 Ma (Barr and MacDonald, 1981). Pailin is the best known centre in Cambodia (Berrangé and Jobbins, 1976). Other centres are in Bokeo and in the Rovieng district (Fontaine and Workman, 1978). In Thailand, the main occurrences are in the Chanthaburi, Trat and Kanchanaburi provinces (Asnachina and Pitragool, 1978). A comprehensive review has been given by Vichit (1987). Most of the rubies come from Trat, which supplies about 70% of the world's production. The gem corundum has been seen in the basalt, and is released by weathering and mined from overlying and contiguous gravels.

Ruby, sapphire and spinel are mined in the Mogok area of north Burma from placers and eluvial sediments derived from marble and calc-silicate migmatites associated with alaskite (The Mogok Gneiss) (Searle and Ba, 1964), but a few mines work the marble source rock itself.

Jadeite is mined in the Lonkin area 150 km SSW of Myitkyina, and provides the world's foremost source of high grade jadeite. Most of it is mined from jadeite-albite bodies associated with peridotite and serpentinite, formed as a result of high pressure metamorphism of an ophiolite assemblage in the Jade Mines area (Bender, 1983). The jadeite at Tawmaw, Myanmar, occurs in association with kosmochlor and chromian sodic amphibole (Mevel and Kienast, 1986).

Nephrite jade is mined at Fengtien, Hualien in Taiwan, to form the basis of an important ornamental stone industry, involving more than 600 lapidary factories. The annual production is around 1000 t. The nephrite occurs as lenses 0.1 to 2.0 m thick, 5 to 50 m long in the contact zones between serpentinite layers and black shales (Wang, 1987). The mineralogy is of tremolite with chromite inclusions.

Mesozoic rhyolitic volcanic rocks in Indochina have been hydrothermally altered to a kaolinite-pyrophyllite rock, known as pagodite (Tran, 1986). Such rocks are quarried in Cambodia, Laos and Vietnam. The best known is the Trassey deposit of Cambodia, due west of Phnom Penh, where the pagodite is veined green, purple, pink, grey and white, and much valued for carving statues and jewellery. The main components are kaolinite, pyrophyllite, alunite, and sometimes the pagodite contains diaspore, andalusite and corundum.

15.2.1 Diamonds

Placer diamonds occur in two regions of South-East Asia: Phuket in SW Thailand, and W and SE Kalimantan. Only in the latter are they commercially significant, but their origin is important in tectonic reconstructions of the region.

15.2.1.1 Phuket, south Thailand

The sea-going dredges are designed to recover cassiterite from the Quaternary alluvium, which is buried beneath a thin layer of marine clay. The heavy minerals are fine grained. The jigs are set to retain the cassiterite, and they also retain tourmaline, garnet, monazite and ilmenite. However, small rounded diamonds are frequently recovered during jig cleaning. Placer diamonds have been found in Tongkah Harbour and Amphoe Kathu, near Phuket, and along the mainland rivers of Khlong Lei, near Takua Pa, and Khlong Phangnga. The tenor is low and is uneconomic as a source of diamond, estimated as one carat from 28 000 m^3 of alluvium. The diamonds are rounded and usually yellowish, with maximum size of 1.9 carats. There are no associated heavy minerals which would suggest a local kimberlite source (Garson *et al.*, 1975). It is therefore concluded that the diamonds have been eroded from the local outcrops of the pebbly mudstones of the Carbo-Permian Phuket Group. These are interpreted to be marine tilloids provenanced from glaciated Gondwanaland (Hutchison, 1993). Accordingly, the ultimate source of the diamonds could have been cratonic Australia or India before the Carboniferous (?) split-up of Gondwanaland.

15.2.1.2 Kalimantan

Placer diamonds have been mined for centuries at three main localities: Landak and Kapuas rivers in the Sanggau Regency of west Kalimantan, Barito River Basin in Central Kalimantan, and Martapura–Pleihari in SE Kalimantan (Van Bemmelen, 1970). The industry flourished in the eighteenth century and continued, with peak productions of 3292 carats in 1940 at Martapura–Pleihari and 6673 carats in 1873 in Landak–Sanggau, but production there has closed-down by 1940. At Tanah Laut, near Gunung Lawak in SE Borneo, there were nearly 4000 diamond diggers in 1836. Most sites closed, with only Martapura continuing to mine, cut and polish diamonds.

The total production, to-date probably more than 1 M carats, has been exclusively from Cainozoic sedimentary sources, predominantly Quaternary river terraces and alluvium of thickness up to 12 metres. The largest diamond ever found weighed 6 carats. The placer diamonds are accompanied by corundum, rutile, brookite, gold, platinum and pyrite. Corundum was traditionally considered a pathfinder mineral, but it is genetically unrelated and results from contact metamorphism of bauxite. The mining is economically attractive because the alluvium contains gold and platinum near Martapura and Pleihari.

The primary source of the Kalimantan diamonds has long been debated and sought. The first reasonable theory was that the Pamali Breccia of the Bobaris ophiolite could be a kimberlite. However the placer deposits contain no kimberlite indicator minerals (Cr-pyrope, Mg-ilmenite). The ophiolitic complexes of SE Kalimantan occur in the Bobaris and the Meratus Mountains, overlain by the Upper Cretaceous Manunggul Group, to which the Pamali Breccia belongs (Sikumbang, 1986).

Sampling of the Pamali Breccia led to a recovery of 13 small diamonds (total 0.35 carats, largest 0.1 carat) from a 10 t sample of residual soil overlying the breccia, however the underlying fresh breccia yielded only 0.05 carats from a 15 t sample. It has now been shown that the Pamali Breccia is in fact composed of a serpentinite sandstone and conglomerate, derived by erosion of the ophiolite (Bergman *et al.*, 1987). Diamonds were also found in Upper Cretaceous basal conglomerate of the Manunggul Formation and in Eocene conglomerate and sandstone (Van Bemmelen, 1970). The diamonds found in hard rocks are accordingly all alluvial in origin, and the primary source remains an enigma.

Anaconda (1985) made an extensive but unsuccessful search for kimberlitic or lamproitic sources for the Kalimantan diamonds, before abandoning their project. Kajanite is a volcanic or hypabyssal rock containing olivine, leucite, nepheline, phlogopite, richterite, magnetite and chromite, occurring in the Kajan River in central Kalimantan (1°30′ N; 115°30′ E). It could be a possible source of diamonds. However, geochemically it lies outside of the lamproite clan (Bergman, 1987).

Small non-diamondiferous mafic to felsic minette dykes, K-Ar dated 7.8 Ma, intrude Lower Tertiary non-marine strata in the Karamu River (0°20′ N; 114°25′E). They are mantle derived rocks but are not of lamproite affinity and are unlikely to be the primary source of the alluvial diamonds (Bergman *et al.*, 1988).

The lack of any classic kimberlite or lamproite indicator minerals in the diamond-bearing alluvium indicates that the diamonds have probably been recycled more than once in their erosional/sedimentation history, so that a primary source is expected to remain elusive and may even lie in some ancestral craton from which western Borneo was rifted. The possibility of an ophiolite source remains unproven and not widely accepted.

15.2.1.3 Sarawak

Diamonds were panned last century from the river gravels in the headwaters of the Sungai Sarawak Kiri. They are interpreted to have been derived from Tertiary sandstones and conglomerate (Plateau and Kayan Sandstone), a source similar to occurrences in the Barito-Meratus area of Kalimantan. The diamonds occur along with pebble corundum and small quantities of gold (Wilford, 1955).

15.3 PHOSPHORITE

The SE region of China contains important Cambrian sedimentary phosphorite deposits (Yang *et al.*, 1986).

Phosphorite occurs as a fissure vein filling in Permian limestone at Tuk Meas, 45 km ENE of Kampot, Cambodia. Production began in 1966 with an annual production of 12 000 tonnes of 20% P_2O_5 ore.

By far the most important locality in Indochina is Lao Cai, in the upper Red River valley of north Vietnam (Tran *et al.*, 1979). The apatite rock occurs as thick lenses and beds in Cambrian Camduong Suite quartzite and marble. The metamorphic complex is derived from a sedimentary phosphorite. Known reserves are quoted at about 300 to 1000 Mt of ore with 10 to 40% P_2O_5. The Lower Cambrian phosphorite-containing section is more than 300 m thick. The main horizon is a grey dolomite-apatite rock of thickness 6 to 12 m containing quartz, pyrite and mica, in which the phosphate occurs as microcrystalline fluor-apatite. The unaltered rock contains 23 to 34% P_2O_5 but leaching of the upper zones has given enrichment to 42% P_2O_5. The richest parts contain 96% granular apatite. In addition there are quartzites containing 16 to 38% apatite and graphitic calc-schists where the apatite content ranges from 6 to 40%. Mining began pre-War but was reopened on a larger scale in 1957 with annual outputs of about 1 Mt of apatite. Opencast mining is carried out by benching to a depth of 120 m (Fontaine and Workman, 1978).

15.4 URANIUM

There are no proven commercially economic uranium deposits in South-East Asia, but several prospects have been investigated.

The Main Range granite batholiths of Peninsular Malaysia, western Thailand and the tin-belt islands of Indonesia are rich in uranium on a world scale (Cobbing *et al.*, 1986; Pitfield, 1987, 1988). The average U value is 12.7 ± 4.5 ppm and Th 47 ± 15 ppm. Locally, uranium contents exceed 20 ppm. Granites at Yala and Songkhla, southern Thailand, locally contain more than 500 ppm U (Shawe, 1984). Radioactive minerals occur disseminated throughout and concentrated in hydrothermally altered zones in granite in Surat Thani Province and the granite at Hua Hin contains monazite (Shawe, 1984).

Rivers flowing over outcrops of such granites should have leached and carried sufficient uranium to result in significant deposition in the Mesozoic and Tertiary intermontane basins. The black shales and lignites are the most likely to have been enriched in uranium.

15.4.1 Sumatra

The uranium content of the Upper Triassic Sibolga granite averages 20 ppm. The overlying Palaeogene Sibolga Formation has been drilled 6.5 km N of Sibolga and shales and carbonaceous siltstones contain 500 to 2500 ppm U (Koesoemadinata and Sastrawiharjo, 1988).

Although the Sopan Tertiary granite is overlain by soil containing anomalies of 71 to 91 ppm U, actual rock samples contain only 14 ppm U. The nearby Tertiary Ombilin intermontane basin was drilled for lignite exploration. There is no uranium enrichment in the non-oxidized sandstone or coaly clays (Barthel, 1988).

15.4.2 Thailand

Mineralization is known in the Jurassic Sao Khua Formation sandstone of the Khorat Group. Analyses gave only 20 to 43 ppm U (Barthel, 1988). At Phu Wiang, on the west edge of the Khorat basin, Lower Jurassic sandstone and conglomerate are mineralized in copper and uranium (Shawe *et al.*, 1975; Shawe, 1984). Radioactive minerals, malachite and azurite are disseminated and coat fractures in strata-bound pod-like bodies up to 1.5 m thick, in a N–S zone 1.5 km long and 150 m wide. The richest part contains 0.03 to 0.08% U and 0.8 to 3.0% Cu. Groundwater may be anomalously high in copper northwards along this zone (Shawe *et al.*, 1975).

The Tertiary intermontane lacustrine basins occur near to basement outcrops such as the Fang granite, which contains an average 12 ppm U and locally up to 128 ppm (Barthel, 1988), so may be favourable for U concentration. No direct observation of mineralization was observed. However, the lignite and lignitic clays should be possible uranium concentration formations.

15.4.3 Kalimantan

The Rirang River, a tributary of the Kalan, 400 km SE of Pontianak, contains mineralized boulders containing 0.6 to 6.7% U, but no *in situ* outcrops have been found. The boulders are of breccia containing monazite, uraninite, molybdenite, pyrite, rutile and tourmaline (Tjokrokardono and Sastratenaya, 1988). The Kalan prospect has been investigated at Remaja. Breccia zones within meta-sediments average 600 ppm U (Sarbini and Wirakusumah, 1988).

Pitchblende is associated with molybdenite and pyrite in Eocene aphanitic rhyolite in the Kawat River of the Upper Mahakam district of central Borneo (Djokolelono and Agoes, 1988).

15.4.4 Malaysia

The Boundary Range granite of NE Peninsular Malaysia has high U values and quartz veins contain rhabdophane, florencite and uraninite in shear zones (Ong and Mior Sallehhuddin, 1988).

15.5 SALT

The Upper Cretaceous Maha Sarakham Formation of the Khorat Basin of Thailand and Cambodia contains redbeds and salt deposits (Utha-Aroon, 1993). The salt beds have been cored. From the base upwards, the thin Basal Anhydrite is succeeded by the 63 m thick Lower Salt, with an overlying 36 m thick potash zone. The 26 m thick Lower Clastic Unit is overlain by the 89 m thick Middle Salt, the 39 m thick Middle Clastic Unit, then the 24 m thick Upper Salt and 209 m thick Upper Clastic Unit.

The Lower Salt is predominantly of halite. The potash zone is mainly of carnallite and sylvite. The Middle and Upper Salt are also of halite (Utha-Aroon, 1993).

16

References

Abdullah Sani b. H. Hashim (1983). Pemetaan geologi kawasan teluk Datuk dan Sepang, Selangor (syit **101** & **102**). *Geological Survey of Malaysia Annual Report 1983*, 163–7.

Abdullah Sani b. H. Hashim (1985). Discovery of an ammonoid (*Agathiceras* sp.) and crinoid stems in the Kenny Hill Formation of Peninsular Malaysia, and its significance. *Newsletter of the Geological Society of Malaysia (Warta Geologi)*, **11** (No. 5), 205–12. Kuala Lumpur.

Achache, J., Courtillot, V. and Besse, J. (1983). Paleomagnetic constraints on the Late Cretaceous and Cenozoic tectonics of Southeast Asia. *Earth and Planetary Letters*, **63**, 123–36.

Achmad, Z and Samuel, L. (1984). Stratigraphy and depositional cycles in the N.E. Kalimantan Basin. *Proceedings of the 13th annual convention*, **1**, 109–20, Indonesian Petroleum Association, Jakarta.

Adam, J. W. H. (1960). On the geology of the primary tin-ore deposits in the sedimentary formation of Billiton. *Geologie en Mijnbouw*, **22**, 406–26.

Adams, C.G. and Haak, R. (1962). The stratigraphical succession in the Batu Gading area, Middle Baram, North Sarawak. In *The geology and mineral resources of the Suai-Baram area, north Sarawak* (ed N.S. Haile), pp. 141–50 Geological Survey Department, British Territories in Borneo Memoir, **13**, Kuching.

Adiwidjaja, P. and De Coster, G.L. (1973). Pre-Tertiary paleotopography and related sedimentation in South Sumatra. *Proceedings of the 2nd annual convention*, 89–103, Indonesian Petroleum Association, Jakarta.

Adnan, A., Sukowitono, and Supriyanto (1991). Jatibarang Sub-Basin—a half graben model in the onshore of northwest Java. *Proceedings of the 20th annual convention*, **1**, 279–87. Indonesian Petroleum Association, Jakarta.

Agostinelli, E., Mohamad Raisuddin bin Ahmad Tajuddin, Antonielli, E. and Mohamad bin Mohd. Aris. (1990). Miocene–Pliocene paleogeographic evolution of a tract of Sarawak offshore between Bintulu and Miri. *Geological Society of Malaysia Bulletin*, **27**, 117 35.

Ahmad bin Jantan (1973). Stratigraphy of the Singa Formation (Upper Paleozoic) in the southwestern part of the Langkawi island group, West Malaysia. *Unpubl. M.Sc. thesis*, University of Malaya, Kuala Lumpur.

Ahrendt, H., Chonglakmani, C., Hansen, B.T., and Helmcke, D. (1993). Geochronological cross section through northern Thailand. *Journal of Southeast Asian Earth Sciences*, **8**, 207–17.

Akiyama, Yoshio (1984). A case history—exploration, evaluation and development of the Mamut porphyry copper deposit. *Geological Society Malaysia Bulletin*, **17**, 237–55

Alderton, D.H.M. and Sudharto, R.T. (1987). Mineralization at Gunung Limbung, West Java: a fluid inclusion and geochemical study. *Proceedings of the Pacific Rim Congress 87*, Australasian Inst. Min. Metal., Parkville, Australia, 5–8.

Alderton, D., Harmon, R., Sloane, H. and Sudharto, T. (1994). Fluid inclusion and stable isotope studies at Gunung Limbung Cu/Pb/Zn deposit, West Java. *Journ. South-East Asian Earth Sciences*, **10**, 25–38.

Aleva, G. J. J. (1973). Aspects of the historical and physical geology of the Sunda Shelf esential to the exploration of submarine tin placers. *Geologie en Mijnbouw*, **52**, 79–91.

Alexander, J.B. and Flinter, B.H. (1965). A note on varlamoffite and associated minerals from the Batang Padang district, Perak, Malaysia. *Mineralogical Magazine*, **35**, 622–7.

Almond, J., Vincent, P. and Williams, L.R. (1990). The application of detailed reservoir geological studies in the D18 Field, Balingian Province, offshore Sarawak. *Geological Society of Malaysia Bulletin*, **27**, 137–59.

Anaconda Indonesia Inc. (1985). *Final report Kalimantan diamond Project*, P.T. Aneka Tambang Arch. (unpublished).

Anderson, B.L., Bon, J. and Wahono, H.E. (1993). Reassessment of the Miocene stratigraphy, paleogeography and petroleum geochemistry of the Langsa Block in the offshore North Sumatra Basin. *Proceedings of the 22nd annual convention*, **1**, 169–89, Indonesian Petroleum Association, Jakarta.

Anonymous 1972: Penrose field conference on ophiolites. *Geotimes*, **17**, 24–5.

Armitage, J.H. and Viotti, C. (1977). Stratigraphic nomenclature—southern end Malay Basin. *Proceedings of the 6th annual convention*, 69–94, Indonesian Petroleum Association, Jakarta.

Arrykul, S., Ishihara, S., Bungbrakearti, N., Busaracome, P. and Sawata, H. (1978). Some characteristics of primary cassiterite and wolframite deposits in southern Thailand. Prinya Nutalaya (ed.) *Proceedings of the 3rd regional conference on geology and mineral resources of S. E. Asia*, Asian Instit. Technology, Bangkok, 415–17.

Asama, K. (1976). *Gigantopteris* flora in Southeast Asia and its phytopalaeogeographic significance. *Geology and Palaeontology of Southeast Asia*, **17**, 191–207, Univ. Tokyo Press.

Asama, K. (1984). *Gigantopteris* flora in China and South-East Asia. *Geology and Palaentology of Southeast Asia*, **25**, 311–25, Univ. of Tokyo Press.

ASCOPE (1981). *Tertiary sedimentary basins of the Gulf of Thailand and South China Sea: stratigraphy, structure and hydrocarbon occurrences*. ASEAN Council on Petroleum, Jakarta.

Asnachina, P. (1981). The stibnite-bearing deposits of northern Thailand. *Proceedings of the 4th regional conference on the geology of Southeast Asia*, Geol. Soc. Philippines, Manila, 627–31.

Asnachina, P and Chantaramee, S. (1986). Regional controls of hydrothermal ore localization in northern Thailand. *Geological Society Malaysia Bulletin*, **19**, 421–9.

Asnachina, P and Pitragool, S. (1978). Review of the non-metallic mineral deposits of Thailand. Prinya Nutalaya (ed.) *Geology and Mineral Resources of Southeast Asia*, Asian Instit. Technology, Bangkok, 795–804.

Austin, G.T. (1992). *Gemstones*. Annual Report for **1991**. U.S. Department of the Interior, Bureau of Mines. Govt. Printing Office, Washington.

Bachman, S.B., Lewis, S.D. and Schweller, W.J. (1983). Evolution of a forearc basin, Luzon Central Valley, Philippines. *American Association of Petroleum Geologists Bulletin*, **67**, 1143–62.

Bacuta, G.C. (1979). *Geology of some alpine-type chromite deposits in the Philippines*, Bureau of Mines Special Publication; Journal Geological Society of the Philippines, **32** (2).

Bacuta, G.C., Gibbs, A.K., Kay, R.W. and Kuoubm, B.R. (1987). Platinum group element abundance in chromite deposits of the Acoje ophiolite block, Zambales ophiolite complex, Philippines [Abstract], *Geo-Platinum 87* (ed. by H.M. Prichard), pp. 381–2. Elsevier Scientific Publications, Essex, England.

Bainton, C.S. (1978). Coal formations in Indonesia. *Proceedings regional conference on geology and mineral resources of Southeast Asia, Jakarta Aug. 1975*, 55–63, (ed. by S. Wiryosutonto and A. Sudradjat), Indonesian Association of Geologists, Jakarta.

Banthay Komphasouk, Somboun Douang, Aphay Many, Saman Aneka and Nguyen Huu Thac (1986). Main structural features and distribution of coal in Laos. *Proceedings of the 1st conference on geology of Indochina*, **2**, 579–89. General Department of Geology, Hanoi, Vietnam.

Barr, S.M. and MacDonald, A.S. (1981). Geochemistry and geochronology of late Cenozoic basalts of Southeast Asia. *Geological Society America Bulletin*, **92**, 1069–142.

Barr, S.M. and MacDonald, A.S. (1987). Nan River suture zone, northern Thailand. *Geology*, **15**, 907–10.

Barr, S.M. and MacDonald, A.S. (1991). Towards a late Palaeozoic–early Mesozoic tectonic model for Thailand. *Journal of Thai Geosciences*, ChiengMai, Thailand, **1**, 11–22.

Barthel, F.H. (1988). Sandstone uranium deposits in central Sumatera, Indonesia, and northern Thailand. *Uranium deposits in Asia and the Pacific: geology and exploration, Proceedings of a technical committee meeting, Jakarta, 16–19 Dec., 1985*, International Atomic Energy Agency, Vienna, IAEA-TC **543**, 179–92.

Basir, J., Tahir, S. and Samsuddin, A.R. (1985). Lower Cretaceous radiolaria from the Chert-Spilite Formation, Kudat, Sabah. *Geological Society of Malaysia Newsletter*, **11**,161–2.

Batchelor, B.C. (1979). Geological characteristics of certain coastal and offshore placers as essential guides for tin exploration in Sundaland, Southeast Asia. *Geological Society Malaysia Bulletin*, **11**, 283–313.

Baumann, P. (1972). Les faunes de foraminiferes de L'Eocene supérieur à la bas du Miocene dans le bassin de Pasir, sud de Kalimantan (Borneo). *Revue de L'Institut Français du Pétrole*, Nov-Dec, 817–29.

Bean, J.H. (1969). The iron-ore deposits of West Malaysia. *Geological Survey West Malaysia, Economic Bulletin*, **2**, Ipoh.

Beddoes, L.R. (1976). The Balabac sub-basin, southwestern Sulu Sea, Philippines. *Offshore Southeast Asia Conference*, Paper 15. Southeast Asian Petroleum Exploration Society, Singapore.

Bell, R.M. and Jessop, R.G.C. (1974). Exploration and geology of the west Sulu Basin, Philippines. *Australian Petroleum Exploration Association Journal*, **14**, 21–8.

Belov, A.A., Gatinsky, Y.G. and Mossakovsky, A.A. (1986). A precis on pre-Alpine tectonic history of Tethyan palaeo-oceans. *Tectonophysics*, **127**, 197–211.

Bender, F. (1983). *Geology of Burma*. Gebrüder Borntraeger, Berlin.

Bergman, S.C. (1987). Lamproites and other potassium-rich igneous rocks: a review of their occurrence, mineralogy and geochemistry. In *Alkaline igneous rocks* (ed. Fitton J.G. and Upton B.G.J.), Geological Society London, Special Publication, **30**, 103–90.

Bergman, S.C., Turner, W.S. and Krol, L.G. (1987). A reassessment of the diamondiferous Pemali Breccia, southeast Kalimantan, Indonesia: intrusive kimberlite breccia or sedimentary conglomerate? *Geological Society of America Special Paper*, **215**, 183–95.

Bergman, S.C., Dunn, D.P. and Krol, L.G. (1988). Rock and mineral chemistry of the Linhaisai minette, central Kalimantan, Indonesia, and the origin of Borneo diamonds. *Canadian Mineralogist*, **26**, 23–43.

Berrangé, J.P. and Jobbins, E.A. (1976). The geology, gemmology, mining methods and economic potential of the Pailin ruby and sapphire gem field, Khmer Republic. *Overseas Division Report*, **35**, 27p. Institute of Geological Sciences, London.

Bignell, J.D. and Snelling, N.J. (1977). K-Ar ages on some basic igneous rocks from Peninsular Malaysia. *Geological Society Malaysia Bulletin*, **8**, 89–93.

Bishop, W.P. (1980). Structure, stratigraphy, and hydrocarbons offshore southern Kalimantan, Indonesia. *American Association Petroleum Geologists Bulletin*, **64**, 37–58.

Blossom, J.W. (1993). *Molybdenum*. Annual Report for **1991**, U.S. Dept. of the Interior, Bureau of Mines. Govt. Printing Office, Washington.

Bol, A.J. and Van Hoorn, B. (1980). Structural styles in western Sabah offshore. *Geological Society Malaysia Bulletin*, **12**, 1–16.

Bowin, C. *et al.* (1980). Arc-continent collision in the Banda Sea region. *American Association of Petroleum Geologists Bulletin*, **64**, 868–915.

Bradford, E.F. (1972). The geology and mineral resources of the Gunung Jerai area, Kedah. *Geological Survey of Malaysia, District Memoir*, **13**.

B.R.G.M. (1982). Geological mapping and mineral exploration in northeast Kalimantan 1979–1982, Final Report.. *Rapport du Bureau de Recherches Géologique et Minières*, **82 RDM 007AO**. Orléans, France.

Briais, A., Tapponnier, P. and Pautot, G. (1989). Constraints of Sea Beam data on crustal fabrics and seafloor spreading in the South China Sea. *Earth and Planetary Science Letters*, **95**, 307–20.

Briais, A., Patriat, P. and Tapponnier, P. (1993). Updated interpretation of magnetic anomalies and seafloor spreading stages in the South China Sea: implications for the Tertiary tectonics of Southeast Asia. *Journal of Geophysical Research*, **98**, 6299–328.

Brinckmann, J. and Hinze, C. (1981). On the geology of the Bawdwin lead-zinc mine, northern Shan State, Burma. *Geologisches Jaarbuch*, Hannover, **D43**, 7–45.

Brondijk, J.F. (1963). A reclassification of a part of the Setap Shale Formation as the Temburong Formation. *British Borneo Geological Survey Annual Report for 1962*, 56–60. Kuching.

Brown, G.F. *et al.* (1951). Geologic reconnaissance of the mineral deposits of Thailand. *U.S. Geological Survey Bulletin*, **984**.

Bryner, L. (1969). Ore deposits of the Philippines—an introduction to their geology. *Economic Geology*, **64**, 644–66.

Buffetaut, E., Helmcke-Ingavat, R., Jaeger, J.J., Jongkanjanasoontorn, Y., Suteethorn, V. and Tong, H. (1989). Fossil vertebrates and the age of the intermontane basins of Thailand. *Intermontane basins: geology and resources*, Proceedings of International Symposium, (ed. T. Thanasuthipitak and P. Ounchanum), 187–95. Chiang Mai University, Thailand.

Bureau of Mines (1976). A review of oil exploration and stratigraphy of sedimentary basins of the Philippines. *United Nations* ESCAP-CCOP Technical Bulletin, Bangkok, **10**, 55–99.

Bureau of Mines (1981). *Geology and mineral resources of the Philippines*, **1**, *(geology)*. Bureau of Mines and Geo-sciences, Manila.

Bureau of Mines (1986). *Geology and mineral resources of the Philippines*, **2**, *(mineral resources)*. Bureau of Mines and Geo-sciences, Metro Manila.

Burgath, K.P. (1987). Platinum group minerals in ophiolitic chromitites and alluvial placer deposits, Meratus-Bobaris area, S.E. Kalimantan. In *Geo-Platinum '87* (ed. H.M. Prichard), Elsevier Scientific Publishers, United Kingdom, 383–404.

Burnaman, M.D., Helm, R.B., and Beeman, C.R. (1985). Discovery of the Cunda Gas Field, Bee Block, North Sumatra: an integrated geologic/seismic stratigraphic case history. In *Proceedings of the 14th annual convention*, Vol. **1**, 453–95. Indonesian Petroleum Association, Jakarta.

Burri, P. (1989). Hydrocarbon potential of Tertiary intermontane basins in Thailand. *Intermontane Basins: geology and resources, Proceedings of the International Symposium*. (ed. T. Thanasuthipitak and P. Ounchanum), pp. 3–12. Chiang Mai University, Thailand.

Burton, C.K. (1973). Geology and mineral resources Johore Bahru–Kulai area, south Johore. *Geological Survey of Malaysia Map Bulletin* **2**, Ipoh.

Burton, C.K. (1983). Observations on the geology of the porphyry copper sub-province of Southwest Negros, Philippines. *Geological Society of Malaysia Bulletin*, **16**, 215–39.

Burton, C.K. (1986). The Kanchanaburi Supergroup of Peninsular and Western Thailand. *Geological Soc. Malaysia Bulletin*, **20**, 311–61.

Callow, K.J. (1967). The geology of the Thanksgiving Mine, Baguio district, Mountain Province, Philippines. *Economic Geology*, **62**, 472–81.

Cameron, N.R., Clarke, M.C.G., Aldiss, D.T., Aspden, J.A. and Djunuddin, A. (1980). The geological evolution of Northern Sumatra. *Proceedings of the 9th annual convention*, 149–87. Indonesian Petroleum Association, Jakarta.

Carlile, J.C., Digdowirogo, S. and Darius, K. (1990). Geologic setting, characteristics and regional exploration for gold in the volcanic arcs of North Sulawesi, Indonesia. *Journal of Geochemical Exploration*, **35**, 105–40.

Carlin, J.F. jr. (1992). *Tin*. Annual Report for **1991**, U.S. Dept. of the Interior, Bureau of Mines. Govt. Printing Office, Washington

Caughey, C.A. and Wahyudi, T. (1993). Gas reservoirs in the Lower Miocene Peutu Formation, Aceh Timur, Sumatra. *Proceedings of the 22nd annual convention*, **1**, 191–218. Indonesian Petroleum Association, Jakarta.

C.C.O.P. (1991). Total sedimentary isopach maps offshore east Asia, 6 sheets, 1:4 000 000. *Technical Bulletin*, **23**. Committee for Co-ordination of Joint Prospecting for Mineral Resources in Asian Offshore Areas, Bangkok.

Chakkaphak, P. and Veeraburus, M. (1982). *Geological map of Thailand, 1:1 000 000*. Department of Mineral Resources, Bangkok.

Chakraborty, K.R. and Metcalfe, I. (1984). Analysis of mesoscopic structures at Mersing and Tanjung Kempit, Johore, Peninsular Malaysia. *Geological Society Malaysia Bulletin*, **17**, 357–71.

Chang Chengfa and Pan Yusheng (1984). A preliminary synthesis of the tectonic evolution of the Qinghai-Xizang (Tibet) Plateau. Tectonics of Asia Colloquium. *27th International Geological Congress, Report* **5**, 190–206. Moscow.

Charoensri, Prachon (1982). Tungsten resources of Thailand. *Symposium on tungsten geology* (ed. J.V. Hepworth) *Jiangxi, China, October 1981*; ESCAP/RMRDC, Bandung Indonesia, and Geological Publishing House, Beijing, China, 33–43.

Charusiri, P., Clark, A.H., Farrar, E., Archibald, D. and Charisuri, B. (1993). Granite belts in Thailand: evidence from the $^{40}Ar/^{39}Ar$ geochronological and geological syntheses. *Journal of Southeast Asian Earth Sciences*, **8**, 127–36.

Chen S. (1987). Magnetic profiles. *Atlas of Geology and Geophysics of the South China Sea, 1:2 000 000.* Second Marine Geology Investigation Brigade of the Ministry of Geology and Mineral Resources, Guangdong.

Chen Sizhong and Pei Cunmin (1993). Geology and geochemistry of source rocks of the eastern Pearl River Mouth Basin, South China Sea. *Journal of Southeast Asian Sciences*, **8**, 393–406.

Chen Xin and Wang Zhitai (1988). Distribution of tin deposits in China and their metallogenic conditions. In *Geology of tin deposits in Asia and the Pacific*, (ed. C S. Hutchison), Springer-Verlag, Heidelberg, 235–44.

Chen Yuchuan, Huang Minzhi, Xu Yu, Ai Yongde, Li Xiangming, Tang Shaohua and Meng Lingku (1988). Geological and metallogenic features and model of the Dachang cassiterite-sulphide polymetallic ore belt, Guangxi, China. In (ed. C.S. Hutchison),*Geology of tin deposits in Asia and the Pacific*. Springer-Verlag, Heidelberg, 358–72.

Chinbunchorn, N., Pradidtan, S. and Sattayarak, N. (1989). Petroleum potential of Tertiary intermontane basins in Thailand. *Intermontane basins: geology and resources*. Proceedings of the international symposium, (ed. T. Thanasuthipitak and P. Ounchanum). Chiang Mai University, Thailand, 29–42.

Chong Nai Hooi (1976). Tungsten occurrences of the Thai-Malay Peninsula. *unpubl. M.Sc. thesis*, University of Malaya, Kuala Lumpur.

Chu Ling Heng, Fateh Chand and D. Santokh Singh (1988). Primary tin mineralization in Malaysia: aspects of geological setting and exploration strategy. In *Geology of tin deposits in Asia and the Pacific*, (ed. C.S. Hutchison), Springer-Verlag, Heidelberg, 593–613.

Clennell, B.N. (1991). The origin and tectonic significance of mélanges in eastern Sabah, Malaysia. *Journal of Southeast Asian Earth Sciences*, **6**, 407–30.

Clennell, B.N. (1992). *The mélanges of Sabah, Malaysia*. Ph.D. thesis, Royal Holloway and Bedford New College, University of London (unpubl.).

Clure, J. (1991). Spreading centres and their effect on oil generation in the Sunda region. *Proceedings of the 20th annual convention*, Indonesian Petroleum Association, Jakarta, **1**, 37–49.

Cobbing, E.J., Mallick, D.I.J., Pitfield, P.E.J. and Teoh, L.H. (1986). The granites of the Southeast Asian tin belt. *Journal of the Geological Society London*, **143**, 537–50.

Cobbing, E.J., Pitfield, P.E.J., Darbyshire, D.P.F. and Mallick, D.I.J. (1992). The granites of the South-East Asian tin belt. *Overseas Memoir*, **10**. British Geological Survey, Keyworth, Nottinghamshire.

Coffield, D.Q., Bergman, S.C., Garrard, R.A., Guritno, N., Robinson, N.M. and Talbot, J. (1993). Tectonic and stratigraphic evolution of the Kalosi PSC area and associated development of a Tertiary petroleum system, South Sulawesi, Indonesia. *Proceedings of the 22nd annual convention*, **1**, 679–706. Indonesian Petroleum Association, Jakarta.

Collenette, P. (1965). The geology and mineral resources of the Pensiangan and Upper Kinabatangan area, Sabah, Malaysia. *Borneo Region Malaysian Geological Survey Memoir*, **12**.

Colley, H. (1976). Classification and exploration guide for Kuroko-type deposits based on occurrences in Fiji. *Transactions Institution Mining and Metallurgy*, **85**, B190–9.

Cooke, D.R. and Bloom, M.S. (1990). Epithermal and subjacent porphyry mineralization, Acupan, Baguio District, Philippines: a fluid-inclusion and paragenetic study. *Journal Geochemical Exploration*, **35**, 297–340.

Cooper, M.A., Herbert, R. and Hill, G.S. (1989). The structural evolution of Triassic intermontane basins in Notheastern Thailand. *Intermontane basins: geology and resources, Proceedings of the international symposium* (ed. T. Thanasuthipitak and P. Ounchanum), 231–42. Chiang Mai University, Thailand.

Counillon, H. (1914). Flore fossile des gîtes de charbon de L'Annam. *Bulletin Service Géologique de L'Indochine*, **1** (no. 2), 21p.

Crawford, M.B. and Searle, M.P. (1992). Field relationships and geochemistry of pre-collisional (India–Asia) granitoid magmatism in the central Karakoram, northern Pakistan, *Tectonophysics*, **206**, 171–92.

Creaney, S., Koch, P.S., and Rudolph, K.W. (1993). Petroleum migration in the Malay Basin—our current understanding. *Petroleum Geology Seminar '93, Programme and Abstracts of Papers*, p. 12. Geological Society of Malaysia, Kuala Lumpur.

Crostella, A. (1977). Geosynclines and plate tectonics in Banda arcs, eastern Indonesia. *American Association of Petroleum Geologists Bulletin*, **61**, 2063–81.

Cunningham, L.D. (1993). *Columbium (niobium) and tantalum*. Annual report for **1991**. U.S. Department of the Interior, Bureau of Mines. Govt. Printing Office, Washington.

Curray, J.R. (1989). The Sunda Arc: a model for oblique plate convergence. *Netherlands Journal of Sea Research*, **24**, 131–40.

Curray, J.R. (1991). Possible greenschist metamorphism at the base of a 22-km sedimentary section, Bay of Bengal. *Geology*, **19**, 1097–100.

Curray, J.R. and Munasinghe, T. (1989). Timing of intraplate deformation, northeastern Indian Ocean. *Earth and Planetary Science Letters*, **94**, 71–7.

Cuthbertson, A.S. (1988). *Regional geology of the central Kalimantan contract of work area*, and 1:250 000 map. P. T. Rio Tinto Indonesia, unpubl. reports **368** and **377**.

Daines, S.R. (1985). Structural history of the W Natuna Basin and the tectonic evolution of the Sunda Region. *Proceedings of the 14th Annual convention*, Vol. **1**, 39–61. Indonesian Petroleum Association, Jakarta.

Daly, M.C., Hooper, B.G.D. and Smith, D.G. (1989). Tertiary plate tectonics and basin evolution in Indonesia. In *Proceedings of the 6th regional conference on the geology, mineral and hydrocarbon resources of Southeast Asia (GEOSEA 6), Jakarta, July 1987* (ed. B. Situmorang), 105–34. Indonesian Association of Geologists, Jakarta.

Davies, P.R. (1984). Tertiary structural evolution and related hydrocarbon occurrences, North Sumatra Basin. *Proceedings of the 13th annual convention*, **1**, 453–95. Indonesian Petroleum Association, Jakarta.

Davis, D.W., Kumalachan, P., Mantajit, N. and Potisat, S. (1987). Thailand—a new focus of Asian gold mining investment. *Proceedings of the 1st international conference on gold mining, Nov. 1987, Vancouver*. A.I.M.E., New York, **5**, 59–73.

Davis, L.L. (1992). *Gypsum*. Annual Report for **1991**. U.S. Department of the Interior, Bureau of Mines. Govt. Publishing Office, Washington.

De Coster, G.L. (1974). The geology of the Central and South Sumatra basins. In *Proceedings of the 3rd annual convention*, 77–110. Indonesian Petroleum Association, Jakarta.

Dewey, J.F. Cande, S. and Pitman, W.C. III (1989). Tectonic evolution of the India/Eurasia collision zone. *Eclogae geologicae Helvetiae*, **82**(3), 717–34.

Divis, A.F. (1983). The geology and geochemistry of Philippine porphyry copper deposits. In *The tectonic and geologic evolution of Southeast Asian seas and islands*, (ed. D.E. Hayes), Geophysical Monograph, **27**, pp. 173–216. American Geophysical Union, Washington.

Djokolelono, Slamet and Agoes, Effentrip (1988). Uranium occurrences in the volcanic rocks of Upper Mahakam, East Kalimantan. *Uranium deposits in Asia and the Pacific: geology and exploration*. Proceedings of a technical committee meeting, Jakarta, Dec 1985, International Atomic Energy Agency, Vienna, IAEA-TC **543**, 109–20.

Djuanda, H (1985). Facies distribution in the Nurbani Carbonate build-up, Sunda Basin. In *Proceedings of the 14th annual convention*, Vol. **2**, 507–33. Indonesian Petroleum Association, Jakarta.

Djuri and Sudjatmiko (1974). *Geological map of the Majene and western part of the Palopo quadrangles, south Sulawesi, scale 1:250,000*, Geological Research and Development Centre, Bandung, **17/X** and **18/X**.

Doust, H. (1981). Geology and exploration history of offshore Central Sarawak. In *Energy resources of the Pacific region*. AAPG Studies in Geology, **12**, (ed. M.T. Halbouty), 117–32; American Association of Petroleum Geologists, Tulsa.

Doutch, H.F. (1992). Aspects of the structural histories of the Tertiary sedimentary basins of East, Central and West Kalimantan and their margins. *B.M.R. Journal of Australian Geology and Geophysics*, **13**, 237–50. Canberra.

Du Bois, E.P. (1985). Review of principal hydrocarbon-bearing basins around the South China Sea. *Geological Society of Malaysia Bulletin,* **18**, 167–209. Kuala Lumpur.

Durkee, E.F. (1993). Oil, geology, and changing concepts in the southwest Philippines (Palawan and the Sulu Sea). *Geological Society of Malaysia Bulletin,* **33**, 241–62.

England, P.C. and Molnar, P. (1990). Right-lateral shear and rotation as an explanation for strike-slip faulting in eastern Tibet. *Nature,* **344**, 140–2.

Enkin, R.J., Yang, Z., Chen, Y. and Courtillot, V. (1992). Paleomagnetic constraints on the geodynamic history of the major blocks of China from Permian to the Present. *Journal of Geophysical Research,* **97** (B10), 13 953–89.

Epting, M. (1980). Sedimentology of Miocene carbonate buildups, Central Luconia, offshore Sarawak. *Geological Society of Malaysia Bulletin,* **12**, 167–209. Kuala Lumpur.

Eubank, R.T. and Makki, A.C. (1981). Structural geology of the Central Sumatra back-arc basin. *Proceedings of the 10th Annual convention,* 153–96. Indonesian Petroleum Association, Jakarta.

Eugster, H.P. and Chou, I-Ming (1979). A model for the deposition of Cornwall-type magnetite deposits. *Economic Geology,* **74**, 763–74.

Fernandez, H.E. (1968). The geology of cinnabar deposits of central Palawan. *Philippine Geologist,* **22** (No. 2), 71–105.

Fernandez, N. S. (1960). Notes on the geology and chromite deposits of the Zambales Range. *Philippine Geologist,* **14**, 1–8.

Fitch, F.H. (1952). *The geology and mineral resources of the neighbourhood of Kuantan, Pahang.* Geological Survey Department Federation of Malaya Memoir, **6** (new series), Ipoh.

Fitch, T.J. (1972). Plate convergence, transcurrent faults and internal deformation adjacent to Southeast Asia and the western Pacific. *Journal Geophysical Research,* **77**, 4432–62.

Flinter, B.H. (1960). The effects of heat, hydrochloric acid and lead chloride on some Malayan mineral grains. *Overseas Geology and Mineral Resources,* London, **8**, 54–6.

Folinsbee, R.E. Kirkland, K., Nekolaichuk, A. and Smejkal, V. (1972). Chinkuashih—gold-pyrite-enargite-barite hydrothermal deposit in Taiwan. *Memoir Geological Society of America,* **135**, 323–35.

Fontaine, H. (1989). Peculiarities of the Permian of Peninsular Thailand. *C.C.O.P. Newsletter,* **14** (No. 1), 15–20. Bangkok.

Fontaine, H. (1990). *Ten years of CCOP research on the pre-Tertiary of East Asia.* C.C.O.P. Technical Secretariat, Bangkok, CCOP/**TP 20**.

Fontaine, H. and Gafoer, S. (1989). *The pre-Tertiary fossils of Sumatra and their environments.* CCOP/TP **19**, C.C.O.P. Technical Secretariat, Bangkok.

Fontaine, H. and Workman, D.R. (1978). Review of the geology and mineral resources of Kampuchea, Laos and Vietnam. In *Proceedings of the 3rd regional conference on geology and mineral resources of Southeast Asia, Nov. 1978,* (ed. Prinya Nutalaya), pp. 541–603. Asian Institute of Technology, Bangkok,

Fontaine, H., Rodziah Daud and Singh Updesh (1990). A Triassic 'reefal' limestone in the basement of the Malay Basin, South China Sea: Regional implications. *Geological Society Malaysia Bulletin,* **27**, 1–25.

Foo, B.N. (1977). Mineral paragenesis, fluid inclusion studies and geochemistry of the Sungei Lembing tin lodes, West Malaysia. *Transactions Institution Mining and Metallurgy,* London, **86**, B163.

Foo Khong Yee (1983). The Palaeozoic sedimentary rocks of Peninsular Malaysia—stratigraphy and correlation. *Proceedings of the workshop on stratigraphic correlation of Thailand and Malaysia,* Geological Society of Thailand, Bangkok; Geological Society of Malaysia, Kuala Lumpur, 1–19.

Fromaget, J. (1927). *Études géologiques sur le nord de L'Indochine centrale.* Bulletin Service Géologique de L'Indochine, **26** (No.2). Hanoi.

Fromaget, J. (1941). L'Indochine française, sa structure géologique, ses roches, ses mines et leurs relations possibles avec la tectonique. *Bulletin Service Géologique de L'Indochine,* **26** (No. 2), 140p.

Fu Jiamo, Pei Cunmin, Sheng Guoying, Liu Dehan and Chen Sizhong (1993). A geochemical investigation of crude oils from eastern Pearl River River Mouth Basin, South China Sea. *Journal of Southeast Asian Earth Sciences,* **8**, 469–86.

Fu, M., Kwak, T.A.P. and Mernagh, T.P. (1993). Fluid inclusion studies of zoning in the Dachang tin-polymetallic ore-field, People's Republic of China. *Economic Geology,* **88**, 283–300.

Fuller, M., Haston, R., Lin Jin-lu, Richter, B., Schmidtke, E. and Almasco, J. (1991). Tertiary paleomagnetism of regions around the South China Sea. *Journal Southeast Asian Earth Sciences,* **6**, 161–84.

Gambogi, J. (1992). *Titanium.* Annual Report for **1991**. U. S. Dept. of the Interior, Bureau of Mines. Govt. Printing Office, Washington.

Gan Lay Chin (1980). Manson Lode, a stratabound base metal-silver deposit in north Kelantan, Malaysia. *Unpubl. Ph.D. thesis. Mining University of Leoben, Austria.*

Garson, M.S., Young, B., Mitchell, A.H.G. and Tait, B.A.R. (1975). The geology of the tinbelt in peninsular Thailand around Phuket, Phangnga and Takkua Pa. *Overseas Memoir*, **1**. Institute of Geological Sciences, London.

Gatinsky, Y.G. and Hutchison, C.S. (1986). Cathaysia, Gondwanaland, and the Paleotethys in the evolution of continental Southeast Asia. *Geological Society Malaysia Bulletin*, **20**, 179–99.

Gervasio, F. C. (1973). Ore deposits of the Philippine mobile belt. In Metallogenic provinces and mineral deposits in the Southwest Pacific, *Bureau Mineral Resources, Geology and Geophysics, Bulletin*, **141**, (ed. N. H. Fisher). 191–207; Canberra.

Ginger, D.C., Ardjakusumah, W.O., Hedley, R.J. and Pothecary, J. (1993). Inversion history of the West Natuna Basin: examples from the Cumi-Cumi PSC. *Proceedings of the 22nd annual convention*, **1**, 635–58. Indonesian Petroleum Association, Jakarta.

Gobbett, D.J. (1973). Upper Paleozoic. In *Geology of the Malay Peninsula*, (ed. D.J. Gobbett and C.S. Hutchison). 61–95. John Wiley-Interscience, New York.

Gobbett, D.J. and Hutchison, C.S. (ed.) (1973). *Geology of the Malay Peninsula*. John Wiley-Interscience, New York.

Gobbett, D.J. and Tjia, H.D. (1973). Tectonic history. In *Geology of the Malay Peninsula*, (ed. D.J. Gobbett and C.S. Hutchison), pp. 305–30, John Wiley-Interscience, New York.

Golightly, J.P. (1979*a*). Nickeliferous laterites: a general description. *International laterite symposium*, 3–23. (ed. D.J.I. Evans, R.S. Shoemaker and H. Veltman), American Institution Mining Metallurgy Petroleum Engineers, New York.

Golightly, J.P. (1979*b*). Geology of Soroako nickeliferous laterite deposits. *International laterite symposium*, 38–56, (ed. D.J.I. Evans, R.S. Shoemaker and H. Veltman), American Institution Mining Metallurgy Petroleum Engineers, New York.

Goossens, P.J. (1978). The metallogenic provinces of Burma, their definitions, geologic relationships and extensions into China, India and Thailand. Prinya Nutalaya (ed.) *Proceedings of the 3rd regional conference on geology and mineral resources of S. E. Asia*, Asian Institute Technology, Bangkok, 431–92.

Gordon, T.L. (1985). Talang Akar coals—Ardjuna subbasin oil source. *Proceedings of the 14th annual convention*, Indonesian Petroleum Association, Jakarta, **2**, 91–120.

Görür, N. and Sengör, A.M.C. (1992). Paleogeography and tectonic evolution of the eastern Tethysides: implications for the northwest Australian margin breakup history. In *Proceedings of the Ocean Drilling Program, scientific results*, (ed. U. von Rad and B.U. Haq *et al.*), **122**, 83–106. College Station, Texas.

Gower, R. J. W. (1990). Early Tertiary plate reconstructions for the South China Sea region: constraints from northwest Borneo. *Journal of Southeast Asian Earth Sciences*, **4**, 29–35.

Grubb, P.L.C. (1968). Geology and bauxite deposits of the Pengerang area, south-east Johore. *Geological survey Malaysia memoir*, **14**, Ipoh.

Grubb, P.L.C. and Hannaford, P. (1966). Magnetism in cassiterite. *Mineralium Deposita*, **2**, 148–77.

Hageman, H. (1987). Palaeobathymetrical changes in NW Sarawak during the Oligocene to Pliocene. *Geological Society of Malaysia Bulletin*, **21**, 91–102.

Haile, N.S. (1954). The geology and mineral resources of the Strap and Sadong valleys, west Sarawak, including the Klinkang coal range. *Geological Survey Department of the British Territories in Borneo, Memoir* **1**, Kuching.

Haile, N.S. (1957). The geology and mineral resources of the Lupar and Saribas valleys, west Sarawak. *British Borneo Geological Survey Memoir*, **7**, Kuching.

Haile, N.S. (1962). The geology and mineral resources of the Suai-Baram area, north Sarawak. *Geological Survey Department, British Territories in Borneo, Memoir*, **13**, Kuching.

Haile, N.S. (1974). Borneo. *Mesozoic–Cenozoic orogenic belts: data for orogenic studies*. Geological Society of London Special Publication **4** (ed. A.M. Spencer), pp. 333–47.

Haile, N.S. (1994). The Rajang accretionary prism and the trans-Borneo Danau Suture. *Tectonic evolution of SE Asia* conference, abstract volume, p. 17, December 1994, Geological Society of London.

Haile, N.S. and Ho, C.K. (1991). *Geological field guide: Sibu-Miri traverse, Sarawak (24 September 1 October 1991)*. Petronas Petroleum Research Institute, Kuala Lumpur.

Haile, N.S. and Hutchison, C.S. (1992). Obituary of Kenneth Frederick George Hosking, *Journal Southeast Asian Earth Sciences*, **7**, 87–8.

Haile, N.S., Wong, N.P.Y. and Nuttall, C.P. (1965). The geology and mineral resources of Dent Peninsula, Sabah. *Geological Survey of the Borneo Region, Malaysia, memoirs*, **16**.

Haile, N.S., McElhinny, M.W. and McDougall, I. (1977). Palaeomagnetic data and radiometric ages from the Cretaceous of West Kalimantan (Borneo), and their significance in interpreting regional structure. *Journal of the Geological Society of London*, **133**, 133–44.

Haile, N.S., Beckinsale, R.D., Chakraborty, K.R., Abdul Hanif Hussein and Tjahjo Harbdjono (1983). Palaeomagnetism, geochronology and petrology of the dolerite dykes and basaltic lavas from Kuantan. *Geological Society Malaysia Bulletin,* **16**, 71–85.

Hamidsyah, H. and Clarke, M. C. G. (1982). Discovery of primary tungsten and tin mineralization in N. Sumatra, Indonesia (A case history). In *Tungsten geology, Jiangxi, China,* ESCAP/RMRDC, (ed. J. V. Hepworth and Y.H. Zhang Yuhong), pp. 49–58, Bandung, Indonesia.

Hamilton, W. (1979). Tectonics of the Indonesian Region. *U.S. Geological Survey Professional Paper,* **1078**, 345p.

Hamilton, W. (1989). Convergent plate tectonics viewed from the Indonesian region. Geologi Indonesia, J.A. Katili Commemorative Volume (60 years), *Journal Indonesian Association of Geologists*, Jakarta, **12** (No. 1), 35–88.

Haq, B.V., Hardenbol, J. and Vail, P.R. (1987). Chronology of fluctuating sea levels since the Triassic. *Science*, **235**, 1156–67.

Harbury, N.A., Jones, M.E., Audley-Charles, M.G., Metcalfe, I., and Mohamed, K.R. (1990). Structural evolution of Mesozoic Peninsular Malaysia. *Geological Society of London Journal,* **147**, 11-26.

Harbury, N.A., Jones, M.E., Audley-Charles, M.G., Metcalfe, I, Mohamed, K.R. and Altermann, W. (1991). Discussion on structural evolution of Mesozoic Peninsular Malaysia. *Geological Society London Journal,* **148**, 417–19.

Hariadi, N. and Soeparjadi, R.A. (1975). Exploration of the Mentawai Block—west Sumatra. *Proceedings of the 4th annual convention,* **1**, 55–66. Indonesian Petroleum Association, Jakarta.

Hartanto, K., Widianto, E. and Satrizal. (1991). Hydrocarbon prospect related to the local unconformities of the Kuang area, South Sumatra Basin. *Proceedings of the 20th annual convention,* **1**, 17–35. Indonesian Petroleum Association, Jakarta.

Hartono, H.M.S. (1990). Late Cenozoic tectonic development of the Southeast Asian continental margin in the Banda Sea area. *Tectonophysics*, **181**, 267–76.

Harun Mohd. Noor. (1987). Tinggi Field—analysing the DHIs. *Geological Society of Malaysia Bulletin*, **21**, 133–49.

Hasan, K. (1991). The Upper Cretaceous flysch succession of the Balangbaru Formation, southwest Sulawesi. *Proceedings 20th annual convention*, Indonesian Petroleum Association, Jakarta, **1**, 190–7.

Hasan, M.M. and Soebandrio, D.S. (1988). The petroleum geology of Tanjung Laban Field, South Sumatra. *Proceedings of the 17th annual convention,* **1**, 257–74. Indonesian Petroleum Association, Jakarta.

Hasan Madjedi, Kamal and Langitan, F.B. (1977/1978). The discovery and development of Minas Field. *SEAPEX Proceedings,* **IV**, 138–57. Southeast Asia Petroleum Exploration Society, Singapore.

Hawkins, J.W. and Evans, C.A. (1983). Geology of the Zambales Range, Luzon, Philippine Islands: ophiolite derived from island arc-back arc basin pair. In *The Tectonic and Geologic Evolution of Southeast Asian Seas and Islands: Part 2,* (ed. D.E. Hayes), Geophysical Monograph, **27**, 95–123. American Geophysical Union, Washington.

Hazebroek, H.P. and Tan, D.N.K. (1993). Tertiary tectonic evolution of the NW Sabah continental margin. *Proceedings of symposium on tectonic framework and energy resources of the western margin of the Pacific Basin, Kuala Lumpur Dec. 1992. Geological Society of Malaysia Bulletin,* **33**, 195–210.

Hedrick, J.B. (1992). *Thorium.* Annual Report for **1991**. U.S. Dept. of the Interior, Bureau of Mines. Govt. Printing Office, Washington.

Hékinian, R., Bonté, P., Pautot, G., Jacques, D., La beyrie, L., Mikkelsen, N. and Reyes, J.L. (1989). Volcanics from the South China Sea ridge system. *Oceanologica Acta,* **12**, 101–15.

Hermes, J.J. (1974). West Irian. *Mesozoic–Cenozoic orogenic belts* Geological Society of London Special Publication **4** (ed. A.M. Spencer), pp. 475–90.

Hinz, K. (1983). Reflection seismic record across the Northwest Sulu Basin. In *Seismic expression of structural styles—a picture and work atlas*, **3**, section 4.2.183, (ed. A.W. Bally). American Association of Petroleum Geologists.

Hinz, K. and Block, M. (1990). Summary of geophysical data from the Sulu and Celebes seas. In *Proceedings of the Ocean Drilling Program, Initial Reports*, (C. Rangin, E. Silver, M.T. Von Breymann *et al.*), Chapter **5**, 124.

Hinz, K., Block, M., Kudrass, H.R. and Meyer, H. (1991). Structural elements of the Sulu Sea, Philippines. *Geologisches Jaarbuch,* **A127**, 483–506.

Hinz, K., Fritsch, J., Kempter, E.H.K., Mohammad, A.M., Meyer, J., Mohamed, D., Vosberg, H., Weber, J. and Benavidez, J. (1989). Thrust tectonics along the northwestern continental margin of Sabah/Borneo. *Geologisches Rundschau,* **78**, 705–30.

Hinz, K., Kempter, E.H.K. and Schlüter, H.U. (1985). The southern Palawan–Balabac area: an accreted or non-accreted terrain?. *Proceedings of the 3rd Asean Council on Petroleum (ASCOPE), Conference and Exhibition*, December 1985, Kuala Lumpur, **2**, 48–72.

Hinz, K. and Schlüter, H.U. (1985). Geology of the Dangerous Grounds South China Sea, and the continental margin off southwest Palawan: results of Sonne cruises SO-23 and SO-27. *Energy*, **10**, 297–315.

Ho K. F. (1978). Stratigraphic framework for oil exploration in Sarawak. *Geological Society of Malaysia Bulletin*, **10**, 1–13.

Hock, M. and Friedrich, G. (1985). Structural features of ophiolitic chromitites in the Zambales Range, Luzon, Philippines, *Mineralium Deposita*, **20**, 290–301.

Holloway, N.H. (1981). The north Palawan Block, Philippines: its relation to the Asian mainland and its role in the evolution of the South China Sea. *Geological Society of Malaysia Bulletin*, **14**, 19–58.

Hosking, K.F.G. (1970*a*). Aspects of the geology of the tin fields of South-East Asia. *Second Technical Conference on Tin, Bangkok 1969*, **1**, 41–80. (ed. W. Fox), International Tin Council, London.

Hosking, K. F. G. (1970*b*). The primary tin deposits of Southeast Asia. *Minerals Science and Engineering*, **2** (No. 4), 24–50. South Africa.

Hosking, K. F. G. (1973). Primary mineral deposits. In *Geology of the Malay Peninsula*, (ed. D.J. Gobbett and C.S. Hutchison), John Wiley-Interscience, New York, 335–90.

Hosking, K. F. G. (1977). Known relationships between the 'hard rock' tin deposits and the granites of Southeast Asia. *Geological Society Malaysia Bulletin*, **9**, 141–57.

Hosking, K.F.G. (1979). Tin distribution patterns. *Geological Society Malaysia Bulletin*, **11**, 1–70.

Hosking, K.F.G. (1988). The world's major types of tin deposit. In *Geology of tin deposits in Asia and the Pacific* (ed. C. S. Hutchison), pp. 3–49. Springer-Verlag, Heidelberg.

Huang, C.K. (1955). Gold-copper deposits of the Chinkuashih Mine, Taiwan, with special reference to the mineralogy. *Acta Geologica Taiwan*, **7**, 1–20.

Hutchinson, R.W. (1981). Lode tin deposits of exhalative origin. In *Complex tin ores and related problems*, Tech. Publication No. **2**, SEATRAD Centre, Ipoh, 81–97.

Hutchinson, R. W. (1986). Massive sulphide deposits and their possible significance to other ores in Southeast Asia. *Geological Society Malaysia Bulletin*, **19**, 1–22.

Hutchison, C. S. (1972). Alpine-type chromite in North Borneo, with special reference to Darvel Bay. *American Mineralogist*, **57**, 835–856.

Hutchison, C.S. (1973). Metamorphism. In *Geology of the Malay Peninsula*. (ed. D.J. Gobbett and C.S. Hutchison), pp. 253–303. John Wiley-Interscience, New York.

Hutchison, C.S. (1975). Ophiolite in Southeast Asia. *Geological Society America Bulletin*, **86**, 797–806.

Hutchison, C.S. (1977). Granite emplacement and tectonic subdivision of Peninsular Malaysia. *Geological Society Malaysia Bulletin*, **9**, 187–207.

Hutchison, C.S. (1982). The various granitoid series and their relationship to W and Sn mineralization. In *Tungsten Geology, Jiangxi, China*, (ed. J. V. Hepworth and Zhang Yuhong), pp. 87–114. ESCAP / RMRDC, Bandung, Indonesia.

Hutchison, C.S. (1983*a*). *Economic deposits and their tectonic setting*. The MacMillan Press, London and Basingstoke.

Hutchison, C.S. (1983*b*). Multiple Mesozoic Sn-W-Sb granitoids of Southeast Asia, *Circum-Pacific plutonism terranes*, (ed. J.A. Roddick), *Geological Society America Memoir*, **59**, 35–60.

Hutchison, C.S. (1986). Tertiary basins of S.E. Asia—their disparate tectonic origins and eustatic stratigraphic similarities. *Geological Society of Malaysia Bulletin*, **19**, 109–22.

Hutchison, C.S. (1987). Continental growth and collision, and mineral prospection in Southeast Asia. *Impact of Science on Society*, UNESCO, Paris, **145**, 33–44.

Hutchison, C.S. (1988*a*). Tectonic settings of tin-tungsten granites in Southeast Asia. In *Exploration and evaluation techniques for tin/tungsten granites in Southeast Asia and the Western Pacific Region*, (ed. C.S. Hutchison), SEATRAD Centre, Ipoh, Malaysia, Tech. Publ., **6**, 1–24.

Hutchison, C.S. (ed.) (1988*b*). *Geology of tin deposits in Asia and the Pacific*. Springer-Verlag, Heidelberg, 718p.

Hutchison, C.S. (1989*a*). *Geological evolution of South-East Asia*. Oxford Monographs on Geology and Geophysics, **13**, Oxford University Press, 368p.

Hutchison, C.S. (1989*b*). The Palaeo-Tethyan Realm and Indosinian Orogenic System of Southeast Asia. In *Tectonic evolution of the Tethyan Region*, (ed. A.M.C. Sengör), NATO ASI Series C, **259**, pp. 585 643. Kluwer Academic Publishers, Dordrecht, Boston and London.

Hutchison, C.S. (1989*c*). Chemical variation of biotite and hornblende in some Malaysian and Sumatran granitoids. *Geological Society Malaysia Bulletin*, **24**, 101–119.

Hutchison, C.S. (1989*d*). Displaced terranes of the Southwest Pacific. In *The evolution of the Pacific Ocean margins*, (ed. Zvi Ben-Avraham), Oxford University Press, New York, 161–75.

Hutchison, C.S. (1991*a*). Neogene arc-continent collision in Sabah, Northern Borneo (Malaysia)—Comment. *Tectonophysics*, **200**, 325–32.

Hutchison, C.S. (ed.) (1991*b*). *Studies in East Asian tectonics and resources (SEATAR) crustal transect VII:*

Jawa–Kalimantan–Sarawak–South China Sea, IOC, Paris, and CCOP, Bangkok, CCOP/TP **26**.

Hutchison, C.S. (1992*a*). The Southeast Sulu Sea, a Neogene marginal basin with outcropping extensions in Sabah. *Geological Society of Malaysia Bulletin*, **32**, 89–108.

Hutchison, (1992*b*). The Eocene unconformity in Southeast and East Sundaland. *Geological Society of Malaysia Bulletin*, **32**, 69–88.

Hutchison, C.S. (1993). Gondwanaland and Cathaysian blocks, Palaeotethys sutures and Cenozoic tectonics in South-east Asia. *Geologisches Rundschau*, **82**, 388–405.

Hutchison, C.S. (1994*a*). Mélange on the Jerudong Line, Brunei Darussalam, and its regional significance. *Geological Society of Malaysia Bulletin* (in press).

Hutchison, C. S. (1994*b*). The unique Baram Delta. *Geological Society of Malaysia—Annual Geological Conference, Trengganu, Programme and Abstracts of Papers,* 8.

Hutchison, C.S. (1996). The 'Rajang accretionary prism' and 'Lupar Line' problem of Borneo. *Tectonic Evolution of Southeast Asia* (ed. R. Hall and D.B. Blundell), special publication **106**, 247–61, Geological Society of London.

Hutchison, C.S. and Chakraborty, K.R. (1979). Tin: a mantle or crustal source? *Geological Society of Malaysia Bulletin,* **11**, 71–9.

Hutchison, C.S. and Sivam, S.P. (1992). Discussion on structural evolution of Mesozoic Peninsular Malaysia, Journal, Vol. 147, 1990, pp. 11–26. *Journal Geological Society of London,* **149**, 679–80.

Hutchison, C.S. and Taylor, D. (1978). Metallogenesis in S.E.Asia. *Journal Geological Society of London,* **135**, 407–28.

Hutchison, C.S. and Tungah Surat, (1991). Sabah serpentinite sandstone and conglomerate. *Geological Society Malaysia Newsletter,* **17** (2), 59–64.

Ikonnikov, A.B. (1975). Mineral Resources of China. *Geological Society of America, Microfilm,* **2**. Boulder, Colorado.

Ingham, F.T. and Bradford, E.F. (1960). *The geology and mineral resources of the Kinta Valley, Perak, Federation of Malaya.* Geological Survey of Malaysia District Memoir, **9**. Ipoh.

Intasopa, S. and Dunn, T. (1994). Petrology and Sr-Nd isotopic systems of the basalts and rhyolites, Loei, Thailand. *Journal of Southeast Asian Earth Sciences,* **9**, 167–80.

Ishihara, S. (1977). The magnetite-series and ilmenite-series granitic rocks. *Mining Geology,* **27**, 293–305.

Ishihara, S., Sawata, H., Arpornsuwan, S., Busaracome, P. and Bungbrakearti, N. (1979). The magnetite-series and ilmenite-series granitoids and their bearing on tin mineralization, particularly in the Malay Peninsula region. *Geological Society Malaysia Bulletin,* **11**, 103–10.

Izman Hamid and Idrus Mohd. Shuhud (1993). Interpretation of seismic attributes in detecting fluid contacts—case history: Tembungo Field. *Geological Society of Malaysia Annual Geological Conference 1993, Langkawi, Programme and Abstracts of Papers,* p. 16.

Jacobson, G. (1970). Gunung Kinabalu area, Sabah, Malaysia. *Geological Survey of Malaysia Report,* **8**.

Jacobson, H.S., Pierson, C.T., Danusawad, T., Japaketr, T., Ithuputi, B., Siriratanamongkol, C., Prapassornkul, S. and Pholphan, N. (1969). Mineral investigations in northeast Thailand. *U.S. Geological Survey Professional Paper,* **618**, 276p.

James, D.M.D. (ed.) (1984). *The geology and hydrocarbon resources of Negara Brunei Darussalam.* Brunei Museum and Brunei Shell Petroleum Co.

Jennings, J.R. and Lee, C.P. (1985). Preliminary note on the occurrence of Carboniferous-age coals and *in-situ* plant fossils in eastern Peninsular Malaysia. *Newsletter of the Geological Society of Malaysia,* **11**, 117–21.

Jin Xiaochi (1994). Sedimentary and paleogeographic significance of Permo-Carboniferous sequences in western Yunnan, China. *Sonderveroeffentlichungen,* **99**, 136 pp. Geological Institute, University of Cologne, Germany.

Johari Dohari and Mohamad Faisal Abdullah (1994). Geology of the Merit–Pila coal basin, Sarawak, Malaysia. *American Association of Petroleum Geologists Bulletin,* **78**, p. 1149.

John, J.U. (1963). Geology and mineral deposits of east-central Balabac island, Palawan Province, Philippines. *Economic Geology*, **58**, 107–30.

Johnson, H.D., Levell, S. and Mohamad, A.H. (1987). Depositional controls of reservoir thickness and quality distribution in Upper Miocene shallow marine sandstones (Stage IVD) of the Erb West Field, offshore Sabah, NW Borneo. *Geological Society of Malaysia Bulletin,* **21**, 195–230.

Jolly, J.H. (1992). *Zinc.* Annual Report for **1991**. U.S. Dept. of the Interior, Bureau of Mines. U.S. Govt. Printing Office, Washington.

Jolly, J. L. W. (1993). *Copper.* Annual Report for **1991**. U.S. Department of the Interior, Bureau of Mines. Govt. Printing Office, Washington.

Jones, C.R. (1978). The geology and mineral resources of Perlis, north Kedah and the Langkawi islands. *Geological Survey Malaysia, District memoir,* **17**.

Jones, T.S. (1992). *Manganese.* Annual Report for **1991**, U.S. Dept. of the Interior, Bureau of Mines. Govt. Printing Office, Washington.

Jongmans, W.J. and Gothan, W. (1935). Die palaeobotanischen ergebnisse der Djambi-Expedition 1925. *Jaarboek van het Mijnwezen in Nederlandsch Oost-Indie,* **2** (1930), 71–201.

Jonklaas, P. (1991). Integration of depth conversion, seismic inversion and modeling over the Belida Field, South Natuna Sea Block 'B". *Proceedings of the 20th annual convention,* **1**, 557–85. Indonesian Petroleum Association, Jakarta.

Karig, D.E. (1971). Origin and development of marginal basins in the Western Pacific. *Journal of Geophysical Research,* **76**, 2542–61.

Karig, D.E., Lawrence, M.B., Moore, G.F. and Curray, J.R. (1980). Structural framework of the fore-arc basin, NW. Sumatra. *Journal Geological Society London,* **137**, 77–91.

Katili, J.A. (1967). Structure and age of the Indonesian tin belt with special reference to Bangka. *Tectonophysics,* **4** (4–6), 403–18.

Katili, J.A. (1970). Naplet structures and transcurrent faults in Sumatra. *Bulletin National Institute of Geology and Mining,* Bandung, **3**(1), 11–28.

Katili, J.A. (1974*a*). Sumatra. In *Mesozoic–Cenozoic orogenic belts,* (ed. A.M. Spencer), Geological Society London, Special Publ., **4**, 317–31.

Katili, J.A. (1974*b*). Geological environment of the Indonesian mineral deposits, a plate tectonic approach. *Direktorat Geologi, Geological Survey Indonesia, Publikasi Teknik-seri Geologi Ekonomi,* **7**.

Katz, B.J. (1991). Controls on lacustrine source rock development: a model for Indonesia. *Proceedings of the 20th annual convention,* Indonesian Petroleum Association, Jakarta, **1**, 587–619.

Kavalieris, I., van Leeuwen, T.M. and Wilson, M. (1992). Geological setting and styles of mineralization, north arm of Sulawesi, Indonesia. *Journal Southeast Asian Earth Sciences,* **7**, 113–29.

Kavalieris, I., Turvey, D.J., and Heesterman, L.J.L. (1987). The geology and mineralization of the Mangani Mine, Sumatra, Indonesia. *Proceedings of the Pacific Rim Congress 87.* Australasian Institution Mining Metallurgy, Parkville, Australia, 221–5.

Kavalieris, I., Walshe, J.L., Halley, S. and Harrold, B.P. (1990). Dome-related gold mineralization in the Pani Volcanic complex, North Sulawesi, Indonesia: a study of geologic relations, fluid inclusions and chlorite compositions. *Economic Geology,* **85**, 1208–25.

Kelly, P.A., Bissada, K.K., Burda, B.H., Elrod, L.W. and Pheifer, R.N. (1985). Petroleum generation potential of coals and organic rich deposits: significance in Tertiary coal rich basins. *Proceedings of the 14th Annual convention,* Vol. **2**, 3–21. Indonesian Petroleum Association, Jakarta.

Kendarsi Roeslan, (1984). *The coal resources of Indonesia and South East Asia.* Paper presented at AGID-ILP workshop, Sydney 1984, Ministry of Mines and Energy, Directorate of Coal, Jakarta, 44p.

Khin Zaw, U. (1978). Fluid inclusion studies on the Hermyingyi tungsten-tin deposit, southern Burma. *Geology and mineral resources of Southeast Asia; Proceedings of the 3rd GEOSEA regional conference,* Asian Institute of Technology, Bangkok, 393–7.

Khin Zaw, U. (1990). Geological, petrological and geochemical characteristics of granitoid rocks in Burma; with special reference to the associated W-Sn mineralization and their tectonic setting. *Journal of Southeast Asian Earth Sciences,* **4**, 293–335.

Khin Zaw, U. and Daw Khin Myo Thet (1983). A note on a fluid inclusion study of tin-tungsten mineralization at Mawchi Mine, Kayah State, Burma. *Economic Geology,* **78**, 530–4.

Khin Zaw, U. and Goossens, P. J. (1989). A brief account of lead mineralization at Phaungdaw prospect, Pyawbwe Township, Mandalay Division, Burma. *Geological Society of Malaysia Bulletin,* **23**, 121–32.

Khin Zaw, U., U. Aung Pwa, and U. Thet Aung Zan, (1984). Lead-zinc mineralization at Theingon Mine, Bawsaing, Southern Shan State, Burma: a Mississippi Valley-type deposit?. *Geological Society Malaysia Bulletin,* **17**, 283–306.

Kho, C.H. (1966). The Bintulu coalfield: prospecting results. *Annual report for 1965 of the Geological Survey, Borneo Region, Malaysia,* Kuching, 40–68.

Khoo, K. K. (1978). Serpentinite occurrence at Telok Mas, Malacca. *Geological Society Malaysia Newsletter (Warta Geologi),* **4** (No. 1), 1–5.

Kieckhefer, R.M., Moore, G.F., Emmel, F.J., and Sugiarta, W. (1981). Crustal structure of the Sunda forearc region west of central Sumatra from gravity data. *Journal of Geophysical Research,* **86**, 7003–12.

Kieft, C. and Oen, I. S. (1973). Ore minerals in the telluride-bearing gold-silver ores of Salida, Indonesia, with special reference to the distribution of selenium. *Mineralium Deposita,* **8**, 312–20.

Kirk, H.J.C. (1957). The geology and mineral resources of the Upper Rajang and adjacent areas. *Geological Survey Department for the British Territories in Borneo, Memoir,* **8**, Kuching.

Kirk, H.J.C. (1968). The igneous rocks of Sarawak and Sabah. *Geological Survey Borneo Region, Bulletin,* **5**, Kuching.

Kirk, W.S. (1993). *Iron ore.* Annual Report for **1991**. U.S. Dept. of the Interior, Bureau of Mines. Govt. Printing Office, Washington.

Klootwijk, C.T., Gee, J.S., Pierce, J.W., Smith, G.M. and McFadden, P.L. (1992). An early India-Asia contact:

paleomagnetic constraints from Ninety-east Ridge, ODP Leg 121. *Geology*, **20**, 395–8.

Koesoemadinata, R.P. (1978). Sedimentary framework of Tertiary coal basins of Indonesia. *Third regional conference on geology and mineral resources of S.E. Asia* (ed. Prinya Nutalaya), 621–40. Asian Institute of Technology, Bangkok.

Koesoemadinata, R.P. and Sastrawiharjo, Sapardi (1988). Uranium prospects in Tertiary sediments in the Sibolga area, north Sumatra. *Uranium deposits in Asia and the Pacific: geology and exploration.* Proceedings of a technical committee meeting, Jakarta Dec. 1985, International Atomic Energy Agency, Vienna, IAEA-TC **543**, 121–40.

Kohar, A. (1985). Seismic expression of Late Eocene carbonate build-up features in the JS 25 and P. Sepanjang trend, Kangean Block. *Proceedings of the 14th annual convention,* **1**, 437–52, Indonesian Petroleum Association, Jakarta.

Koning, T. (1985). Petroleum geology of the Ombilin intermontane basin, West Sumatra. In *Proceedings of the 14th annual convention,* **1**, 117–37. Indonesian Petroleum Association, Jakarta.

Koning, T. and Darmono, F.X. (1984). The geology of the Beruk Northeast Field, Central Sumatra: Oil production from Pre-Tertiary basement rocks. *Proceedings of the 13th Annual convention,* 385–406, Indonesian Petroleum Association, Jakarta.

Kon'no, E. and Asama, K. (1970). Some Permian plants from the Jengka Pass, Pahang, West Malaysia. *Geology and Palaeontology of SE Asia*, **8**, 97–132. Univ. of Tokyo Press.

Kon'no, E., Asama, K. and Rajah, S.S. (1971). The Late Permian Linggiu flora from the Gunung Blumut area, Johore, Malaysia. *Geology and Palaeontology of SE Asia,* **9**, 1–85. Univ. of Tokyo Press.

Koopmans, B.N. (1968). The Tembeling Formation—lithostratigraphic description (West Malaysia). *Geological Society Malaysia Bulletin,* **1**, 23–43.

Kosaka, H. and Wakita, K. (1978). Some geologic features of the Mamut porphyry copper deposit, Sabah, Malaysia. *Economic geology*, **73**, 618–27.

Kostick, D.S. (1992). *Salt.* Annual Report for **1991**. U.S. Dept. of the Interior, Bureau of Mines. Govt. Printing Office, Washington.

Krähenbuhl, R. (1991). Magmatism, tin mineralization and tectonics of the Main Range, Malaysian Peninsula: consequences for the plate tectonic model of Southeast Asia based on Rb-Sr, K-Ar and fission track data. *Geological Society of Malaysia Bulletin,* **29**, 1–100.

Kuck, P.H. (1993). *Nickel.* Annual Report. for **1991**, U.S. Dept. of the Interior, Bureau of Mines. Govt. Printing Office, Washington.

Kudrass, H.R., Wiedicke, M., Cepek, P., Kreuzer, H. and Muller, P. (1986). Mesozoic and Cainozoic rocks dredged from the South China Sea (Reed Bank area) and Sulu Sea and their significance for plate-tectonic reconstructions. *Marine and Petroleum Geology,* **3**, 19–30.

Kusuma, I. and Darin, T. (1989). The hydrocarbon potential of the Lower Tanjung Formation, Barito Basin, S.E. Kalimantan. *Proceedings of the 18th annual convention, Oct. 1989,* Indonesian Petroleum Association, Jakarta, **1**, 107–38.

Land, D.H. and Jones, C.M. (1987). Coal geology and exploration of part of the Tertiary Kutei basin in east Kalimantan, Indonesia. In Coal and coal-bearing strata: recent advances, (ed. A.C. Scott), *Geological Society of America, Special Publication*, **32**, 235–55.

Lapham, D.M. (1968). Triassic magnetite and diabase at Cornwall, Pennsylvania. *Ore deposits of the United States, 1933–1967. The Graton-Sales Volume*, **1**, (ed. J. D. Ridge), 72–94. American Institution Mining Metallurgy Petroleum Engineers, New York.

Lapouille, A., Haryono, H., Larue, M., Pramuwijowo, S. and Lardy, M. (1985). Age and origin of the seafloor of the Banda Sea (eastern Indonesia). *Oceanologica Acta*, **8**, 379–89.

Le Van Khy (1986). The structure of the Mekong Trough. *International Geology Review*, **28**, 87–95.

Leblanc, M. and Violette, J. F. (1983). Distribution of aluminium-rich and chromium-rich pods in ophiolite peridotites. *Economic Geology*, **78**, 293–301.

Lee, C. P. and Azhar Hj. Hussin. (1991). The Wang Kelian redbeds, a possible extension of the unamed Devonian unit (Rebanggun Beds) into Perlis? *Geological Society of Malaysia, Annual Geological Conference '91, programme and abstracts of papers*, 50.

Lee, C.S. and McCabe, R. (1986). The Banda–Celebes–Sulu Basin: a trapped piece of Cretaceous–Eocene oceanic crust ? *Nature*, **322**, 51–4.

Lee, D. T. C. and Weber, H. S. (1986). Base metal exploration in Sabah. *Geological Society Malaysia Bulletin*, **19**, 405–19.

Lee, R.A. (1982). Petroleum geology of the Malacca Strait contract area (Central Sumatra Basin). *Proceedings of the 11th annual convention,* 243–63. Indonesian Petroleum Association, Jakarta.

Lehmann, B. and Harmanto (1990). Large-scale tin depletion in the Tanjungpandan tin granite, Belitung Island, Indonesia. *Economic Geology*, **85**, 99–111.

Lehmann, B. and Mahawat C. (1989). Metallogeny of tin in central Thailand; a genetic concept. *Geology,* **17**, 426–9.

Lenz, H. and Muller, P. (1981). Rb/Sr age determinations (total rock) of rocks of the Bawdwin Volcanic

Formation / Northern Shan State, Burma. *Geologisches Jaarbuch*, **D43**, 47–52.

Letouzey, J., Warner, P. and Marty, A. (1990). Fault reactivation and structural inversion. Back arc and intraplate compressive deformations, example of the eastern Sunda Shelf (Indonesia). *Tectonophysics*, **183**, 341–62.

Levell, B. K. (1987). The nature and significance of regional unconformities in the hydrocarbon-bearing Neogene sequences offshore West Sabah. *Geological Society of Malaysia Bulletin*, **21**, 55–90.

Levell, B. K. and Kasumajaya, A. (1985). Slumping at the late Miocene shelf-edge offshore West Sabah: a view of a turbidite basin margin. *Geological Society of Malaysia Bulletin,* **18**, 1–29.

Lewis, S.D. and Hayes, D.E. (1983). The tectonics of northward propagating subduction along eastern Luzon, Philippine islands. In *The tectonic and geologic evolution of South-east Asian seas and islands, part 2*, (ed. D.E.Hayes), Geophysical Monograph, **27**, 57–78. American Geophysical Union, Washington.

Li Chongyou and Xu Jing (1982). Genetic classification of tungsten deposits of Jiangxi and neighbouring provences. Hepworth, J. V. and Zhang Yuhong (eds.) *Tungsten Geology, Jiangxi, China,* ESCAP/RMRDC, Bandung, Indonesia, 287–99.

Li Desheng (1984). Geologic evolution of petroliferous basins on continental shelf of China. *American Association of Petroleum Geologists Bulletin*, **68**, 993–1003.

Liechti, P., Roe, F.W. and Haile, N.S. (1960). The geology of Sarawak, Brunei and the western part of North Borneo. *Geological Survey department for the British Territories in Borneo, Bulletin,* **3**. Kuching.

Liew Kit Kong (1993). Structural development of the west-central margin of the Malay Basin (basement of blocks PM 2 and PM 7). *Geological Society Malaysia Annual Geology Conference '93, Programme and abstracts of papers*, 37.

Liew Kit Kong (1994). Structural patterns within the Tertiary basement of the Strait of Malacca. *Geological Society of Malaysia, Annual Geological Conference, Trengganu 1994, Programme and Abstracts of Papers,* 68–9.

Lim, P.S. (compiler) (1985). *Geological Map of Sabah, 1:500,000.* 3rd edition. Geological Survey of Malaysia, Kota Kinabalu.

Liu, Char-Shine, Curray, J.R. and McDonald, J.M. (1983). New constraints on the tectonic evolution of the eastern Indian Ocean. *Earth and Planetary Science Letters*, **65**, 331–42.

Liu, Menggen (1982). Some characteristics of vein-type tungsten ore deposits in southern China and their metallogenic pattern. In *Tungsten geology, Jiangxi, China,* (ed. J.V. Hepworth and Zhang Yuhong), ESCAP/RMRDC, Bandung, Indonesia, 327–38.

Llewellyn, T.O. (1992). *Antimony.* Annual Report for **1991**. U.S. Department of the Interior, Bureau of Mines. Govt. Printing Office, Washington.

Loganathan, P. (1977). The geology and mineral resources of the Segamat area (sheet **115**), Johore. *Geological Survey of Malaysia annual report 1977*, 104–7.

Loganathan, P. (1978). The Ma'Okil Formation—an outlier of the Tahan Supergroup. *Geological Survey of Malaysia Annual Report 1978,* 119–31.

Loiret, B. and Mugniot, J.F. (1982). Seismic sequences interpretation, a contribution to the stratigraphical framework of the Mahakam area. *Proceedings of the 11th annual convention*, **1**, 323–34. Indonesian Petroleum Association, Jakarta.

Lowdner, G.G. and Dow, J.A.S. (1977). Porphyry copper mineralization at the Tapadaa Prospect, northern Sulawesi, Indonesia. *Society Mining Engineers AIME Transactions.* **262**, 191–8.

Lucas, J.M. (1992). *Gold.* Annual Report for **1991**. U.S. Department of the Interior, Bureau of Mines. Govt. Printing Office, Washington.

Lum Hoi Kon (1979). The geology of Gemas area (sheet 105). *Geological Survey of Malaysia annual report 1979*, 131–8.

Malaysia (1985). *Geological map of Peninsular Malaysia,* 8th edn, 1:500 000. Geological Survey of Malaysia, Kuala Lumpur and Ipoh.

Maliki, M.A. and Soenarwi, S. (1991). South Lho Sukon-D1 discovery, North Sumatra. *Proceedings of the 20th annual convention*, **1**, 235–54. Indonesian Petroleum Association, Jakarta.

Malihan, T.D. (1987). The gold-rich Dizon porphyry copper mine in the western Central Luzon island, Philippines: its geology and tectonic setting. *Proceedings Pacific Rim Congress 87.* Australasian Institution of Mining and Metallurgy, Parkville, Australia, 303–7.

Marks, E., Sujatmiko, Samuel, L., Dhanutirto, H., Ismoyowati, T. and Sidik, B.B. (1982). Cenozoic stratigraphic nomenclature in east Kutei Basin, Kalimantan. *Proceedings of the 11th annual convention,* **1**, 145–79. Indonesian Petroleum Association, Jakarta.

Masters, C.D., Root, D.H. and Attanasi, E.D. (1991). Resource constraints in petroleum production potential. *Science*, **253**, 146–52.

Matthews, S.J. and Todd, S.P. (1993). A tectonostratigraphic model for the southern Nam Con Son Basin, offshore Vietnam. *Programme and abstracts of papers, Dec. 1993*, 22–3. Geological Society of Malaysia, Kuala Lumpur.

May, J.A. and Eyles, D.R. (1985). Well log and seismic character of Tertiary Terumbu Carbonate, South China Sea, Indonesia. *American Association of Petroleum Geologists Bulletin*, **69**, 1339–58.

Mazlan bin Haji Madon (1991). Sedimentology and petrography of the Upper Miocene reservoirs in the Jerneh Field, Malay Basin, offshore Peninsular Malaysia. *Unpubl. M.Sc. thesis, Univ. of Malaya*, 210 pages.

Mazlan bin Haji Madon (1992). Note on dolerite, rhyolite and granophyre in the basement of the Tenggol Arch, offshore Terengganu. *Geological Society of Malaysia Bulletin*, **31**, 133–43.

Mercado, A.C., Estoque, J., Domingo, B., Acenas, M. and Abaigar, J. (1987*a*). Geology and ore genesis of the Siana gold-silver deposit, Surigao Del Norte, Philippines. *Proc. Pacific Rim Congress '87*, Australasian Institution Mining Metallurgy, Parkville, Australia, 325–8.

Mercado, A.C., Estoque, J., Lodrigueza, L. and Rebillon, F. (1987*b*). Geology and ore deposits of Masara Mine, Davao Del Norte, Philippines. *Proc. Pacific Rim Congress '87*, Australasian Institution Mining Metallurgy, Parkville, Australia, 329–32.

Mertosono, S. and Nayoan, G.A.S. (1974). The Tertiary basinal area of Central Sumatra. *Proceedings of the 3rd annual convention*, 63–76. Indonesian Petroleum Association, Jakarta.

Metcalfe, I. (1986). Late Palaeozoic palaeogeography of Southeast Asia: some stratigraphical, palaeontological and palaeomagnetic constraints. *Geological Society Malaysia Bulletin*, **19**, 153–64.

Metcalfe, I. (1990*a*). Allochthonous terrane processes in Southeast Asia. *Phil. Trans. Royal Soc. London*, **A331**, 625–40.

Metcalfe, I. (1990*b*). Stratigraphic and tectonic implications of Triassic conodonts from northwest Peninsular Malaysia. *Geological Magazine*, **127**, 567–78.

Metcalfe, I. (1991). Late Palaeozoic and Mesozoic palaeography of Southeast Asia. *Palaeogeography, Palaeoclimatology, Palaeoecology*, **87**, 211–21.

Metcalfe, I., Idris, M. and Tan, J.T. (1980). Stratigraphy and palaeontology of the Carboniferous sediments in the Panching area, Pahang, West Malaysia. *Geological Society Malaysia Bulletin*, **13**, 1–26.

Mevel, C. and Kienast, J.R. (1986). Jadeite-kosmochlor solid solution and chromian sodic amphiboles in jadeites and associated rocks from Tawmaw (Burma). *Bulletin Mineralogie*, **109** (No. 6), 657–66.

Minerals Yearbook (1993). *Minerals Yearbook for 1990*, Vol. **1**, *Metals and Minerals*. U.S. Department of the Interior, Bureau of Mines. Govt. Printing Office, Washington.

Mining (1992). *Mining Annual Review 1992*, Far East. pages 84–106, June 1992, Publ. By Mining Journal, London.

Mitchell, A.H.G. (1992). Late Permian–Mesozoic events and the Mergui Group nappe in Myanmar and Thailand. *Journal Southeast Asian Earth Sciences*, **7**, 165–78.

Mitchell, A.H.G. and Balce, G.R. (1990). Geological features of some epithermal gold systems, Philippines. *Journal Geochemical Exploration*, **35**, 241–96.

Mitchell, A.H.G. and Carlile, J.C. (1994). Arc-related gold-copper mineralization, basement domes and crustal extension. *Journal of Southeast Asian Earth Sciences*, **10**, 39–50.

Mitchell, A.H.G. and Leach, T.M. (1991). *Epithermal gold in the Philippines: island arc metallogenesis, geothermal systems and geology*. Academic Press, London, 457p.

Mitchell, A.H.G., Marshall, J.R., Skinner, A.C., Baker, M.D., Amos, B.J. and Bateson, J.H. (1977). Geology and exploration geochemistry of the Yadanatheingi and of the Kyaukme–Longtawkno areas, Northern Shan States, Burma. *Overseas Geology and Mineral Resources*, London, **51**, 34p.

Moh, G. H. (1986). Complex tin-bearing sulphides of the south Chinese ore type. *Geological Society Malaysia Bulletin*, **19**, 369–74.

Mohammad Yamin Ali and Amita Mohd. Ali. (1993). Reactivated tectonic structural controls on morphological development of the Central Luconia carbonates. *Geological Society of Malaysia Annual Geological Conference 1993, Langkawi, Programme and Abstracts of Papers*, 36.

Mohd. Idrus Ismail (1989). Interim report on the final phase of stress analysis and hydrofracturing directions based on borehole breakout studies. *Proceedings of the 24th session C.C.O.P.*, Bangkok, 227–52.

Mohd. Nazri Ramli. (1988). Stratigraphy and palaeofacies development of Carigali's operating areas in the Malay Basin, South China Sea. *Geological Society of Malaysia Bulletin*, **22**, 153–87.

Mohd. Nor Mansor and Rudolph, K.W. (1993). Seismic recognition of incised valleys in the Malay Basin. *Programme and Abstracts of Papers*, p. 27, Petroleum Geology Seminar '93. Geological Society of Malaysia, Kuala Lumpur.

Mohd. Tahir Ismail and Rudolph, K.W. (1992). Structural trap styles of the Malay Basin. *Symposium on tectonic framework and energy resources of the western margin of the Pacific Basin, Programme and Abstracts of Papers*, Circum Pacific Council and Geological Society of Malaysia, Kuala Lumpur, p. 44.

Molengraaff, G.A.F. (1902). *Geological Explorations in Central Borneo (1893–94)*. E. J. Brill, Leyden.

Molina, J. (1985). Petroleum geochemistry of the Sunda Basin. *Proceedings of the 14th annual convention*, **2**, 143–79. Indonesian Petroleum Association, Jakarta.

Molnar, P. and Tapponnier, P. (1975). Cenozoic tectonics of Asia: effects of continental collision. *Science*, **189**, 419–26.

Moore, G.F. and Karig, D.E. (1980). Structural geology of Nias island, Indonesia: implications for subduction and tectonics. *American Journal of Science*, **280**, 193–223.

Moore, G.F., Curray, J.R., Moore, D.G. and Karig, D.E. (1980). Variations in geologic structure along the Sunda fore-arc, northeastern Indian Ocean. In The tectonic and geologic evolution of Southeast Asian seas and islands, (ed. D.E. Hayes). *Geophysical Monograph*, **23**, 145–60. Geophysical Union, Washington.

Morley, R.J. (1991). Tertiary stratigraphic palynology in Southeast Asia: current status and new directions. *Geological Society of Malaysia Bulletin*, **28**, 1–36.

Morton, A.C., Humphreys, B., Dharmayanti, D.A. and Sundoro (1994). Palaeogeographic implications of the heavy mineral distribution in Miocene sandstones of the North Sumatra Basin. *Journ. South-East Asian Earth Sciences*, **10**, 177–90.

Moulds, P.J. (1989). Development of the Bengkalis Depression, central Sumatra, and its subsequent deformation—a model for other Sumatran grabens? *Proceedings of the 18th annual convention*, Indonesian Petroleum Association, Jakarta, **1**, 217–45.

Muff, R. (1990). Geological, geochemical, and ore microscopic investigations of the massive cupriferous sulfide occurrences at West Sualog, in the Bidu Bidu ophiolite complex, Sabah, East Malaysia. *Geologisches Jaarbuch*, **B74**, 65–95.

Muller, P. and Weiser, T. (1981). Studies of the petrology and ore mineralogy in the area of the Bawdwin ore deposit, northern Shan State, Burma. *Geologisches Jaarbuch*, **D43**, 53–64.

Murphy, R.W. (1989). Petroleum industry usage. In *Inversion Tectonics* (ed. by M.A. Cooper and G.D. Williams), Special Publications Classics, Geological Society of London, 336–7.

Mylius, H-G. (1990). The geological setting of a cupriferous sulfide mineralization in the Kiabau area of the Bidu Bidu Hills, Sabah, East Malaysia. *Geologisches Jaarbuch*, **B74**, 97–114.

Nakapadungrat, Somchai and Maneenai, Decha (1993). The Phuket, Phangnga and Takua Pa tin-field, Thailand. *Journal of Southeast Asian Earth Sciences*, **8**, 359–68.

Newton-Smith, J. (1967). Bidu-Bidu Hills area, Sabah, East Malaysia. *Geological Survey Borneo Region Malaysia, Report*, **4**, Kuching.

Ng Tong San (1987). Trap styles of the Tenggol Arch and the southern part of the Malay Basin. *Geological Society of Malaysia Bulletin*, **21**, 177–93.

Nguyen Duc Tien (1989). Lower Permian foraminifera of Sumatra. In *The pre-Tertiary fossils of Sumatra and their environments* (ed. H. Fontaine and S. Gafoer). C.C.O.P. Technical Publication, **19**, 71–93. Bangkok.

Nik Ramli (1986). Depositional model of a Miocene barred wave- and storm-dominated shoreface and shelf, southeastern Malay Basin, offshore West Malaysia. *American Association. Petroleum Geologists Bulletin*, **70**, 34–47.

Nik Ramli (1988). Characteristics of J sandstone (Tapis Formation) reservoirs in the southeastern part of the Malay Basin, offshore West Malaysia. *Southeast Asian Petroleum Exploration Society (SEAPEX), Singapore, Proceedings*, **8**, 239–48.

Noor Azim Ibrahim and Mazlan Madon (1990). Depositional environments, diagenesis, and porosity of reservoir sandstones in the Malong Field, offshore West Malaysia. *Geological Society of Malaysia Bulletin*, **27**, 27–55.

Norton, I.O. (1994). Tertiary relative plate motions in the north Pacific: the 43 Ma non-event. *Tectonic evolution of SE Asia, December meeting, Abstract Volume*, p. 41. Geological Society of London.

Nuay, E.S., Astarita, A.M., and Edwards, K. (1985). Early Middle Miocene progradation in the southern Kutai Basin. In *Proceedings of the 14th annual convention*, **1**, 63–81. Indonesian Petroleum Association, Jakarta.

Ober, J.A. (1992). *Sulfur*. Annual Report for **1991**: U.S. Dept. of the Interior, Bureau of Mines. Govt. Printing Office, Washington.

Ogura, Y., Ito, K., Koide, K. and Shimosaka, K. (1981). Geochemistry and mineralogy of nickel oxide ores in the Southwest Pacific area. In *Lateritisation processes*, (ed. M.K. Roy *et al.*), 58–67, A. A. Balkema, Rotterdam.

O'Leary, J. and Hill, G.S. (1989). Tertiary basin development in the southern central plains, Thailand. In *Intermontane Basins: geology and resources*. (ed. T. Thanasuthipitak and P. Ounchanum), pp. 254–64. Proceedings of the International Symposium, Chiang Mai University, Thailand.

Ong Yeoh Han and Mior Sallehhuddin bin Mior Jadid (1988). Promising uranium prospects in the central belt area, Peninsular Malaysia. *Uranium deposits in Asia and the Pacific: geology and exploration*. Proceedings of a technical committee meeting, Jakarta, Dec. 1985, IAEA-TC **543**, 97–107. International Atomic Energy Agency, Vienna.

Orberger, B., Friedrich, G. and Woermann, E. (1987). Platinum group element mineralization in the ultramafic sequence of the Acoje ophiolite block, Zambales, Philippines. In *Geo-Platinum 87,* (ed. H.M. Prichard), pp. 361–80. Elsevier Scientific Publications, Essex, England.

Orberger, B and Alleweldt, J. (1994). Contribution to the petrogenesis of the palladium-bearing black dunites of the Acoje block, Zambales ophiolite, Philippines; selenium and sulfur-contents, clinopyroxenes, initial weathering. *Journal of Southeast Asian Earth Sciences,* **9**, 229–39.

O'Sullivan, T., Pegum, D., and Tarigan, J. (1985). Seram oil search, past discoveries and future oil potential. In *Proceedings of the 14th annual convention,* **1**, 3–20. Indonesian Petroleum Association, Jakarta.

Packham, G.H. (1990). Plate motions and Southeast Asia: some tectonic consequences for basin development. *SEAPEX Proceedings,* **9**, 55–68.

Packham, G.H. (1993). Plate tectonics and the development of sedimentary basins of the dextral regime in western Southeast Asia. *Journal of Southeast Asian Earth Sciences,* **8**, 497–511.

Park, C. F. jr. (1972). The iron ore deposits of the Pacific Basin. *Economic Geology,* **67**, 339–49.

Parkinson, C.D. (1991). The petrology, structure and geologic history of the metamorphic rocks of Central Sulawesi. *Unpubl. Ph.D. thesis. University of London.*

Pautot, G., Rangin, C., Briais, A., Tapponnier, P., Beuzart, P., Lericolais, G., Mathieu, X., Wu, J., Han, S., Li, H., Lu, Y. and Zhao, J. (1986). Spreading direction in the Central South China Sea. *Nature,* **321**, 150–4.

Peng Chengdian and Cheng Shuxi (1988). The history of exploration over the past 30 years in the Gejiu tin deposit, Yunnan. In *Geology of tin deposits in Asia and the Pacific,* (ed. C.S. Hutchison), pp. 465–72. Springer-Verlag, Heidelberg.

Phillips, T.L., Noble, R.A. and Sinartio, F.F. (1991). Origin of hydrocarbons, Kangean Block northern platform, offshore N.E. Java Sea. *Proceedings of the 20th annual convention,* **1**, 637–61. Indonesian Petroleum Association, Jakarta.

Pieters, P.E. and S. Supriatna (1989). *Preliminary geological map of the west, central and east Kalimantan area, 1:1 000 000.* Geological Research and Development Centre, Bandung.

Pieters, P.E. and Supriatna, S. (1990). *Geological map of the west, central and east Kalimantan area (colour version), 1:1,000,000.* Geological Research and Development Centre, Bandung; Bureau of Mineral Resources, Canberra.

Pieters, P.E., Trail, D. S. and Supriatna, S. (1987). Correlation of Early Tertiary rocks across Kalimantan. *Proceedings of the 16th annual convention,* **1**, 291–306. Indonesian Petroleum Association, Jakarta.

Piggott, J.D. and Sattayarak, N. (1993). Aspects of sedimentary basin evolution assessed through tectonic subsidence analysis. Example: Northern Gulf of Thailand. *Journal of Southeast Asian Earth Sciences,* **8**, 407–20.

Pigram, C.J. and Panggabean, H. (1984). Rifting of the northern margin of the Australian continent and the origin of some microcontinents in eastern Indonesia. *Tectonophysics,* **107**, 331–53.

Pigram, C.J., Robinson, G.P. and Tobing, S.L. (1982). Late Cainozoic origin for the Bintuni Basin and adjacent Lengguru fold belt, Irian Jaya. *Proceedings of the 11th annual convention,* 109–26. Indonesian Petroleum Association, Jakarta.

Pimm, A.C. (1967). Bau Mining district, West Sarawak. II: Krokong. *Geological Survey Borneo Region Malaysia, Bulletin,* **7**. Kuching.

Pitfield, P.E.J. (1987). *Report on the geochemistry of the Tin Island granites of Indonesia.* British Geological Survey, Overseas Directorate, Report **WC/87/9/R**, 1–52.

Pitfield, P.E.J. (1988). *Report on the geochemistry of the granites of Thailand.* British Geological Survey, Overseas Directorate, Report **WC/88/6**, 1–102.

Pitrakul, S. and Tantisukrit, C. (1973). Review of economic mineral deposits in Thailand. *Conference on the geology of Thailand,* Chiang, Thailand, 51pp.

Plunkert, P.A. and Sehnke, E.D. (1993). *Aluminum, bauxite and alumina.* Annual Report for **1991**. U.S. Dept. of the Interior, Bureau of Mines. Govt. Printing Office, Washington.

Polachan, S. and Racey, A. (1993). Lower Miocene larger foraminifera and petroleum potential of the Tai Formation, Mergui Group, Andaman Sea. *Journal of Southeast Asian Earth Sciences,* **8**, 487–96.

Polachan, S. and Sattayarak, N. (1989). Strike-slip tectonics and the development of Tertiary basins in Thailand. In *Intermontane Basins: geology and resources.* (ed. T. Thanasuthipitak and P. Ounchanum) Proceedings of the international symposium., pp. 243–53. Chiang Mai University, Thailand.

Pollock, R.E., Hayes, J.B., Williams, K.P. and Young, R.A. (1984). The petroleum geology of the KH Field, Kakap, Indonesia. *Proceedings of the 13th annual convention,* **1**, 407–23. Indonesian Petroleum Association, Jakarta.

Pulunggono, A. (1986). Tertiary structural features related to extensional and compressive tectonics in the Palembang Basin, South Sumatra. *Proceedings of the 15th annual convention,* **1**, 187–213. Indonesian Petroleum Association, Jakarta.

Pulunggono, A. and Cameron, N.R. (1984). Sumatran microplates, their characteristics and their role in the evolution of the Central and South Sumatra Basins. *Proceedings of the 13th Annual convention,* **1**, 121–44. Indonesian Petroleum Association, Jakarta.

Putthapiban, P. and Gray, C.M. (1983). Age and tin-tungsten mineralization of the Phuket granites, Thailand. In *Geology and mineral resources of Thailand,* (ed. S. Suensilpong), 1–10. Department of Mineral Resources, Bangkok.

Rajah, S.S. (1979). The Kinta tin field, Malaysia. *Geological Society of Malaysia Bulletin,* **11**, 111–36.

Rajah, S.S. (1982). Tungsten mineralization in Peninsular Malaysia. In *Tungsten geology, Jiangxi, China,* (ed. J.V. Hepworth and Y.H Zhang, pp. 179–95. ESCAP/RMRDC, Bandung, Indonesia.

Rajah, S.S. (1987). Geology and mineral resources of the Gunung Belumut area, Johore, Peninsular Malaysia. *Geological Survey of Malaysia district memoir,* **19**.

Rangin, C. (1989). The Sulu Sea, a back-arc basin setting within a Neogene collision zone. *Tectonophysics,* **161**, 119–41.

Rangin, C., Bellon, H., Benard, F., Letouzey, J., Muller, C. and Sanudin, T. (1990). Neogene arc-continent collision in Sabah, Northern Borneo (Malaysia). In Geodynamic evolution of the Eurasian margin (ed. J. Angelier)., *Tectonophysics,* **183**, 305–19.

Ratanasthien, B. (1986). Factors concerning spontaneous fires in northern Thailand coals. *Geological Society Malaysia Bulletin,* **19**, 393–404.

Redfern, J. (1991). Glacial facies—their sedimentology, distribution and hydrocarbon potential (Abstract). *Geological Society of Malaysia Newsletter (Warta Geologi),* **17**, p. 194.

Réhault, J.P., Malod, J.A., Larue, M., Burhanuddin, S. and Sarmili, L. (1991). A new sketch of the central North Banda Sea, Eastern Indonesia. *Journal of Southeast Asian Earth Sciences,* **6**, 329–34.

Réhault, J.P., Villeneuve, M., Honthaas, C., Bellon, H., Malod, J.A., Cornée, J.J., Butterlin, J., Sarmili, L. and Burhanuddin, S. (1994). New geological samplings along a transect across the south east Banda Sea Basin (Indonesia). *Tectonic evolution of SE Asia, Dec. meeting, Abstract volume,* Geological Society of London, p. 49.

Reminton, C.H. and Pranyoto, U. (1985). A hydrocarbon generation analysis in northwest Java Basin using Lopatin's method. *Proceedings of the 14th annual convention,* **2**, 121–41, Indonesian Petroleum Association, Jakarta.

Reynolds, C.D., Havryluk, I., Bastaman, S. and Atmowidjojo, S. (1973). The exploration of the nickel laterite deposits in Irian Barat, Indonesia. *Geological Society Malaysia Bulletin,* **6**, 309–23.

Rice-Oxley, E.D. (1991). Palaeoenvironments of the Lower Miocene to Pliocene sediments in offshore NW Sabah area. *Geological Society Malaysia Bulletin,* **28**, 165–94.

Richardson, J.A. (1939). *The geology and mineral resources of the neighbourhood of Raub, Pahang, Federated Malay States, with an account of the geology of the Raub Australian Gold Mine.* Geol. Surv. Dept. Federated Malay States, memoir, **3** (new series). Ipoh.

Richter, B., Fuller, M., Schmidtke, E., U Tin Myint, U Tin Ngwe, U Mya Win and Bunopas, S. (1993). Paleomagnetic results from Thailand and Myanmar: implications for the interpretation of tectonic rotations in Southeast Asia. *Journal of Southeast Asian Earth Sciences,* **8**, 247–55.

Ridd, M.F. (1980). Possible Palaeozoic drift of SE Asia and Triassic collision with China. *Journal Geological Society of London,* **137**, 635–40.

Rijks, E.J.H. (1981). Baram Delta geology and hydrocarbon occurrence. *Geological Society of Malaysia Bulletin,* **14**, 1–18. Kuala Lumpur.

Robertson Research (1975). *Burma—an appraisal of mineral resources,* **1**, report, 2, atlas. Robertson Research (Australia) Pty. Ltd.

Rose, R. (1983). Miocene carbonate rocks of Sibolga Basin, Northwest Sumatra. *Proceedings of the 12th annual convention,* **1**, 105–25. Indonesian Petroleum Association, Jakarta.

Rosly Mohd. Nor (1979). Geology of Kenny Hill Formation, Selangor, Peninsular Malaysia. Unpubl. B.Sc. thesis, Dept. of Geology. Univ of Malaya, Kuala Lumpur.

Roure, F., Brun, J.P., Colleta, B. and Van Den Driesse, J. (1992). Geology and kinematics of extensional structures in the Alpine Foreland Basin of Southeastern France. *Journal of Structural Geology*, **14** (15), 503–19.

Ru, Ke and Pigott, J.D. (1986). Episodic rifting and subsidence in the South China Sea. *American Association of Petroleum Geologists Bulletin,* **70**, 1136–55.

Rudolph, K.W. and Lehmann, P.J. (1989). Platform evolution and sequence stratigraphy of the Natuna Platform, South China Sea. In *Controls on carbonate platform and basin development.* The Society of Economic Paleontologists and Mineralogists, Special Publication, **44**, 353–61.

Rung Songsiri (1978). Pa-Daeng, the largest known zinc deposit of Thailand; exploration and mining assessment. Geology and Mineral Resources of Thailand, Special Issue for GEOSEA III, *Journal Geological Society of Thailand,* **3** (No. 1), E 4–1.

Rustandi, E. Nila, E.S. and Sanyoto, P. (1981). *Geologic map of Kotabaru Quadrangle, South Kalimantan, 1:250,000 (sheet 1812)*. Geological Research and Development Centre, Bandung.

Santos-Ynigo, L. and Pacis, M. G. (1981). Bauxitic ores and clays developed in karstic limestone terrains, central Samar island, Philippines. *Proceedings of the 4th regional conference on the geology of Southeast Asia,* Geol. Soc. of the Philippines, Manila, 735–55.

Sarbini, Sumantri Achmad and Wirakusumah, Wahlan (1988). Uranium model for estimation of ore reserves in the Remaja area, west Kalimantan. *Uranium deposits in Asia and the Pacific: geology and exploration.* Proceedings of Technical committee meeting, Jakarta Dec. 1985, International Atomic Energy Agency, Vienna, IACA-TC **543**, 155–65.

Sardjito, Fadianto, E., Djumiati and Hansen, S. (1991). Hydrocarbon prospect of pre-Tertiary basement in Kuang area, South Sumatra. *Proceedings of the 20th annual convention,* **1**, 255–77. Indonesian Petroleum Association, Jakarta.

Sattayarak, N. (1983). Review of the continental Mesozoic stratigraphy of Thailand. *Stratigraphical correlation of Thailand and Malaysia,* **1**, *technical papers*, 127–48. Geological Society of Thailand, Bangkok/ Geological Society of Malaysia, Kuala Lumpur.

Sattayarak, N., Srilulwong, S. and Pum-Im, S. (1989). Petroleum potential of the Triassic pre-Khorat intermontane basin in northeastern Thailand. *Proceedings of the international symposium on intermontane basins: geology and resources*, 43–57. Chiang Mai University, Thailand.

Schaar, G. (1976). The occurrence of hydrocarbons in overpressured reservoirs of the Baram Delta (offshore Sarawak, Malaysia). *Proceedings of the 5th annual convention,* **2**, 163–69. Indonesian Petroleum Association, Jakarta.

Schiller, D.M., Garrard, R.A. and Prasetyo, L. (1991). Eocene submarine fan sediments in southwest Java. *Proceedings of the 20th annual convention*, **1**, 125–81. Indonesian Petroleum Association, Jakarta.

Schmidtke, E.A., Fuller, M.D. and Haston, R.B. (1990). Paleomagnetic data from Sarawak, Malaysia (Borneo) and the Late Mesozoic and Cenozoic tectonics of Sundaland. *Tectonics,* **9**, 123–40.

Schoell, M., Teschner, M., Wehner, H., Durand, B., and Oudin, J.L. (1983). Maturity related biomarker and stable isotope variations and their application to oil/source rock correlation in the Mahakam Delta/Kalimantan. In *Advances in organic geochemistry 1981*, (ed. M. Bjoroy), pp. 156–63. John Wiley, New York.

Schoell, M., Durand, B., and Oudin, J.L. (1985). Migration of oil and gas in the Mahakam Delta, Kalimantan: evidence and quantitative estimate from isotope and biomarker studies. *Proceedings of the 14th annual convention*, **2**, 49–56. Indonesian Petroleum Association, Jakarta.

Schwartz, M.O. and Askury, A.K. (1989). Geologic, geochemical, and fluid inclusion studies of the tin granites from the Bujang Melaka Pluton, Kinta Valley, Malaysia. *Economic Geology,* **84**, 751–79.

Schwartz, M.O. and Surjono (1990). The strata-bound tin deposit Nam Salu, Kelapa Kampit, Indonesia. *Economic Geology,* **85**, 76–98.

Schweller, W.J., Roth, P.H., Karig, D.E. and Bachman, S.B. (1984). Sedimentation history and biostratigraphy of ophiolite-related Tertiary sediments, Luzon, Philippines. *Geological Society of America Bulletin,* **95**, 1333–42.

Scrivenor, J.B. (1928). *The geology of the Malayan ore deposits.* MacMillan Press, London.

Searle, D.L. and Ba Than Haq (1964). The Mogok Belt of Burma and its relationship to the Himalayan Orogeny. *Proceedings of the 22nd international geology congress*, Delhi, **11**. 132–61.

Searls, J.P. (1992). *Barite*. Annual Report for **1991**. U.S. Department of the Interior, Bureau of Mines. Govt. Printing Office, Washington.

Sengör, A.M. Celâl (1984). The Cimmeride Orogenic System and the tectonics of Eurasia. *Geological Society of America Special Paper* **195**, 82pp.

Sengör, A.M.C., Altiner, D., Cin, A., Ustaomer, T., and Hsu, K.J. (1988). Origin and assembly of the Tethyside orogenic collage at the expense of Gondwana-Land. In *Gondwana and Tethys*, (ed. M.G. Audley-Charles and A. Hallam), pp. 119–81. Geol. Soc. London, special publication **37**.

Shawe, D.R. (1984). *Geology and mineral deposits of Thailand.* U. S. Geological Survey, Denver, Colorado. Open File Report **84–403**, 190 p. + maps.

Shawe, D.R., Hite, R.J. and Inthuputi, B. (1975). Potential for sandstone-type uranium deposits in Jurassic rocks, Khorat Plateau, Thailand. *Economic Geology,* **70**, 538–41.

Shi, G.R. and Waterhouse, J.B. (1991). Early Permian brachiopods from Perak, west Malaysia. *Journal of Southeast Asian Earth Sciences*, **6**, 25–39.

Sikumbang, N. (1986). Geology and tectonics of the pre-Tertiary rocks in the Meratus Mountains; south-east Kalimantan, Indonesia. Unpubl. Ph.D. thesis. Royal Holloway and Bedford New College, University of London.

Sillitoe, R.H. (1973). The tops and bottoms of porphyry copper deposits. *Economic Geology,* **68**, 799–815.

Sillitoe, R.H. and Bonham, H.F. (1990). Sediment-hosted gold deposits: distal products of magmatic-hydrothermal systems. *Geology*, **18**, 157–61.

Sillitoe, R.H. and Gappe, I.M. jr. (1984). Philippine porphyry copper deposits: geologic setting and characteristics. *Technical Publication,* **TP 14**, C.C.O.P., Bangkok, 89p.

Silver, E.A., Yoko Joyodiwiryo and McCaffrey, R. (1978). Gravity results and emplacement geometry of the Sulawesi ultramafic belt, Indonesia. *Geology,* **6**, 527–31.

Silver, E.A., Gill, J.B., Schwartz, D., Prasetyo, H. and Duncan, R.A. (1985). Evidence for a submerged and displaced continental borderland, North Banda Sea, Indonesia. *Geology,* **13**, 687–91.

Silver, E.A. *et al.* (1989*a*). Origins of marginal basins. *Nature,* **338**, 380–1.

Silver, E.A. *et al.* (1989*b*). Leg 124 researchers drill marginal basins. *Geotimes,* **May 1989**, 15–7.

Simmons, S.F. and Browne, P.R.L. (1990). Mineralogic, alteration and fluid-inclusion studies of epithermal gold-bearing veins at Mt Muro Prospect, Central Kalimantan (Borneo), Indonesia. *Journal of Geochemical Exploration,* **35**, 63–103.

Singapore (1976). *Geology of the Republic of Singapore,* Public Works Department, Singapore., 79 pages + 10 map sheets.

Siregar, M.S. and Sunaryo, R. (1980). Depositional environments and hydrocarbon prospects, Tanjung Formation, Barito Basin, Kalimantan, Indonesia. *Proceedings of the 9th annual convention,* **1**, 379–400. Indonesian Petroleum Association, Jakarta.

Situmorang, B. and Yulihanto, B. (1985). The role of strike slip faulting in structural development of the North Sumatra Basin. In *Proceedings of the 14th annual convention,* **1**, 21–38. Indonesian Petroleum Association, Jakarta.

Situmorang, B., Yulihanto, B., Guntur, A., Himawan, R. and Jacob, T.G. (1991). Structural development of the Ombilin Basin, West Sumatra. In *Proceedings of the 20th annual convention,* **1**, 1–15. Indonesian Petroleum Association, Jakarta, .

Sivam, S.P. (1969). Quaternary alluvial deposits in the north Kinta Valley, Perak. Unpubl. M.Sc. thesis. University of Malaya, Kuala Lumpur.

Smith, G.R. (1992). *Tungsten.* Annual Report for **1991**. U.S. Dept. of the Interior, Bureau of Mines. U.S. Govt. Printing Office, Washington.

Soeparjadi, R.A., Nayoan, G.A.S., Beddoes, L.R. and James, W.V. (1975). Exploration play concepts in Indonesia. *Proceedings of the 9th world petroleum congress, exploration and transportation,* **3**, 51–64. Applied Science Publishers, London.

Soeparyono, N. and Lennox, P.G. (1989). Structural development of hydrocarbon traps in the Cepu field, northeast Java, Indonesia. *Proceedings of the 18th annual convention,* **1**, 139–56. Indonesian Petroleum Association.

Sorem, R.K. (1956). Geology of three major manganese districts in the Philippines. *Proceedings of the 8th Pacific science congress, 1953,* Vol. **2a**, 661–8. Quezon City; National Research Council.

Staritskiy, Y.G., Maymin, Y.S. and Trofimov, V.A. (1973). Tectonic development of north Vietnam. *International Geology Review,* **15**, 1381–90.

Stauffer, P.H. (1968). Studies in the Crocker Formation, Sabah. *Borneo Region of Malaysia Geological Survey Bulletin,* **8**, 1–13.

Stauffer, P.H. (1973*a*). Kenny Hill Formation. In *Geology of the Malay Peninsula,* (ed. D.J. Gobbett and C.S. Hutchison), pp. 87–91. John Wiley-Interscience, New York.

Stauffer, P.H. (1973*b*). Cenozoic. In *Geology of the Malay Peninsula: West Malaysia and Singapore,* (ed. D.J. Gobbett and C.S. Hutchison), pp. 143–76. John Wiley-Interscience, New York.

Stauffer, P.H. (1978). Anatomy of the Australasian tektite strewnfield and probable site of its source crater. *Proceedings of the 3rd regional conference on the geology and mineral resources of Southeast Asia* (ed. Prinya Nutalaya), pp. 285–9. Asian Institute of Technology, Bangkok.

Stauffer, P.H. and Lee Chai Peng (1986). Late Paleozoic glacial marine facies in Southeast Asia and its implications. GEOSEA V *Proceedings,* Geological Society Malaysia Bulletin, **20**, 363–97.

Stauffer, P.H. and Mantajit, Nopadom (1981). Late Palaeozoic tilloids of Malaya, Thailand and Burma. In *Earth's pre-Pleistocene glacial record,* (eds. M.J. Hambrey and W.B. Harland), pp. 331–7. Cambridge University Press.

Stoll, W. C. (1962). Geology and petrology of the Masinloc chromite deposit, Zambales, Luzon, Philippine islands. *Philippine geologist,* **16**, Manila, 1–41.

Stowasser, W.F. (1992). *Phosphate rock.* Annual Report for **1991**. U.S. Department of the Interior, Bureau of Mines. Govt. Printing Office, Washington.

Su Daquan, White, N. and MacKenzie, D. (1989). Extension and subsidence of the Pearl River Mouth Basin, northern South China Sea. *Basin Research,* **2**, 205–22.

Sukamto, R., Supriatna, S., Trail, D. and Pieters, P. (1988). Collaborative regional geochemical survey in Kalimantan: results of gold analyses. *Research Newsletter,* **8**, 4–6. Bureau of Mineral Resources, Canberra.

Sumii, T., Wakita, K., Matsubayashi, O., Fujii, K., Moritani, T. and Gryc, G. (1992). *Energy-Resources map of the Circum-Pacific region northwest quadrant, 1:10 000 000,* Map CP **40**. U. S. Geological Survey.

Sundararaman, P. and Teerman, S.C. (1991). Porphyrins in extracts as indicators of maturity and depositional environment of Tertiary source rocks from Central Sumatra. *Proceedings of the 20th annual convention,* Indonesian Petroleum Association, Jakarta, **1**, 621–35.

Suntharalingam, T. (1968). Upper Palaeozoic stratigraphy of the area west of Kampar, Perak. *Geological Society of Malaysia Bulletin,* **1**, 1–15.

Supriatna, S., Rustandi, E. and Heryanto, R. (1980). *Geological map of Sampanahan quadrangle, South Kalimantan, 1:250,000 (sheet 1813).* Geological Research and Development Centre, Bandung.

Suryono and Clarke, M.C.G. (1982). Primary tungsten occurrences in Sumatra and the Indonesian tin islands. In *Tungsten geology,* (ed. J.V. Hepworth and Y.H Zhang, pp. 217–31. *Jiangxi, China,* ESCAP/RMRDC, Bandung, Indonesia.

Sutoto, A. (1991). Reservoir geology of the Belida Field South Natuna Sea, Block 'B". *Proceedings of the 20th annual convention,* **1**, 453–78. Indonesian Petroleum Association, Jakarta.

Swinburn, P.M. (1993). Tectonic styles of the Balingian Province. *Petroleum Geology Seminar '93, Programme and Abstracts of Papers,* p. 15. Geological Society of Malaysia, Kuala Lumpur.

Tan, D.N.K. (1979). Lupar Valley, West Sarawak, Malaysia. *Geological Survey of Malaysia Report,* **13**, Kuching, Sarawak.

Tan, D.N.K. (compiler) (1982). *Geological map of Sarawak, 1:500 000,* 1st edn. Geological Survey of Malaysia, Kuching.

Tan, D.N.K. (1986). Palaeogeographic development of West Sarawak. *Geological Society of Malaysia Bulletin,* **19**, 39–49.

Tan, D.N.K. and Lamy, J.M. (1990). Tectonic evolution of the NW Sabah continental margin since the Late Eocene. *Geological Society of Malaysia Bulletin,* **27**, 241–60.

Tan, L. P. (1976). Taiwan copper mineralization in relation to plate tectonics. *Geological Survey of Taiwan Bulletin,* **25** (May), 1–18.

Tan Say Biow (1989). The Wang Phar tungsten deposits. *Geological Society of Malaysia Bulletin,* **23**, 133–45.

Tantisukrit, C. (1978). Review of the metallic mineral deposits of Thailand. In *Proceedings of the 3rd regional conference on geology and mineral resources of SE Asia,* (ed. Prinya Nutalaya), pp. 783–93. Asian Institute of Technology, Bangkok.

Tapponnier, P., Peltzer, G., Le Dain, A.Y., Armijo, R. and Cobbold, P. (1982). Propagating extrusion tectonics in Asia, new insights from simple experiments with plasticene. *Geology,* **10**, 611–16.

Tapponnier, P., Peltzer, G. and Armijo, R. (1986). On the mechanics of the collision between India and Asia. In *Collision Tectonics,* (ed. M.P. Coward and A.C. Ries), Geological Society London special publication, **19**, 115–57.

Tate, R.B. (1976). Palaeo-environment studies in Brunei. *Proceedings of the Southeast Asia Petroleum Exploration Society,* Singapore, **3**, 102–24.

Tate, R.B. (1991). Cross-border correlation of geological formations in Sarawak and Kalimantan. *Geological Society of Malaysia Bulletin,* **28**, 63–95.

Taylor, B. and Hayes, D.E. (1980). The tectonic evolution of the South China Basin. In *The tectonic and geologic evolution of Southeast Asian seas and islands,* (ed. D.E. Hayes), Geophysical Monograph, **23**, 89–104. American Geophysical Union, Washington.

Taylor, B. and Hayes, D.E. (1983). Origin and history of the South China Sea Basin. In *The tectonic and geologic evolution of Southeast Asian seas and islands: Part 2,* (ed. D.E. Hayes), Geophysical Monograph, **27**, 23–56. American Geophysical Union, Washington.

Taylor, D. (1971). An outline of the geology of Bukit Ibam orebody, Rompin, Pahang. *Geological Society Malaysia Bulletin,* **4**, 71–89.

Taylor, D. (1986). Some thoughts on the developments of the alluvial tinfields of the Malay-Thai Peninsula. *Geological Society Malaysia Bulletin,* **19**, 375–92.

Taylor, D. and Hutchison, C. S. (1979). Patterns of mineralization in Southeast Asia, their relationship to broad-scale geological features and the relevance of plate-tectonic concepts to their understanding. (ed. M.J. Jones) *Proceedings of the 11th Commonwealth mining and metallurgy congress, HongKong,* Institution Mining and Metallurgy, London, 93–107.

Taylor, D. and Toh Swee Cheng (1981). Regional mineral exploration, Pahang Tenggara: a case history of base-metal exploration project in the Malay Peninsula. Asia Mining, *Proceedings of the 1981 meeting,* 67–78.

Taylor, D. and Van Leeuwen, T. (1980). Porphyry-type deposits in Southeast Asia. *Mining Geology,* Tokyo, Special Issue, **8**, 95–116.

Teh Guan Hoe (1981). The Tekka tin deposit, Perak, Peninsular Malaysia. *Geological Society of Malaysia Bulletin,* **14**, 101–18.

Teh, Guan Hoe (1994). Barite and associated massive sulphide and Fe-Mn mineralization in the Central Belt of Peninsular Malaysia. *Geological Society of Malaysia Newsletter,* **20**, 101–2.

Templeton, D.A. (1993). *Zirconium and Hafnium.* Annual Report for **1991**. U.S. Dept. of the Interior, Bureau of Mines, U.S. Govt. Printing Office, Washington.

Ting, C. S. (1992). Jurassic–Cretaceous palaeogeography of the Jagoi–Serikin area as indicated by the Bau Limestone Formation. *Geological Society of Malaysia Bulletin*, **31**, 21–38.

Tjia, H.D. (1964). Topographic lineaments in Riau and Lingga archipelagos, Indonesia: their structural significance. *International geological congress, Report of the 22nd Session*, New Delhi, **4**, 566–81.

Tjia, H.D. (1985). Gaya struktur Selat Melaka: suatu pembandingan. *Sains Malaysiana*, **14** (no. 1), 37–64.

Tjia, H.D. (1989). Tectonic history of the Bentong–Bengkalis suture. Geologi Indonesia. J.A. Katili Commemorative Volume (60 years), *Journal of Indonesian Association of Geologists*, **12** (No. 1), July 1989, 89–111, Jakarta.

Tjia, H.D., Ibraham Komoo, Lim, P.S. and Tungah Surat (1990). The Maliau Basin, Sabah: geology and tectonic setting. *Geological Society of Malaysia Bulletin*, **27**, 261–92.

Tjokrokardono, Soeprapto and Sastratenaya, A. Sariyana (1988). Rich mineralized boulders of the Rirang River, West Kalimantan. *Uranium Deposits in Asia and the Pacific: geology and exploration*. Proceedings of a technical committee meeting, Jakarta Dec 1985, IAEA-TC **543**, 79–95. International Atomic Energy Agency, Vienna.

Tobler, A. (1922). Djambi verslag. Uitkomsten van het geol. mijnb. onderzolk in de Residentie Djambi 1906–12. *Jaarbuch Mijnwezen*, 1919, Verh. III + atlas. Batavia (Jakarta).

Tongkul, F. (1990). Structural style and tectonics of Western and Northern Sabah. *Geological Society of Malaysia Bulletin*, *27*, 227–39.

Tran Dzuy and Tran Xuan Dzuc (1986). Iron skarn mineralization in north Trungbo and its perspective. *Proceedings of the 1st conference on geology of Indochina*, **2**, 591–7. General Department of Geology, Hanoi, Vietnam.

Tran Van Tri, Tran Kim Thach and Truong Cam Bao (1979). *Geology of Vietnam (north part)*. General Department of Geology, Research Institute of Geology and Mineral Resources, Hanoi.

Tran Xuan Dzuc and Nguyen Van Tung (1986). Mineralization of barite in the north Trungson region. *Proceedings of the 1st conference on geology of Indochina*, **2**, 707–14. General Department of Geology, Hanoi, Vietnam.

Tran Xuan Toan (1986). Pagodite deposits in Indochina. *Proceedings of the 1st conference on geology of Indochina*, **2**, 679–89. General Department of Geology, Hanoi, Vietnam.

Trevena, A.S. and Clark, R.A. (1986). Diagenesis of sandstone reservoirs of Pattani Basin, Gulf of Thailand. *American Association of Petroleum Geologists Bulletin*, **70**, 299–308.

Tyrrell, W.W. jr. and Christian, H.E. jr. (1992). Exploration history of Liuhua 11–1 Field, Pearl River Mouth Basin, China. *American Association of Petroleum Geologists Bulletin*, **76** (No. 8), 1209–23.

U Ko Ko (1986). Preliminary synthesis of the geology of Bangka Island, Indonesia. GEOSEA V Proceedings, *Geological Society of Malaysia Bulletin*, **20**, 81–96.

Umar, I., Yasin, A. and Koesoemadinata, S. (1982). *Geologic map of the Balikpapan Quadrangle, East Kalimantan*, 1:250,000 (sheet 1814). Geological Research and Development Centre, Bandung.

Utha-Aroon, Cherdsak (1993). Continental origin of the Maha Sarakham evaporites, northeastern Thailand. *Journal of Southeast Asian Earth Sciences*, **8**, 193–203.

Vachard, D. (1989). A rich algal micro flora from the Lower Permian of Jambi Province. In *The Pre-Tertiary fossils of Sumatra and their environments*, (H. Fontaine and S. Gafoer), C.C.O.P. Technical Publication, **19**, Bangkok.

Vallet, J. (1983). Seismic facies study in the Sepasu area of East Kalimantan. *Proceedings of the 12th annual convention*, **1**, 359–72. Indonesian Petroleum Association, Jakarta.

Van Bemmelen, R. W. (1949 and 1970). *The geology of Indonesia*, **1A**: *General geology of Indonesia and adjacent archipelagos*, **2**: *Economic geology*, **1B**: *Portfolio and index*, 1st and 2nd edn. Martinus Nijhoff, The Hague.

Van de Weerd, A.A. and Armin, R.A. (1992). Origin and evolution of the Tertiary hydrocarbon-bearing basins in Kalimantan (Borneo), Indonesia. *American Association of Petroleum Geologists Bulletin*, **76**, 1778–803.

Van de Weerd, A.A., Armin, R.A., Mahadi, S. and Ware, P.L.B. (1987). Geologic setting of the Kerendan gas and condensate discovery, Tertiary sedimentation and paleogeography of the north-western part of the Kutei Basin, Kalimantan, Indonesia. *Proceedings of the 16th annual convention*, **1**, 317–38. Indonesian Petroleum Association, Jakarta.

Van Gool, M., Huson, W.J., Prawirasasra, R. and Owen, T.R. (1987). Heat flow and seismic observations in the Northwestern Banda Arc. *Journal of Geophysical Research*, **92**, 2581–6.

Van Leeuwen, T.M., Leach, T., Hawke, A.A. and Hawke, M.M. (1990). The Kelian disseminated gold deposit, East Kalimantan, Indonesia. *Journal of Geochemical Exploration*, **35**, 1–61.

Veevers, J.J. (ed.) (1984). *Phanerozoic earth history of Australia*. Oxford monographs on geology and geophysics, **2**. Oxford University Press.

Vichit, Pongsak (1987). Gemstones in Thailand. *Journal of the Geological society of Thailand*, **9**, 108–33.

Vincelette, R.R. (1973). Reef exploration in Irian Jaya, Indonesia. *Proceedings of the 2nd annual convention,* 243–77. Indonesian Petroleum Association, Jakarta.

Virta, R.L. (1992). *Talc and pyrophyllite.* Annual Report for **1991**. U.S. Dept. of the Interior, Bureau of Mines. Govt. Printing Office, Washington.

Virta, R.L. (1993). *Clays.* Annual Report for **1991**. U.S. Department of the Interior, Bureau of Mines. Govt. Printing Office, Washington.

Vozenin-Serra, C. (1989). Lower Permian continental flora of Sumatra. In *The Pre-Tertiary fossils of Sumatra and their environments,* (ed. H. Fontaine and S. Gafoer), C.C.O.P. Technical Publication **19**, 53–7.

Wahab, A. and Martono, D. (1985). Application of oil geochemistry for hydrocarbon exploration in northwest Java. In *Proceedings of the 14th annual convention,* **2**, 657–82. Indonesian Petroleum Association, Jakarta.

Wain, T. and Berod, B. (1989). The tectonic framework and paleogeographic evolution of the Upper Kutei Basin. *Proceedings of the 18th annual convention,* **1**, 55–78. Indonesian Petroleum Association, Jakarta.

Wajzer, M.R., Barber, A.J., Hidayat, S. and Suharsono (1991). Accretion, collision and strike-slip faulting: the Woyla Group as a key to the tectonic evolution of North Sumatra. *Journal Southeast Asian Earth Sciences*, **6**, 447–61.

Walker, T.R., Williams, A.F., Wong, D., Kadir, M.K.A. and Wong, R.H.F. (1992). Structural framework and hydrocarbon potential of the southern Sandakan Basin, Eastern Sabah, Malaysia. *Symposium on tectonic framework and energy resources of the western margin of the Pacific Basin, Kuala Lumpur, Nov-Dec 1992, Programme and Abstracts of Papers,* p. 53. Circum Pacific Council and Geological Society of Malaysia.

Wan Ismail Wan Yusoff (1990). Heat flow in offshore Malaysian basins. *C.C.O.P. Technical Bulletin,* **23**, 39–54. Bangkok.

Wang Wen-Sheng (1987). Nephrite deposits of Taiwan. *Proceedings of the Pacific Rim Congress 87,* Australasian Institution of Mining and Metallurgy, Parkville, Australia, 639–40.

Wang Zhifen (1983). Some problems on the mineralization of tin deposits in Gejiu, Yunnan. *Acta Geologica Sinica,* **2**, 154–63.

Waterhouse, J.B. (1982). An Early Permian cool-water fauna from pebbly mudstones in South Thailand. *Geological Magazine,* **119**, 337–54.

Waples, D.W. (1980). Time and temperature in petroleum formation: application of Lopatin's method to petroleum exploration. *American Association of Petroleum Geologists Bulletin,* **64**, 916–26.

Weber, H.S. and Lee, D.T.C. (1990). Mineral resources investigation in Sabah, East Malaysia, 1980–1984, selected results and conclusions. *Geologisches Jaarbuch,* **B74**, 3–29.

Wee, G. (1993). Structural framework and trap styles in southwestern offshore Sabah. *Geological Society of Malaysia, Petroleum Geology Seminar 1993, Programme and Abstracts of Papers,* 21.

Weissel, J. K. (1980). Evidence for Eocene oceanic crust in the Celebes Basin. In *The tectonic and geologic evolution of Southeast Asian seas and islands.* (ed. D.E. Hayes), Geophysical Monograph, pp. **23**, 37–47. American Geophysical Union, Washington.

White, J.M. jr. and Wing, R.S. (1978). Structural development of the South China Sea with particular reference to Indonesia. *Proceedings of the 7th annual convention,* 159–77. Indonesian Petroleum Association, Jakarta.

Whittle, A.P. and Short, G.A. (1978). The Petroleum Geology of the Tembungo Field, East Malaysia. *Offshore South-East Asia Conference (Feb. 1978),* 1–11. South East Asian Petroleum Exploration Society, Singapore.

Wight, A.W.R. and Hardian, D. (1982). Importance of diagenesis in carbonate exploration and production, lower Batu Raja carbonates, Krisna Field, Java Sea. *Proceedings of the 11th annual convention,* 211–36. Indonesian Petroleum Association, Jakarta.

Wight, A.W.R., Hare, L.H. and Reynolds, J.R. (1993). Tarakan Basin, NE Kalimantan, Indonesia: a century of exploration and future hydrocarbon potential. *Proceedings of symposium on tectonic framework and energy resources of the western margin of the Pacific basin, Kuala Lumpur, Nov-Dec 1992,* Geological Society of Malaysia Bulletin, **33**, 263–88.

Wikarno, D.A.D., Suyatna, and Sukardi, D. (1988). Granitoids of Sumatra and the tin islands. In *Geology of Tin Deposits in Asia and the Pacific,* (ed. C.S. Hutchison), pp. 571–89. Springer-Verlag, Heidelberg.

Wilford, G.E. (1955). The geology and mineral resources of the Kuching–Lundu area, West Sarawak, including the Bau Mining District. *Geological Survey Department British Territories in Borneo memoir,* **3**.

Wilford, G.E. (1961). The geology and mineral resources of Brunei and adjacent parts of Sarawak with descriptions of Seria and Miri oilfields. *Geological Survey Department British Territories in Borneo memoir,* **10**.

Willbourn, E.S. and Ingham, F.T. (1933). Geology of the scheelite mine, Kramat Pulai, F.M.S. *Quarterly Journal of the Geological Society of London,* **89**, 449–79.

Williams, H.H., Kelley, P.A., Janks, J.S. and Christensen, R.M. (1985). The Paleogene rift basin source rocks of

Central Sumatra. In *Proceedings of the 14th annual convention,* **2,** 57–90. Indonesian Petroleum Association, Jakarta.

Williams, P.R. and Harahap, B.H. (1987). Preliminary geochemical and age data for post-subduction intrusive rocks, northwest Borneo. *Australian Journal of Earth Sciences,* **34,** 405–15.

Williams, P.R., Johnston, C.R., Almond, R.A. and Simamora, W.H. (1988). Late Cretaceous to Early Tertiary structural elements of West Kalimantan. *Tectonophysics,* **148,** 279–97.

Wilson, J.G. (1987). Structural control of gold mineralization at a plate boundary. The Kyauk Pahto photogeological case history, north-central Burma. *Proceedings of the Pacific Rim Congress '87,* Australasian Institution Mining Metallurgy, Parkville, Australia, 487–93.

Wilson, R.A.M. (1964). The geology and mineral resources of the Labuan and Padas Valley area, Sabah, Malaysia. *Geological Survey of the Borneo Region of Malaysia, memoir,* **17.**

Wing-Easton, N. (1937). The tin-ore of Bangka, Billiton and Singkep, Malay Archipelago. *Economic Geology,* **32,** 1–29 and 154–82.

Wolfe, J.A. (1972). K:Ar dating in the Philippines. *Journal of the Geological Society of the Philippines,* **26,** 11–2.

Wolfe, J.A. (1986). Breccias related to explosive volcanism. *Journal Southeast Asian earth sciences,* **1,** 63–79.

Wolfenden, E.B. (1960). The geology and mineral resources of the Lower Rajang Valley and adjoining areas, Sarawak. *Geological Survey Department for the British Territories in Borneo, Memoir,* **11.** Kuching.

Wolfenden, E.B. (1965). Bau Mining District, West Sarawak, Malaysia, Part I: Bau. *Geological Survey Borneo Region, Malaysia, Bulletin,* **7.** Kuching.

Wong, R.H.F. (1993). Sequence stratigraphy of the Middle Miocene–Pliocene southern offshore Sandakan Basin, East Sabah. *Geological Society of Malaysia Bulletin,* **33,** 129–42.

Wong, W.S. and Lee, A.K. (1986). Gold investigations in the Sungai Luit area, Mengapur Base Metal District, Pahang. *Geological Society of Malaysia Bulletin,* **20,** 667–700.

Wongsosantiko, A. and Wirojudo, G.K. (1984). Tertiary tectonic evolution and related hydrocarbon potential in the Natuna area. *Proceedings of the 13th annual convention,* **1,** 161–83. Indonesian Petroleum Association, Jakarta.

Wood, B.G.M. (1985). The mechanics of progressive deformation in crustal plates—a working model for Southeast Asia. *Geological Society of Malaysia Bulletin,* **18,** 55–99.

Woodbury, W.D. (1992). *Lead.* Annual Report for 1991. U.S. Department of the Interior, Bureau of Mines. Govt. Printing Office, Washington.

Woodroof, P.B. and Carr, A.D. (1992). Source rock and hydrocarbon geochemistry, offshore NW Sabah, Malaysia. *Symposium on tectonic framework and energy resources of the western margin of the Pacific Basin, Kuala Lumpur Nov–Dec. 1992, Programme and Abstracts of Papers,* p. 64. Geological Society of Malaysia and Circum Pacific Council.

Woollands, M.A. and Haw, D. (1976). Tertiary stratigraphy and sedimentation in the Gulf of Thailand. *Proceedings of the Offshore Southeast Asia Conference, Feb. 1976,* paper **7,** 1–22. Southeast Asia Petroleum Exploration Society, Singapore.

Wopfner, H. and Jin, X. (1993). Baoshan and Tengchong blocks of western Yunnan (China) in the late Palaeozoic mosaic of the eastern Tethys. In *Gondwana dispersion and Asian accretion; International Geological Correlation Program Project 321, 3rd international symposium,* Department of Geology, University of Malaya, Kuala Lumpur, 47–9.

Xia KangYuan and Zhou Di (1993). The geophysical characteristics and evolution of northern and southern margins of the South China Sea. *Proceedings of symposium on tectonic framework and energy resources of the western margin of the Pacific Basin, Kuala Lumpur 1992,* Geological Society of Malaysia Bulletin, **33,** 223–40.

Xu Keqin, Hu Shouci, Sun Minzhi, Zhang Jingrong, Ye Jun and Li Huiping (1982). Regional factors controlling the formation of tungsten deposits in south China. In *Tungsten geology,* (ed. J.V. Hepworth and Zhang Yuhong), pp. 473–88. *Jiangxi, China,* ESCAP/RMRDC, Bandung, Indonesia.

Xu Keqin and Zhu Jinchu (1988). Time-space distribution of tin/tungsten deposits in south China and controlling factors of mineralization. In *Geology of tin deposits in Asia and the Pacific,* (ed. C.S. Hutchison), pp. 265–77. Springer-Verlag, Heidelberg.

Yan MeiZhong, Wu YongLe and Li ChongYou (1980). Metallogenic systems of tungsten in southeast China and their mineralization characteristics. In *Granitic magmatism and related mineralization,* (ed. S. Ishihara and S.Takenouchi), Mining geology, Special Issue **8,** 215–21.

Yang Minggui and Lu Dekui (1982). Structural characteristics, forms of arrangement and composition of the vein-type tungsten ore in Xihuashan–Piotang district. In *Tungsten geology,* (ed. J.V. Hepworth and Y.H Zhang, pp. 521–31. *Jiangxi, China,* ESCAP/RMRDC, Bandung, Indonesia.

Yang Zunyi, Cheng Yuqi and Wang Hongzhen (1986). *The geology of China*, Oxford Monographs on Geology and Geophysics, **3**. Clarendon Press, Oxford.

Yeap, E. B. (1993). Tin and gold mineralizations in Peninsular Malaysia and their relationships to the tectonic development. *Journal of Southeast Asian Earth Sciences*, **8**, 329–48.

Yumul, G.P. jr. and Balce, G.R. (1994). Supra-subduction zone ophiolites as favorable hosts for chromitite, platinum and massive sulfide deposits. *Journal of Southeast Asian Earth Sciences*, **10**, 65–79.

Zeijlmans van Emmichoven, C.P.A. (1935). De Geologie van het centrale en oostelijke Deel van de Westerafdeeling van Borneo. *Jaarboek van het Mijnwezen in Nederlandsch Oost-Indie 'Verlandelingen'*.

Zeiller, R. (1903). Sur la flore des gîtes de charbon du Tonkin. *Étude des gîtes mineraux de La France*, Imprinterie Nationale, Paris, 328.

Zeliff, C.W., Trollope, S.W., and Maulana, E. (1985). Exploration cycles in the Corridor Block, South Sumatra. In *Proceedings of the 14th annual convention*, **1**, 379–400. Indonesian Petroleum Association, Jakarta.

Zhang Qiming and Zhang Quanxing (1991). A distinctive hydrocarbon basin—Yinggehai Basin, South China Sea. *Journal Southeast Asian Earth Sciences*, **6** (2), 69–74.

Zillman, N.J. and Paten, R.J. (1975*a*). Exploration and petroleum prospects Bula Basin, Seram, Indonesia. In *Proceedings of the 4th annual convention*, 129–48. Indonesian Petroleum Association, Jakarta.

Zillman, N.J. and Paten, R.J. (1975*b*). Geology and petroleum prospects of Seram Island, eastern Indonesia. *Australian Petroleum Exploration Association Journal*, **15** (1), 73–80.

Zwierzycki, J. (1930). Geologische overzichtskaart van de Nederlandsch Indische Archipel, schaal 1:1,000,000, toelichting bij blad VIII (Midden Sumatra, Bangka en de Riauw eilanden). *Jaarboek van het Mijnwezen in Nederlandsch Oost-Indie*, **38** (1929), 73–157.

Zwierzycki, J. (1935). Die geologische ergebnisse der palaeobotanische Djambi-Expedition 1925. *Jaarboek van het Mijnwezen in Nederlandsch Oost-Indie*, **2** (1930), 1–70 + map.

Index

figures and tables—§ after the page entry
major section—in bold numbers

Made in the USA
Monee, IL
11 November 2025

34346731R00155